D0793607

# Apoptosis, Genomic Integrity, and Cancer

## Jones and Bartlett Titles in Biological Science

*AIDS: Science and Society, Fourth Edition*
Fan/Conner/Villarreal

*AIDS: The Biological Basis, Fourth Edition*
Weeks/Alcamo

*Alcamo's Fundamentals of Microbiology, Seventh Edition*
Pommerville

*Alcamo's Laboratory Fundamentals of Microbiology, Seventh Edition*
Pommerville

*Anatomy and Physiology: Understanding the Human Body*
Clark

*Apoptosis, Genomic Integrity, and Cancer*
Van Lancker

*Aquatic Entomology*
McCafferty/Provonsha

*Botany, An Introduction to Plant Biology, Third Edition*
Mauseth

*The Cancer Book*
Cooper

*Cell Biology: Organelle Structure and Function*
Sadava

*Cells*
Lewin/Cassimeris/Lingappa/Plopper

*Defending Evolution: A Guide to the Evolution/Creation Controversy*
Alters

*Diversity of Life: The Illustrated Guide to the Five Kingdoms*
Margulis/Schwartz/Dolan

*DNA Sequencing: Optimizing the Process and Analysis*
Kieleczawa

*DNA Sequencing II: Optimizing Preparation and Cleanup*
Kieleczawa

*Early Life, Evolution on the Precambrian Earth, Second Edition*
Margulis/Dolan

*Electron Microscopy, Second Edition*
Bozzola/Russell

*Elements of Human Cancer*
Cooper

*Encounters in Microbiology*
Alcamo

*Essential Genetics: A Genomic Perspective, Fourth Edition*
Hartl/Jones

*Essential Medical Terminology, Second Edition*
Stanfield

*Essentials of Molecular Biology, Fourth Edition*
Malacinski

*Evolution, Third Edition*
Strickberger

*Experimental Techniques in Bacterial Genetics*
Maloy

*Exploring the Way Life Works: The Science of Biology*
Hoagland/Dodson/Hauck

*Genetics, Analysis of Genes and Genomes, Sixth Edition*
Hartl/Jones

*Genetics of Populations, Third Edition*
Hedrick

*Genomic and Molecular Neuro-Oncology*
Zhang/Fuller

*Guide to Infectious Diseases by Body System*
Pommerville

*Grant Application Writer's Handbook*
Reif-Lehrer

*How Cancer Works*
Sompayrac

*How Pathogenic Viruses Work*
Sompayrac

*How the Human Genome Works*
McConkey

*Human Anatomy and Physiology Coloring Workbook and Study Guide, Second Edition*
Anderson/Spitzer

*Human Anatomy Flash Cards*
Clark

*Human Biology, Fifth Edition*
Chiras

*Human Biology Laboratory Manual*
Welsh

*Human Body Systems: Structure, Function, and Environment*
Chiras

*Human Embryonic Stem Cells: An Introduction to the Science and Therapeutic Potential*
Kiessling/Anderson

*Human Genetics: The Molecular Revolution*
McConkey

*The Illustrated Glossary of Protoctista*
Margulis/McKhann/Olendzenski

*Introduction to Human Disease, Sixth Edition*
Crowley

*Introduction to the Biology of Marine Life, Eighth Edition*
Sumich/Morrissey

*Laboratory and Field Investigations in Marine Life, Eighth Edition*
Sumich/Gordon

*Laboratory Research Notebooks*
Jones and Bartlett Publishers

*A Laboratory Textbook of Anatomy and Physiology, Cat Version, Eighth Edition*
Donnersberger/Lesak Scott

*A Laboratory Textbook of Anatomy and Physiology, Fetal Pig Version, Eighth Edition*
Donnersberger/Lesak Scott

*Major Events in the History of Life*
Schopf

*Medical Biochemistry: Pearls of Wisdom*
Eichler

*Medical Genetics: Pearls of Wisdom*
Sanger

*Microbes and Society*
Alcamo

*Microbial Genetics, Second Edition*
Maloy/Cronan/Freifelder

*Microbiology: Pearls of Wisdom*
Booth

*Missing Links: Evolutionary Concepts and Transitions Through Time*
Martin

*Oncogenes, Second Edition*
Cooper

*100 Years Exploring Life, 1888–1988, The Marine Biological Laboratory at Woods Hole*
Maienschein

*Origin and Evolution of Intelligence*
Schopf

*Plant Cell Biology: Structure and Function*
Gunning/Steer

*Plants, Genes, and Crop Biotechnology, Second Edition*
Chrispeels/Sadava

*Population Biology*
Hedrick

*Protein Microarrays*
Schena

*Statistics: Concepts and Applications for Science*
LeBlanc

*Teaching Biological Evolution in Higher Education*
Alters

*Vertebrates: A Laboratory Text*
Wessels

# Apoptosis, Genomic Integrity, and Cancer

Julien L. Van Lancker
University of California at Los Angeles

JONES AND BARTLETT PUBLISHERS
*Sudbury, Massachusetts*
BOSTON TORONTO LONDON SINGAPORE

*World Headquarters*
Jones and Bartlett Publishers
40 Tall Pine Drive
Sudbury, MA 01776
978-443-5000
info@jbpub.com
www.jbpub.com

Jones and Bartlett Publishers
Canada
6339 Ormindale Way
Mississauga, Ontario
L5V 1J2
CANADA

Jones and Bartlett Publishers
International
Barb House, Barb Mews
London W6 7PA
UK

Jones and Bartlett's books and products are available through most bookstores and online booksellers. To contact Jones and Bartlett Publishers directly, call 800-832-0034, fax 978-443-8000, or visit our website www.jbpub.com.

Substantial discounts on bulk quantities of Jones and Bartlett's publications are available to corporations, professional associations, and other qualified organizations. For details and specific discount information, contact the special sales department at Jones and Bartlett via the above contact information or send an email to specialsales@jbpub.com.



**Production Credits**
Chief Executive Officer: Clayton Jones
Chief Operating Officer: Don W. Jones, Jr.
President, Higher Education and Professional Publishing: Robert W. Holland, Jr.
V.P., Design and Production: Anne Spencer
V.P., Sales and Marketing: William Kane
V.P., Manufacturing and Inventory Control: Therese Connell
Executive Editor, Science: Stephen L. Weaver
Acquisitions Editor, Science: Cathleen Sether
Managing Editor, Science: Dean W. DeChambeau
Editorial Assitant: Molly Steinbach
Senior Production Editor: Louis C. Bruno, Jr.
Marketing Manager: Andrea DeFronzo
Text and Cover Design: Anne Spencer
Illustrations: Lynne Olson
Composition: SPI Publisher Services
Printing and Binding: Malloy
Cover Printing: Malloy
Cover Photo: Microworks/Phototake

**Library of Congress Cataloging-in-Publication Data**
Van Lancker, Julien L., 1924–
Apoptosis, genomic integrity, and cancer: an introdution to interacting molecules/ Julien L. Van Lancker.
p. cm.
Includes bibliographical references and index.
ISBN 0-7637-4541-3 (alk. paper)
1. Apoptosis. 2. Cancer cells. 3. Genomes. 4. DNA repair. I. Title.
Qh671.V36 2006
571.9′36—dc22 2005049240

Printed in the United States of America
09 08 07 06 05 10 9 8 7 6 5 4 3 2 1

To all my teachers and my students
and
to my wife and family

# Brief Contents

## Part I. Introduction to Apoptosis 1



## Part II. Integrity of the Genome 131



## Part III. The Life and Death of the Cancer Cell 257

# Contents

# Preface

Remarkable progress was made in cancer diagnosis, prognosis, and therapy during the last decades of the twentieth century. Diagnosis and prognosis were improved by refining morphopathological criteria and by developing more detailed imaging procedures with ultrasound, computer tomography, and magnetic resonance imaging (MRI).

All three traditional methods of cancer therapy have become more efficient in various ways: surgery by more effectively eradicating all parts of the tumor from healthy tissues; radiation therapy by better targeting the cancer mass and, thereby, causing less damage to surrounding normal tissue; and chemotherapy by the development of a large pharmacopoeia of antimetabolites thereby allowing for more selective and less destructive treatments. Cooperation between surgeons, radiation oncologists, and chemotherapists has led to the development of combined therapy protocols often based on new and more precise understanding of cancer's pathogenesis. By comparing the protocols regionally, nationally, or internationally, their effectiveness in securing disease-free survival may be evaluated. This multipronged attack on cancer helped to cure many traditionally incurable malignancies, raised the quality of life, and secured longer disease-free lives for many patients. Yet, it would be unfair to past and future victims of cancer to declare that all cancers have been conquered.

Unless unexpected and wondrous discoveries give us surprising effective new tools for the cure or diagnosis of cancers at an early stage, the hope of conquering cancer will most likely rest on better understanding of the molecular pathogenesis of each cancer type and on targeting the special molecules that cause it.

This book reviews the new insight into cancer pathogenesis that emerged during the last decades of the twentieth century with special focus on apoptosis and loss of genomic integrity. It is not intended to be, or could not be, a comprehensive description of the pathways of cell replication, DNA repair, or programmed cell death; it is instead an earnest attempt to explore the ways by which this new knowledge has had an impact on modern concepts of cancer pathogenesis. Investigations of cell proliferation have led to the identification of cell-cycle checkpoints at the G1/S and the G2/M transitions and during anaphase and metaphase. The discovery of the protein P53 has contributed to the clarification of controls of genomic integrity, and molecular investigations of the pathways to programmed cell death has revealed apoptosis' importance in the homeostasis of cell populations.

Xeroderma pigmentosum, ataxia telangiectasia, Fanconi anemia, Nijmegen breakage, and Bloom syndromes are known to be associated with a predisposition to cancer. The characterization of those genes' loci and the identification of their encoded proteins helped to uncover the genes' role in DNA repair. Later, the products of two other genes associated with predisposition to cancer of the breast, *BRCA1* and *BRCA2*, were also found to function in DNA repair.

The progress in cancer research further demonstrates that cancer cells not only proliferate but also fail to die. In addition some of the proliferating cells undergo sequential somatic mutations that are at the source of new clones with survival advantages (acceleration of cell proliferation, loss of differentiation, loss of apoptosis, invasion, and metastasis). Although the exact cause of the hypermutable phenotype is still debated, failure of integral DNA repair and persistent chromosomal anomalies are likely to be significant contributors to hypermutability.

The identification of the mutated products in the cancer cell's phenotype has already, at least in some cases, facilitated prevention and the early diagnosis of several cancers and has disclosed new targets for therapy for some cancers. I hope that this work will provide the reader not only with an introduction to present knowledge but also with a bridge to information to come.

## Acknowledgments

Many persons have encouraged, made suggestions, and assisted me at various stages of this endeavor. I am grateful for the support of Drs. Jonathan Braun, Jonathan Said, and Rodney Withers. I also wish to express my appreciation for the helpful suggestions offered to me by my colleagues and friends who read portions of the manuscripts: Drs Michael Collins, Wayne Grody, Andre Nel, and Henry Pitot. My thanks to Dr. Ahmed El-Said and Geoffrey Robertson who contributed to the retrieving of references from PubMed, to Fernando Casimiro who typed the original manuscript, to Susan Stehn for her insightful review of the manuscript and her constructive suggestions, to Dorina Gui for her competent assistance with the computer generated graphics, to Lynne Olson for several drawings, to Carol Appleton for her photographs, and to Linda Escobar for her gracious assistance. Special thanks to the staff of Jones and Bartlett who provided professional assistance in copywriting (Shellie Newell), indexing (Deborah Patton), and production (Lou Bruno). In the end this work could not have been completed without the patience and the never failing support and assistance of my wife, Jill. Part of this work was supported by the Anna Memorial Cancer Foundation.

Julien L. Van Lancker
Sherman Oaks, California
October, 2005

# Foreword

In *Apoptosis, Genomic Integrity, and Cancer*, Dr. Julien Van Lancker has provided the reader and the biomedical establishment with an entirely new perspective on cell growth and death, the *yin* and *yang* at the core of all biological processes. So central are these concepts that the treatise required a synthesis of studies from many disparate fields of biomedical research and clinical medicine across many decades of history. Dr. Van Lancker has accomplished this task with a mastery that is both expansive and inclusive. His critical eye draws from theories and discoveries in molecular biology, cell biology, pathology, biochemistry, pharmacology, oncology, and even philosophy. The reader is presented with not only the most current knowledge of these complex interacting molecules but also the history of how we got to this point and how our understanding has changed over time; see for example the description of the tumor suppressor gene *p53* in Chapter 1 or the discussion of DNA repair in Chapter 2, which goes back to the classic Avery and McLeod experiments and builds logically from there. I can think of no one more apt to lead the reader on this historic journey than the author, whose scientific career has spanned more than six decades and interwoven with all of these fields. Dr. Van Lancker's formal clinical training in the specialties of pathology, radiation oncology, and nuclear medicine, and his productive research career in the molecular biology of DNA repair and radiation damage, furnish precisely the requisite combination of expertise and perspective to address the subject.

There are many books on cancer biology and oncology and a growing number on apoptosis, but the combination of cancer and apoptosis in this book is quite unique. Texts on cancer tend to focus on cell proliferation and control, but as Dr. Van Lancker convincingly argues, that is only half the story. All of biology represents a balance between life and death, an "indissoluble marriage" as he calls it, and this is especially demonstrable at the individual cell and tissue levels, where it is fundamental to the processes of embryogenesis, hematopoiesis, homeostasis, morphogenesis, and immune function. This balance is also central to abnormal cell processes, including cancer, and the function of apoptosis. Its derangement may be just as important in determining tumor behavior, cancer prognosis, and response to therapy as the pro-growth/loss-of-control factors on which we usually focus. In that sense, this book opens a new door for us and imparts a novel conceptualization of neoplasia.

In ecological systems, as in the single organism, life and death are both essential elements for maintaining homeostasis and renewal. Just as organismal death is elemental to the life cycle of the natural world, programmed cell death (apoptosis) is essential for the life cycle of the organism. This begins in embryonic development, where some tissues must involute to form fetal structures, and occurs throughout life as terminally differentiated cells must "fall away" (the literal meaning of the Greek root for *apoptosis*) to make room for new replacements from progenitor or stem cells. This process occurs continually in the blood and bone marrow, the immune system, the skin, the bronchial epithelium, and in solid organs such as the liver.

It seems counterintuitive, in the Darwinian sense, that a living cell would have evolved a mechanism for promoting its own death, but at the whole-organism level, such pathways obviously confer advantages and may well have furnished the key to the evolution of complex multicellular species. Furthermore, when the pathway is defective, it can have severe adverse consequences for the whole organism. Such is the case with cancer, which we now know involves more than merely increased cell proliferation or faster mitotic rate. In fact, in some tumors the rate of cell division is actually lower than that in the native tissue; therefore, some loss of capacity for programmed cell death is clearly responsible.

Moreover, the degree to which tumors respond to therapy, particularly chemotherapy and radiation, is dependent to a large extent on their characteristic ability to respond to apoptotic signals. Assessment of this property may have both prognostic and therapeutic value, and there have been efforts to measure the "apoptotic index" of individual tumors as a complement to the more generally utilized mitotic index. Just as cancer is in part a manifestation of defective apoptosis, a semblance of functional apoptosis may be induced in the therapeutic death of cancer cells through chemotherapeutic, radiotherapeutic, hormonal, hyperthermal, or small-molecule targeted approaches. Enhancing this response, therefore, becomes an obvious target for newer cancer therapies. Aside from the original Greek definition of *apoptosis,* its similarity to the word *mitosis* also drives home the *yin-yang* relationship of functional opposites. Indeed, at the molecular level, the same oncogenes and tumor suppressor genes involved in promoting or suppressing cell growth are also directly involved in the apoptotic pathway (*p53, bcl-2, c-fos*, etc.). Because these responses are both determined and effected at the genomic level, the subtext running throughout the book, "genomic integrity" is of utmost relevance.

This book, then, ties together these mutually opposing yet interdependent life processes in a way that will be entirely new to most readers. But this should come as no surprise to those who know the author, whose entire career has comprised new ways to synthesize seemingly disparate problems. This is even true in my field of DNA diagnostics where, to the best of my knowledge, Dr. Van Lancker was the first person to use the now popular term for our discipline, *molecular pathology*, and even included it in the title of an influential journal article as far back as 1976—long before the field was born, let alone recognized. Herein the reader will find equivalent far-reaching insight to the still largely uncharted relationship of apoptosis and cancer at the genomic level. It is an ambitious inquiry and a remarkable journey.

Wayne W. Grody, M.D., Ph.D.
UCLA School of Medicine

PART

# Introduction to Apoptosis

# I

BECAUSE OF FEAR OF OUR OWN DEATH, or because of the grief suffered by those who remain alive, the quick and the dead are at antipodes in the minds of humans. Yet life and death continuously interact in the living world. Humans kill and consume cattle that eat grass in green pastures, salmons swim upstream to the Kodiak shallow lakes to spawn and ultimately become the prey of bears, eagles, and foxes. Animal life thrives on other forms of life. Even the harmony of the living organism depends on the death of its parts. Nature has built-in cycles of life and death. Outside of the tropics, every autumn almost all leaves die slowly, and it is no wonder that old age is referred to as the winter of our lives. Growth and maintenance of the animal organism result from the delicate balance of cellular proliferation, maturation, and elimination. During embryogenesis millions of cells are killed or commit suicide in preselected areas. After their elimination, the dead cells are replaced by new cells, sometimes of a different type.

Morphogenesis of bone tissue results from repeated destruction of preexisting histological structures with replacement by new ones. Red cells, polymorphonuclear cells, and lymphocytes of the blood all have finite lifespans, yet the total circulating number of these cells remains constant in the healthy. Superficial layers of skin and mucosae are shed periodically, but never are surfaces denuded in the healthy organism. These few examples suggest that cellular death is no less important to organismal health and survival than cellular proliferation.

Because of this indissoluble marriage between life and death, the cells, which since Theodor Schwann (1810–1882) (1) have been considered the unit of life and since Virchow (1821–1902) (2) the target of injury, preserve the harmony of the whole by balancing cell proliferation and cell death.

Cell death can be brutal if caused by trauma or burns, whether accidental or voluntarily inflicted. Open wounds or mutilations are visible to all; they cause pain in the victim and fright, horror, and pity in the bystander. If not controlled by various defense mechanisms, immunity, inflammation, and healing by cell proliferation, these trauma can kill or handicap the victim for life. The blockage of an artery of the heart is not directly visible, but its consequences are tragic. The massive destruction of well integrated but different populations of cells (endothelial cells, fibroblasts, skin, or heart muscle cells) is called necrosis. The latter is almost always associated with inflammation.

The term *necrosis* comes from the Greek word νεκροσ, meaning "the dead," "death," and "dead body." Every organ in the body is subject to necrosis, but some organs are affected preferentially because they are more accessible to trauma, because the life of the cells of the organ depends on the presence of enzymes or other biochemical compounds particularly sensitive to specific toxins, or because the vascular tree that nourishes that organ, for some mechanical or metabolic reason, is prone to occlusion. Since necrosis is an acute process leading to the simultaneous death of a large number of cells, it is recognizable macroscopically. The gross appearance is naturally a function of the agent responsible for necrosis and of the organ involved. Pathologists have described a great variety of necrotic forms in detail. Although a great deal has been learned about the molecular mechanisms of inflammation and wound healing, that extent of knowledge is still wanting for necrosis.

In contrast to necrosis, programmed cell death does not involve inflammation and a multicellular healing process. Programmed cell death is present through the entire lifespan of the whole organism, from small invertebrates to humans. The birth of a butterfly fascinates our children. It happens in three stages. The larva is a caterpillar, which turns into a pupa after molting several times. The amazing metamorphosis of a caterpillar into a butterfly takes place during the pupal stage. Various appendages, including the wings and the limbs, start as internal buds and by a combination of cell death and cell growth are molded into the glorious shape and colors of a monarch. The spinal ganglia of the early chick embryo contain more neurons than are needed for adequate innervation of most of the adult body except for those parts of the chick's body that include the wings and the limbs. Thus, cells die in large numbers in parts of the brain where they have been overproduced.

The development of the wing buds of the chick's embryo also involves selective and rigidly scheduled cellular death. At a critical stage of the chick embryo's development, some cells of the wings bud die to sculpt the shape of the wings. If the "condemned cells" are transplanted into another embryo, which has reached a stage of development different from that of the donor embryo, the grafted cells die on schedule as if never transplanted.

Two embryonic structures of mesenchymal origin are involved in the development of the genitourinary tract of mammals: the primitive excretory tract (mesonephric or wolffian) and the paramesonephric or müllerian ducts. Both types of ducts are found in embryos of both sexes at early stages of development. In the female, the development of the excretory urinary tract develops from the mesonephric duct, and the müllerian duct grows and differentiates to yield the uterus, fallopian tubes, and the vagina. In contrast, in the male the müllerian ducts regress. The mesonephric ducts eventually connect with the testis, and both the genital and the urinary systems use a common pathway for excretion. In the human embryo the separation of fingers, toes, and lips results from the death of connecting cells.

The significance of these and other observations and experiments was summarized by Saunders (3). They concluded that during development cells do not die by chance; on the contrary, their death is programmed in space and time, and it is monitored by a "death clock." Perhaps because programmed cell death is discrete and quick, as was pointed out by Horvitz (4), and also because the tools of molecular biology were, at the time, not sharp enough to dissect the molecular mechanisms of programmed cell death, the process was ignored. However, the morphological observations of Kerr and associates (5, 6) and the discovery, in Horvitz's laboratory, of genes responsible for identifying and executing doomed cells in the small worm *Caenorhabditis elegans*, were a molecular treasure chest full of surprises. Many are remarkable but one stands out: the evolutionary conservation of the molecular process from worm to humans. But, like most if not all physiological phenomena, programmed cell death or apoptosis is not without its Damoclean counterparts. If apoptosis fails, too many cells grow and compromise homeostasis in the entire organism. It is now clear that cancer is not only caused by uncontrolled growth but also by defective apoptosis. When apoptosis appears unscheduled, it can contribute to discrete cell death as observed in Parkinson's disease or other neurodegenerative disease. The studies on programmed cell death have taught us lessons in evolution, physiology, and pathology. The purpose of the first part of this endeavor is a modest introduction to some of the molecular events known to contribute to programmed cell death.

## References and Recommended Reading


1. Metler, C. and Metler, F.S. *History of Medicine.* Philadelphia: Blakinston, 1947. pp. 90–100.
2. Virchow, R. Cellular pathology as based upon physiological and pathological histology. Twenty lectures delivered in the Pathological Institute of Berlin during the months of February, March, and April, 1858. Translated from the second edition by Chance, F. in *The Classics of Medicine Library.* Birmingham, Alabama: Griphon Editions, Ltd., 1978.
3. Saunders, J.W. Death in embryonic systems. *Science.* 1966;154:604–42.
4. Horvitz, H.R. Worms, life, and death (Nobel lectures). *Chembiochem.* 4(8).697–711.
5. Kerr, J.F. History of the events leading to the formulation of the apoptosis concept. *Toxicology.* 2002(Dec. 27);181–2, 471–4 (review).
6. Kerr, J.F, Wylie, A.H., and Currie, A.R. Apoptosis: a basic biological phenomenon with wide-ranging implications in tissue kinetics. *Br J Cancer.* 1972;26(4)239–57 (review).

CHAPTER

# MORPHOLOGY AND OCCURRENCE OF APOPTOSIS

# 1

Programmed cell death was on the mind of biologists long ago (1–3). However, it is the persistent work of Kerr and his associates that laid the groundwork for programmed cell death and demonstrated its role in embryology, in the regulation of healthy cell populations, and in cell pathology. Still, it took much longer for investigators to discover the molecular mechanisms of programmed cell death, and that search is far from complete.

Programmed cell death contrasts with necrosis, which is often extensive, caused by internal or external nefarious agents, and followed by inflammatory reactions and scarring. Programmed cell death is discrete and an important part of development and homeostasis. In 1972 Kerr et al. named this process "apoptosis," a composite word borrowed from the Greek απο–πτοσισ, which refers to falling autumn leaves (4). Apoptosis can be triggered by highly specific molecules such as members of the tumor necrosis family or even noxious agents such as radiation or chemotherapy, and through the intervening years we have learned that the choice between cell life and cell death are carefully regulated by balancing the action of intracellular survival agents (e.g., Bcl-2 and Bclx and survivin) and cell death-promoting agents (e.g., Bad and Bax). Surprisingly, many of the antiapoptotic and proapoptotic agents belong to the same protein family and are conserved during evolution from nematodes to humans. If it could be anticipated that apoptosis at some stage would involve proteolytic enzymes, it was not immediately obvious that a special family of enzymes, now called caspases, is responsible for the sequential proteolytic steps involved in apoptosis. Even more surprising was the finding that the release of a molecule that is part of the electron transport chain, cytochrome c, triggers the caspase cascade. Thus, mitochondrial alterations are central to our understanding of apoptosis. However, in some cases calcium release from the endoplasmic reticulum, stimulation of sphingomyelinase, and release of ceramides from the cell membrane can either be the first event or part of the apoptotic sequence. While some of the proteins involved in apoptosis are constitutive, others are synthesized de novo. In either case, some critical genes control the process and their selective mutations may either abolish or accelerate apoptosis, thereby causing disease, among them cancer or neurodegenerative diseases. Well-regulated apoptosis is critical for the whole organism's harmonious development and its survival in good health (5).

## DNA Degradation in Apoptosis

Apoptosis is a carefully regulated, self-destructive cellular process with distinctive, but variable with the circumstances, morphological and biochemical characteristics. The pathway to death is programmed in the course of development. It can also be triggered by association of ligands and death receptors or by various toxins during the entire lifespan of plants and animals. The process is completed without causing damage to neighboring cells.

Wyllie and associates summarized the differences between apoptosis and necrosis (6). The most characteristic differences are:

1. Apoptosis concerns an individual cell, while necrosis affects a multitude of cells often of different embryologic origin (epithelial, mesenchymal, etc.).
2. Necrosis is never physiologic.
3. Necrosis is associated with systemic inflammation and distortion of the tissue (permanent ulceration, scarring, etc.).

Critical observations of apoptosis were made with the electron microscope. The earliest events in the process involve the chromatin. The chromatin segregates into compact masses that rest on the nuclear envelope, and at the same time the cytoplasm condenses and the membrane assumes a more convoluted shape. Pathologists previously referred to this type of cell death as shrinkage necrosis. Within minutes the nucleus disintegrates, the membrane becomes more convoluted, and it engulfs segments of cytoplasm including intact organelles such as mitochondria and endoplasmic reticulum and fragments of the nucleus. These apoptotic bodies are soon phagocytized by adjacent cells in which they blend with lysosomes and are digested[1] (**FIGURE 1-1**) (7).

In contrast to necrosis, there is no inflammation in apoptosis and the architecture of the tissue (e.g., skin or intestine) is not altered. From the start the process was believed to require activation of gene expression and energy. **TABLE 1-1** lists various stages of life in which apoptosis may occur.

### Chromatin Degradation in Apoptosis

DNA is degraded in many—if not most—forms of cell death. In some cases, for example, after exposure to ionizing radiations it occurs early, in part because of unrepaired damage. DNA fragments are potentially dangerous because they may trigger autoimmune reactions or parasitize the genome of neighboring cells. Therefore, it is imperative that fragmented DNA of dead cells be degraded. In necrosis degradation of DNA is disorderly. In apoptosis it takes place at specific sites yielding single nucleosomes or multiples of nucleosomes.

Thus, a characteristic DNA fragmentation that assumes a ladder configuration on gel electrophoresis resulting from internucleosomal cleavage is in most cases considered to be the biochemical hallmark of apoptosis (6–9). In 1998, Enari and coworkers discovered an enzyme that cleaves chromatin at the internucleosomal site (9). Two years later the same group, working in Nagata's laboratory, used mouse lymphoma cells to show that the agent responsible for the cleavage of chromatin is a deoxyribonuclease activated by a caspase (8), a member of a family of proteolytic enzymes (described in Chapter 6) that plays a critical role in apoptosis. The deoxyribonuclease (40,000 Da,

[1]Note the similarity between the apoptotic bodies and the foci of cytoplasmic degradation described by the Chicago School. The latter, however, generally assumed that the foci were sites of autodigestion. Hewson Swift suggested that they might be fragments picked up by the phagocytic cells (personal communication made in 1964).

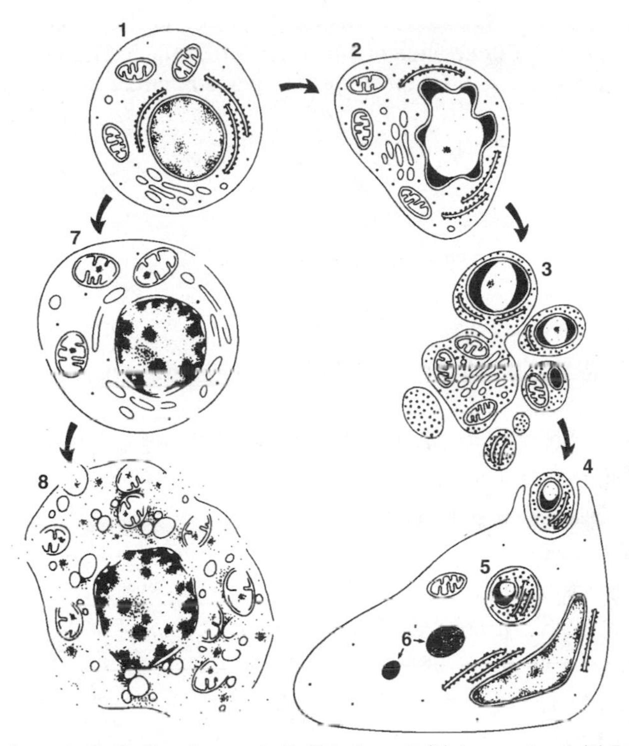


**FIGURE 1-1** Process of apoptosis (2–6) and necrosis (7, 8) in the cell. (1) A normal cell. (2) Early apoptosis is characterized by compaction and segregation of chromatin in sharply circumscribed masses that abut on the inner surface of the nuclear envelope, convolution of the nuclear outline, condensation of the cytoplasm with preservation of the integrity of organelles, and the beginning of convolution of the cell surface. (3) The nucleus fragments, and further condensation of the cytoplasm is associated with extensive cell surface protrusion, followed by separation of the surface protuberances to produce membrane-bounded apoptotic bodies of varying size and composition. These bodies are phagocytized (4) by nearby cells and are degraded by lysosomal enzymes (5) and are rapidly reduced to nondescript residues within telolysosomes (6). In the irreversibly injured cell, the onset of necrosis (7) is manifest as irregular clumping of chromatin without radical change in its distribution, gross swelling of mitochondria with the appearance of flocculent densities in their matrices, dissolution of ribosomes, and focal rupture of membranes. At a more advanced stage of this process (8), all cellular components disintegrate. In tissues, the overall configuration of the cell is reasonably maintained until it is removed by mononuclear phagocytes, but in cell cultures dissolution eventually ensues. (From Kerr, J.F.R., et al. "Anatomical methods in cell death," Chapter 1, *Methods in Cell Biology,* Vol. 46 *Cell Death,* L.M. Schwartz and B.A. Osborne, eds., San Diego: Academic Press, 1995, with permission.)

343 amino acids) is a basic protein that carries a nuclear localization signal. It binds to the nucleus by its carboxyterminal. Caspase activated deoxyribonuclease (CAD) is at first located in the cytoplasm where it is maintained inactive by its binding to an inhibitor, ICAD. This inhibitor has been identified as an acid protein that complexes with CAD, thereby inhibiting its activation and translocation to the nucleus. The combination CAD-ICAD protects the chromatin from DNAse cleavage during the cell's life. The inhibitor ICAD is cleaved at two sites: asparagines 117 and 224. The cleavage of ICAD releases CAD and once free, CAD degrades DNA. ICAD is cleaved from the complex (CAD-ICAD) by the proteolytic action of caspase 3 (**TABLE 1-2, FIGURE 1-2**).

ICAD, the inhibitor exists in two forms, generated by alternative splicing, a long (331 amino acids, 30,000 Da) and a short (226 amino acids, 29,000 Da). Gel electrophoresis of the products of CAD reveals the appearance of large nucleotide fragments before the typical nucleosomal ladder is generated, suggesting that the same enzyme is responsible for the two modes of cleavage. Overexpression of ICAD blocks chromatin degradation, but does not prevent other events associated with apoptosis such as mitochondrial and other organelle degradation, or membrane blebbing (8, 9).

**Table 1-1 Occurrence of apoptosis (programmed cell death) in organs at different stages of the mammalian life cycle**

| |
|---|
| Cell death during development (e.g., in brain) |
| Cell death in normal adult tissues (e.g., in liver) |
| Elimination of autoreactive T cell clones |
| Involution processes |
| Ovarian follicular atresia |
| Lactating breast |
| Adrenal in neonates or after suppression of ACTH secretion |
| Prostate after castration |
| Exocrine pancreas after ductular obstruction |
| Renal tissue after ureteric ligation |
| Reversion of artificially induced hyperplasia[a] |
| Apoptosis in altruistic suicide |
| Cells exposed to radiation |
| Spontaneous cell death in tumors |
| Response to carcinogens in target tissues |
| In graft versus host response |
| Cell death caused by T cells |

[a]Hepatocytes after cyproterone acetate cessation.
ACTH = adrenocorticotrophic hormone.

In summary, DNA fragmentation is caused by an enzyme called CAD, also called DFF40, which is DNA fragmentation factor 40. In the non-apoptic cell, CAD is bound to an inhibitor ICAD (also DFF45). In cells that do not undergo apoptosis, ICAD is bound to CAD through two binding sites. After an apoptotic stimulation, ICAD is cleaved into three fragments by caspase 3, and CAD and ICAD separate (**FIGURE 1-2**).

There are two forms of ICAD, a short one of 265 amino acids and a long one of 331 amino acids. It is the long ICAD that associates with CAD.

CAD possesses two domains: a C terminal, the nuclease catalytic domain, and an N terminal or regulatory domain (10). ICAD contains a sequence homologous to the N domain of CAD. CAD's amino acid sequence is known and its C terminal contains six histidine residues (11). Site mutations or chemical modifications of some of these histidine demonstrate that they are needed for DNA binding (12) and possibly for enzymatic cleavage by CAD.

**Table 1-2 Nomenclature of DNA fragmentation enzymes in apoptosis**

| | |
|---|---|
| CAD | Caspase-activated DNase |
| ICAD | Inhibitor of caspase-activated DNase |
| DFF40 | DNA fragmentation factor 40 (equal to CAD) |
| DFF45 | DNA fragmentation factor 45 (equal to ICAD) |

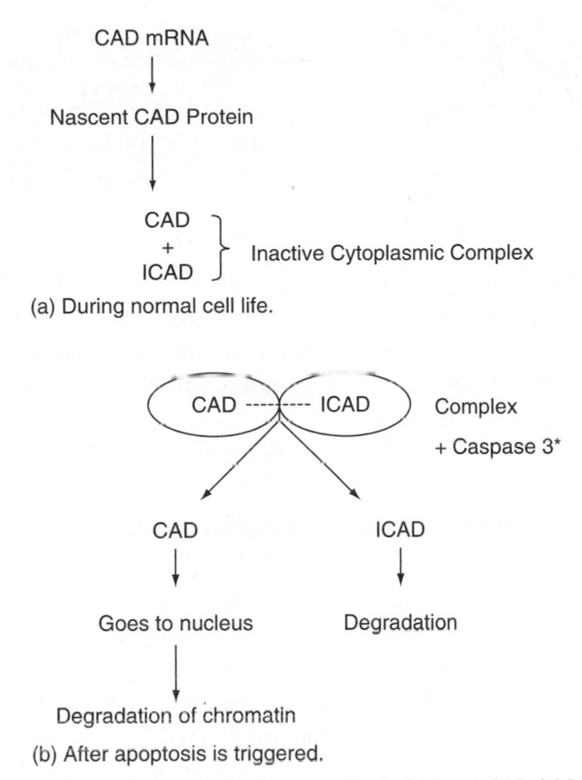


**FIGURE 1-2** The pathway to chromatin degradation in apoptosis via CAD and ICAD. (a) During the normal cell life. (b) After apoptosis is triggered. Cleavage of ICAD releases CAD, and once free, CAD degrades DNA. ICAD is cleaved from the complex (CAD-ICAD) by the proteolytic action of caspase 3.
*See references 9–11.

The potential roles of the amino acids necessary for DNA cleavage were investigated, and substitution of lysine by glutamine and tyrosine by phenylalanine demonstrate that the conserved lysines and tyrosines of the active site are essential for cleavage, but not for DNA binding (13). On the basis of these and other studies it was ultimately established that:

1. The CAD-ICAD interacting site includes amino-acid residues in the CAD sequence up to amino acid 241.
2. The DNA binding site is located between Lys 155 and His 313. All residues that block the catalytic activity are located between Lys 155 and His 313.

## CIDE and CAD-ICAD Interaction

In an attempt to understand the molecular function of ICAD and DFF45, Inohara et al. searched available gene libraries for genes susceptible to encode proteins homologous to DFF45. They discovered two such mammalian genes that function in apoptosis, one gene whose product is associated with terminal differentiation of fat cells, *ESP27*, and another found in *Drosophila melanogaster*.

Because the products of these genes were found to contribute to apoptosis, they were named CIDEs for cell death-inducing DFF45-like effectors. The CIDEs have two domains, a C terminal (called CIDE C) and an N terminal (called CIDE N).

We focus on the findings made on a mouse gene that possesses two initiation sites and encodes two highly homologous proteins, one 217 amino acid long and another shorter by 17 amino acids in its N terminal sequence. These two proteins are referred to as CIDE A and CIDE B.

CIDE A and CIDE B share homology to each other, and they are also to an extent homologous to DFF45 9, the CAD inhibitor. However, the homology between CIDE A and CIDE B and that of DFF45 is restricted to the N terminal region of the latter (39 and 29% amino acid identity for A and B, respectively) (**FIGURE 1-3**). The C terminal regions of CIDE A and CIDE B, referred to as CIDE C, are highly homologous to each other, but they present no homology with the C domain of DFF45. Human CIDE A and B have been identified. Several properties of mouse and human CIDE A and B are critical to the understanding of their function. The various CIDEs have different functions that are sometimes related to DNA fragmentation, but here their relevance resides in the interaction of their CIDE N domains.

It was stated earlier that CAD has two distinct domains, a C terminal enzymatic (residues 290–345 amino acids), and an N terminal regulatory (residues 1–83) domain. The regulatory domain is homologous to the N domains of the CIDEs, referred to as CIDE N (14). Lugovskoy and associates (15) reasoned that the high level of homology between the CIDE N of DFF40 (CAD), DFF45 (ICAD), and the N terminal domains of other CIDE Ns suggest a common regulatory function for CIDE Ns in DNA fragmentation. This led to the investigation of the solution structure of the CIDE N domain of the CIDE B, which is the short CIDE encoded by the mouse gene. The results of these studies led to a model for the interaction of one CIDE N with the CIDE N of another molecule, hoping that the model may reflect the nature of the in vivo molecular interactions.

The CIDE N structure consists of a twisted, five-stranded β sheet and two α helices arranged in an α/β roll fold (16):

- CIDE N of CIDE B interacts with the CIDE N of CAD (DFF40) and ICAD (DFF45) in a homophilic interaction.
- The interaction face of each CIDE N has two binding sites: a positively and a negatively charged one.
- The cluster of positive charges is convex and that of the negative charges is concave. When the acidic site of one CIDE N points North and the acidic site of the second CIDE N points South, each of the basic sites of CIDE are facing the acidic sites of the companion CIDE. Consequently, the two adjacent CIDE Ns will attract to each other through a cluster of opposite charges and form electrostatic bonds between the two CIDE Ns (**FIGURE 1-4**).

On the basis of these findings, a model was proposed for the activation and inhibition of CAD, the enzyme by ICAD the inhibitor. In absence of apoptosis, CAD and ICAD bind each other through their mutual CIDE N, and as a result the enzyme is inhibited. In apoptosis, caspase 3 splits the ICAD into three domains including the CIDE N of ICAD, and each of the three domains of ICAD dissociate from CAD. However, as long as the milieu contains an excess of CIDE N, the catalytic site of CAD

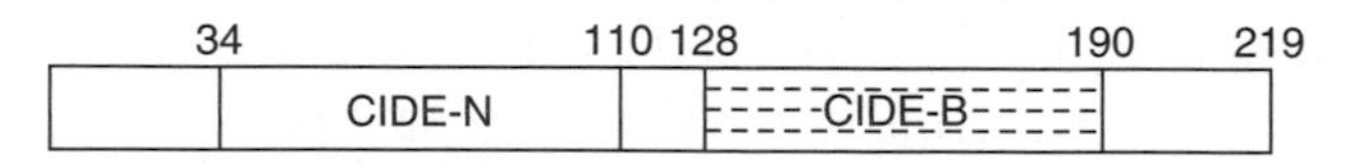


**FIGURE 1-3** Domain structure of CIDE-B. The homology region is limited to the *N* terminal (CIDE-N).

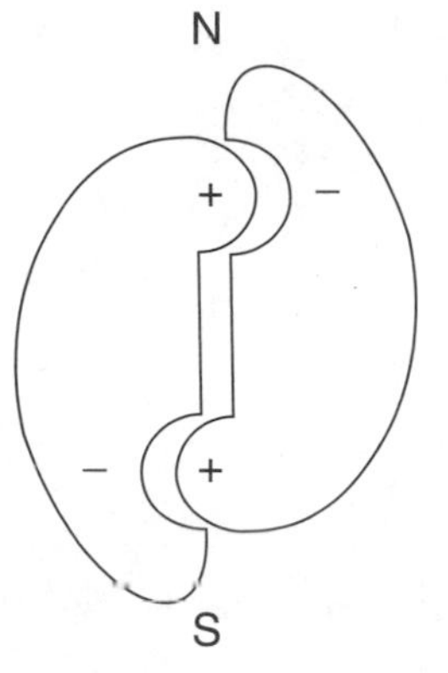


**FIGURE 1-4** Highly schematic representation of the yin/yang interaction between two N CIDES.

remains, but it is not excluded that the CIDE N of the pool may neutralize each other by the homophlilic interaction.

In addition to providing an elegant model for the interaction between the DNAse, CAD, and its inhibitor, ICAD, these findings also constitute a model for homophilic reaction, a type of reaction often seen among proteins that interact in apoptosis. The breakdown of the cell's chromatin into either single nucleosomes or chains of multiples of nucleosomes is responsible for the ladder structure observed on electrophoresis. The molecular mechanism by which chromatin is degraded by CAD was clarified by Woo and associates, who solved the crystal structure of activated CAD/ICAD (16).

A vector containing the appropriate promoters and the coding sequences for CAD and ICAD was constructed, and CAD (residues 1–139) and ICAD (residues 1–312) sequences (both devoid of their NSL) were expressed simultaneously in *Escherichia coli*. The CAD ICAD complex was purified and CAD separated from the CAD/ICAD complex after degradation of ICAD by caspase 3. The recombinant monomeric CAD was made of 321 residues. The folded sequence was found to contain three separated domains: C1, C2, and C3. C1 corresponded to the N-terminal CAD domain (residues 1–85), and its electron density was practically nonexistent as shown previously (11). It Is suggested that C1 functions as a mass capable to move around within the molecular field. C2 is small (residues 86–131) and its sequence folds into three alpha helices. C3, the largest domain, is divided in three components or alpha helices (α1–α8) that are assembled to emerge as a stretched out structure.

The active enzyme is a dimer in which the monomers are held together through weak interaction between the helices of the monomeric C2 domains. It is suggested that the weak attachment of the monomers, because it makes the energy demands low, may explain the easy separation of the dimers. The C3 domains of the dimeric enzyme are kept separate and have been compared to the two blades of a pair of scissors in their open mode. The width of the gap created by the scissor-like enzyme is apt to grab a DNA double strand, which, as it reaches the bottom of the gap, finds it sequence cut as it meets the catalytic portion of the enzyme (in particular the invariant His 263, His 308, Lys 310, or His 313). Woo and associates concluded that the molecular structure of the enzyme, CAD, "with this unique substrate-recognition mode," is well fit for cleavage of the nucleosomal linked double-stranded DNA at the exclusion of the DNA wrapped around the core histones (16) (**FIGURE 1-5**).

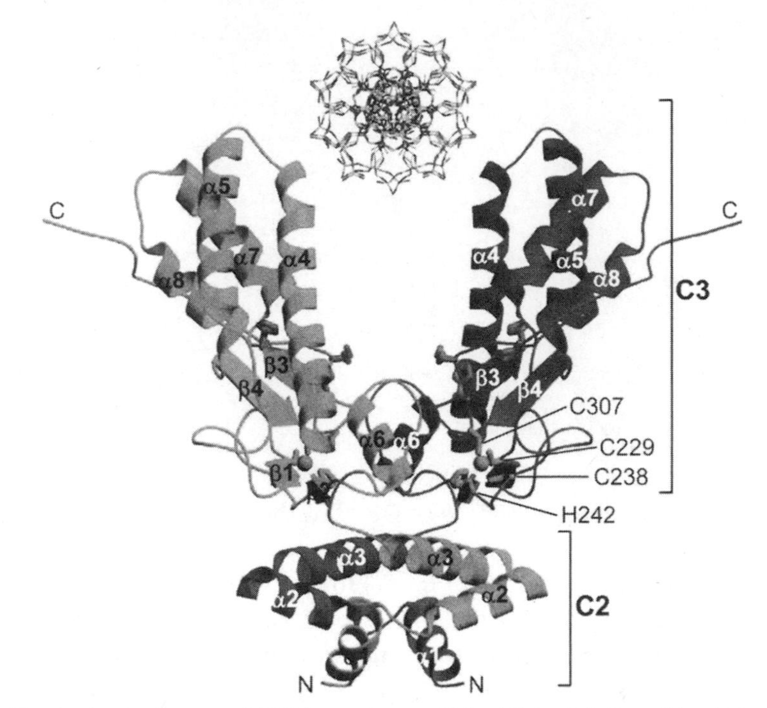


**FIGURE 1-5** Dimeric structural aspect of CAD (see reference 16 for full details). (From Woo, E-J., et al. Structural mechanism for inactivation and activation of CAD/DFF40 in the apoptotic pathway. *Molecular Cell* 14:531–539, 2004. With permission from Cell Press.)

## Chaperones and Their Contribution to the Modulation of Apoptosis and DNA Degradation

Apoptosis is an important homeostatic component in the life of organs. Not only does it contribute to the control of the normal cell population, but also it is needed for the elimination of cells with undesirable activities, among them immunologic cells that attack the self or cells that have acquired threatening survival advantages over their neighbors. Molecular chaperones are among those molecules that modulate apoptosis by either stimulating or inhibiting it. For example, heat shock protein 27 (HSP27), HSP70, and HSP90 function as antiapoptotic proteins, the first by binding to cytochrome C and the two others by binding to Apaf-1, thereby inhibiting the formation of the apoptosome. In contrast, HSP60 can function as a proapototic agent by stimulating the maturation of caspase 3. That HSP 60 sustains apoptosis by stabilizing CAD is of particular interest. CAD's activity is normally lost approximately one hour after its release from ICAD, but it is maintained longer in the presence of HSP70 (9).

The extent to which HSPs participate in the regulation of apoptosis in general and the degradation of DNA in particular is still under investigation, but already chaperones appear to be important contributors to the homeostasis of the cell population. Heath shock proteins may not be the only factors that modulate the activation of CAD, as at the end of apoptosis ICAD can also function as a protector of CAD's functional folding. This role of ICAD as a chaperone for CAD has been attributed to a high density cluster found in the C terminal of DFF45 (ICAD-L) (17, 18).

The role of the chaperone in protecting against apoptosis is not without pathophysiologic importance. Indeed, short-term hypoxia increases CAD activity in cardiomyocytes derived from failing hearts through separation of the CAD-ICAD complex by caspase 3. In contrast, normal cardiomyocytes resist the activation of CAD caused by short periods of hypoxia.

### Non-CAD Endonucleases and Apoptosis

In the face of the rich population of endonucleases found in the cell, it is not surprising that investigators looked for enzymic mechanisms for DNA degradation other than CAD-ICAD. The apoptosis-inducing factor (AIF) is one of them.

AIF is a highly conserved mammalian intramembrane mitochondrial oxireductase that is released from mitochondria during mitochondrial membrane permeabilization (MMP). AIF is transferred to cytoplasm and nucleus, where it functions as a DNAse. The three-dimensional crystal structure of AIF has been described; human AIF contains a catalytic site (an oxidoreductase) and a putative DNA binding site. It has been suggested that while AIF is in the mitochondria, the redox status prevents apoptosis, and once AIF has leaked from mitochondria to the cytoplasm and nucleus, the redox mode favors apoptosis (19).

Endonuclease G is another mitochondrial intramembrane protein that is released by permeabilization of the mitochondria. Like AIF, endonuclease G is a DNA degradation agent when added to nuclei. *C. elegans* harbors the AIF homolog WAN-1 that is transferred to the nucleus during apoptosis. Its release is associated with that of endonuclease G and the two appear to degrade DNA. At first these events were believed to be caspase-independent. But there is evidence that caspase activation is required to release these two mitochondrial proteins (20).

### Lysosomes and DNA Degradation

When caspase 3-resistant ICAD mutants (ICAD-Sdm) are expressed in cells exposed to apoptotic stimuli, DNA is not fragmented into multiples of, or single nucleosomal fragments. Thus, CAD is the principal enzyme involved in DNA fragmentation in cells subjected to apoptotic stimuli; it occurs in the apoptotic cell without the help of other cells (e.g., macrophages).

However, if nondigested DNA is potentially harmful, is it conceivable that a backup pathway for DNA fragmentation might exist? McIlroy and associates tested this hypothesis (21, 22). Thymocytes prepared from ICAD-Sdm transgenic mice and from wild-type mice were exposed to different apoptotic stimuli ($\gamma$ radiation, dexamethasone and FAS antibody). The cells were tested for DNA fragmentation by TUNEL (terminal deoxy transferase uridine triphosphate nick end-labeling) (23) and for their aptitude to be phagocytized by annexin V binding (see Chapter 2). While the wild-type cells were both positive for annexin V, a measure of the aptitude of the cells to be phagocytized, and TUNEL, a measure of DNA degradation, the ICAD-Sdm transgenic mice were found to be annexin V positive only (**TABLE 1-3**).

**Table 1-3 Thymocyte exposed to apoptotic stimuli in vitro**

| | Wild | Transgenic |
|---|---|---|
| TUNEL | + | – |
| Annexin V | + | + |

Sakahira extended their studies in vivo. Apoptosis is normally observed in the thymus, the corpus luteum, and the ovarian follicle of mice, and it can be detected by the TUNEL method[2] (18). Thymocytes were prepared from thymus of irradiated mice at various times after exposure for analysis by flow cytometry of: TUNEL positivity, annexin V binding capacity, and caspase 3 activation. While the three parameters increased almost linearly and simultaneously in the thymocytes derived from wild mice, the three curves were widely separated in thymocytes derived from transgenic mice. The thymocytes of wild-type mice are prone to be phagocytized and have their caspase 3 activated, and therefore, CAD is freed from ICAD and the cells are TUNEL-positive. In contrast, in the thymocytes derived from transgenic mice (caspase resistant) annexin V binding is remarkable, caspase release is slow, and both TUNEL and DNA degradation are even slower. Thus, cells containing unfragmented DNA are phagocytized.

Can the DNA of the prey be degraded by the phagocyte's lysosmal deoxyribonucleases? Thymocytes exposed to an apoptotic stimulus (such as glucocorticoid hormone) are presented to peritoneal macrophages; they are engulfed and the ICAD-Sdm cells become TUNEL-positive. Proof that lysosomes (DNase(s)) are involved was provided by inhibiting lysosomal acidification by chloraquine (19, 20).

*Caenorhabditis elegans* have yielded additional information about DNA degradation. The process involves the product of several genes. *ced-ced-1* is required for the formation of TUNEL+ fragments with 5′ $PO_4$ and 3′ OH termini, and *ced-7* facilitates the process. Thus, mutations of *ced-ced-1* arrest the process and mutations of *ced-7* slow that process down. The *nuc-1* gene encodes a protein with properties similar to that of mammalian DNase II. NUC-1 is active at an acidic pH, and does not require $Ca^{++}$ or $Mg^{++}$ for activity. It generates fragments with 5′ OH and 3′ $PO_4$ ends and, therefore, they are not TUNEL+. *ced-2, ced-5,* and *ced-10* are involved in the formation of pseudopodia that expand around the dying cells.

On the basis of these laborious and elegant studies, a three-step model for DNA degradation in programmed cell death was proposed (21, 22) In the first step, an unknown endonuclease generates TUNEL reactive fragments. The products of *ced-1* and *ced-7* are required. The products of *ced-1* and *ced-7* contribute to signaling the readiness of the dying cells for engulfment. The products of *ced-2*, *ced*-5, *ced-6*, and *ced-10* are required for pseudopodal expansion (21). Clearly, the work of Wu et al. in *C. elegans* (24) and that of McIlroy et al. (21, 22) suggests that the process of DNA degradation is conserved from *C. elegans* to mammals, as are other critical aspects of programmed cell death. Moreover, in *C. elegans* the process signals the need for phagocytosis; however, the molecular nature of that signal remains unclear.

In the second step, DNase II, the product of the *nuc-1* gene, generates TUNEL-unreactive fragments. The mechanism of the focal or general acidification of the dying cell, which is required for DNase II activity, is still unclear.

In the third step, the TUNEL unreactive fragments are further degraded and activity of the *nuc-1* gene is required for the process, but it is also likely that the DNase(s)

[2]One of the many methods used to detect apoptosis includes the labeling of the free 3′ end of the DNA fragments generated in the nuclei of apoptotic cells. A terminal deoxynucleotide transferase is used to bind a biotin-uridine complex to the free 3′ ends. Streptavidin is a compound that avidly binds to the biotin that is added to cells in culture. The biotin-streptavidin remains colorless in living cells, but stains the apoptotic cells (TUNEL method).

secreted by the phagocytes are involved. In any case, the last step is associated with expansion of the pseudopodia followed by engulfment of the dead cell.[3]

It should be added that the manifestations of apoptosis have been observed in cytoblasts and enucleated cells. Moreover, thioacetamide produces apoptosis in L37 cells (a cell line derived from a rat liver cell line transgenic for the albumin SV40-T antigen) without causing DNA fragmentation (25). These findings suggest that the mechanisms that control apoptosis can function independently of nucleosomal degradation.

## Conclusion

The principal enzymatic machinery that regulates nuclear degradation involves a nuclease CAD and its inhibitor ICAD. CAD-ICAD is present as a cytoplasmic complex. The association of the two molecules is likely to be secured by electrostatic complementation. In apoptosis, CAD is released from ICAD through cleavage by caspase 3. CAD functions a molecular "scissor" constructed to attack the molecular chain that links nucleosomes. Other enzymic mechanisms have been proposed. They include the caspase independent release of AIF, an intramembrane oxyreductase, and that of a mitochondrial endonuclease (endonuclease G). However, it is still not clear if the CAD-ICAD and the AIF DNA degradation occur simultaneously or independently, and how critical the AIF release is to DNA degradation.

Is DNA fragmentation always a requirement for cell death by apoptosis? Apoptotic cytoblasts provide a convincing argument against the absolute necessity for DNA fragmentation (25). These observations suggest that apoptosis could occur in absence of nuclear fragmentation, if apoptosis can be defined in terms of cytoplasmic events (such as loss of membrane integrity and decrease in mitochondrial functions) induced by caspase activation and inhibited by antiapoptotic agents (Bcl-2). However, in apoptotic nucleated cells, early destruction of DNA appears to be most desirable, if genomic instability and autoimmunity are to be avoided (25).

## References and Recommended Reading


1. Glucksman, A. Cell death in normal vertebrate ontogeny. *Biol Rev.* 1951;24:58–86.
2. Saunders, J.W. Death in embryonic systems. *Science.* 1966;154:604–42.
3. Kerr, J.F.R. Shrinkage necrosis: a distinct mode of cellular death. *J Pathol.* 1971;105:13–20.
4. Kerr, J.F.R., Wyllie, A.H., and Currie, A.R. Apoptosis: a basic biological phenomenon with wide ranging application in cell kinetics. *Br J Cancer.* 1972;26:239–57.
5. Rich, T., Watson, C.J., and Wyllie, A. Apoptosis: the germs of death. *Nat Cell Biol.* 1999;1:E69–E71.
6. Wyllie, A.H., Kerr, J.F.R., and Currie, A.R. The significance of apoptosis. *Int Rev Cytol.* 1980;68:251–356.


[3]In the absence of added stimuli, the levels of apoptosis were the same in wild-type and transgenic mice (ICAD-Sdm). However, when these mice were exposed to 1300 cGy of $\gamma$ radiation, there was a great difference between the wild-type and transgenic mice. A large number of cells were TUNEL-positive in the former and few clusters were TUNEL-positive in the latter.

7. Kerr, J.F.R., Gobè, G.C., Winterford, C.M., and Harmond, B.V. Anatomical methods in cell death. In: Schwartz, L.M., and Osborne, B.A., eds. *Cell Death* (from Methods in Cell Biology Series). 1995;46:1–27, San Diego, Academic Press.
8. Mukae, N., Yokoyama, H., Yokokura, T., Sakoyama, Y., Sakahira, H., and Nagata, S. Identification and developmental expression of inhibitor of caspase-activated DNAse (ICAD) in *Drosophila melanogaster. J Biol Chem.* 2000;275:21402–8.
9. Enari, M., Sakahira, H., Yokoyama, H., Okawa, K., Iwamatsu, A., and Nagata, S. A caspase-activated DNAse that degrades DNA during apoptosis, and its inhibitor ICAD. *Nature.* 1998;391:43–50.
10. Inohara, N., Koseki, T., Chen, S., Wu, X., and Nunez, G. CIDE, a novel family of cell death activators with homology to the 45 kDa subunit of the DNA fragmentation factor. *EMBO J.* 1998;17:2526–33.
11. Uegaki, K., Otomo, T., Sakahira, H., Shimizu, M., Yumoto, N., Kyogoku, Y., and Nagata, S. Structure of the CAD domain of caspase-activated DNAse and interaction with the CAD domain of its inhibitor. *M Mol Biol.* 2000;297:1121–8.
12. Meiss, N., Enari, M., Sakahira, H., Fukuda, Y., Inazawa, J., Toh, H., and Nagata, S. Molecular cloning and characterization of human caspase-activated DNAse. *Proc Natl Acad Sci USA* 1998;95:9123–8.
13. Korn, C., Scholz, S.R., Gimadutdinow, O., Pingoud, A., and Meiss, G. Involvement of conserved histidine, lysine and tyrosine residues in the mechanism of DNA cleavage by the caspase-3 activated DNAse CAD. *Nucleic Acids Res.* 2002; 30:1325–32.
14. Inohara, N., Koseki, T., Chen, S., Benedict, M.A., and Nunez, G. Identification of regulatory and catalytic domains in the apoptosis nuclease DFF40/CAD. *J Biol Chem.* 1999;274:270–4.
15. Lugovskoy, A.A., Zhou, P., Chou, J.J., McCarty, J.S., Li, P., and Wagner, G. Solution structure of the CIDE-N domain of CIDE-B and a model for CIDE-N/CIDE-N interactions in the DNA fragmentation pathway of apoptosis. *Cell.* 1999; 99:747–55.
16. Woo, E.J., Kim, Y.G., Kim, M.S., Hhang, W.D., Shin S, Robinson, H., Park S.Y., and Oh, B.H. Structural mechanism for inactivation and activation A of CAD/DF40 in the apoptotic Pathway. *Mol Cell.* 2004;14:531–9.
17. Fukushima, K., Kikuchi, J., Koshiba, S., Kigawa, T., Kuroda, Y., and Yokoyama, S. Solution structure of the DFF-C domain of DFF45/ICAD. A structural basis for the regulation of apoptic DNA fragmentation. *J Mol Biol.* 2002;321:317–27.
18. Sakahira, H., Iwamatsu, A., and Nagata, S. Specific chaperone-like activity of inhibitor of caspase-activated DNAse for caspase-activated DNAse. *J Biol Chem.* 2000;275:8091–6.
19. Lipton, S.A., and Bossy-Wetzel, E. Dueling activities of AIF in cell death versus survival: DNA binding and redox activity. *Cell.* 2002;111:147–50. (review).
20. Wang, X., Yang, C., Chai, J., Shi, Y., and Xue, D. Mechanisms of AIF-mediated apoptotic DNA degradation in *Caenorhabditis elegans. Science.* 2002;298:1587–92.
21. McIlroy, D., Sakahira, H., Talanian, R.V., and Nagata, S. Involvement of caspase 3-activated DNAse in internucleosomal DNA cleavage induced by diverse apoptotic stimuli. *Oncogene.* 1999;18:4401–8.

22. McIlroy, D., Tanaka, M., Sakahira, H., Fukuyama, H., Suzuki, M., Yamamura, K., Ohsawa, Y., Uchiyama, Y., and Nagata, S. An auxiliary mode of apoptotic DNA fragmentation provided by phagocytes. *Genes Dev.* 2000;14:549–58.
23. Gavrieli, Y., Sherman, Y., and Ben-Sasson, S.A. Identification of programmed cell death *in situ* via specific labeling DNA fragmentation. *J Cell Biol.* 1992; 119:493–501.
24. Wu, Y-C., Stanfield, G.M., Horvitz, H.R. NUC-1, a *Caenorhabditis elegans* DNAse II homolog, functions in an intermediate step of DNA degradation during apoptosis. *Genes Dev.* 2000;14:536–48.
25. Bulera, S.J., Sattler, C.A., Gast, W.L., Heath, S., Festerling, T.A., Pitot, H.C. The mechanism of thioacetamide-induced apoptosis in the L37 albumin-SV40-T-antigen transgenic rat hepatocyte-derived cell line occurs without DNA fragmentation. *In Vitro Cell Dev Biol Anim.* 1998;34:685–93.

CHAPTER

# 2 RECOGNITION AND DISPOSAL OF THE APOPTOTIC BODIES

Apoptosis generates membrane-surrounded apoptotic bodies that contain potentially threatening molecules. Among them, but not exclusively, are hydrolases operating at physiological or acidic pHs, such as lipases, amylases, proteases, and nucleases. Clearly, to avoid an inflammatory reaction or an autoimmune response, apoptotic bodies need to be removed from the tissues before they lyse. Moreover, it is essential that the cells that engulf the apoptotic bodies do not release cytokines that might trigger inflammation. The elimination takes place by phagocytosis. Professional phagocytes (monocytes, macrophages), semiprofessional phagocytes (dendritic cells, glomerular mesangial cells, Kupffer cells of the liver, vascular endothelial cell, Sertoli cells), and even resident cells in the tissue and cancer cells may acquire phagocytic properties.

Although a great deal has been learned about the morphology, physiology, and molecular biology of phagocytosis of foreign agents including parasites, bacteria, and viruses, our knowledge of phagocytosis of the self, albeit that the prey is in a dying process, is still limited. From the start, changes in the molecular rearrangement in the cell membrane of the apoptotic bodies were suspected of signaling the need for engulfment of the phagocyte. To be receptive, the phagocyte must make some sort of a receptor available that allows binding between the apoptotic bodies and the phagocyte. The recognition sites include appearance at the surface of senescent cell or apoptotic bodies of phosphatidylacetylserine or sugar groups that are hidden inside the membrane in normal cells. Integrins, located at the surface of the macrophage, in another sequence of events, may bind to the extracellular products secreted by the macrophage (e.g., thrombospondin) and use the latter to form a bridge between a molecule in the dying body (e.g., vitronectin) and the phagocyte. This chapter briefly reviews the relevant properties of these most investigated molecules and their participation in the engulfment of the apoptotic bodies.

## Phosphatidylserine

The lipid bilayer of the eukaryotic cell membrane contains several phospholipids and phosphatidylcholine sphingomyelin, phosphatidylserine, and phosphatidyl etanolamine. Among them, only phosphatidylserine carries a negative charge. Phosphatidylserine (**FIGURE 2-1**) is located in the cytoplasmic face of the membrane where its content varies

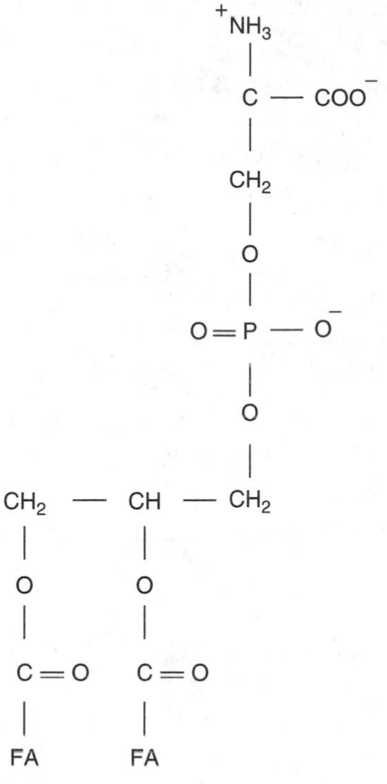

**FIGURE 2-1** Formula of phosphatidylserine.

with the cell type, from 4% of the total lipid weight in the liver to 7% in the erythrocyte plasma membrane.

Protein kinase C, after it has been translocated to the cytoplasmic aspect of the plasma membrane, is activated by diglycerol, $Ca^{2+}$, and the negatively charged phosphatidylserine. The activated protein C kinase phosphorylates serine or threonine residues, thereby triggering various signals such as activation of transcription and phosphorylation of activator or inhibitor proteins.

In apoptotic cells, the negatively charged phosphatidylserine, by unknown mechanisms, is transferred from the cytoplasmic face of the membrane to the cell surface, where it has been reported to signal the apoptotic body's readiness for phagocytosis to the macrophage (1).

The exteriorization of phosphatidylserine must be associated with structural membrane changes. The latter are manifested by disruption of the symmetry of the phospholipids within the membrane. The integrity of the cell membrane requires the persistent activity of a special group of enzymes:

- A translocase (amino-phospholipid-translocase)
- An ATPase that contributes to the recruitment of the phospholipids to the membrane
- A scramblase that catalyzes the movements of lipids in and out of the membrane
- A flopase that secures lipid exit also prevents leakage from the apoptotic cell that would cause inflammation (2).

While the scramblase is adenosine 5-triphosphate (ATP) independent, the flopase requires ATP (3). It has been shown that in central nervous system (CNS) cells, inhibition of the translocase is sufficient for the disruption of the symmetry of the phospholipids within the membrane. To prevent the release of toxic and proinflammatory molecules, the leakage of that asymmetry needs to be strictly controlled (3).

In cultured cell lines in which apoptosis is triggered by Fas ligation (see below), phosphatidylserine is externalized prior to other apoptotic events, including the increase of the plasma membrane permeability. When cells are exposed to N-etylmaleimide, phosphatidylserine is externalized but apoptosis does not take place; still, these cells are phagocytized (4). Adayev and associates, using a different cell line and a different trigger for apoptosis, have confirmed that the externalization of phosphatidylserine is followed by phagocytosis, but they also showed that the externalization occurs late in the sequence of events that characterize apoptosis (5).

The study by Zhuang et al. may explain why cell surface alteration appears early in some studies and late in others (6). They exposed monocytes to mitochondrial inhibitors (antimycin A and oligomycin) to induce apoptosis. Their study showed that (1) the inhibitors interfered with plasma membrane alterations, including phosphatidylserine externalization as well as the thrombospondin recognition pathway; and (2) this took place without interfering with other characteristic manifestations of apoptosis resulting from caspase activation, such as poly (ADP)-ribose polymerase (PARP) breakdown[1] (from 116 kDa to 86 kDa) and DNA degradation. The authors concluded that surface alterations and nuclear and cytoplasmic changes associated with apoptosis are independent events, possibly initiated by different transducing pathways (6).

If phosphatidylserine exterioralization and nuclear and cytoplasmic alterations are independent manifestations of apoptosis, it is conceivable that they also can be temporally dissociated. Consequently, it is not impossible that phosphatidylserine exteriorization may occur early or late in apoptosis depending upon the variable circumstances. In any event, there is little doubt that the exteriorization of phosphatidylserine in apoptotic cells or apoptotic bodies signals their readiness for phagocytosis to macrophages. This does not exclude the possibility, however, that other molecules also may be required for successful phagocytosis. Transfection of melanoma cells with CD36 enhanced their ability to phagocytize neutrophils, lymphocytes, and fibroblast (7).

Monocytes, the precursors of macrophages, are inefficient phagocytes. In contrast to the indifferentiated products, the macrophages are efficient phagocytes. Indeed, the appearance of phosphatidylserine at the cell surface must occur in both phagocytes and the death cell for efficient engulfment of the apoptotic bodies. The appearance of phosphatidylserine at the surface of macrophages is linked to its differentiation and is not limited to the presences of apoptotic bodies.

Van den Eijnde and associates injected a combination of biotinylated annexin V and a calcium-dependent phosphatidylserine binding protein into healthy mouse and chick embryos and into *Drosophila* pupae. The apoptotic cells that emerged during development were observed to bind to the annexin, suggesting that the exposure of phosphatidylserine prior to engulfment of apoptotic cells is phylogenetically conserved (8).

Transfection of CD36 to poorly phagocytic cells enables them to phagocytize. Thus, CD36 can convert an occasional phagocyte into a professional phagocyte. CD36 (molecular wt 88 kDa) is an adhesion molecule elaborated by platelets, monocytes, and endothelial cells. It contributes to the recognition of the readiness of apoptotic cells for

[1]PARP is a caspase substrate.

phagocytosis. It is involved in the $\alpha_V\beta_3$ integrin as well as the phosphatidylserine recognition pathways (9).

The β2 glycoprotein I (β2GP1) is a plasma protein that binds negatively charged phospholipids, among them phosphatidylserine, and also contributes to phagocytosis. The macrophage membrane possesses a receptor that binds β2GP1. Therefore, it seems reasonable that β2GP1 may play an important role in the uptake of phosphatidylserine–β2GP1 complex and phagocytosis of apoptotic cells (10). Antibodies to other potential phosphatidylserine receptors such as CD6 did not inhibit the phosphatidylserine–β2GP1 uptake, leading to the conclusion that CD6 or other potential macrophage receptors are not responsible for the formation of phosphatidylserine–β2 glycoprotein 1 complex.

Finally, does phosphatidylserine externalization only occur in one form of death—apoptosis—at the exclusion of necrosis? Brouckaert and colleagues reported that phagocytosis by macrophages of dead cells whether by necrosis or apoptosis uses a phosphatidylserine dependent pathway (11). It is of interest that the apoptotic cells are engulfed faster than necrotic cells and, moreover, there is no inflammation associated with the clearance of either type of death cells (12).

## Potential Receptors for Phosphatidylserine

Although no specific receptor for phosphatidylserine is known, several potential receptors have been proposed: CD36, CD68, CD14, Lox 1, and B2GPI. The problem with these putative receptors is that they are promiscuous, and researchers are unable to distinguish between different anionic phospholipids found in the cell membrane. As a result, several phospholipids may compete for the same receptor. For example, overloading of the medium with phosphoinositol blocks the interaction of most anionic phospholipids with the above receptors. However, Fadock and colleagues demonstrated in 1992 that the binding of phosphatidylserine to phagocytes is not inhibited by other anionic phospholipids, suggesting that the phagocytes harbor specific receptors for phosphatidylserine (PSR) (9). The same group later cloned the gene that encodes the "phosphatidylserine receptor" and transfected the gene to B and T cells. The latter cells specifically recognized phosphatidylserine on the surface of apoptotic cells and phagocytized them. The receptor is present on macrophages, fibroblasts, and epithelial cells. Indeed, antibodies to the receptor as well as phosphatidylserine liposomes inhibited phagocytosis of apoptotic cells by the cells mentioned. Genes of unknown function but with sequence homology to the gene that encodes the PSR have been found in *Caenorhabditis elegans* and *Drosophila,* suggesting that phagocytosis may be conserved throughout phylogeny.

Death cells accumulate in PSR defective mice, causing abnormal development of brain and lungs and neonatal death. A finding indicating that the presence of PSR is indispensable to harmonious organogenesis and that the absence of PSR may be associated with brain malformation and respiratory distress.

### ABC and Phagocytosis

Glyburide is an inhibitor of the ATP binding cassette (ABC). Such a cassette is believed to be involved in engulfment by macrophages. Marguet and associates (13) investigated the effects of glyburide (and other inhibitors of the ABC cassette, e.g., oligomycin) on the redistribution of phosphatidylserine at the surface of thymocytes exposed to a trigger of apoptosis (e.g., gamma irradiation) as well as on the phagocytic macrophages. Glyburide inhibited the uptake of apoptotic bodies, whether the irradiated thymocytes or the phagocytic macrophages were exposed to the inhibitor. Inhibition of both irradiated thymocytes and phagocytes had an additive effect on the impairment of the uptake of the apoptotic bodies. Thus, during the process of phagocytosis, membrane changes take place in the target and the phagocyte, and the same pharmacologic agent inhibits the changes in both. The authors further attempted to clarify the nature of the changes in thymocyte's undergoing apoptosis (10). Gamma irradiated thymocytes were exposed to glyburide and the redistribution of phosphatidylserine on the surface was determined by annexing binding. Glyburide interfered with the redistribution of phosphatidylserine. Two enzymes are required for the redistribution of phosphatidylserine, a translocase, and a scramblase. The scramblase can be activated by a calcium ionophore. The glyburide inhibits the activation of the scramblase. Marguet and Chimini (14) concluded that:

- Exposure of phosphatidylserine on the phagocytic prey is required for phagocytosis.
- Phospholipid scrambling in the cell membrane contributes to the exteriorization of the phosphatidylserine.
- The ABC1 (mouse cassette) is likely to be involved in maintaining the molecular arrangements of lipid molecules in the cell membrane.

## Lectins

The molecular composition of the plasma membrane of eukaryotic cells includes 2 to 10% of carbohydrates in the form of either glycolipids[2] or glycoproteins. In glycoproteins the sugars form either O or N linkages. In the O type, the sugar binds to serine or threonine residues (e.g., N-acetylgalactosamine) and in the N type, the sugar protein linkage involves an asparagenic residue (e.g., N-acetylglucosamine).

Lectins are proteins that bind carbohydrates.[3] Discovered in bacteria over a 100 years ago, it took over 70 years to show that they were also present in mammalian cells, where they recognize oligosaccharides whether they are located on soluble glycoproteins, a component of the cellular matrix, or on the plasma membrane. Binding of lectins to monosaccharides can be monovalent when it involves an individual lectin site or multivalent as a result of clustering of binding sites. The first type is weak and the second is strong.

The biological processes performed by lectins include cell-to-cell adhesion, phagocytosis, and intracellular glycoprotein transport. Over 100 animal lectins have been identified, and in some cases their genes have been cloned (15, 16).

In acute inflammation neutrophils are recruited from blood to the inflamed tissue. The blood neutrophils first adhere weakly to the vascular endothelial cells; however, in a second step, the neutrophils and the endothelial cells bind to each other more tightly. Finally, the neutrophils migrate into the tissue by crawling between the endothelial cells (diapedesis). Adhesion between neutrophil and endothelium is mediated by protein carbohydrate binding, a protein that is expressed in the plasma membrane of

[2]In higher organisms, glycolipids are derivatives of sphingosine molecules linked to one or two sugar residues.

[3]Several plant lectins have proved to be useful tools for localization of specific sugars, among them the jack bean–derived concanavalin A, phytoagglutinin derived from red kidney beans, the wheat germ agglutinin, and the peanut lectin.

the endothelial cell. It is a transmembrane protein referred to as ρ selectin. The most distal portion of the extracellular segment of the protein is a lectin domain that serves as a receptor for an oligosaccharide (galactose, N-acetylglucosamine, fucose, or sialic acid) present in glycolipids and glycoproteins lodged in the neutrophil plasma membrane.

The C type[4] carbohydrate recognition domain is linked distally to an endothelial growth factor (EGF)-like domain, a complement binding repeat, the transmembrane segment, and the C terminal sequence (14).[5] The natural killer (NK) cells, a group of lymphocytes distinct from T and B cells, kill virus-infected and transformed cells without prior sensitization. The NK cell needs to distinguish target cells (e.g., infected cells) from normal cells (e.g., non-infected cells) and the molecular mechanism of recognition involves two types of receptors: one that inhibits activation of the NK cell and another that activates it. The inhibitory receptors recognize major histocompatibility complex (MHC) Class I components. The genes encoding several of these receptors have been cloned, and the products belong to two protein families: C lectins and Ig immunoglobulins.[6] The human genes encoding the C lectin type of receptor are located on chromosome 12p12.3-p13.1. The exact nature of the activating receptor, which operates only after the inhibitory receptors have given the permission to kill, remains unknown, but there are several candidates (17).

The two examples, that of polymorphonuclear ligation and the mode of action of NK cells, reveal the importance of lectins in cell biology as well as the paucity of knowledge in the field. Yet it cannot be excluded that lectin may prove to play a role in phagocytosis of apoptotic cells.

## Integrins

Except for cancer cells, cells can only proliferate in vitro if the serum contains extracellular components (e.g., fibronectin). Integrins mediate interactions between cells and components of the extracellular matrix. At first they were considered to be only adhesion molecules, but further investigations revealed that integrins, like other extracellular matrix components such as laminins (18)[7] contribute intracellular signals and thereby participate in cell proliferations, cell differentiation, and phagocytosis (19).

Integrins bind to matrix molecules including collagen, actin, fibronectin, and laminin. Integrins constitute a large family of protein transmembrane heterodimers made of two types of subunits referred to as α or β chains. Each subunit is a transmembrane glycoprotein. In mammals, 16 α and 8 β chains have been identified that combine to form at least 22 receptors. The dimer is composed of an extracellular globular head that includes the $NH_2$ terminals of the α and β chains. An α and a β stalk links the globular head to the transmembrane segment, itself connected to the intracellular COOH terminal (**FIGURE 2-2**). The extracellular portion of the α subunit contains up to 1114 amino acids and that of the β subunit contains up to 678 residues. The globular component of the α chain contains multiple divalent cations binding domains reminiscent of the $Ca^{2+}$ binding domains in calmodulin and troponin C. The divalent cation-binding domain acts in ligand binding. Most of the α N-terminal

[4]C type implies that the lectin is calcium dependent for function.

[5]Those interested in the structural basis for lectin carbohydrate recognition can read Weiss and Drichamer, *Annu Rev Biochem.* 1996 (16).

[6]The Ig types inhibitory receptors are referred to as killer-cell inhibitory receptors (KIRs).

[7]Myoloblasts growing on fibronectin proliferate without any sign of differentiation. In contrast, while grown in presence of laminin, their growth ceases and they fuse to form myotubules.

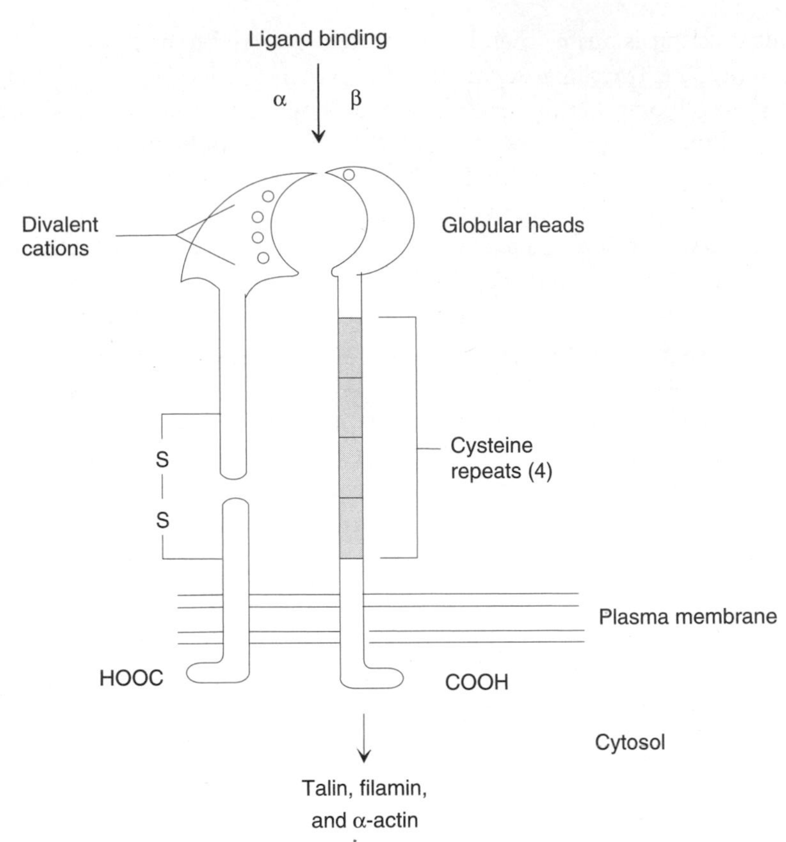


**FIGURE 2-2** Schematic representation of integrin dimers (modified from Alberts, B., Bray, D., Lewis, J., Raff, M., Keith Roberts, K., and Watson, J.D. *Molecular Biology of the Cell,* 4th ed., New York: Garland Science, 2002, p. 1113).

binding domain involves an amino acid sequence starting with residue 1 and ending approximately with residue 440. In some cases there is an additional domain in the N-terminal α chain, the I or insertion domain.[8] In such cases the sequence of the α N-terminal reads as follows: (a) 1–130 residues, (b) 190 residues, and (c) 320–600 residues. The sequence of all these segments of the α chain can be divided into seven conserved repeats, each comprising 60 amino acids. Each repeat forms a four-stranded β sheet and the combination of the seven repeats assumes a shape reminiscent of a propeller. Repeats 5, 6, and 7 have a $Ca^{2+}$ binding site in all α chains and there is also a $Mg^{2+}$ site at the center of the structure. The ligand binding is reduced by mutations in sequence 153–165 and 187–190.

The stalk of the α chain is cleaved by proteolysis and an S-S bridge links the two segments. The stalk of the β chain includes four cysteine-rich repeated sequences. In some integrins, the intracellular chain has alternatively spliced domains (e.g., $\beta_3$, $\beta_4$, $\alpha_6$, $\alpha_3$). The different chains most likely determine the nature of the binding between the target and the integrin. The cytosolic domain of most integrins connects to actin in the cytoskeleton.[9] Overlapping domains of the glomerular portion of the α and β chains form the ligand binding site of the receptor.

As mentioned earlier, the α and the β chains form 22 combinations, and among them, $\beta_1$ is involved with nine α chains that bind to collagens, laminin fibronectin, and

[8]The latter is seen, for example, in lymphocytes and the I domain is homologous to a similar structure in the von Willebrand factor.

[9]An exception is the hemidesmosomal α6β4integrin, which binds to intermediate filaments.

vascular cellular adhesive molecule (VCAM):

- $\beta_1\alpha_V$ binds vitronectin and fibronectin, thus heterodimers that include a $\beta_1$ chain tend to bind to extracellular units.
- The $\beta_2$ family heterodimerizes with special $\alpha$ chains ($\alpha_L$, $\alpha_M$, and $\alpha_X$), which are found only in leukocytes and are mainly involved in cell to cell binding.
- The $\beta_3$ group forms heterodimers with $\alpha_{HQ}$ and $\alpha_V$; they mainly bind to the extracellular matrix components and the von Willebrand factor.
- The $\beta_4$, $\beta_5$, $\beta_6$, and $\beta_8$ groups are less promiscuous and combine to form only one type of heterodimer with one of five $\alpha$ isoforms: $\alpha_6$, $\alpha_V$, $\alpha_V$, $\alpha_4$, and $\alpha_V$, respectively. The $\beta_7$ heterodimers heterodimerize with two $\alpha$ chains ($\alpha_4$ and $\alpha_{IEL}$). The $\beta_4$, $\beta_5$, and $\beta_6$ heterodimers bind laminin, vibronectin, and fibronectin, respectively. $\beta_7\alpha_4$ binds fibronectin, or VCAM-1. The binding target of $\beta_7\alpha_{IEL}$ and $\beta_8$ are unknown. Thus, in many cases different integrins bind the same ligand, but all may not be activated at the same time.

In contrast to hormonal receptors, which have high affinities, integrins have relatively low affinities for their ligands. However, the cell surface usually possesses a large number of integrins that may act simultaneously, thereby securing a firm attachment with cellular matrix components. Inactivation of the links between integrin and ligand is easier if the binding affinity is low. For example, cell migration can be facilitated by modulating the integrin's affinity to the nature of the ligand as required.

Integrins extend their receptor outside the cells and their tail inside the cytosol. The ability of integrins ($\beta_1$, $\beta_2$, and $\beta_3$ families) to become competent for ligand binding is activated through signals emerging in the cytoplasm, and this is referred to as inside out signaling. Two regions in the sequence of the $\beta_1$ chain of the platelet integrin $\alpha_{II}\beta_3$ are critical. One is located in the cytoplasmic region (the highly conserved NPXY) and another is in the region of the sequence proximal to the cell membrane (KLLXXXD) in the $\beta$ chain of the integrin. All $\alpha$ chains also contain highly conserved motifs (KXGFFKR) needed to modulate integrin affinity. Mutations, deletion, or truncation in these regions reduce or abolish activation.

The inside out signaling pathway regulates chemokinin growth factor. The full length of the $\beta_2$ complement receptor type 3 (CR3; CD11b/C18) is required for an effective phagocytic signal. The outside in signaling emerges from the extracellular matrix (ECM) and triggers transduction signals. The latter function is coordinated with that of actin and contributes to the modulation of the balance between cell survival and apoptosis and differentiation. Further details are in the study of Dedhar and Hannigan (19). Although a great deal remains to be learned about the nature of the signals activated by the integrins, an important kinase already has been found to be involved in outside signaling: the focal adhesion-associated kinase (FAK). FAK is a nonreceptor tyrosine kinase with a central catalytic domain flanked by $NH_2$ and COOH terminal sequences, which are uncharacteristic for tyrosine kinases. Activation of FAK is associated with focal adhesion. The model for activation involves:

1. Binding of the ligand to the extracellular domain of the integrin's $\alpha$ and $\beta$ chains.
2. Interaction of the tail of the $\beta$ chain with talin,[10] possibly through a conformational change of the $\beta$ chain.

3. Interaction with vinculin[10] and paxillin[10] via talin.
4. Interaction of talin and vinculin with actin in the cytoplasm.
5. FAK then undergoes conformational changes and its amino terminal reacts with $\beta$ chain sequences located close to the plasma membrane.
6. Different FAK molecules are brought together, FAK is autophosphorylated, and actin contracts; this presumably leads to signaling via the Rho family of GTPases(18).

Integrins have been shown to be involved in phagocytosis of bacteria (e.g., *Escherichia coli*, *Streptococcus pyogenes*, and *Mycobacterium tuberculosis*) by macrophages. Integrins are also involved in the phagocytosis of degenerative axon fragments and rod outer segments by retinal pigment epithelial cells. Therefore, a role of integrin in phagocytosis of apoptotic cells cannot be excluded.

*Escherichia coli* phagocytosis by insect hematocytes requires cytoskeletal rearrangement and integrin receptors. However, some pathogenic *E. coli* elaborate a necrotizing factor I (CNF1) that impairs phagocytosis and activates Rho GTPases. This leads in vitro to membrane alterations in monocytes, including the formation of lamellipodia and cell spreading of macrophages. These changes are associated with reorganization of the actin cytoskeleton and prevent integrin activation on which phagocytosis depends.

Adequate retinal function requires that the outer segments of photoreceptor rods that are shed be phagocytized by retinal pigment epithelial cells. In vitro, vitronectin stimulates the rod fragments uptake by retinal pigmented cells. That uptake depends on the activation of $\alpha_V\beta_5$ integrin. The integrin-vibronectin receptor is expressed on the apical membrane of the human retinal pigment epithelial cell and contributes to the engulfment of the rod segment (20).

Consider the phagocytosis of apoptotic neutrophils. Although neutrophil invasion is useful in the early stages of inflammation, neutrophils have a short lifespan and the release of their contents, among them acid hydrolases, damage tissues. Therefore, it is imperative that the apoptotic neutrophils be removed by macrophages or resident cells by phagocytosis.[11] Thioglycolate injections in the peritoneum of experimental animals cause acute inflammation with accumulation of neutrophils. Thus, neutrophils escape from the blood vessels and accumulate in the extravascular space. The apoptosis and phagocytosis of neutrophils is delayed most likely because of low level of oxygenation. Moreover, phagocytosis of apoptotic neutrophils can be blocked by antibodies to CDIIb/CD18, suggesting that CDIIb/18[12] contributes to the elimination of the extravasated neutrophils (21). Multiple receptors, including integrin receptors, are likely to cooperate in phagocytizing extravasated polymorphonuclear cells (22).

Polymorphonuclear (PMN) phagocytosis alone does not stop the inflammatory reaction. The reason for the arrest of the inflammation is not clear. Kobayashi and associates investigated whether the apoptotic polymorphonuclear contributes to the anti-inflammatory process (23). Using microarrays, they studied the genes that are either activated (~98) or repressed (~100) in polymorphonuclear cells prior to apoptosis. While several metabolic pathways were up-regulated, seven genes involved in fatty acid regu-

[10] Talin is a phosphoprotein that connects the cytoskeleton to the plasma membrane, it binds strongly to vinculin. Vinculin is a cytoskeleton protein present at the cytoplasmic phase of the plasma membrane at the site of adhesion plaques, which anchor actin microfilaments to the plasma membrane. Paxillin are adhesion molecules associated with vinculin. They become phosphorylated under certain stimuli.

[11] Note that with rare exceptions, most experiments on apoptotic cell phagocytosis have been done in vitro.

[12] CIIb is the $\alpha_M$ subunit of integrin C3, and C18 is a $\beta$ chain that heterodimerizes with C11, a, b, c and d (see Janeway 1999, Appendix I 33).

lation, which contribute to lipid peroxidation, were down-regulated. It is not excluded that the combination of these events may cooperate in inhibiting inflammation.

No antibody antigen complex linking apoptotic neutrophils to macrophages was identified until Jersmann et al. (24, 25) reported that the antigen on apoptotic neutrophil is fetuin, a scialoglycoprotein that binds the antigen BOB93.

In the early stages of wound healing fibroblasts actively proliferate. If the wound is to heal without scarring or cheloids, the excess fibroblasts must be eliminated without causing inflammation. This occurs through apoptosis and phagocytosis of the death cell. Prior to engulfment of the fibroblast, CD36 appears on their surface and substantial amounts of thrombospondin are released. Inhibition of either of these events as well as blockade of integrin $\alpha_v\beta_3$ impairs the phagocytosis of fibroblasts. Thus, the CD36 macrophage receptor, thrombospondin, and the integrin $\alpha_v\beta_3$ tether the fibroblast possibly by reacting with phosphatidylserine (26).

Glomerular mesangial cells are semiprofessional phagocytes that ingest apoptotic leukocytes and apoptotic resident cells in the course of glomerular inflammation, thereby contributing to the recovery of glomerulonephritis. With the help of inhibitors of various potential mechanisms used by other known receptors for phagocytosis (such as phosphatidylserine recognition), it became clear that the vitronectin receptor $\alpha_V/\beta_3$ integrin mediates the phagocytosis of apoptotic cells and that a rise in free intracellular calcium is required for the process. In contrast to semiprofessional macrophages, the professional phagocytes use all known receptors for phagocytosis to engulf apoptotic bodies with different degrees of efficiency, and the process does not require mobilization of calcium. Thus, the reception of the signal for the apoptotic body's phagocytosis in glomerular mesangial cells is different from the response in macrophages; diverse multiple receptors are involved in the case of macrophages and only the $\alpha_5/\beta_3$, vitronectin, and integrins are required in dendritic cells (27).

CD14 is a bacterial receptor for the bacterial endotoxin lipopolysaccharide (LPS). The binding of LPS and the receptor causes inflammation. It is, therefore, surprising that CD14 also functions to remove apoptotic cells (28). CD14's cloned gene product is a 356 amino acid glycoprotein (55 kDa) found primarily in macrocytes and monocytes, and in small amounts in PMN leukocytes. The protein has been detected in numerous other cells by antibody labeling (in B, endothelial, and epithelial cells). The molecule exists in two forms: bound and soluble. The molecule is embedded in the plasma membrane through a glycosyl-phosphatidylinositol anchor. CD14 functions as an inflammatory and immune receptor and is also involved in the clearance of apoptotic cells. In this chapter we focus on the latter function.

The CD14 of the phagocyte likely functions as a receptor for one or more specific ligands located at the surface of the apoptotic cells. But as pointed out by Dewitt et al., this does not exclude a role for the soluble CD14, which may function as a bridge between receptor and ligand molecules (28). The preferred mode of action of CD14 in apoptosis is that of a bridge between phagocyte and dying cell, and this bridge is able to bring several macrophage receptors in contact with the ligand on the apoptic cell (29).

## Engulfment and Autoimmunity

The importance of the phagocytosis of apoptotic bodies is emphasized by situations in which the process is suppressed. The phagocytic clearance of apoptotic cells is impaired in lupus erythematosus. To determine whether this impairment may contribute to the autoimmune reactions associated with lupus, mice injected intravenously with syngeneic apoptotic thymocytes were found to develop autoantibodies (antinuclear, anti-SSDNA, and anticardiolipin). In contrast, apoptotic thymocytes are rapidly cleared in normal mice. These observations suggest that the presence of apoptotic bodies can induce a transient autoimmune response even in normal mice. Consequently, it is quite conceivable that when the rate of apoptosis is increased and phagocytosis concomitantly defective, as is the case in lupus erythematosus, remnant apoptotic bodies might contribute to the autoimmune response.

When macrophages are cocultured with dying CD4+ T cells and infected with *Trypanosoma cruzi,* replications of the parasite increase. Freire-de-Lima and associates exposed murine resident peritoneal macrophages to T cells at three different stages of their life cycles: apoptotic T cells, necrotic T cells, and viable T cells (30). After five days of exposure, each cell mixture was washed, infected with *T. cruzi,* and the rate of infection of the mixture determined by counting the parasites. When macrophages were mixed with apoptotic cells, the rate of infection was more than twice that in untreated cells and almost three times that in macrophages mixed with necrotic cells. Moreover, a single injection of apoptotic cells in mice seven days after infection caused a doubling of the parasitemia compared to the controls (noninjected mice). Keeping in mind that vitronectin has been reported to be involved in apoptosis; it was further shown that the blockade of vitronectin with specific antibodies was sufficient to prevent parasites replication.[13] Apoptosis of T cell and phagocytosis of the apoptotic cells causes exacerbation of the parasitemia. Blockade of vitronectin prevents the exacerbation (30). The association of apoptosis and a rise in parasitemia has also been observed with *Entamoeba histolytica* (31) and in *Streptococci iniae* (32).

## Apoptosis and Inflammation

The absence of inflammation in apoptosis compared to necrosis determines the fate of the surrounding tissues. Inflammation is a complex response to injury involving humoral factors (e.g., cytokinins) and several different cell reactions. The presence of unphagocytized death cells caused by trauma, toxins, or parasites is a major cause of inflammation. Unless the apoptotic cells are opsonized, apoptosis does not cause inflammation because the cells that die by apoptosis generate an anti-inflammatory environment in part resulting from the release of transforming growth factor (TGF-β) (9). It should be added that the engulfment of the apoptotic bodies also protects against inflammation. However, questions that need answers include: Is the engulfment sufficient to destroy the apoptotic bodies and could the apoptotic bodies kill the phagocyte? What happens if the phagocyte betrays its victim and uses fragments of it to activate other cells that may cause inflammation? Filaci et al. investigated the effects of apoptosis on macrophages (33). The engulfed apoptotic bodies did not generate an immune response because the phagocytic macrophage was unable to present antigen to the T cells from an autologous cell line. Antigen presentation is inhibited by phagocytosis of apoptotic bodies as a result of the binding of apoptotic DNA to the human lymphocyte antigen (HLA) Class II molecules of the macrophages. There is also impairment of inflammatory stimulators (e.g., reduction in tumor necrosis factor [TNF-α] secretion) and increased secretion of TGF-β interferes with antigen presentation to neighboring dendritic cells (33).

[13] *Trypanosoma cruzi* causes Chagas disease, a condition that leads to severe, often fatal cardiac illness.

The molecular mechanism(s) that initiates the panoply of events that inhibits inflammation after engulfment of apoptotic bodies is unknown. It has been proposed that exposure of phosphatidylserine at the surface of the apoptotic cell (e.g., in microglia) keeps them from triggering an inflammatory response (34).

## Significance of Clearance of Apoptotic Bodies

Assuming that the engulfment fails entirely or even partially, inflammation is the price. If engulfment is successful but digestion of cellular components is incomplete, there is a risk for autoimmune disease and cancer. Incomplete digestion of the cell's content is of significance particularly if the cell harbors parasites (viruses, bacteria, or ameba), as the new environment may provide shelter for the parasite.

Limitations of our present knowledge of the recognition and phagocytosis of the apoptotic bodies should not lead to an underestimation of the importance of the process. Consider the complexity of the events. In contrast to necrosis, which is associated with inflammation and pain, apoptosis is silent and, except when massive, hardly detectable with the light microscope. When completed with the formation of apoptotic bodies, the membranes of the latter must signal their readiness to be engulfed and the molecular signal must be recognized by a receptor on the phagocyte. A further complication is the frequent building of molecular bridges between the prey and the phagocytes. To date, one major (if not the only) signal appears at the surface of the apoptotic body, phosphatidylserine. The tethering devices between phosphatidylserine on the apoptotic body to the phagocyte involve:

Phosphatidylserine — Cdc 42-Rac

Phosphatidylserine — Gaz 6 and Mer (receptor tyrosine kinase)

Phosphatidylserine — MF-GE-8 $\alpha_v\beta$

Phosphatidylserine — ICAM 3 CD14.

Other connections between apoptotic bodies and phagocyte involve putative signals on the surface of the apoptotic cell, such as:

C1q (member of complement pathway) — Cal reticulin

Thrombospondin $\alpha_v\beta_3$ (integrin) CD36

The cells that undergo apoptosis vary considerably. Most, if not all, cells can undergo programmed cell death albeit under variable circumstances: embryonic development, involutions in adults' exposure to reactive oxygen species (ROS), genotoxic agents, or hormonal exposure (glucocorticoid), etc. Different phagocytes can be involved in the clearance of apoptotic cells and they may be professional, semiprofessional fixed or mobile, or they may be occasional (neighboring cells). Consequently, it is not surprising that a variety of molecular connections between apoptotic cells and phagocytes may exist. The identification of those that operate in vivo is far from simple, since most experiments on engulfment of mammalian cells were done in vitro. The consequences of failing engulfment are a threat to the cellular environment in vivo and include: awakening of an inflammation response; triggering of autoimmune reactions; innocent sheltering of surviving parasites; and even the possibility of contributing to cancer through transduction of oncogenic DNA (9,12).[14]

[14]Exceptions may exist, but at present the evidence is extremely weak.

## Conclusion

There are remarkable differences between traditional phagocytosis of foreign invaders and phagocytosis of the self by apoptotic bodies. Since traditional phagocytosis is facilitated mainly by penetration of the animal body by either living organisms (viruses, bacteria, or parasites) or inert material (asbestos or suture remnants), engulfment of the apoptotic bodies is constant and involves the self.

Apoptosis ends by cellular fragmentation and yields apoptotic bodies. These apoptotic bodies need to be phagocytized before they release their content into the tissues, because such events would be deleterious to neighboring healthy cells. Traditional phagocytosis mainly involves professional phagocytes. In contrast, the apoptotic bodies are engulfed by professional phagocytes, semiprofessional phagocytes, or even resident cells. To date, the most comprehensive studies of the genetic control of the engulfment of the apoptotic corpse were done in *C. elegans*, and evidence is emerging that the process may be phylogenetically conserved. In mammalian cells, several molecular structures have been involved in the recognition of the target oligosaccharides, such as phosphatidylserine, lectin, and integrins. At present it is impossible to give a comprehensive picture of phagocytosis of the apoptotic bodies in vivo, since most of the studies were done in vitro, and it cannot be excluded that several different mechanisms might operate simultaneously or in sequence. In any event, it seems established that phagocytosis is not the primary cause of death. Indeed, it can occur in absence of the phagocytotic pathway. (For further details refer to the excellent reviews by de Almeida and Linden and that of Fadeel [35, 36].)

## References and Recommended Reading


1. Fadok, V.A., Voelker, D.R., Campbell, P.A., Cohen, J.J., Bratton, D.L., and Henson, P.M. Exposure of phosphatidylserine on the surface of apoptotic lymphocytes triggers specific recognition and removal by macrophages. *J Immunol.* 1992; 148:2207–16.
2. Williamson, P., and Schlegel, R.A. Transbilayer phospholipid movement and the clearance of apoptotic cells. *Biochim Biophys Acta.* 2002;1585:53–63.
3. Das, P., Estephan, R., and Banerjee, P. Apoptosis is associated with an inhibition of aminophospholipid translocase (APTL) in CNS-derived HN2-5 and HOG cells and phosphatidylserine is a recognition molecule in microglial uptake of the apoptotic HN2-5 cells. *Life Sci.* 2003;72:2617–27.
4. Shiratsuchi, A., Osada, S, Kanazawa, S., and Nakanishi, Y. Essential role of phosphatidylserine externalization in apoptosing cell phagocytosis by macrophages. *Biochem Biophys Res Commun.* 1998;246:549–55.
5. Adayev, T., Estephan, R., Meserole, S., Mazza, B., Yurkow, E.J., and Banerjee, P. Externalization of phosphatidylserine may not be an early signal of apoptosis in neuronal cells, but only the phosphatidylserine-displaying apoptotic cells are phagocytosed by microglia. *J Neurochem.* 1998;71:1854–64.
6. Zhuang, J., Ren, Y., Snowden, R.T., Zhu, H., Gogvadze, V., Savill, J.S., and Cohen, G.M. Dissociation of phagocyte recognition of cells undergoing apoptosis from other features of the apoptotic program. *J Biol Chem.* 1998;273:15628–32.

7. Savill, J., Hogg, N., Ren, Y., and Haslett, C. Thrombospondin cooperates with CD36 and the vitronectin receptor in macrophage recognition of neutrophils undergoing apoptosis. *J Clin Invest.* 1992;90:1513–22.
8. Van den Eijnde, S.M., Boshart, L., Baehrecke, E.H., De Zeeuw, C.I., Reutelingsperger, C.P., and Vermeij-Keers, C. Cell surface exposure of phosphatidylserine during apoptosis is phylogenetically conserved. *Apoptosis.* 1998;3:9–16.
9. Fadok, V.A., Warner, M.L., Bratton, D.L., and Henson, P.M. CD36 is required for phagocytosis of apoptotic cells by human macrophages that use either a phosphatidylserine receptor or the vitronectin receptor (alpha v beta 3). *J. Immunol.* 1998;161:6250–7.
10. Balasubramanian, K., and Schroit, A.J. Characterization of phosphatidylserine-dependent beta2-glycoprotein I macrophage interactions. Implications for apoptotic cell clearance by phagocytes. *J Biol Chem.* 1998;273:29272–7.
11. Brouckaert, G., Kalai, M., Krysko, D.V., Saelens, X., Vercammen, D., Haegeman, G., D'Herde, K., and Vandenabeele, P. Phagocytosis of necrotic cells by macrophages in phosphatidylserine dependent and does not induce inflammatory cytokine production. *Mol Biol Cell.* 2004;15(3):1089–1100.
12. Krieser, R.J. and White, K. Engulfment mechanism of apoptotic cells. *Curr Opin Cell Biol.* 2002;14:734–8.
13. Chambenoit, O., Hamon, Y., Marguet, D., Rigneault, H., Rosseneu, M., and Chimini, G. Specific docking of apolipoprotein A-1 at the cell surface requires a functional ABCA1 transporter. *J Biol Chem.* 2001;276(13):9955–60.
14. Marguet, D., and Chimini, G. The ABCA1 transporter and ApoA-1: obligate or facultative partners? *Trends Cardiovasc Med.* 2002;12(7):294–8 (review).
15. Rini, J., and Drickamer, K. Carbohydrates and glycoconjugates. *Curr Opin Struct Biol.* 1997;7:615–6.
16. Weis, W.I., and Drickamer, K. Structural basis of lectin-carbohydrate recognition. *Annu Rev Biochem.* 1996;65:441–73.
17. Timonen, T., and Helander, T.S. Natural killer cell target cell interactions. *Curr Opin Cell Biol.* 1997;9:667–73.
18. Giancotti, F.G. Integrin signaling: specificity and control of cell survival and cell cycle progression. *Curr Opin Cell Biol.* 1997;9:691–700.
19. Dedhar, S., and Hannigan, G.E. Integrin cytoplasmic interactions and bidirectional transmembrane signaling. *Curr Opin Cell Biol.* 1996;8:657–69.
20. Clegg, D.O., Mullick, L.H., Wingerd, K.L., Lin, H., Atienza, J.W., Bradshaw, A.D., Gervin, D.B., and Cann, G.M. Adhesive events in retinal development and function: the role of integrin receptors. *Results Probl Cell Differ.* 2000;31:141–56.
21. Coxon, A., Rieu, P., Barkalow, F.J., Askari, S., Sharpe, A.H., von Andrian, U.H. Arnout, M.A., and Mayadas, T.N. A novel role for the beta 2 integrin CD11b/CD18 in neutrophil apoptosis: a homeostatic mechanism in inflammation. *Immunity.* 1996;5:653–66.
22. McCutcheon, J.C., Hart, S.P., Canning, M., Ross, K., Humphries, M.J., and Dransfield, I. Regulation of macrophage phagocytosis of apoptotic neutrophils by adhesion to fibronectin. *J Leukoc Biol.* 1998;64:600–7.

23. Kobayashi, S.D., Voyich, J.M., Somerville, G.A., Braughton, K.R., Malech, H.L., Musser, J.M., and DeLeo, F.R. An apoptosis-differentiation program in human polymorphonuclear leukocytes facilitates resolution of inflammation. *J Leukoc Biol.* 2003;73:315–22.

24. Jersmann, H.P., Dransfield, I., and Hart, S.P. Fetuin/alpha2-HS glycoprotein enhances phagocytosis of apoptotic cells and macropinocytosis by human macrophages. *Clin Sci* (Lond). 2003;105(3):273–8.

25. Lord, J.M. A physiological role for alpha2-HS glycoprotein: stimulation of macrophage uptake of apoptotic cells. *Clin Sci* (Lond). 2003;105(3):267–8.

26. Moodley, Y., Rigby, P., Bundell, C., Bunt, S., Hayashi, H., Misso, N., McAnulty, R., Laurent, G., Scaffidi, A., Thompson, P., and Knight, D. Macrophage recognition and phagocytosis of apoptotic fibroblasts is critically dependent on fibroblast-derived thrombospondin 1 and CD36. *Am J Pathol.* 2003;162:771–9.

27. Rubartelli, A., Poggi, A., Zocchi, M.R. The selective engulfment of apoptotic bodies by dendritic cells is mediated by the alpha(v)beta3 integrin and requires intracellular and extracellular calcium. *Eur J Immunol.* 1997;27:1893–900.

28. Devitt, A., Moffatt, O.D., Raykundalia, C., Capra, J.D., Simmons, D.L., and Gregory, C.D. Human CD14 mediates recognition and phagocytosis of apoptotic cells. *Nature.* 1998;392:505–9.

29. Savill, J. Apoptosis. Phagocytic docking without shocking. *Nature.* 1998;392:442–3.

30. Freire-de-Lima, C.G., Nascimento, D.O., Soares, M.B., Bozza, P.T., Castro-Faria-Neto, H.C., de Mello, F.G., Dos Reis, G.A., and Lopes, M.F. Uptake of apoptotic cells drives the growth of a pathogenic trypanosome in macrophages. *Nature.* 2000;403:199–203.

31. Huston, C.D., Boettner, D.R., Miller-Sims, V., and Petri, W.A. Jr. Apoptotic killing and phagocytosis of host cells by the parasite *Entamoeba histolytica. Infect Immun.* 2003;71:964–72.

32. Zlotkin, A., Chilmonczyk, S., Eyngor, M., Hurvitz, A., Ghittino, C., and Eldar, A. Trojan horse effect: phagocyte-mediated *Streptococcus iniae* infection of fish. *Infect Immun.* 2003;71:2318–25.

33. Filaci, G., Contini, P., Fravega, M., Fenoglio, D., Azzarone, B., Julien-Giron, M., Fiocca, R., Boggio, M., Necchi, V., De Lerma Barbaro, A., Merlo, A., Rizzi, M., Ghio, M., Setti, M., Puppo, F., Zanetti, M., and Indiveri, F. Apoptotic DNA binds to HLA class II molecules inhibiting antigen presentation and participating in the development of anti-inflammatory functional behavior of phagocytic macrophages. *Hum Immunol.* 2003;64:9–20.

34. De Simone, R., Ajmone-Cat, M.A., Tirassa, P., and Minghetti, L. Apoptotic PC12 cells exposing phosphatidylserine promote the production of anti-inflammatory and neuroprotective molecules by microglial cells. *J Neuropathol Exp Neurol.* 2003;62:208–16.

35. de Almeida, C.J., and Linden, R. Phagocytosis of apoptotic cells: a matter of balance. *Cell Mol Life Sci.* 2005;62(14):1532–46.

36. Fadeel, B. Programmed cell clearance. *Cell Mol Life Sci.* 2003;60(12):2575–85 (review).

CHAPTER

# AGENTS THAT TRIGGER APOPTOSIS

# 3

The list of agents that cause apoptosis is large and still growing. The selective programmed cell death of a few or even a large population of cells, without causing inflammation scarring or other organ distortions (ulcers, fistulas, etc.), is triggered by genes that operate under physiologic or pathologic conditions. Gene products are released during development or at some specific time of the life cycle that cause selective cell death. Withdrawal of survival factors leaves no alternative but death. In the course of the organ's lifespan, some cells die either because the organ is returned to a lower level of function or because of aging (atrophy). Cells respond by apoptosis or programmed cell death to endogenous triggers such as the cytokines of the tumor necrosis factor-alpha (TNF-α) family. Agents that damage DNA beyond repair cause cell death, but if the cells don't die, the DNA insult may be replicated and cause cancer. Common triggers of apoptosis are listed in **TABLE 3-1**. This chapter focuses on the trigger mechanisms of the tumor necrosis factor family.

## Tumor Necrosis Factor

### Historical Overview

The history of the discovery of the tumor necrosis factor (TNF) is interesting because it is linked to early observations of spontaneous tumor regression. Around 400 BC, the Hipppocratic scrolls are perhaps the first to record such spontaneous tumor regression in patients suffering from erysipelas. In 1884, William B. Coley, a surgeon working at the New York Hospital (now the Memorial), observed that a recurrent round cell sarcoma of the neck in a 21-year-old male, who had been operated on five times over a period of three years, regressed spontaneously after the patient suffered an episode of erysipelas (1–3). The observation was not lost on Coley, who concluded that erysipelas produced a powerful substance that killed the sarcoma cells. Coley again infected patients with the erysipelas agent, observed tumor regression, and concluded that the curative effect of this agent was systemic. Further experiments using heat-killed bacterial cultures established that the culture medium contained a soluble toxin that, when injected in humans, was almost as effective in triggering tumor regression as erysipelas itself. Extracts prepared from *Bacillus prodigiosus* and *Streptococcus erysipelatis* were

**Table 3-1 Agents causing apoptosis**

1. Agents that directly interact with the genome: they include the factors that trigger cell death in *C. elegans*; hormones such as thyroxine, glucocorticoid, and retinoic acid; and growth factors or local factors that modulate embryogenesis and development in mammals.
2. Withdrawal of hormonal trophic signals leads to atrophy of the breast, the prostate, and the adrenal gland and causes atresia of the ovarian follicle; withdrawal of EGF or NGF can lead to apoptosis in cells in culture.
3. Molecules that react with specific receptors, among them: TNF and FAS (APO I, CD 95), the Mullerian inhibitor factor.
4. Physiological or nonphysiological (when in excess) agents that damage DNA: radiation, free radical or reactive oxygen species, mutagens.
5. Some pathological agents such as viruses.

equally effective. Assuming that the diagnosis was correct in all treated cases, Coley is said to have cured 10% of his cancer patients by such means.

Although the notion that erysipelas could lead to regression of tumors was not completely forgotten by clinicians, the Coley toxin was abandoned in favor of radiotherapy and improved surgery. Before World War II, Coley's daughter Helen discovered that the tumors in mice injected with killed strains of Gram negative bacteria started to bleed, turned black, and regressed. Histologic examination revealed hemorrhagic necrosis in the tumor but not in normal tissues.

In 1943, Murray Shear extracted an endotoxin from Gram negative bacteria that contributed to hemorrhagic necrosis in experimental tumors. The active compound was purified and it proved to be an outer wall lipopolysaccharide, now referred to as LPS. All effects of LPS are not nefarious; the endotoxin confers resistance to repeated bacterial infections and whole body irradiation. At the core of the molecule, the trisaccharide 2-keto-3-deoxyoctonate is linked to lipid A(1,4), that is responsible for the endotoxin's effect. The bovine bacillus of tuberculosis, discovered by Calmette-Guérin and referred to as BCG (bacillus Calmette-Guérin), rendered mice more resistant to the growth of implanted or induced tumors. The combination of BCG and LPS was even more effective in inducing tumor necrosis than either one alone. It soon became obvious that BCG did not act in the same way as LPS. BCG is only active before tumor implantation, it confers immunity against a number of different antigens, and its effect is eliminated by immunosuppression. In contrast, LPS causes mainly hemorrhagic necrosis in tumor implants, an effect that is not dependent upon an immune response, and it survives immunosuppression (5).

In 1975, Elizabeth Carswell and Lloyd Old demonstrated that the serum of mice infected with BCG and LPS contained a soluble factor that not only causes hemorrhagic necrosis in tumors in vivo, but also was cytostatic and cytotoxic for cell cultured in vitro (6). A biological assay was set up and the factor named TNF was soon purified. It was found to be present in the serum of several species of challenged animals (mice, rabbits, guinea pigs, and humans) and was shown not to be species-specific. It was further demonstrated that the tumor necrosis-induced sarcoma in mouse resulting from intramuscular methylcholantrene injection was due to TNF production rather than to

LPS (6). Serum from mice treated with either BCG or LPS alone did not contain TNF and complement was not involved in cell killing in vivo or in vitro.

## Discovery of TNF Cytokines

The addition of TNF to serum of cultured cells proved to be both cytostatic ($H^3$ thymidine incorporation) and cytotoxic by toluidine blue exclusion or phase contrast microscopy. These two effects could not be separated even by extensive purification. In addition to its cytostatic and cytotoxic effects, TNF serum stimulates the differentiation of B cell lymphocytes; it substitutes for helper T cells, induces the formation of complement receptors on the surface, and the emergence of immunoglobulin.

TNF is a dimer with a molecular weight of 40,000 (gel filtration) that drops to 18,000 after sodium dodecyl sulfate (SDS) gel filtration (7). A human TNF was purified from a cloned histiocytoma line, which produces large amounts of TNF after stimulation with LPS and from a B cell line (Luk 11) stimulated by 4β-phorbol-12β-myristate 13α-acetate (PMA), a tumor promoter. The purified TNF contained no interferon, an important feature because of the synergistic effects of interferon and TNF, two classes of leukokinins produced by the hematopoietic system. After purification of the human TNF (molecular wt 45,000 on gel filtration, 17,000 on polyacrylamide gels containing SDS), the molecule was found to contain two cysteines per subunit that form an intrachain disulfide.

After partial sequencing and complementary DNA (cDNA) selection, a plasmid construction was performed. The precursor mRNA encodes 233 amino acids, and contains 4 exons. When the leader sequence was removed the mature DNA sequences code for 157 amino acids. There was good correlation between mouse recombinant TNF (157 amino acids) and human recombinant TNF: 76% homology in the respective sequences (157 amino acids). Recombinant TNF injected in humans is rapidly cleared from the serum (half-life 15–30 minutes) and is no longer detectable in the serum 12 hours after injection. The recombinant TNF, however, causes severe local inflammations.

TNF receptors with high affinity for recombinant TNF exist at the surface of sensitive cells (1000 to 10,000 copies). However, cellular response is not dependent upon the number of receptors and, therefore, the sensitivity to TNF depends not only on the number of receptors, but also upon the mechanism of transduction of information from membrane to cytoplasm and nucleus.

The immune response requires the cooperation of several cell types, among them T cells, B cells, neutrophils, macrophages, and others. Such wide cooperation depends on timely and appropriate signals between the various cell types. This is achieved in part by special molecules called cytokines. Not unlike growth factors and some hormones, cytokines bind to cell surface receptors, trigger tyrosine phosphorylation of transcription factors, and thereby trigger a cascade of biochemical events.

Cytokines are either peptides or glycoproteins that fall into at least four categories—interferons, interleukins, colony-stimulating factors, and the TNF family—and each of these groups has defined, limited functions. There is, however, one more class with multiple and varied functions. The latter includes the tumor necrosis factors α and β, and the transforming growth factors (TGF-β). The function of TNF and TGF-β includes the induction of apoptosis. In summary, the interest in and, indeed, the name tumor necrosis factor emerged from the factor's capacity to induce hemorrhagic necrosis in tumors after induction by LPS and the killing of tumor cells by activated macrophages. The discovery that TNF is a potent inducer of apoptosis further stimulated research on the function of TNF. This section will focus on the role of TNF family of proteins in apoptosis.

## TNF Family and Apoptosis

### Lymphotoxin, TNF-α, TNF-β, and Cachectin

Investigations of the tumor necrosis factor paralleled that of lymphotoxin and cachectin. The first proved to be a compound homologous to the TNF and the second is similar to the TNF. About the same time, LPS endotoxin was found to induce the factor causing necrosis in rodent tumor. It was independently discovered that lymphocytes, when stimulated by interleukin-2 (IL-2), secrete a lymphotoxin. In contrast to the TNF that is produced by macrophages, other hematopoietic cells, and cells unrelated to the hematopoietic system, lymphotoxin is only elaborated by lymphocytes. Moreover, TNF can be induced by numerous agents and, in contrast, lymphotoxin induction is limited to fewer agents (phorbol-ester, thymosin α, and IL-1α/β).

Lymphotoxin was purified and sequenced, the DNA was cloned, and its gene mapped. The gene is located in the major histocompatibility complex (MHC) region on chromosome 6, and has 1000 nucleotides of that of the tumor necrosis factor. The amino acid sequence is 28% identical and 52% homologous to that of the TNF. Moreover, both the TNF and the lymphotoxin use the same receptor(s). In view of all the similarities between the TNF and lymphotoxin, they were renamed TNF-α and TNF-β, respectively (8).

If weight loss is not an absolute diagnostic criterion, it is often a sign of the existence of cancer. Sometimes even a relatively small cancer may cause substantial weight loss and, depending upon the cancer, severe emaciation and weakness are associated with its progression. This phenomenon is named cachexia from the Greek κακοσ (bad) and ηχισ (a bad situation). Cachexia is not unique to cancer, and many chronic or acute infections lead to cachexia or the so-called consumption. Novelists past and present often have given moving description of the consumption experienced by their tuberculous characters. Cachexia is not unique to humans. A case at point is that of cattle in Africa suffering from sleeping sickness, a disease caused by *Trypanosomes brucei*; the syndrome includes anorexia (but not always), weight loss, muscle wasting, and anemia. In cancer patients, there is no correlation between the development of cachexia and food intake or size and histology of the primary cancer. Many components of the pathogenic mechanisms of cachexia are still unknown (9), but earlier theories included elaboration of a toxin and the nitrogen trap. It cannot be excluded that both may contribute concomitantly or independently to cachexia. However, the involvement of a toxin is established on solid scientific grounds.

Studies in Cerami's laboratory, based on earlier observations on animals and humans victims of trypanosomiasis, demonstrate that LPS[1] induces macrophages to elaborate a substance that inhibits adipocyte lipoprotein lipase and ultimately leads to the cachectic state. The factor, a protein named cachectin, was purified and sequenced.

[1]LPS is a lipopolysaccharide found in the cell wall of Gram negative bacteria that is also a potent stimulator of TNF production.

Cachectin proved to have substantial sequence homology with TNF and it was established that TNF was the cytokine that mediated the cachectic effects of LPS. Antibodies to TNF abolished these effects, at least in part (10). The elevation of plasma triglycerides in septic or cancerous patients results in part from reduced lipoprotein lipase activity and stimulation of their hepatic secretion. TNF also induces other metabolic alterations including a marked increase in glucose utilization in most organs (liver, kidney, intestine, and skin), but not in skeletal muscle. In some experiments, it also causes muscle protein degradation. In summary, TNF causes hypertriglyceridemia, but it is still impossible to outline a consistent pathogenic mechanism for cachexia, and it is likely that septic shock and cachexia are caused by the release of several different cytokines acting in sequence or in concert.

### TNF's Structure and Its Genes

TNF-α and TNF-β are respectively made of 157 amino acids (17,000 Da) and of 172 amino acids (25,000 Da). TNF-α is non-glycosylated and TNF-β is glycosylated. The 157 amino acid subunit assembles with two of its homologues to form a trimer. Consider the structure of the single subunit; it forms a single chain, and its amino acid sequence is completely known (**FIGURE 3-1**) (3). Cysteine 69 and 101 form the single disulfide bond found in the peptide. Further studies revealed that the 157 amino acid chain forms two antiparallel β-pleated sheets[2] generating a β sandwich structure. The N and C terminal are located at the lower end of the structure; the C terminal is mostly included in the β structure. The N terminal extends out the main structure and up to 8 terminal amino acids can be deleted without affecting biological activity. The trimer is formed through a simple edge to face packing of the β sheet sandwich. There is a remarkable structural homology between TNF monomer and that of the capsid of several small spherical plant viruses (the jelly roll motif),[3] for example, the satellite tobacco necrosis virus. This led to the suggestion that TNF and the viral protein are both products of an ancestral gene that encodes a protein with a jelly roll configuration. Moreover, the trimeric arrangement in the adenovirus capsid is similar to the packing of the monomer in TNF. In contrast to the similarities described between TNF and viral capsid, there is no such analogy with other lymphokinines, for example, IL-2 and IL-1β, despite their overlapping biological roles (10,11).

TNF genes were mapped in mouse and humans, and are located on chromosome 17 in mice and the short arm of chromosome 6 in humans. Both mice and human TNF genes are flanked by Class III and Class I MHC genes. The *TNF-α* gene is located 5′ to that of *TNF-β*; each gene is 3 kb pairs long and each has four exons and three introns. TNF-α is inducible in a greater variety of cells than TNF-β. There is 80 to 89% homology between the *TNF-α* and *TNF-β* genes, except in the 5′ flanking region, which is believed to respond to transcription factors (13).

[2]Antiparallel β structures, the second largest group of protein domain structure after the α structure, are found in various proteins such as enzymes, transport protein, antibodies, and virus capsids. They are made of a variable number of β strands (4 to 10), and two β sheets are usually packed against each other in an antiparallel fashion (12).

[3]The jelly roll motif is frequently found in nature's polypeptides. It is a complicated structure, essentially a peptide wrapped around a barrel core like a jellyroll (details are in Branden and Tootze, pp. 70–71) (12).

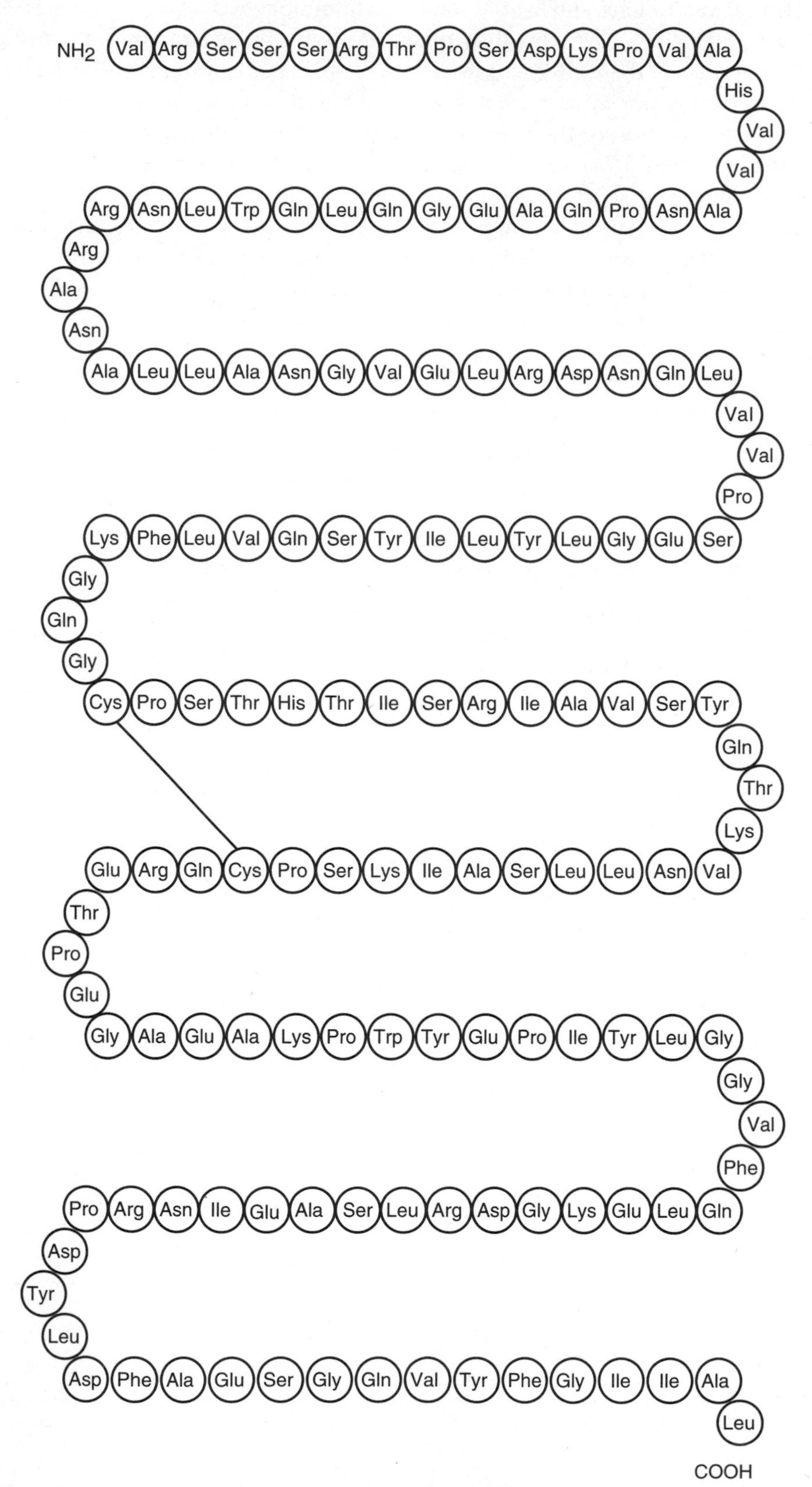


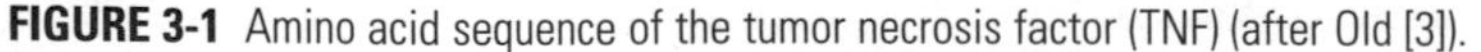
**FIGURE 3-1** Amino acid sequence of the tumor necrosis factor (TNF) (after Old [3]).

## TNF Ligands

At least 11 members in the TNF family of ligands[4] have been identified. Many, but not all, of them induce apoptosis. Among those that can induce apoptosis are TNF-α, lymphotoxin B heteromer, and the Fas ligand, which induce apoptosis when bound to their receptors. All the members of the TNF family of ligands are trimers and all except lymphotoxin B have identical subunits.

Lymphotoxin B is composed of two types of subunits: one lymphotoxin α and two β subunits. The ligands can be entirely secreted by the cell of origin, partially secreted, or remain protein bound. Lymphotoxin α is entirely secreted. TNF is predominantly secreted, but also acts as a transmembrane protein that triggers cell-to-cell contacts.

## TNF Receptors

There are numerous TNF receptors for the TNF family of ligand and new ones are still being discovered. As a whole, TNF receptors are pleiotrophic in the signals they send inside the cells and, thus, they trigger numerous and sometimes paradoxical biological responses, including leukocytic activation and proliferation, neural growth, bone development of inflammation, and apoptosis. If the hope is to control apoptosis by either arresting it in degenerative diseases or by promoting it in cancer, then it is imperative to understand the nature of the message that emerges from the receptor to reach its target. As the TNF receptor family grew (at least 20 members of the family are known), it became a superfamily divided into two classes depending upon whether the receptor contains a death domain (14). The former are sometimes referred to as death receptors. TNF receptors are surface receptors in that they have an extracellular, intramembrane, and intracellular component. In contrast to kinase receptors whose intracellular COOH terminal possesses tyrosine kinase activity, the TNF receptors have no catalytic function.

The TNF receptors are made of three subunits; the extracellular region is composed of three to six cysteine-rich pseudorepeats containing six cysteine residues and approximately 40 amino acids, with variation in size and numbers of the amino acids among the repeats. A single polypeptidic chain forms the intramembrane domain, which is short (40–221 amino acids) and lacks sequence homology (14, 15). As mentioned earlier, some members of the TNF receptor family, among them TNFR1 and FAS, contain a 60-residue conserved sequence in their cytoplasmic domains, the so-called "death domain" because the mutation in that sequence abolishes apoptosis (15–17).

The TNF receptor family is divided into two major groups, one that includes a death domain in its cytoplasmic COOH terminal and another devoid of the death domain. Among the members of the family that contain a death domain are TNFR1, FAS, TRAMP, TRAIL-R1, and TRAIL R-2. In contrast, TNFR II, TRAIL R3, and TRAIL R4 are devoid of a death domain. As explained later in this chapter, the presence of the death domain is critical for the transduction of the signal from the cell surface to the target molecules, a signal that triggers the biological response, for example, inflammation or apoptosis (14, 18). Mutations in the sequence of the death domain abolish the ability of the receptor to cause apoptosis after binding of the ligand. The complete sequence of events that links the activation of the receptor to the target biological response is not entirely understood, but a good deal is known about the immediate events that follow the binding of the ligand to the receptor. The intramembrane monomers trimerize.

[4]Members of the TNF family include: TNF, lymphotoxin α, lymphotoxin β heteromes, FAS or Apo I, CD40, CD27, CD30, OX40 antigen, T2 antigen of Shope, and 4-1BB ligands. NGF is a dimeric protein that is not a member of the TNF ligand family, but its receptors are members of the TNF receptors family.

## TNF Receptor Trimerization

It is generally assumed that the binding of ligands activates surface receptors by inducing conformational changes that may lead to clustering, and in keeping with this notion, it was believed that ligand triggered TNF receptors to trimerize and that trimerization is one of the first steps in sending the transduction signal (19).

Chan and colleagues showed that TNF receptors trimerize prior to ligand binding. As mentioned earlier, the extracellular segment of TNFs (60–80 amino acids) contains four cysteine-rich domains (CRDs): there are well ordered $CRD_1$, $CRD_2$, and $CRD_3$, all located immediately downstream of the N terminal, and a less conserved one located in the sequence adjacent to the cell membrane. $CRD_2$ and $CRD_3$ are known to contribute to the formation of the ligand pocket. Chan and colleagues showed that a domain, possibly including $CRD_1$, but distinct from $CRD_2$ and $CRD_3$, is necessary and sufficient for TNF receptor trimerization. This domain is referred to as the preligand activating domain (PLAD). Thus, in contrast to the previously generally accepted view—that ligand binding triggers trimerization—it appears that TNF receptors are preformed complexes (20).

Similarly, FAS homotrimerizes before the ligand is present and its pre-assembly is required to bind the FAS ligand and trigger apoptosis. As in the case of TNF, the death domain of FAS is not required for trimerization, only the PLAD is required. Surprisingly, PLAD allows trimerization of receptors that are defective because of mutations in the FAS monomer (e.g., mutation or deletion of the death domain in splice variants) (18). Participation of death domain deficient mutants confirms that the death domain is not required for trimerization.

TNF receptor 1 (TNFR1) and FAS trimerize prior to ligand binding (18, 19). These observations are seminal to the relative roles of receptors and ligands in signal transduction. What is the role of the ligand if it is not required for trimerization? The traditional view has it that the ligand induces conformational changes in the monomer that leads to trimerization (20), but it cannot be excluded that ligand binding also can induce conformational changes in the trimer. As pointed out by Goldstein, pre-assembled dimeric erythropoietin receptors are unable to transduce signal (21). Binding of the ligand causes a conformational change that allows signaling (21, 22). As pointed out by Chan and associates, the inhibition of PLADs function could prevent activation of TNF receptors and interfere with their signal whether it be apoptosis or activation of nuclear factor-κB (NF-κB) (20).

Huang et al. showed that polymerization of FAS receptors by cross linking enhances the signal for apoptosis (23). Is it possible that the FAS ligand facilitates homotrimeric receptors to cluster?

After ligand binding and receptor trimerization the trimer binds to non-catalytic proteins, such as TRADD and FADD, that contribute to the transfer of the signal. Thus, in contrast to receptors that transduce their signal by an enzymic reaction (by activating a second messenger, e.g., cyclic AMP or triggering a chain of phosphorylation that leads to the release of transcription factors), the TNF receptors build a protein chain or scaffold to transduce their message.

## Molecules Associated with the TNF and FAS Receptors

There are numerous proteins that interact directly or indirectly with the TNF receptors. Those interacting with TNFRs 1 and 2, FAS, and to an extent TRAIL have been investigated most often. The monomer of TNFR1 is encoded by a locus in chromosome 12p13. TNFR1 (55 kDa) is ubiquitous, and after trimerization it transduces its message through a protein chain.

TNF1 receptor-associated death domain (TRADD) is a 34 kDa molecule that contains, as its name implies, a death domain in its C terminal. The structure of the death domain and that of the death effector domains (discussed later in this chapter) are both made of six antiparallel amphipathic alpha helices (24). TRADD is possibly the first molecule recruited by TNFR1 after activation of the receptor. The death domain of TRADD associates with that of TNFR1 and acts as an upstream transducer of apoptosis (25). The death domain of TNFR1, an approximately 80 amino acid sequence, is found in the intracellular segment, the COOH terminal sequence.

TRADD has no enzymic activity, and thus its death domain is a protein interaction motif. TRADD recruits another death domain containing the FAS-associated death domain (FADD) protein. FADD (or MORT) associates with TNFR, again by the homophilic reaction of their respective death domains.

Human FADD[5] (23.3 kDa) is a 208 amino acid protein containing two functional domains, the death domain in its C terminal (70 amino acids) and a death effector domain (76 amino acids) in its N terminal. When overexpressed it leads to FAS-mediated apoptosis. There is 68% amino acid identity and 80% similarity between the human and the mouse FADD (205 amino acids), indicating that FADD is conserved among mammalian species.

TRADD and FADD serve as signal transducing agents to link the signal received by TNFR1 after ligand binding and trimerization to the target: caspase 8. The molecular scaffolds that include TNFR, TRADD, FADD, and caspase 8 (or FAS-TRADD-FADD) are referred to as the death-inducing signal complex (DISC).

The death effector domain associates with one of the two death effector domains of procaspase 8 (see Chapter 6) (26). Thus, TRADD and FADD serve as an upstream signal-transducing molecule that leads to apoptosis through activation of caspases (**FIGURE 3-2**).

However, the TNF receptors have two major targets: activation of caspases and activation of NF-κB. The first leads to apoptosis and the second to inflammation and survival.

What determines the selection between life and death? The TRAF family of proteins (6 members, TRAF 1–6) is likely to be a critical element in the choice. These proteins react with several of the TNF receptors superfamily. The proteins of this family lack a death domain, and are characterized by their unique but conserved C terminal domain, the TRAF domain, which is divided between the TRAF N and TRAF C subdomains (27–29).[6] The TRAF N terminal domain (except in TRAF1) contains an amino terminal RING finger region that is followed by characteristic cysteines and histidines linked to zinc by coordinated bonds (a cluster of 5 to 7 zinc fingers) (30). As for the TRAF domain (C terminal region), it interacts with TRADD. Thus, in addition to triggering the proapoptotic pathway (TNFR1-TRADD-FADD-Caspase), TNFR1 and TNFR2 can engage another pathway TNFR-TRADD-TRAF→NF-κB, and the latter pathway is believed to be involved in cell survival.

Clearly, TRAFs are important adapter molecules in the TNFR downstream transduction pathway, and so the intracellular distribution and the tissue specificity of the

---

[5]FADD is encoded on the 11q13.3 locus, a region associated with diabetes susceptibility. Moreover, the FADD locus is amplified in some breast adenocarcinomas. The closeness of FADD to these other genes suggests that FADD may be associated with autoimmunity and carcinogenesis. The *FADD/MORT* gene is made of two exons, exon 1 (286 bp) and exon 2 (341 bp), separated by a 2 kb intron. Patients with autoimmune lymphoproliferative disorders are resistant to Fas-induced apoptosis because of a mutation of the *FAS* gene.

[6]TRAF 2 to 6 contain an N terminal zinc finger region, a ring finger, and five tandem zinc fingers. The function of the zinc finger is still not clear, but it is believed to activate downstream signaling, in particular for NK-κB action.

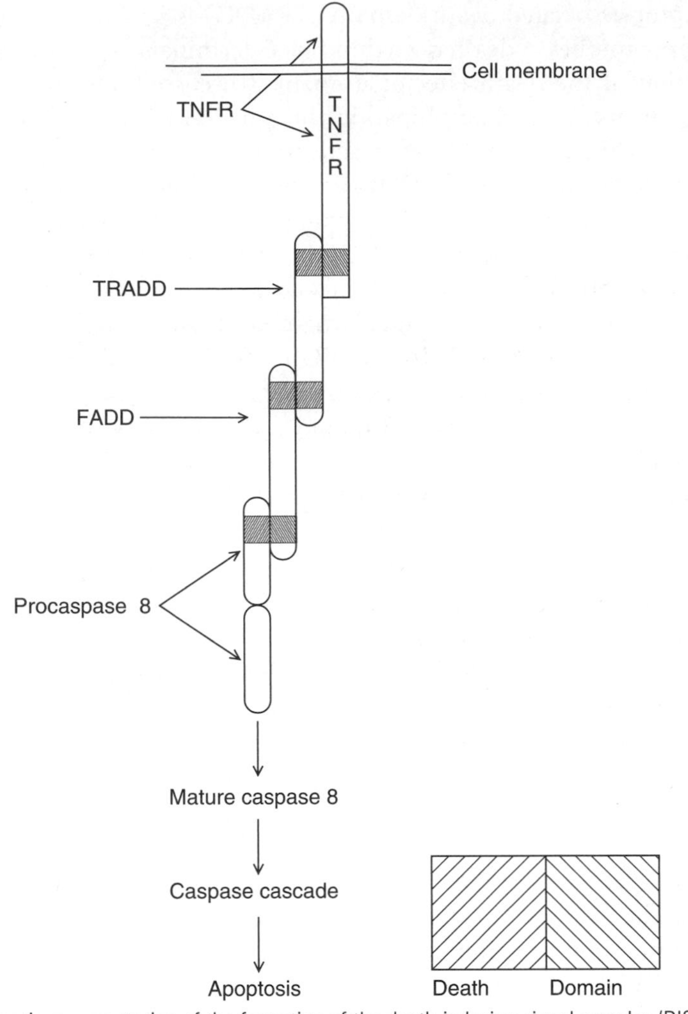


**FIGURE 3-2** Schematic representation of the formation of the death-inducing signal complex (DISC) by the interaction of the death domains of TRADD, FADD, and procaspase 8.

TRAFs are most intriguing. TRAF1, TRAF2, TRAF3, and TRAF5 are located in the cytosol. They bind to receptors and recruit other proteins including other TRAF proteins by homo- or heterodimerization through their C terminal region, the TRAF domain (31). While TRAF2, TRAF3, and TRAF6 are ubiquitous, TRAF5 is expressed the most in lung, spleen, and thymus, and high expression of TRAF6 occurs only during embryogenesis, for instance, in the hippocampus and the olfactory regions. TRAF4, mainly a nuclear protein, was also found in the cytosol of breast cancer. Its expression ceases with differentiation and malignant transformation (32). The presence of the zinc finger domain determines whether TRAF4 localizes to the nucleus or the cytoplasm. Deletion of the zinc finger causes nuclear localization and transcription deregulation. The *TRAF3* gene encodes four messenger RNAs that result from alternative polyadenylation, initiation, or splicing (33).

Although much more needs to be learned about the type of TRAF molecules that associates with TRADD and the downstream transduction pathway that leads to NF-κB activation, it seems that the TRADD-TRAF association may provide a molecular mechanism for selection between the two pathways; apoptosis or NF-κB activation. If so, it is conceivable that sites of control for divergence from the activation of apoptosis to that of the transcription factor NF-κB might be discovered. For example, molecular control sites might exist at the level of the receptor by selection of TNFR2 over TNFR1. Or molecular controls may stimulate TRADD to bind with different TRAFs or (FADD) depending upon which pathway needs to be entered (apoptosis or NF-κB).

## FAS and the FAS Ligand

Some of the properties of FAS and its ligand were discussed in the section devoted to the TNF family and its family of receptors. The FAS ligand, a member of the TNF family, binds to the 95 kDa cell membrane protein (APO-1, CD95) or the FAS receptor (itself a member of the TNF receptor family) and triggers apoptosis in target cells. FAS is a 317-amino-acid transmembrane glycoprotein, with three cysteine-rich sequences in the extracellular domains. FAS is expressed on the cell surface of a variety of cells, including epithelial, hepatocyte, and hematopoietic cells, and its role in triggering apoptosis of cells regulating immunity has been a major focus of investigation (18, 34), in part because two recessive autosomal diseases in mice have helped to understand the role of the FAS receptor and the FAS ligand in immunology and autoimmune diseases.

MRL/*lpr/lpr* mice (*lpr* for lymphoproliferative gene) suffer a condition that mimics human lupus erythematosus. The manifestations include vasculitis, lymphadenopathy, uncontrolled proliferation of T cells, and severe glomerulonephritis caused by autoantibodies (among them, antiDNA, anticardiolipin) and formation of autoimmune complexes. The disease is hereditary, transmitted in an autosomal recessive mode, and caused by a mutation of the FAS locus leading to a retrotransposon insertion. The responsible gene is *lpr*, located on chromosome 19 in the mouse. The mutation causes premature termination of the transcription of the gene that encodes a FAS receptor protein. Similarly, mutations of the gene encoding the FAS ligand (*Fas6*) cause a lupus-like systemic autoimmune disease in C57BL/6-*lpr/gdl* mice and rarely in humans. In mice it is caused by a point mutation in the gene encoding the FAS ligand, the *gdl*, located on chromosome 9 in the mouse. A new strain of mice homozygous for both the *lpr* and the *gdl* genes (B6/*lpr/gdl* mice) has been genetically produced.

Thus, apoptosis is inhibited in mice homozygous for the *lpr* and *gdl* genes. These findings helped to establish that FAS-FAS ligand interaction triggers apoptosis of T and B cells, and thereby it plays a critical role in eliminating lymphocytes that react to the self. The implications for understanding autoimmune diseases and graft rejection are obvious.

## TRAIL and TRAIL Receptors

The TNF-related apoptosis-inducing ligand (TRAIL) was a latecomer in the TNF family of ligands constitutively expressed in several human tissues. TRAIL or Apo-2 L induces apoptosis after interaction with its receptor TRAIL R1 and TRAIL R2. The binding to the receptor triggers a death signal pathway similar to that triggered by the FAS ligand (35, 36). Like TNF, TRAIL contains a death domain (a requirement for downstream transduction), the proper receptor is activated and the molecules needed to form the traditional disc are recruited. Special interest in TRAIL emerged from the observation that it preferentially causes apoptosis in cancer cells, thus providing a potential tool for the destruction of these cells. What makes normal cells resistant to TRAIL? One possible mechanism is based on the existence of decoy receptors (TRAIL R3 and TRAIL R4) that compete with the active receptors (TRAIL R1 and TRAIL R2) for the small amount of TRAIL available in a normal cell. However, there is a poor correlation between receptors expression and cell sensitivity to TRAIL and so other factors must be involved to secure cell sensitivity to TRAIL. It has also been proposed that the proapoptotic signal triggered by the binding of TRAIL to its receptor is too weak in cancer cells to interfere with the anti-apoptotic properties of Bcl-2 (37).

An 1146 bp cDNA encoding a mouse homologue of human TRAIL receptor has been cloned that encodes a protein 381 amino acids long. The receptor contains an extracellular cysteine-rich domain, a transmembrane domain, and a cytoplasmic death domain. It thus resembles the human TNFRs, FAS, and TRAIL R2 (38). Hopefully, these studies of TRAIL's mode of action in cancer will help to clarify the role(s) of TRAIL in humans. Indeed, if TRAIL is found to induce apoptosis in most or even in a few cancer cell types, it may also be useful as a potential antineoplastic agent alone or in combination with chemotherapy or radiotherapy. Unfortunately, experiments of Jo et al. suggest that TRAIL causes apoptosis in normal hepatocytes isolated from human, mouse, rat, and rhesus monkey (39). If these data are confirmed, the finding is critical for the use of TRAIL as an antineoplastic agent. Could TRAIL, in addition to killing the cancer cell, cause massive liver necrosis or does it also kill other normal cells? Studies of Smyth et al. have opened new horizons on the potential role of TRAIL (40). In immunosurveillance, TRAIL seems to exert its apoptotic properties through both T and NK cells. More about TRAIL is in the section on cancer.

## RIP

Receptor interacting protein (RIP) was originally identified by its association with TNF induction of apoptosis. RIP binds to the cytoplasmic domain of FAS-TNFR1 and is a 74-kDa protein that possesses a death domain homologous in sequence, in its COOH terminal region, and a tyrosine kinase in its amino terminal sequence. RIP was first discovered in yeast; it contains a tyrosine kinase in its N terminal sequence, a dead domain, and a C terminal. The C terminal is required for cell death, but the kinase and the dead domain are involved in the binding to the adaptor protein (41).

RIP associates with TNFR1 through the adaptor molecule TRADD (42), so if the death domain is necessary, that of the kinase is not. RIP-deficient mice have a markedly reduced ability to activate NF-κB. RIP is believed to bind to TRAF2 not through its death domain, but through an intermediary molecule. In keeping with the suggested role of NF-κB pathway as an activator of survival factors, it is not surprising that RIP-deficient mice are hypersensitive to apoptosis (43).

Cell lines that lack RIP are unable to activate NF-κB in response to TNF and this defect can be corrected by RIP restitution. The chaperone HSP90 regulates the stability and the population of signal transduction molecules and HSP90 has been found to bind to RIP. Inhibition of the HSP90 binding to RIP by geldamicin leads to degradation of RIP and blockage of the NF-κB pathway. In contrast, proteosomal inhibition (by MG 132) prevents RIP degradation, but for reasons that are as yet not clear, the inhibition does not restore NF-κB activation (43, 44). Available data strongly suggest that RIP contributes to the cell's decision to activate the NF-κB pathway or cause apoptosis, a decision between life and death. Lin et al's study sheds some light on the mechanism that leads to the selection of survival over apoptosis.

TNF-induced apoptosis leads to cleavage of RIP at the level of aspartic acid 324, yielding two cleavage products. In particular, one referred to as RIPc, which facilitates interaction between TRADD and FADD, thereby increasing the sensitivity of cells to

TNF. Moreover, RIP mutants that are resistant to cleavage by caspase 8 prevent TNF-induced apoptosis (45). The significance of this study resides in that it contributes clues for the mechanism of control of the passage from apoptosis to NF-κB, an important crossroad between apoptosis and inflammation or apoptosis and cell proliferation (46–49).When TNF's interaction with its receptor is cytotoxic in some cells but not in others, what determines the cell's choice between life and death? The intricacies of the molecular mechanisms that determine the choice between apoptosis and survival are not completely understood. For example, Legler et al. show that TNF-a and TNFR1 need to be recruited to a lipid raft for proper functioning (50). The translocation of TNF-α and TNFR1 to the raft is associated with the binding of RIP. TRADD and TRAF2 are further recruited through their dead domain (Figure 3-2). The complex engages NF-κB activation and is later ubiquitylated. In any event, failure to reach the raft impairs NF-κB activation. Because activation of NF-κB protects against apoptosis (46–49), while inhibition of the pathway enhances apoptosis in response to various inducing agents, it seems clear that the selection of survival over apoptosis depends largely on the engagement of the NF-κB pathway. The work shows that RIP is critical to this selection (45). When RIP is bound to the activated TNFR1 and to TRADD, FADD, TRAF2, and caspase 8 (the DISC), it signals the activation of the NF-κB pathway. RIP is believed to be the kinase that operates upstream of IKK. When RIP is cleaved by caspase 8, RIPc enhances TRADD and FADD association and promotes apoptosis (**FIGURE 3-3**). The function of NF-κB in cancer is discussed further in Chapter 26 (see Figure 26-4).

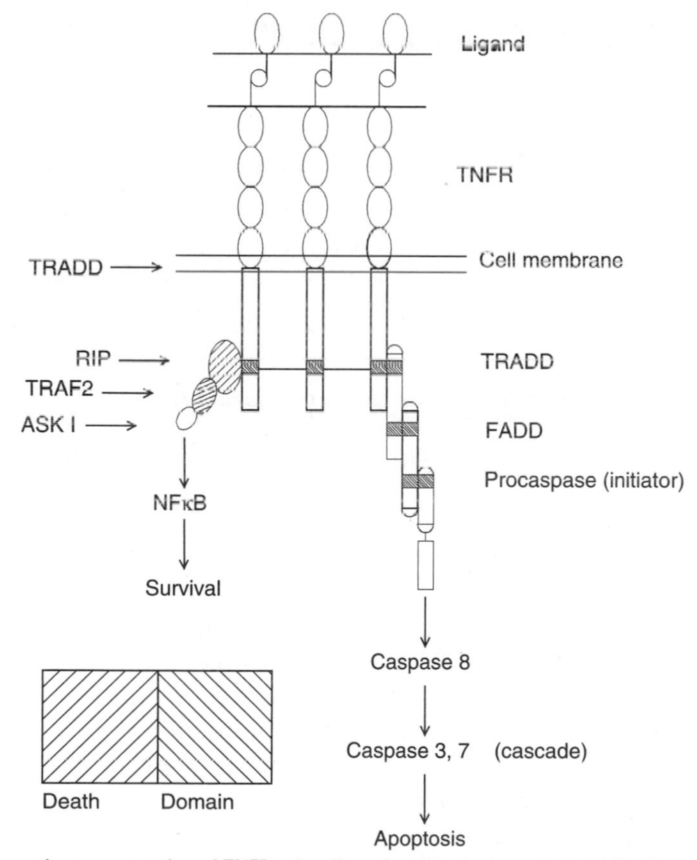

**FIGURE 3-3** Schematic representation of TNFR1 signaling of apoptosis or survival of NF-κB.

## DAXX

DAXX, a nuclear protein, a transcription modulator, and a receptor-associated protein that contributes to apoptosis, was first identified as a protein that binds to the cytosolic tail of FAS and activates the c jun NH(2)-terminal kinase (51). The physiologic role of the death domain-associated protein (DAXX) is still unclear. It is believed to be a proapoptotic protein. DAXX-deficient mice present an hyperproliferative disorder (52, 53). DAXX seems to exert its proapoptotic effect from the nucleus by a molecular mechanism that is far from clear (54). No conclusive mode of action for DAXX has been reached despite active research in the field.

It seems established that DAXX binds to TNFR1 and CD95 receptors in yeast, engages the JNK signal pathway, and induces apoptosis. The C terminal of DAXX is required but is not indispensable for activation of the pathway (52). In promyelocytic leukemia (PML), DAXX is found in the nucleus within PML oncogenic bodies (PODs). DAXX is required during embryogenic development. DAXX-deficient mice reveal complex congenital anomalies and are hypersensitive to apoptosis, suggesting that DAXX modulates the apoptotic process, but its function varies with the circumstances proapoptotic in some and antiapoptotic in others (55). DAXX is a transcription regulator whose mode of action and targets must still be identified. However, it appears that DAXX is a transcription repressor or corepressor (56) whose repression can, at least in some cases, be relieved by phosphorylation, for example, by the homeodomain interacting protein kinase (HIPK1) that phosphorylates DAXX on Ser 669.

In some cases the corepressor activity of DAXX can result from its sequestration. DAXX directly binds to the glucorticoid receptor and prevents its activation, possibly including the activation of its proapoptotic effect. The binding between receptor and DAXX involves the C terminal sequence 501–740 of DAXX). The promyelocytic protein, PML relieves the repression of the glucocorticoid receptor by translocating DAXX to PML oncogenic domains or POD's (57); however, PML sumoylation interferes with DAXX translocation. In summary the transfer of DAXX from the glucocorticoid receptor to PODs sequesters DAXX and allows for activation of the glucocorticoid receptor; in contrast sumoylation of PML blocks the sequestration of DAXX and allows for its binding to the glucorticoid receptor and its repression. As pointed out by Lin et al., these results illustrate the role of PODs in regulating gene transcription by sequestering corepressors in PODs (58).

## Negative Switches of Apoptosis: FAP1, IAPS, and SODD

Numerous inhibitors of apoptosis have been discovered, only those most relevant to the pathway to apoptosis triggered by the TNF family of ligands are briefly listed here.

Human FAS-associated phosphatase-1, FAP-1, is a long, membrane-bound protein (2485 amino acids) that interacts with 15 amino acids of the carboxyl-terminal of FAS. FAP-1 prevents the access of the ligands, FASL, to its cognate receptor, by inhibiting its export to the cell surface (59) FAP-1 contributes the cell resistance apoptosis in colon carcinoma (60).

Inhibitors of apoptosis proteins (IAPs) have been found in insects and in humans. IAPs interact with the positive switches for apoptosis. They are discussed in further detail in the section on inhibitors of caspases (Chapter 6).

A silencer of the dead domain (SODD) was discovered by Jiang and associates in 1999 (61). SODD is associated with the dead domain of the TNFR1 monomer, and inactivates it by preventing trimerization. It does not interact with other components of the activated TNFR1 complex (TRADD, FADD, and RIP). Moreover, SODD seems to interact specifically with the dead domain of TNFR1 and does not bind to other members of the TNF receptors family, including FAS and TNFR2 (61).

### Disposal of TNF Receptors

When the TNF ligand binds to receptors (TNFR1 or TNFR2), it forms an extracellular complex with components of the receptors. The complex activates the cytoplasmic domain and triggers its biological effect. The extracellular complex then further aggregates, is endocytized, and dissociates in a low pH environment. The latter changes appear to result from changes in configuration of the receptor extracellular segment, which at low pH buries the TNF interaction surface (62).

Physiologically, two circulating forms of TNF exist, one that is bioactive and the other that is immunodetectable but inactive. The latter is believed to be bound to inactivating proteins. The inactivated TNF can bind to circulating TNF receptors cleaved from the cell membrane by proteases. Then the soluble receptors (55 and 75 kDa) inactivate TNF and facilitate its clearance mainly by kidney and liver (63).

## Conclusion

The existence of so called "factors" causing hemorrhagic necrosis in cancers has been known for over a century, but the effective agent was not identified until the middle 1970s. The interest in and, indeed, the name of tumor necrosis factor (TNF) emerged from this factor's effect on different cancers in vivo. TNF is a cytokine with multiple functions, among which is that of triggering apoptosis. TNF exists in a secreted or a transmembrane form and the cytokine plays multiple roles in pathology, for example, in autoimmune and inflammatory diseases, transplantation, and cancer. It damages endothelial cells and triggers oxidative damage in many cells. This chapter focuses on its role in apoptosis, much of which has been investigated in vitro on cells in culture (64).

TNF is a member of a large family of cytokines that includes the FAS ligand. TNF's subunits form a trimer that binds to one or more receptors, for example, TNFR1, TNFR2, FAS, TRAIL, etc. The TNF receptors are also members of a large superfamily of receptors, some of which possess a special sequence referred to as the death domain (~80 amino acids located in the C terminal region), while others are devoid of the death domain. The ligands exist in either soluble or a transmembrane form, but the most investigated death receptors are of TNF and FAS. TNF can either induce apoptosis or activate gene transcription that encodes survival proteins; FAS mainly cause cell death.

The TNF receptors are characterized by a variable number of cysteine-rich clusters in their extracellular domain and their cytoplasmic domain. Interaction of the trimeric ligand with the receptor activates the cytoplasmic domain, which possesses no enzymic activity. Thus, signal transduction involves the building of a protein scaffold including TRADD, FADD/MORT, and TRAFs caspase 8 to form the death-inducing signal complex (DISC). A number of positive and negative modulators bind to the cytoplasmic domain of the receptor: RIP and DAXX are positive effectors while SODD is a negative effector. Clearance of TNF can take place in at least two ways: inhibition by binding to inactivating proteins or endocytosis of TNF bound to the receptor.

## References and Recommended Reading


1. Oettgen, H.F., Carswell, E.A., Kassel, R.L., Fiore, N., Williamson, B., Hoffman, M.K., Haranaka, K., and Old, L.J. Endotoxin-induced tumor necrosis factor. *Recent Results Cancer Res.* 1980;75:207–12.
2. Oettgen, H.F., and Old, L.J. Tumor necrosis factor. *Important Adv Oncol.* 1987;105–30.
3. Old, L.J. Tumor necrosis factor. *Sci Am.* 1988;258:59–60, 69–75.
4. Luderitz, O., Galanos, C., Lehmann, V., Nurminen, M., Rietschel, E.T., Rosenfelder, G., Simon, M., and Westphal, O. Lipid A: chemical structure and biological activity. *J Infect Dis.* 1973;128(Suppl):17–29.
5. Zbar, B., Bernstein, I.D., and Rapp, H.J. Suppression of tumor growth at the site of infection with living Bacillus Calmette-Guerin. *J Natl Cancer Inst.* 1971;46:831–9.
6. Carswell, E.A., Old, L.J., Kassel, R.L., Green, S., Fiore, N., and Williamson, B. An endotoxin-induced serum factor that causes necrosis of tumors. *Proc Natl Acad Sci USA.* 1975;72:3666–70.
7. Haranaka, K., Satomi, N., Sakurai, A., and Nariuchi, H. Purification and partial amino acid sequence of rabbit tumor necrosis factor. *Int J Cancer.* 1985;36:395–400.
8. Aggarwal, B.B. Comparative analysis of the structure and function of TNF-alpha and TNF-beta. *Immunol Ser.* 1992;56:61–78.
9. Souba, W. Clinical manifestations of cachexia. In: DeVita, T.D., Jr., Helman, S., Rosenberg, A.S., eds. *Cancer: Principles and Practice of Oncology.* Philadelphia, PA: Lippincott and Raven; 1997;2481–2844.
10. Bazzoni, F., and Beutler, B. The tumor necrosis factor ligand and receptor families. *N Engl J Med.* 1996;334:1717–25.
11. Eskay, R.L., Grino, M., and Chen, H. Interleukins, signal transduction, and the immune system-mediated stress response. *Adv Exp Med Biol.*1990;272:331–44.
12. Branden, C., and Tooze, J. *Introduction to Protein Structure.* New York: Garland, 1991.
13. Vilcek, J., and Lee, T.H. Tumor necrosis factor. New insights into the molecular mechanisms of its multiple actions. *J Biol Chem.* 1991;266:7313–6.
14. Ashkenazi, A., and Dixit, V.M. Death receptors signaling and modulation. *Science* 1998;281:1305–1308.
15. Hill, C.M., and Lunec, J. The TNF-ligand and receptor superfamilies: controllers of immunity and the Trojan horses of autoimmune disease? *Mol Aspects Med.* 1996;17:455–509.
16. Smith, C.A., Farrah, T., and Goodwin, R.G. The TNF receptor superfamily of cellular and viral proteins: activation, costimulation, and death. *Cell.* 1994;76:959–62.
17. Tewari, M., and Dixit, V.M. Recent advances in tumor necrosis factor and CD40 signaling. *Curr Opin Genet Dev.* 1996;6:39–44.
18. Siegel, R.M., Frederiksen, J.K., Zacharias, D.A., Chen, F.K., Johnson, M., Lynch, D., Tsien, R.Y., and Lenardo, M.J. Fas preassociation required for apoptosis signaling and dominant inhibition by pathogenic mutations. *Science.* 2000;288:2354–7.

19. Banner, D.W., D'Arch, A., Janes, W., Gentz, R., Schoenfeld, H.J., Broger, C., Loetscher, H., and Lesslauer, W. Crystal structure of the soluble human 55 kd TNF receptor-human TNF beta complex: implications for TNF receptor activation. *Cell.* 1993;73:431–45.

20. Chan, F.K., Chun, H.J., Zheng, L., Siegel, R.M., Bui, K.L., and Lenardo, M.J. A domain in TNF receptors that mediates ligand-independent receptor assembly and signaling. *Science.* 2000;288:2351–4.

21. Golstein, P. Signal transduction. FasL binds preassembled Fas. *Science.* 2000; 288:2328–9.

22. Remy, I., Wilson, I.A., and Michnick, S.W. Erythropoietin receptor activation by a ligand-induced conformation change. *Science.* 1999;283:990–3.

23. Huang, D.C., Hahne, M., Schroeter, M., Frei, K., Fontana, A., Villunger, A., Newton, K., Tschopp, J., and Strasser, A. Activation of Fas by FasL induces apoptosis by a mechanism that cannot be blocked by Bcl-2 or Bcl-x(L). *Proc Natl Acad Sci USA.* 1999;96:14871–6.

24. Huang, B., Eberstadt, M., Olejniczak, E.T., Meadows, R.P., and Fesik, S.W. NMR structure and mutagenesis of the Fas (APO-1/CD95) death domain. *Nature.* 1996;384:638–41.

25. Hsu, H., Xiong, J., and Goeddel, D.V. The TNF receptor 1-associated protein TRADD signals cell death and NF-kappa B activation. *Cell.* 1995;81:495–504.

26. Strasser, A., and Newton, K. FADD/MORT1, a signal transducer that can promote cell death or cell growth. *Int J Biochem Cell Biol.* 1999;31:533–7.

27. Rothe, M., Wong, S.C., Henzel, W.J., and Goeddel, D.V. A novel family of putative signal transducers associated with the cytoplasmic domain of the 75 kDa tumor necrosis factor receptor. *Cell.* 1994;78:681–92.

28. Takeuchi, M., Rothe, M., and Goeddel, D.V. Anatomy of TRAF2 distinct domains for nuclear factor-kappa B activation and association with tumor necrosis factor signaling proteins. *J Biol Chem.* 1996;271:19935–42.

29. Dadgostar, H., and Cheng, G. An intact zinc ring finger is required for tumor necrosis factor receptor-associated factor-mediated nuclear factor-kappa B activation but is dispensable for c-Jun N-terminal kinase signaling. *J Biol Chem.* 1998;273:24775–80.

30. Arch, R.H., Gedrich, R.W., and Thompson, C.B. Tumor necrosis factor receptor associated factors (TRAFs) a family of adapter proteins that regulate life and death. *Genes Dev.* 1998;12:2821–30.

31. Park, Y.C., Burkitt, V., Villa, A.R., Tong, L., and Wu, H. Structural basis for self-association and receptor recognition of human TRAF2. *Nature.* 1999;398: 533–8.

32. Krajewska, M., Krajewski, S., Zapata, J.M., Van Arsdale, T., Gascoyne, R.D., Berem, K., McFadden, D., Shabaik, A., Hugh, J., Reynolds, A., Clevenger, C.V., and Reed, J.C. TRAF-4 expression in epithelial progenitor cells. Analysis in normal adult, fetal, and tumor issues. *Am J Pathol.* 1998;152:1549–61.

33. Arch, R.H., Gedrich, R.W., and Thompson, C.B. Translocation of TRAF proteins regulates apoptotic threshold of cells. *Biochem Biophys Res Commun.* 2000;272:936–45.

34. Van Parijs, L., and Abbas, A.K. Role of Fas-mediated cell death in the regulation of immune responses. *Curr Opin Immunol.* 1996;8:355–61.
35. Bodmer, J.L., Holler, N., Reynard, S., Vinciguerra, P., Schneider, P., Juo, P., Blenis, J., and Tschopp, J. TRAIL receptor-2 signals apoptosis through FADD and caspase-8. *Nat Cell Biol.* 2000;2:241–3.
36. Bodmer, J.L., Meier, P., Tschopp, J., and Schneider, P. Cysteine 230 is essential for the structure and activity of the cytotoxic ligand TRAIL. *J Biol Chem.* 2000; 275:20632–7.
37. MacFarlane, M. TRAIL-induced signaling and apoptosis. *Toxicol Lett.* 2003; 139:89–97.
38. Wu, G.S., Burns, T.F., Zhan, Y., Alnemri, E.S., and El-Deiry, W.S. Molecular cloning and functional analysis of the mouse homologue of the KILLER/DR5 tumor necrosis factor-related apoptosis-inducing ligand (TRAIL) death receptor. *Cancer Res.* 1999;59:2770–5.
39. Jo, M., Kim, T-H., Seol, D-W., Esplen, J.E., Dorko, K., Billiar, T.R., and Strom, S.C. Apoptosis induced in normal human hepatocytes by tumor necrosis factor-related apoptosis-inducing ligand. *Nat Med.* 2000;6:564–7.
40. Smyth, M.J., Takeda, K., Hayakawa, Y., Peschon, J.J., van den Brink, M.R., and Yagita, H. Nature's TRAIL—on a patch to cancer immunotherapy. *Immunity.* 2003;18:1–6.
41. Kelliher, M.A., Grimm, S., Ishida, Y., Kuo, F., Stanger, B.Z., and Leder, P. The death domain kinase RIP mediates the TNF-induced NF-kappa B signal. *Immunity.* 1998; 8:297–303.
42. Hsu, H., Huang, J., Shu, H.B., Baichwal, V., and Goeddel, D.V. TNF-dependent recruitment of the protein kinase RIP to the TNF receptor–1 signaling complex. *Immunity.* 1996;4:387–96.
43. Zhao, C., and Wang, E. Heat shock protein 90 suppressor tumor necrosis factor alpha induced apoptosis by preventing the cleavage of Bid in NH3T3 fibroblast. *Cell Signal.* 2004;16:313–21.
44. Lewis, J., Devin, A., Miller, A., Lin, Y., Rodriguez, Y., Neckers, L, and Liu, Z.G. Disruption of hsp 90 function results in degradation of the death domain kinase, receptor-interacting protein (RIP), and blockage of tumor necrosis factor-induced nuclear factor-kappa B activation. *J Biol Chem.* 2000;275:10519–26.
45. Lin, Y., Devin, A., Rodriguez, Y., and Liu, Z.G. Cleavage of the death domain kinase RIP by caspase-8 prompts TNF-induced apoptosis. *Genes Dev.* 1999;13:2514–26.
46. Beg, A.A., and Baltimore, D. An essential role for NF-kappa B in preventing TNF-alpha-induced cell death. *Science.* 1996;274:782–4.
47. Wang, C.Y. Mayo, M.W., and Baldwin, A.S. Jr. TNF- and cancer therapy-induced apoptosis: potentiation by inhibition of NF-kappa B. *Science.* 1996;274:784–7.
48. Barinaga, M. Life-death balance within the cell. *Science.* 1996;274:724.
49. Van Antwerp, D.J., Martin, S.J., Verma, I.M., and Green, D.R. Inhibition of TNF-induced apoptosis by NF-kappa B. *Trends Cell Biol.* 1998;8:107–11.
50. Legler, D.F., Micheau, O., Doucey, M.A., Tschopp, J., and Bron, C. Recruitment of TNF receptor 1 to lipid rafts is essential for TNF alpha-mediated NF-kappa B activation. *Immunity.* 2003;18(5):655-64.

51. Chang, H.Y., Nishitoh, H., Yang, X., Ichijo, H., and Baltimore, D. Activation of apoptosis signal-regulating kinase 1 (ASK 1) by the adapter protein Daxx. *Science.* 1998;281:1860–3.

52. Michaelson, J.S., Bader, D., Kuo, F., Kozak, C., and Leder, P. Loss of Daxx, a promiscuously interacting protein, results in extensive apoptosis in early mouse development. *Genes Dev.* 1999;13:1918–23.

53. Michaelson, J.S. The Daxx enigma. *Apoptosis.* 2000;5:217–20.

54. Torii, S., Egan, D.A., Evans, R.A., and Reed, J.C. Human Daxx regulates Fas-induced apoptosis from nuclear PML oncogenic domains (PODs). *EMBO J.* 1999;18:6037–49.

55. Michaelson, J.S., and Leder, P. RNAi reveals anti-apoptotic and transcriptionally repressive activities of DAXX. *J Cell Sci.* 2003;116:345–52.

56. Emelyanov, A.V., Kovac, C.R., Sepulveda, M.A., and Birshtein, B.K. The interaction of Pax5 (BSAP) with Daxx can result in transcriptional activation in B cells. *J Biol Chem.* 2002;277:56–64.

57. Salomoni, P., and Pandolfi, P.P. The role of PML in tumor suppression. *Cell.* 2002;108:165–70.

58. Lin, D.Y., Lai, M.Z., Ann, D.K., and Shih, H.M. Promyelocytic leukemia protein (PML) functions as a glucocorticoid receptor co-activator by sequestering Daxx to the PML oncogenic domains (PODs) to enhance its transactivation potential. *J Biol Chem.* 2003;278:15958–65.

59. Ivanov, V.N., Lopez-Bergami, P., Maulit, G., Sato, T.A., Sassoon, D., and Ronai, Z. FAP-1 association with Fas (Apo-1) inhibits Fas expression on the cell surface. *Mol Cell Biol.* 2003;23(10)3623–35.

60. Yao, H., Song, E., Chen, J., and Hamar, P. Expression of FAP-1 by human colon adenocarcinoma: implication for resistance against Fas-mediated apoptosis in cancer. *Br J Cancer.* 2004;91:1718–25.

61. Jiang, Y., Woronicz, J.D., Liu, W., and Goeddel, D.V. Prevention of constitutive TNF receptor 1 signaling by silencer of death domains. *Science.* 1999;283:543–6.

62. Naismith, J.H., Devine, T.Q., Kohno, T., and Sprang, S.R. Structures of the extra cellular domain of the type I tumor necrosis factor receptor. *Structure.* 1996;4:1251–62

63. Beutler, B.A., Milsark, I.W., and Cerami, A. Cachectin/tumor necrosis factor: production, distribution, and metabolic fate in vivo. *J Immunol.* 1985;135:3972–7.

64. Rink, L., and Kirchner, H. Recent progress in the tumor necrosis factor-alpha filed. *Int Arch Allergy Immunol.* 1996;111:199–209.

CHAPTER

# 4 GENES AND GENE PRODUCTS IN APOPTOSIS OF *C. ELEGANS*

Programmed cell death or apoptosis for many cells is the natural exit from their life cycle. Like cell proliferation, it is critical for homeostasis and when deregulated it causes disease. Therefore, it is not surprising that batteries of genes that tightly regulate the process appeared during evolution. Many of these genes survived from worm to human, and the products of these genes fall in five categories:

1. The first group functions as brakes for the entire process (CED-9, and the anti-apoptotic members of the Bcl-2 family, IAPs and survivin).
2. The second group, when released, engages the process (CED-4 and the proapoptotic agents of the Bcl-2 family).
3. The third group initiates and executes the sequence of events that culminate into apoptosis (caspases).
4. The fourth group acts as guardians of the genome and determines whether the cell, depending upon the integrity of its genome, should or should not enter the cell cycle or engage in apoptosis (P53 family).
5. The fifth group transduces the signal.

## Apoptosis in *Caenorhabditis elegans*

Each of these families of gene products are discussed separately in this Chapter, first in *C. elegans* and second in mammalian cells. *Caenorhabditis elegans,* which is a tiny worm that is a nematode, proved to be a most useful tool for the study of programmed cell death and identification of genes associated with apoptosis. The special properties that make this creature so appealing to biologists are: (a) its transparency; (b) it has a small genome that is completely sequenced; (c) its biologic simplicity; and (d) its short life cycle (**TABLE 4-1**).

After reproduction in a hermaphrodite mode, 3 days for development, and 17 days of adulthood, the worms die. But they can be grown easily on agar plates by feeding them *Escherichia coli.*

**Table 4-1 Some properties of the *Caenorhabditis elegans***

| |
|---|
| Small transparent nematode (1.5 mm) |
| Feeds on *Escherichia coli* |
| Reproduces in a hermaphrodite mode |
| Morphologically composed of 1090 cells, of which 131 cells die in a constant manner in each worm (same cells, same number) |
| Haploid genomes $8 \times 10^7$ nucleotide base pairs |
| Six chromosomes (5 autosomes and 1 X chromosome) |

Transparency makes it easy to observe the development of the worm under special microscopes and the small genome ($40 \times 10^7$ for *Saccharomyces cerevisiae* genome) facilitates cloning of the gene. Moreover, defective mutants are readily generated by exposing the worms to methane sulfonate or ionizing radiation. In a milestone study in cell biology, students of *C. elegans* led by Horvitz traced the fate or lineage of each of the original embryonic cells though innumerable cell divisions and determined the location and the differentiation of each cell in the worm. They observed that among the total number of cells, some are destined to die, mainly neurons that secrete serotonin; there are 105/131 such cells in the hermaphrodite. The constant deaths (in number, time, and location) of 105 of 131 (among the 1090) somatic cells offered a unique advantage for the study of programmed cell death (1). Programmed cell death is not limited, however, to neurons. Programmed cell death during the development of *C. elegans* is not described here because the focus is on the genes involved in apoptosis.

Most genes of the *C. elegans* were cloned and their role investigated by microinjection into the mitotic germline. Several of these genes play a role in programmed cell death (2).

The exact number of genes that operate in programmed cell death of *C. elegans* is not completely known, but among these genes are *ces-1* and *ces-2*, *ced1-2-3-4-5-6-7-9-10*, *egl-1*, and *nuc1*. *ces-1* and *ces-2* determine when cells must die. *ced-3*, *ced-4*, and *ced-9* are responsible for the killing of the cell. Engulfment of the apoptotic body requires two groups of genes: *ced-1*, *ced-6*, *ced-7*, and *ced-2*, *ced-5*, and *ced-10*. Finally *nuc-1* encodes protein(s) that degrade DNA. Thus, the complete apoptotic process in *C. elegans* involves five components: (1) the determination of whether to keep the cell alive or to kill it; (2) a delicate regulatory mechanism leading to the process of apoptosis; (3) execution of the cell; (4) engulfment of the apoptotic body; and (5) DNA degradation.

## *ces-7* and *ces-2* Genes

The products of *ces-1* and *ces-2* genes both contribute to the indictment of the cell, where indictment, when carried to its ultimate consequence, determines whether the cell lives or dies. The genes have been cloned (3). In wild-type animals some special cells (neurosecretory motor neurons [NSM]) located in the lateral face of the pharyngeal wall of the worm inevitably die. When the functions of the proteins encoded by the *ces-2* gene are reduced, the NMS cells survive and differentiate into neurons that secrete serotonin. The finding implies that the fully functional wild-type *ces-2* gene promotes cell death. In contrast, loss of function of the *ces-1* gene preserves the wild-type phenotype, suggesting that the wild-type *ces-1* tends to interfere with the wild-type *ces-2*'s decision to kill the cell. Further experiments led to a model in which the wild-type *ces-2* down-regulates the activity of the wild-type *ces-1* and thereby causes the cells to die. *ces-2* and *ces-1* are located on the same chromosome on which *ces-2* is found upstream of *ces-1*. In summary, *ces-2* causes cell death, *ces-1* prevents cell death, and the determination for the cell to die results from the down-regulation of *ces-1*. It is not clear whether mammalian homologues of *ces-2* and *ces-1* exist; however, it has been shown the *ces-2* product belongs to a family of BZIP proteins that includes the hepatic leukemia factor (HLF) and BZIP transcription factor.

## *ced-3, ced-4,* and *ced-9* Genes

The products of three genes—*ced-3, ced-4,* and *ced-9*—are involved in the second step of apoptosis, the decision of whether the cell must die or live.

### *ced-9*

*Ced-9* is the advocate for the defense and its product, a protein composed of 280 amino acids, serves to prevent cell death in most cells. Its mammalian homologue is Bcl2 (**TABLE 4-2**), which when expressed in *C. elegans,* inhibits normal apoptosis. Mutations in critical domains of CED-9 ($BH_1$, $BH_2$, or $BH_3$) interfere with its ability to block the function of CED-4.

### *Ced-3*

*Ced-3* is the executor of cell death. It is a cysteine protease with functions analogous to mammalian interleukin β-converting enzyme (ICE or caspase 1). Its amino acid sequence is 43% similar to that of ICE and this similarity is most striking in a region of the sequence that includes the active cysteine. However, studies suggest that CED-3 resembles caspase 3 more than it resembles caspase 1. Although most of the substrates of CED-3 remain unknown, poly (ADP)-ribose-polymerase is among the substrates identified. The completion of apoptosis also requires the activation of *ced4* (4).

### *ced-4*

The interpretation of the role of *ced-4* and its product is complicated for at least two reasons. First, the gene for CED-4 encodes two transcripts through alternative splicing: a short one, CED-4S, and a long one, CED-4L. The short transcript is approximately 20 times more frequent than the long one. Second, the proteins derived from *ced-4S* and *ced-4L* have antagonistic functions; the first activates cell death and the second protects against cell death.

CED-4 induces cell death but does not kill the cells. It is a large protein containing four domains including two potential calcium binding sites and an adenosine

**Table 4-2 Genes and gene products involved in whether the cell must die or live**

| Gene | Protein (*N* amino acids) | Homologue | Function |
|---|---|---|---|
| *ced-3* | 503 | Ice | Proapoptotic |
| *ced-4* | 544 | Apaf1 | Proapoptotic |
| *ced-9* | 280 | Bcl-2 | Antiapoptotic |

5′-triphosphate (ATP) binding site. Calcium is not required for CED-4 activity (4–7). CED-4 is a homologue of the mammalian APAF1 (Table 4-2).

The products of *ced-3* and *ced-4* are necessary for cell death and, in contrast, the product of *ced-9* prevents cellular death. *ced-9* encodes a protein (280 amino acids), a homologue of the mammalian antiapoptotic gene *bcl-2*. CED-9 functions upstream of CED-3. CED-9 is anchored to cellular membranes, including the nuclear membrane. In contrast, CED-4 is located in the cytosol and the mitochondria, mainly in the outer mitochondrial membrane. It appears that the product of *ced-4* (CED-4) binds to that of *ced-3* (CED-3). The first serves as a trigger of apoptosis and the second as the executioner of apoptosis. The product of *ced-9* (CED-9) binds to CED-4; as a result the latter is sequestered and unable to bind CED-3 (**FIGURE 4-1**), and the triggering of apoptosis is thereby prevented. Mutations that inhibit the binding of *ced-9* to *ced-3* activate cell killing and, moreover, overexpression of *ced-9* prevents apoptosis.

In 1998, Conradt and Horvitz discovered a protein, Egl-1, that interacts with CED-9 (the homologue of Bcl-2) and is required for cell death. Mutations that inactivate the gene prevent cell death (8), and overexpression induces cell death. Further experiments established that the product of *egl-1* acts upstream of those of the *ced-3*, *ced-4*, and *ced-9* genes. The product of the *egl-1* gene is a protein homologous to the BH-3 proteins found in mammalian cells and is a proapoptotic protein that binds to antiapoptotic proteins.

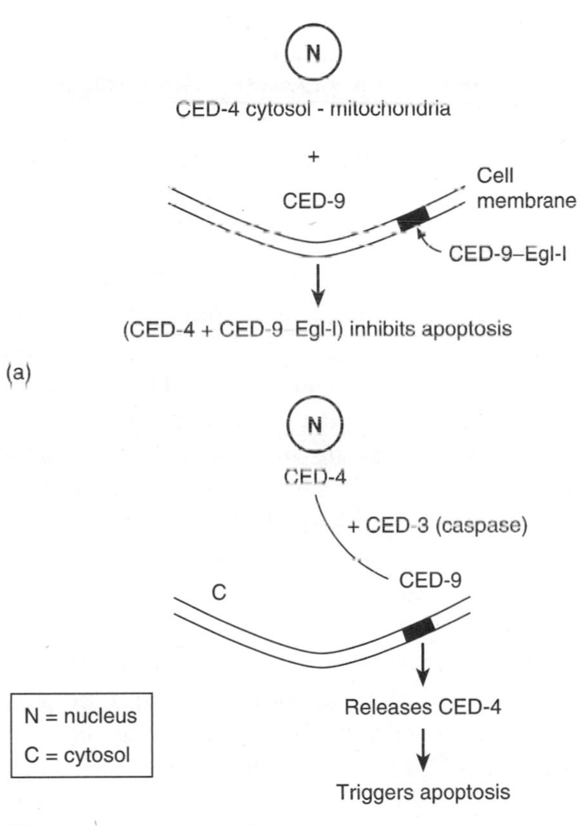

**FIGURE 4-1** Mode of action of CED-3, CED-4, and CED-9. (a) Before apoptosis, CED-4, which is a cytosolic mitochondria protein is bound to CED-9, a membrane bound protein. (b) During apoptosis, CED-9 in the presence of CED-3 releases CED-4. The latter interacts with CED-3, and apoptosis follows.

The discovery of Egl-1 and its reaction with CED-9 further supported the analogy between the pathways for programmed cell death in the worm and mammal.

## Engulfment of the Corpses

The discrete phagocytosis of cell corpses after programmed cell death prevents inflammation and is therefore essential for the protection of the cellular environment. This clearance process requires that the death cell be recognized, engulfed, and then phagocytized by neighboring cells. These events most likely involve complex interactions between the membrane of the cellular corpse and the phagocyte. Identification of the genes that encode molecules involved in the process in the worm has seriously begun. In mammals, dead cells signal their presence to the phagocytes by the emergence of phosphatidylserine on the surface of their membrane. Whether the cells of the worm emit the same signal is not certain. However, it has been established that in *C. elegans,* several genes are involved in the phagocytosis of the apoptotic cells.

Three genes are known to participate in engulfment: *ced-5, ced-6,* and *ced-7.* Workers in Horvitz's laboratory showed that *ced-5* is necessary for engulfment, and that *ced-5* mutants are not only unable to engulf, but they also impair migration of some gonadal cells. Note that engulfment and migration must in some way modulate the function of the cell membrane.

*Ced-5* encodes a 1781-amino-acid protein that contains a proline-rich region and shares 26% identity with the human DOCK-180 and a *Drosophila* myeloblast proteins, a family of proteins (referred to as CDM) that functions in the extension of cell surfaces (9). Overexpression of DOCK-180 restores migration but not engulfment in *C. elegans* worms deficient in their ability to engulf corpses. Whether CED-5 and DOCK-180 are functional homologs is uncertain, although studies suggest that DOCK-180 is found in most cells except in hematopoietic cells, where the homologue M-DOCK has been identified and is associated with focal adhesions (10).[1]

## DOCK-180 and Other Proteins

A brief pause to consider the role of DOCK-180 and other associated proteins first found in mammals may help to clarify the description of the role of CED-5 in engulfment. DOCK-180 binds to the adaptor protein CrkII, which is involved in many signaling cascades including those stimulated by integrin. The Crk's adaptors, among them CrkII, are phosphotyrosine proteins that contain Src homologous domains ($SH_2$, $SH_3$). They recruit other proteins through the $SH_2$ domain and thereby become linked to their sites of action. Integrin activation causes Crk to bind to phosphorylated $p130^{cas}$. $P130^{cas}$ (P130 Crk associate substrate) is a 130 kD protein first found in v-Crk (an oncogene produced by the CT 10 retrovirus) and v-src transformed cells, where it is heavily phosphorylated and functions as a signal transduction factor. The protein includes multiple $SH_2$ and $SH_3$ binding sites; because of this special structure, the protein readily functions as a docking protein. For example, $p130^{cas}$ recruits CrkII to focal adhesions. The exact function of $p130^{cas}$ in focal adhesions is not known, but it has been suggested that it may transduce cell adhesion-generated signals needed for cell survival. Indeed, dephosphorylation of $p130^{cas}$ destabilizes the protein and is associated with

[1]Focal adhesions are a family of adherins, which are junctions, each with a different morphology and function. They include desmosomes, hemodesmosomes, and zonula adherens, which form specialized membrane attachment sites composed of integrins, bundles of actin filaments, and other cytoskeletal proteins. Focal adhesions are involved in cell adhesion, migration, and morphology.

apoptosis (11). NH3T3 cells stimulated with integrin phosphorylates DOCK-180, which in turn binds to CrkII, which then binds to P130$^{cas}$. Coexpression of these three proteins—P130$^{cas}$, CrkII, and DOCK-180—causes membrane spreading and accumulation of the [DOCK-180 CrkL, P130$^{cas}$] complex to focal adhesions.

In conclusion, it appears that DOCK-180 acts downstream to integrin stimulation and complexes with the adaptor CrkII and P130$^{cas}$. DOCK-180 further binds to the small GTPase Rac (a regulator of cytoskeletal actin) and activates several membrane functions, including cell migration and phagocytosis of apoptotic cells (12, 13). A special domain has been identified in DOCK-180: the DOCK homology region 2, DNR-2, which specifically binds to the nucleotide-free Rac and activates it in vitro, and is needed for Rac activation in vivo. The domain is found in several related proteins, suggesting the existence of an evolutionary conserved superfamily of DOCK-180 related proteins.

## *ced-5* and Other Genes in Engulfment

Returning to the function of the products of *ced-5* and other *C. elegans* genes in the engulfment of apopoptic cells, to better understand the function of CED-5 and CED-3, Reddien and Horvitz (14) sought to clarify the role of CED-5 (the orthologue of DOCK-180) in relation to CED-2 and CED-10 in engulfment of apoptotic bodies. Their seminal studies found the following:

- Appropriate mutation experiments established that the products of those two genes are required for engulfment.
- The *ced-2* gene encodes a protein homologous to the human CrkII protein and, moreover, that the *ced-10* gene encodes a protein homologous to the human Rac GTPase.
- They showed that the product of *ced-2* (CrkII) interacts with that of *ced-5* (DOCK-180).
- They provided evidence that *ced-2* and *ced-10* are expressed in the engulfing of the cell rather than during the killing of the cell.
- They proposed that CED-10 (= RGTPase) functions downstream of *ced-2* (= CrkII) and *ced-5* (= DOCK-180).

On the basis of their findings, they proposed a model in which the dying cells send a signal to the engulfing cell, and as a result CED-2 (CrkII) binds to CED-5 (DOCK-180), the latter complex localizes at the cell membrane and activates *ced-10* (Rac GTPase), which then directs the membranes cytoskeleton to emit the engulfing pseudopodia (**FIGURE 4-2**).

As pointed out earlier, the homologies between the products of *ced-2, ced-5,* and *ced-10* with the human proteins Crk-2, DOCK-180, and Rac GTPase suggest that the mechanism for phagocytosis proposed above may have survived during evolution (**TABLE 4-3**).

The exact role of the products of the *ced-6* and *ced-7* genes is not clear. *ced-6* is required for engulfment of apoptotic cells; it encodes a new protein that contains phosphoryltyrosine binding sites in its N terminus and $SH_3$ domains in its C terminus. The phosphoryltyrosine binding site reacts specifically with polypeptides containing a MPXY (p) consensus motif. Overexpression of *ced-6* can partially correct incapacitation of the engulfing process caused by defective *ced-1* and *ced-7,* but not that caused by *ced-2, ced-5,* and *ced-10.* This led to the conclusion that *ced-6* functions downstream of *ced-1*

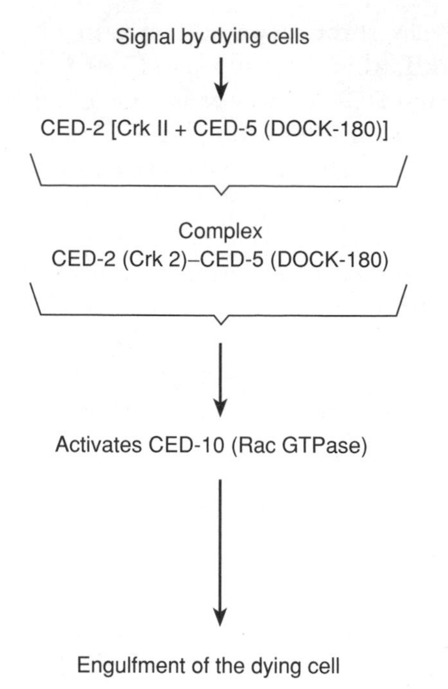


**FIGURE 4-2** Engulfment of dying cells in *Caenorhabditis elegans* (see references 14 and 15).

**Table 4-3 Gene product and the mammalian homologue**

| *C. elegans* Gene Product | Mammalian Homologue(s) |
|---|---|
| CED-3 | ICE, caspase-3 |
| CED-4 | APAF-1 |
| CED-9 | Bcl-2 family |
| CED-5 | DOCK-180 |
| CED-2 | Crk II |
| CED-10 | Rac GTPase |

and *ced*-7. The *ced*-7 gene encodes an ABC transporter, the ATP binding cassette.[2] ABC transporters possess four core domains: two transmembrane or T domains that form the passageway, and two cytosolic ATP binding domains (A domains). ABC transporters constitute a superfamily of approximately a 100 proteins; they are found in organisms ranging from bacteria (permeases) to humans, for example, the multiple drug resistance transport protein (MDR) and the cystic fibrosis transmembrane regulator. Although it is not known at what point the *ced-6* and *ced*-7 gene products might function in engulfment of the dead or dying cells, it can be expected that an ATP cas-

[2]ABC transporters use energy from ATP hydrolysis to pump substrates (ions, sugars, peptides, polysaccharides, drugs, proteins) across cell membranes against concentration gradients either to import or export molecules. However, they only function in one direction. The human ATP binding cassette transporter (*hABC1*) gene is an open reading frame 6603 bp long that encodes a polypeptide made of 2207 amino acids (~220 kDa).

sette transporter and a protein containing phosphoryltyrosine binding sites will find a place in the chain of events that lead to engulfment.

Finally a *nuc-1,* which cleaves DNA at specific sites, is required to conclude apoptosis, although it is not required for cell death. However, in the absence of *nuc-1,* the apoptotic bodies persist even after engulfment.

## Conclusion

Tracking the fate or lineage of each of the embryonic cells through their differentiation in adulthood and the identification of 131 cells of *C. elegans* destined to die provided a unique model for the study of programmed cell death from initiation of the process to engulfment of the corpses. Identification of the genes directly involved in the triggering and execution of cellular death provided a logical, albeit still incomplete, mechanism for apoptosis. Finally the homologies between CED-3, CED-9, and the mammalian ICE and Bcl-2; between CED-2 and CrkII; between CED-5 and Dock-180; and between CED-10 and the Rac GTPase have provided dramatic confirmation of the conservation of the regulation of programmed cell death, phagocytosis, and caspases from nematodes to humans (Table 4-3).

Despite the elegance of the methods and the enormous amount of work devoted to the study of programmed cell death in *C. elegans,* some questions remain. Among them are the exact functions of the products of *ces-1* and *ces-2,* the role of the *ced*-4L transcript, and the detailed mechanism of engulfment of the apoptotic bodies. However, if the findings made in *C. elegans* are a prologue in the evolution of knowledge, these findings are bound to help scientists complete our understanding of apoptosis in mammals. The reader is referred to the excellent reviews of Reddien and Horvitz (14 and 15) for further details.

## References and Recommended Reading


1. Gumienny, T.L., Lambie, E., Hartwieg, E., Horvitz, H.R., and Hengartner, M.O. Genetic control of progammed cell death in the *Caenorhabtitis elangans* hermophrodite germline. *Development* 1999;126:1011–22.
2. Xue, D., Wu, Y., and Shok, M. Programmed cell death in *C. elegans*: the genetic framework. In: Jacobson, M., McCarthy, N., eds. *Apoptosis. The Molecular Biology of Programmed Cell Death.* Oxford, UK: Oxford University Press; 2002:23–85.
3. Metzstein, M., Hengartner, M.O., Tsung, N., Ellis, R.E., and Horvitz, H.R. Transcriptional regulator of programmed cell death by *C. elegans* gene. *Nature* 1996;382:545–7.
4. Yuan, J.Y., and Horvitz, H.R. The *Caenorhabditis elegans* genes Ced-3 and Ced-4 act autonomously to cause programmed cell death. *Dev Biol.* 1990;138:33–41.
5. Yuan, J.Y., Shaham, S., Ledoux, S., Ellis, H.M., and Horvitz, H.R. The *C. elegans* cell death gene Ced-3 encodes a protein similar mammalian interleukin-1-B converting enzyme. *Cell* 1993;75:641–52.
6. Yuan, J., and Horvitz, H.R. The *Caenorhabditis elegans* cell death gene Ced-4 encodes a novel protein and is expressed during the period of extensive programmed cell death. *Development* 1996;116:309–20.

7. Yang, X., Chang, H.Y., and Baltimore, D. Essential role of Ced-4 oligomerization in CED-1 activation and apoptosis. *Science* 1998;281:1356–7.
8. Conradt, B., and Horvitz, H.R. The *C. elegans* protein EGL–1 is required for programmed cell death and interacts with the Bcl2 like protein CED-9.*Cell* 1998;93:519–20.
9. Wu, Y., and Horvitz, H.R. *C. elegans* phagocytosis and cell-migration protein CED-5 is similar to human DOCK180. *Nature.* 1998;392:501–4.
10. Thomas, S.M., and Brugge, J.S. Cellular functions regulated by SRC family kinases. *Ann Rev Cell Dev Biol.* 1997;13:513–610.
11. Weng, L.P., Wang, X., and Yu, O. Transmembrane tyrosine phosphatase LAR induces apoptosis by dephosphorylating and destabilizing p130Cas. *Genes Cells.* 1999;4:185–96.
12. Cote, J.F., and Vuori, K. Identification of an evolutionarily conserved superfamily of DOCK180-related proteins with guanine nucleotide exchange activity. *J Cell Sci.* 2002;115:4901–13.
13. Nishihara, H., Maeda, M., Oda, A., Tsuda, M., Sawa, H., Nagashima, K., and Tanaka, S. DOCK2 associates with CrkL and regulates Rac1 in human leukemia cell lines. *Blood.* 2002;100:3968–74.
14. Reddien, P.W., and Horvitz, H.R. CED-2/CrkII and CED-10/Rac control phagocytosis and cell migration in *Caenorhabditis elegans. Nat Cell Biol.* 2000;2(3):131–6.
15. Reddien, P.W., and Horvitz, H.R. The engulfment process of programmed cell death in *Caenorhabditis elegans. Annu Rev Cell Biol.* 2004;20:193–221 (review).

CHAPTER

# THE BCL-2 FAMILY 5

## Structure and Role

The product of the *bcl-2* gene was the first of a large family of inhibitors and activators of apoptosis to be discovered. Bcl-2 is a 26 kDa protein found first in the mitochondrial membrane, but later also found in the nuclear membrane and the endoplasmic reticulum. The Bcl-2 protein is coded for by a 230-kb gene that is translocated (t14;18) to the heavy chain locus in certain B cell malignancies (follicular lymphoma and chronic lymphocytic leukemia). The translocation of the *bcl-2* gene to chromosome 18 leads to its overexpression. Therefore, it was at first believed that the product of the gene (Bcl-2) stimulated cell proliferation. It was later established that *bcl-2* is an oncogene that is constantly activated in some forms of malignancies and protects the cells from undergoing apoptosis. There is 23% identity and 46% similarity between the *ced-9* gene and the *bcl-2* gene (1, 2). Moreover, *bcl-2* rescues mutant yeasts deficient in *ced-9* and prevents programmed cell death in *Caenorhabditis elegans.* Thus, the genes protecting against apoptosis can be found from nematodes to mammals and may even be present in unicellular organisms as well.[1]

The products of the *bcl-2* gene family are critical to the manifestation of apoptosis. Bcl-2-defective mice are abnormal during development and after birth. Development of the kidney is impaired and it becomes polycystic, there are distortions of the small intestine and abnormal hair pigmentation. After birth the mice are immunodeficient because of the loss of B and T cells (4).

The Bcl-2 family includes both protectors (Bcl-2, Bcl-X1, Bcl-W, Bfl-1, Brag-1, Mcl-1, A1, Boo) and stimulators (Bax, Bak, Bcl-X5, Bad, Bid, Bik, and Hrk) of apoptosis. There are some structural common denominators among the members of the Bcl-2 family. Bcl-2 and other members of the family (Bcl-x1, Bcl-w, and Mcl-, A1) contain four homologous sequence sites referred to as BH1, BH2, BH3, and BH4, each of which constitutes an α helical segment of the molecule (5). Each BH domain is highly conserved in the antiapoptotic members of the family, at least in mammals and *C. elegans.* The main characteristics for the Bcl-2 family include:

- Sequence conservation is not as rigid for the proapoptotic members of the family except in the BH3 domain. Few possess a BH4 domain, and many of the proapoptotic

[1]Jean Claude Ameisen (1996) pointed out that the process of apoptosis is not limited to multicellular organisms, but is also found in unicellular organisms such as trypanosomes (*cruzy* and *brucei rhodiense*) as well as the slime mold (*Dyctyostellium dicoideum*) and the ciliate (*Tetrahymena thermophila*) (3).

molecules contain only one of the conserved BH domains, BH3 (BH3-only subfamily). The viral proapoptotic members of the family are devoid of the BH4 domain.

- Most members of the antiapoptic and proapoptotic of the Bcl-2 family possess a hydrophobic COOH terminal. Notable exceptions are the proapoptotic Bid and Bad. At least in the case of Bcl-2, the COOH terminal (TM) is needed for mitochondrial targeting. Deletion of the COOH terminal reduces the death inhibitory capacity as well as the binding of some proteins (such as Raf1, a protein kinase, and the mammalian equivalent of CED-4, see below) to the BH4 region.
- $Bclx_L$ and Bid contain a flexible loop domain between residues 26 and 76 and residues 31 and 70, respectively. Some of these sites are missing in certain members of the Bcl-2 family (BH1 in some or BH4 in others, and both in still others).

Many of the products of the *bcl-2* gene family are mitochondrial proteins (e.g., Bcl-2, Bax, $Bclx_L$ , and BHRF), but as pointed out earlier, Bcl-2 also can be found in the endoplasmic reticulum and the nucleus. In mitochondria the Bcl-2 sites are distributed irregularly between the inner and outer membranes, and they are stoichiometrically related to a transition pore.

Available evidence suggests that in some (if not in all) cases, the ratio of antiapoptotic to proapoptotic agents determines whether the cell will live or die.[2] This so-called "death–life rheostat" results from the competitive binding between antiapoptotic and proapoptotic proteins. It is the ratio of proapoptotic to antiapoptotic proteins that determines whether the cell lives or dies (6), and the conserved BH regions are important for these interactions.

1. BH1 and BH2 regions are indispensable for the formation of Bcl-2 homodimers.
2. The BH1 and BH2 regions of Bcl-2 and $Bclx_L$ are required for the heterodimerization of the antiapoptotic protein and the proapoptotic protein (e.g., Bax).
3. Similarly, the BH3 region of Bax is needed for its heterodimerization with the $Bclx_L$. Therefore, the BH3 region is sometimes referred to the death region.
4. BH4 is absent in poorly antiapoptotic molecules and in proapoptotic molecules.

The fatal role of the BH3 domain is further supported by the existence of short proteins containing the BH3, TM, and small lateral sequence that antagonize survival (7) (**FIGURES 5-1** and **5-2**).

If an appropriate balance between proapoptotic and antiapoptotic agents of the Bcl-2 family seem to be required for the cell to determine whether it enters apoptosis, binding between these opposing forces may not always be necessary. Thus, the $BH_1$, $BH_2$, and $BH_3$ domains of $Bclx_L$ form a cleft in which the dead domain of the proapoptotic agents lodges. Minn and associates constructed mutants of $Bclx_L$ that remained antiapoptotic but were unable to bind Bax (8). These findings raised a new problem, namely that of the antiapoptotic role of $Bclx_L$ in the absence of binding to apoptotic

[2]Mutation of $Bclx_L$ that prevents its binding to Bax do not interfere with the antiapoptotic action of $Bclx_L$. $Bclx_L$ binds caspase 8 via CED4 or its mammalian equivalent, thereby sequestering the caspase and preventing it from exerting its function as a death agonist. As a result, two mechanisms for the antiapoptotic effect of Bcl-2 and $Bclx_L$ have been contemplated: binding to proapoptotic members of the family (e.g., Bax or Bak) or direct sequestration of the death executor (caspase 8) by the proapoptotic agent.

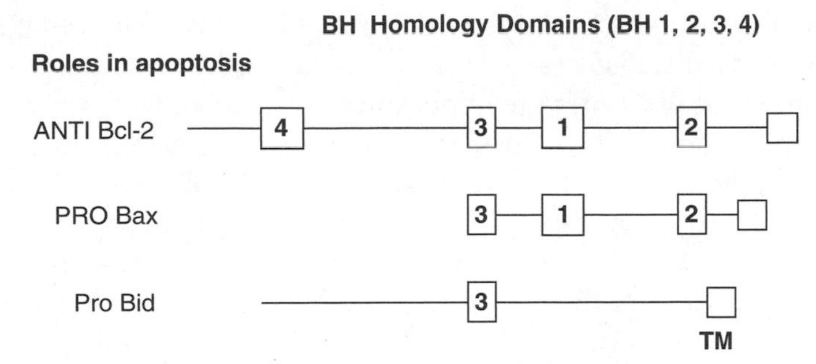

**FIGURE 5-1** Location and numbers of BH homology domains in Bcl2, Bax, and Bid.

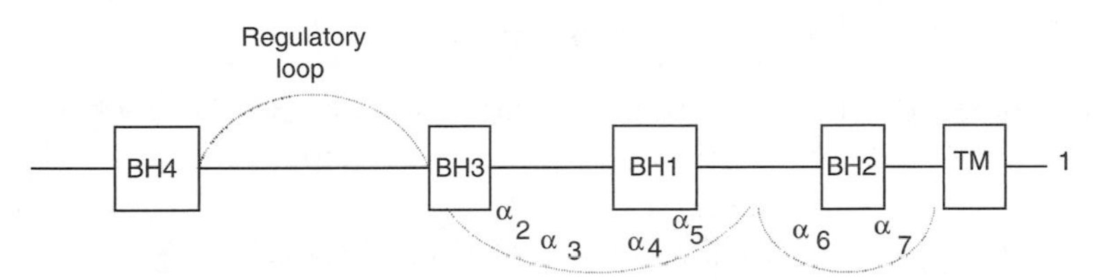


**FIGURE 5-2** Relative position of BH homology domains, regulatory loop, and alpha helices in Bcl-2. (Modified from Tsujimoto [7].)

agonists. Recognizing that $Bclx_L$ forms channels within synthetic membranes, Minn et al. wondered if that property could, by itself, contribute to the proapoptotic function of $Bclx_L$ without requiring heterodimerization. To answer this question, they constructed a new set of mutants by replacing certain amino acids within the $Bclx_L$ sequences forming the channel. The chimera remained able to heterodimerize and inhibit apoptosis, but the normal membrane potential was not maintained. Moreover, heterodimerization with Bax was not required for the membrane potential alteration. The study concluded that $Bclx_L$ modulates cell survival in at least two ways, heterodimerization with a proapoptotic agent and its ability to maintain an appropriate potential channel (8).

Muchmore and associates investigated the structure of $Bclx_L$ (9). The molecule is composed of seven antiparallel central helices. A long flexible region links the first and second helices. The flexible loop exhibits the greatest variability in amino acid sequence and can be deleted without a loss of the antiapoptotic effect. The BH4 domain comprises most of the first helix. The COOH terminal is part of the remaining components of the molecule, namely the three homology regions: BH1, BH2, and BH3. The cleft contains a highly conserved glycine 138 located in the BH1 domain. The structure of $Bclx_L$ is similar to that of colicin and the diphtheria toxins, especially with respect to their membrane-binding domain (9).

Bcl-2 is a tail-anchored protein inserted in the mitochondrial membrane through its COOH terminus. However, during translation on the ribosome the $NH_2$ terminus is translated first and the COOH tail last. Egan et al. showed that the targeting of Bcl-2 to the ribosome is cotranslational and determined by 13 amino acids (204 to 218 in B $Bclx_L$), a flexible flanking domain that shields the emerging hydrophobic COOH terminal until it can be properly inserted into the membrane (10). Thus, the insertion of the COOH terminal until it can be properly inserted into the membrane. The insertion of the COOH terminal is likely to occur posttranslationally (9).

A physiologic event illustrates the importance of the interaction between antiapoptotic and proapoptotic agents. Consider the fate of human ovarian follicles. At the onset of puberty, the ovary contains approximately 400,000 follicles, but only 400 of them mature and lead to ovulation. After menopause 99% of the follicles disappear through apoptosis (11). In transgenic mice in which Bcl-2 is overexpressed, follicular apoptosis is suppressed and the size of the follicle is increased. Moreover, teratomas develop in older animals (12). In contrast, Bcl-2 null mice have a decreased number of primordial follicles. Bax null mice present defective apoptosis in the follicles and the latter are conserved.

## Posttranslational Modification of Bcl-2

At least two modes of posttranslational modifications of Bcl-2 have been described: phosphorylation-dephosphorylation and proteolysis. If the ratio of antiapoptotic to proapoptotic proteins is an essential part of the regulation of apoptosis, this regulatory mechanism is further modulated by posttranslational events such as serine phosphorylation, binding to growth factors, and possibly proteolysis. Serine phosphorylation has been described for Bcl-2 and Bad. Posttranslational phosphorylation is likely to contribute to the transduction of signals that lead to the consumption of apoptosis in the case of Bcl-2, and the inhibition of apoptosis in the case of Bad. A regulatory loop exists between the BH4 and the BH3 domains of Bcl-2 that contains the serine phosphorylating site(s). Phosphorylation of those sites abolishes the antiapoptotic activity of Bcl-2. Taxol, vincristine, and vinblastine are believed to assist the phosphorylation of those sites and thereby accelerate cell death through apoptosis. Site specific mutations revealed that serine 70, an evolutionary conserved site, is required for phosphorylation. Interleukin-3 (IL-3) and the nerve growth factor, agents that stimulate the growth of cells by activating cellular proliferation and inhibiting apoptosis, also increase the phosphorylation of serine 70. Therefore, it appears that Bcl-2 phosphorylation at this specific site is critical for suppression of apoptosis. The phosphorylating enzyme is suspected to be protein kinase C (11, 13).

Since in vivo or in vitro maintenance of the balance of the cell population is often secured by a combination of proliferation and apoptosis, it is expected that Bcl-2 could be rapidly dephosphorylated. Deng et al., using (among other experiments) specific inhibitors of phosphatase 2A, demonstrated that this enzyme plays a regulatory role of the function of Bcl-2 in apoptosis (**FIGURE 5-3**) (13).

Regulation of Bcl-2 phosphorylation is critical to its capacity to prevent apoptosis. Although the specific enzymes involved in phosphorylation and dephosphorylation of Bcl-2 have not been fully identified, two candidates have been discovered in mice myeloid cells by Deng and colleagues: protein kinase C for phosphorylation of Bcl-2 and suppression of apoptosis, and protein phosphatase 2A for dephosphorylation and activation of apoptosis. As the study emphasized, phosphatase might be a new target for inducing apoptosis in leukemic cells resistant to chemotherapy (13, 14).

Of course, a phosphorylation-dephosphorylation off and on switch does not exclude other mechanisms of inactivation of the proapoptotic activity of Bcl-2 that might act independently or in sequence. Other mechanisms of amplification of the caspase cascade that lead to apoptosis have been described. The proteosomal degradation of Bcl-2

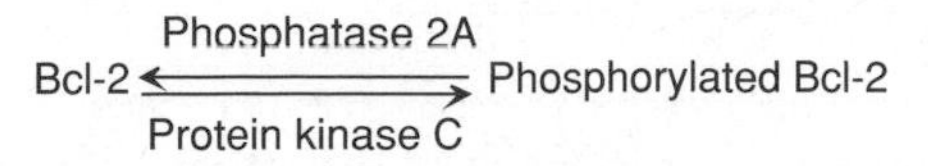


**FIGURE 5-3** Bcl-2 phosphorylation.

may also contribute to the activation of Bcl-2. Apoptosis induced by tumor necrosis factor-alpha (TNF-α) is associated with a decrease in the level of Bcl-2. Mutation of Bcl-2 amino acid susceptible to bind ubiquitin maintains high levels of Bcl-2. Inhibitors of the proteosomal attack of Bcl-2 prevent the induction of apoptosis (15).

A link seems to exist between Bcl-2 dephosphorylation and Bcl-2 ubiquination and proteosomal degradation. TNF-α induces dephosphorylation and subsequent ubiquination, and degradation of Bcl-2 in endothelial cells. Inactivation of mitogen activated protein (MAP) kinase contributes to ubiquitination and degradation of Bcl-2. In contrast, inactivation of the protein kinase B/akt does not affect Bcl-2 stability. Therefore, it appears that MAP kinase contributes to the phosphorylation of Bcl-2 (most likely on serine 87) and protein against degradation in endothelial cells (16).

Inasmuch as the Bcl-2 molecule contains a BH3 domain, it is not inconceivable that the Bcl-2 proteolytic breakdown might activate the BH3 domain and convert Bcl-2 to a dead effector similar to Bax. After induction of apoptosis, caspase 3 cleaves the loop domain of Bcl-2 at asp 34 to yield a proapoptotic compound. The latter function is linked to the BH3 sequence of the transmembrane domain and further triggers the caspase cascade. A mechanism by which the cleavage of Bcl-2 leaves the cell without protection against cell death can therefore readily be imagined.

Bcl-2 is believed to heterodimerize with a proapoptotic agent (e.g., Bax) that neutralizes its proapoptotic activity; cytochrome c is released and activates caspase 9. The latter activates the caspase cascade including caspase 3, which in turn cleaves Bcl-2 and destroys at least one protector against apoptosis (**FIGURE 5-4**) (17).

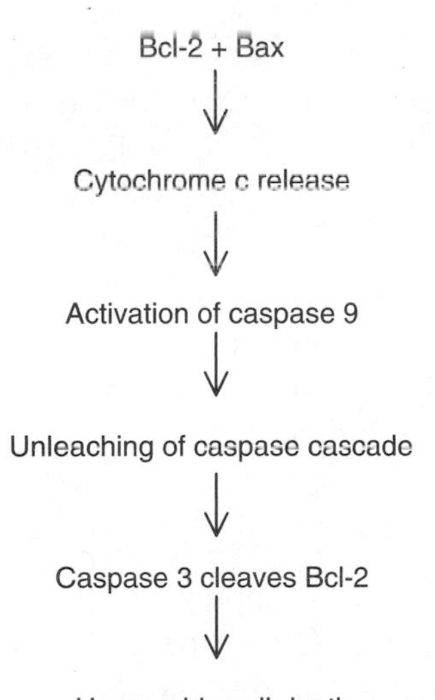


**FIGURE 5-4** The pathway from cytochrome c release to apoptosis.

## Bax

Bax, a member of the Bcl-2 family, is an agonist of apoptosis. The molecule contains three of the BH domains: BH1, BH2, and BH3. Bax resides in the mitochondrial membrane. The BH3 domain is necessary for heterodimerization with antagonistic members of the Bcl-2 family: Bcl-2, $Bclx_L$, and Mcl-1. In contrast, BH1, BH2, and the N terminal are not needed for binding (18). However, it is still not always clear why heterodimerization of Bax and Bcl-2 are required for induction of apoptosis.

Mice with a deficient *bax* gene develop lymphoid hyperplasia of T and B cells, but are also infertile because of defective testicular development and impaired spermatogenesis. Thus, the cell damage caused by Bax deficiency in knockout mice varies with the cell lineage. These differences may be the result of variations in cellular expression and redundancy of the members of the Bcl-2 family (19).

Although apoptosis is regulated by the interaction between proapoptotic and antiapoptotic agents, and heterodimerization between proapoptotic and antiapoptotic agents can be demonstrated in vitro and even in intact cells, it is not clear whether heterodimerization is always required for function. Mutations of the BH1 and BH2 domains of Bcl-2 hinder its protective effect against apoptosis. Thus, the substitution of alanine for the highly conserved glycine 145 in Bcl-2 or its homologue glycine 138 in $Bclx_L$ prevents homodimerization and heterodimerization, and inactivates the protein (20). While these findings suggest that heterodimerization is needed to regulate apoptosis, $Bclx_L$ mutants that are unable to heterodimerize but still can block apoptosis have been reported (21). Mutants of Bcl-2 (G145A) and $Bclx_L$138 are unable to homodimerize with full length wild-type Bax, but can dimerize with truncated mutants containing the BH3 domain and block the apoptosis caused by the truncated Bax mutants. The findings emphasize the importance of BH3 in securing dimerization, and the coupling of cell death and dimerization (22).

In contrast, studies of Zha and Reed suggest that dimerization of Bcl-2 and Bax is not an absolute requirement for apoptosis (23). The BH3 domain contains a conserved motif, the IGDE sequence (residues 66–69). Zha and Reed (23) investigated the role of the IGDE motif in dimerization by introducing alanine substitutions in it. In yeast, the Bax mutants D68A and E69A, which are able to homodimerize, cannot heterodimerize with Bcl-2. In both yeast and human cells, overexpression of the E69A mutant causes apoptosis that is suppressed by co-expression of Bcl-2, despite the inability of Bax to heterodimerize. However, complete deletion of IGDE (Bax Δ IGDE) from the Bax sequence blocks homo- or heterodimerization and the proapoptotic properties are lost in yeast or mammals. Bax and the two mutants bind Bid, a member of the Bcl-2 family believed to activate Bax. Zha and Reed conclude that:

1. Bax can cause apoptosis without heterodimerizing with Bcl-2. However, the possibility of binding of the mutants to redundant antiapoptotic members of the Bcl-2 family cannot be excluded.
2. Bcl-2 rescues yeast and mammalian cells from Bax-induced apoptosis.
3. Bax is likely to interact with Bid to exert its death agonistic function (21).

The possibility that Bax exerts its proapoptotic function differently in different cells cannot be excluded. Thus, two different mechanisms have been proposed: (1) The BH3 of Bax binds to Bcl-2 (heterodimerization) and thereby sequesters it; and (2) Bax activates cytotoxic channels that unbalance the homeostasis of ions, proteins, and other molecules.

Wang and colleagues further investigated the amino acid substitutions in the amphipathic α helix located outside of the IGDE motif (24). Amino acids 63, 67, 70, and 74 of the wild-type Bax were replaced by alanine. Located in the hydrophobic face of the BH3 domain, these amino acids yielded a mutant containing the following four amino acid substitutions: L63A, G67A, L70A, and M74A (Bax mutant mIII-I). These mutations did not affect the amphipathic nature of the sequence or the proapoptotic functions of Bax, but they abolished the capacity of Bax to homo- or heterodimerize. This led Wang et al. to conclude that the manifestation of the proapoptotic properties of Bax requires that it engage in an intramembranous conformation (24).

There is evidence that recombinant Bax protein added to isolated mitochondria induces cytochrome C release and activation of cytosolic caspases. However, when Bax is released by exposing mitochondria to calcium, Bax does not cause mitochondrial swelling induced by permeability transition (25).

If Bax can exert its proapoptotic activities without heterodimerization, is homodimerization necessary for function? Mutations, such as the deletion of IGDE motif in the BH3 domain of Bax, which prevents homodimerization and also cancels its proapoptotic function, suggest that cell death and homodimerization are linked.

In conclusion, it appears that, depending upon the circumstances, Bax[3] exerts its proapoptotic properties by two different mechanisms: heterodimerization and activation of the permeability transition pore. Both mechanisms are likely to lead to cytochrome c and cytosolic caspase activation.

The existence of several modes of interaction between proapoptotic and antiapoptotic proteins does not negate the notion of a molecular rheostat that balances the operations of these antagonistic proteins. This balance may be affected by other mechanisms such as posttranslational events (e.g., phosphorylation and dephosphorylation of Bcl-2 and Bax, and the relative amounts of these proteins and calcium). Hence, Bcl-2 expression in the brain rapidly decreases after birth except in portions that continue to develop (e.g., the dentate nucleus). However, the expression of $Bclx_L$ in the brain (possibly only in neurons) is maintained in the fetus after birth, when it is found in pyramidal and granular neurons of the hippocampus (26, 27). In contrast, the proapoptotic splice variant $Bclx_S$ is barely detectable in premature brain. Because of these changes in distribution of Bcl-2 and $Bclx_L$ as the brain matures, it has been proposed the $Bclx_L$ gradually replaces Bcl-2 and may secure survival of postnatal neurons (28). The hippocampus is rich in receptors for both mineralocorticoids and glucocorticoids. Cortisone activates both receptors in rats, but the affinity for mineralocorticoid is 10 times greater than that for corticosteroids. Stimulation of glucocorticoid receptors increases Bax synthesis in neurons and leads to apoptosis in the dentate granular cell population. Aged subjects are more vulnerable to glucocorticosteroid than younger ones and are less able to up-regulate $Bclx_L$. This leads to an imbalance between the antiapoptotic $Bclx_L$ and the proapoptotic Bax in favor of apoptosis. Such findings clearly suggest that the balance between anti- and proapoptotic proteins is critical to the survival of granular cells of the dentate nucleus (29). The discoveries of other proapoptotic homologues of Bcl-2 have further confirmed the existence of more than one

[3]In addition to the classical Bax, a number of variants of that molecule exists, some are derived from alternate splicing and are discussed later in this chapter.

mechanism involved in unleashing the caspase cascade; consider the roles of Diva, Molt, and Bok.

## Diva, Molt, and Bok

### Diva

Diva is a proapoptotic homologue of Bcl-2 found in different parts of the embryo and in several organs in mice up to four weeks after birth. In adult mice, Diva is found only in the testicle and the ovary. Diva contains homologues of the four domains (BH1, BH2, BH3, and BH4) of Bcl-2 as well as the carboxyterminal hydrophobic domain. Like Bax, Diva inhibits the binding of $Bclx_L$ to Apaf-1. It does not heterodimerize with $Bcl\text{-}x_L$. Thus, Diva binds directly to Apaf-1 (see below) and thereby unleashes the caspase cascade. The finding emphasizes the ability of the cell to use different pathways to induce apoptosis. When cell death occurrs, Diva acts as a proapoptotic agent that bypasses heterodimerization with proapoptotic Bcl-2 or $Bclx_L$.

### Molt

The proapoptotic Molt ($BH_3$, $BH_1$, $BH_2TM$), a member of the Bcl-2 family with a highly conserved $BH_3$ domain, does not heterodimerize with Bcl-2 or $Bclx_L$, but its proapoptotic activity is inhibited by caspase inhibitors. Molt mRNA is detectable in liver, thymus, lung, and in intestinal epithelia (30).

### Bok

Bok is a proapoptotic member of the Bcl-2 family similar to Mtd found in the ovary. It dimerizes with the proapoptotic Mcl-1 probably through its BH3 domain. The gene has five exons, and splicing out of exon 3 during transcription yields a shorter form of the *Bok* gene. The splicing variant (Bok-S) loses 43 amino acid from the full-length amino acid Bok-L sequence. The deletion is associated with fusion between the N terminal half of the $BH_3$ domain with the C terminal half of the BH1 domain, but Bok-S is still proapoptotic but unable to heterodimerize.

### Mtd

Mtd is a proapoptotic member of the Bcl-2 family that contains the four BH conserved domains and the hydrophobic COOH terminal. Messenger RNA (mRNA) of Mtd have been found in embryonic (liver, thymus, lung, and intestine) and adult tissues (brain, liver, and lymphoid tissue). The molecule does not heterodimerize and mutations of conserved amino acids in BH3 do not abolish apoptosis (31). It is a natural member of the Bcl-2 family that does not dimerize.

## Proteins with a $BH_3$ Only Domain

In addition to the now classical members of the Bcl-2 family, there is a group of molecules that contain the conserved BH3 domain only (BH3 domain only) in the amphipathic helical domain. The BH3 only group includes: Bid, Bad, Blk, Bim, Bok, Bnips, Rad 9, and Noxa in mammals, and in *C. elegans*, Egl-1. The remaining sequence of these molecules has no homology with the BH1, BH2, and BH4 domains. The exact role of each of the "death domain only" proteins is still not clear.

## Bad

Bad is a member of the "BH3 only" subfamily of the broader Bcl-2 superfamily. The molecule is devoid of the COOH terminal hydrophobic sequence, which usually serves to target the mitochondria, although Bad is found in both mitochondria and cytosol. In the mitochondria, it heterodimerizes with Bcl-2 or $Bclx_L$ and thereby inactivates their proapoptotic activity. In the cytosol Bad is bound to protein. Kelekar et al. identified, by deletion analysis, a 26-amino-acid sequence that constitutes a minimal domain for the binding of murine Bad to Bclx, thereby inhibiting the antiapoptic function of Bclx (32).

Bad contains two binding domains for the 14-3-3 protein. A yeast mutation in the 14-3-3 binding site (S137A) of Bad impairs the binding of Bad to the protein. However, the mutation does not impair heterodimerization with Bcl-2. Coexpression of 14-3-3 with the S113A mutant did not prevent cell killing, probably because the mutant cannot bind 14-3-3 but can still heterodimerize.

In cells stimulated by a survival factor (e.g., IL-3), Bad is phosphorylated at a consensus site; when the survival factor is withdrawn, Bad is dephosphorylated and moves from cytosol to mitochondria where it heterodimerizes with antiapoptotic agents (Bcl-2 or $Bclx_L$). Thus, the phosphorylation–dephosphorylation processes are critical to the function of Bad as a proapoptotic agent; nonphosphorylated Bad inhibits antiapoptotic molecules (Bcl-2 or $Bclx_L$). Phosphorylated Bad binds to 14-3-3 molecules in the cytoplasm where it is sequestered and loses its proapoptotic properties (33).

### Bad Phosphorylation

In lymphoid cells stimulated to proliferate by IL-3, Bad is phosphorylated at two sites: 112 and 136. Phosphorylation at either site facilitates the binding of Bad to the 14-3-3 protein rather than to $Bclx_L$, thus allowing the antiapoptotic protein of the latter to keep the cell alive. Mutations at the Bad phosphorylation site generate a proapoptotic mutant more active than the wild-type Bad.

Two kinases have been suspected to be involved in the phosphorylation of Bad protein kinase B/akt: PKB/Akt and protein kinase A (PKA). PKB/akt is an effective antiapoptotic agent in many cells. It phosphorylates Bad on serine 136. However, the antiapoptotic effect of PKD/akt may not be uniquely linked to the phosphorylation of Bad because PKB/akt phosphorylates other proteins functioning in apoptosis, among them caspase 9. Neither can it be excluded that other kinases can also phosphorylate Bad. Indeed, the cyclic adenosine monophosphate (cAMP)-dependent PKA was shown to phosphorylate Bad in vitro at serine 112 (34). In summary, in the pathway for prevention of apoptosis Bad's phosphorylation is critical. It leads to the sequestration of Bad by association with protein 14-3-3, but the definitive mechanism of phosphorylation

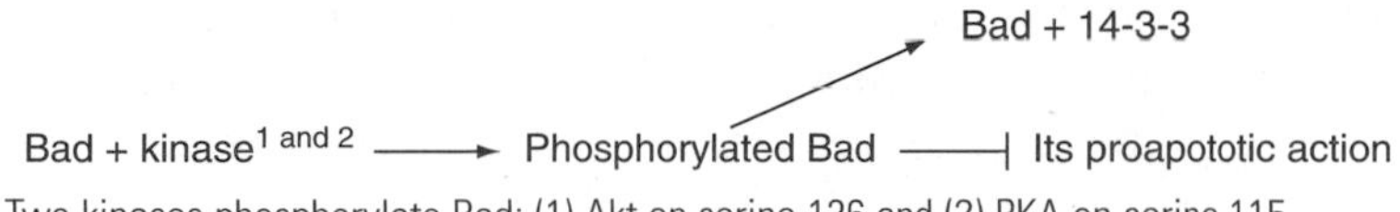

**FIGURE 5-5** Two kinases phosphorylate Bad: (1) Akt on serine 136 and (2) PKA on serine 115.

remains somewhat elusive. Further opportunities to return the role of Bad's phosphorylation in apoptosis are found later in this chapter (**FIGURE 5-5**).

After purification of a protein kinase that phosphorylates Bad at S112 in IL-3 stimulated cells, it was established that the kinase was primarily found in mitochondria (34) Phosphorylation of endogenous Bad in IL-3 stimulated cells was impaired by PKA inhibitors. This finding led to the assumption that inhibition of the mitochondrial S112 Bad kinase activity must involve a mechanism that targets the protein kinase to the mitochondria. Indeed, it is known that the type II cyclin adenosine 5-triphosphate (ATP)–dependent kinase holoenzyme is a molecular complex that includes an inhibitory subunit (R). The R subunit is a dimer (RII) that keeps two catalytic subunits (c subunits) inactive. Moreover, the RII directs the subcellular location of the holoenzyme by associating with a special member of a family of signal proteins, kinase anchoring proteins (AKAP). AKAP contains a conserved domain that anchors it to RII. Finally, binding of the R subunit and AKAP can be prevented by a competitive inhibition of the binding. The inhibition was found to also impair the S112 phosphorylation. The combination of these events led to the conclusion that the S112 kinase phosphorylates Bad only when it is properly anchored to the mitochondrial membrane, and that once phosphorylated, Bad loses its proapoptotic function through sequestration in the cytoplasm where it binds to 14-3-3 proteins with high affinity (34).

### Bad Dephosphorylation

Wang and associates endeavored to identify the phosphatase that dephosphorylates Bad. They postulated that calcineurin, a serine threonine phosphatase that can induce apoptosis, might be a logical candidate. They transfected 393T cells with a plasmid encoding constitutively active calcineurin devoid of its negative regulatory domain and demonstrated that Bad was dephosphorylated, and transferred to mitochondria where it formed an heterodimer with $Bclx_L$. These findings provide an explanation for the apoptotic effect of the increased release of cytosolic $Ca^{2+}$. The release activates calcineurin, which in turn activates Bad dephosphorylation and its binding to the antiapoptotic molecule (**FIGURES 5-6A** and **B**) (35). In conclusion, regulation of Bad phosphorylation and dephosphorylation involves a specific mitochondrial protein kinase anchored at the phosphorylation site and calcineurin at the dephosphorylation site. In mitochondria, unphosphorylated Bad is bound to antiapoptotic proteins and the latter are inhibited, allowing for apoptosis. Activation of the protein phosphokinase phosphorylates Bad, and Bad moves to the cytoplasm where it binds to 14-3-3 proteins and becomes inactive, permitting cell survival.

Note that $Bclx_L$ bound Bad can also be phosphorylated by PKB/akt at the S136 site, which may operate like the phosphorylation of S112. Both sites are susceptible to be dephosphorylated by calcineurin serine phosphatase (**FIGURE 5-6**).

## Bid

After triggering of apoptosis by TNF-α or Fas the activation of caspase 8 causes the proteolytic attack of p22 Bid, another BH3 only protein, to yield a truncated form (tBid). The latter is translocated from cytosol to mitochondria where it becomes an integral membrane protein. Three mechanisms have been postulated for the proapoptotic activity of Bid (36):

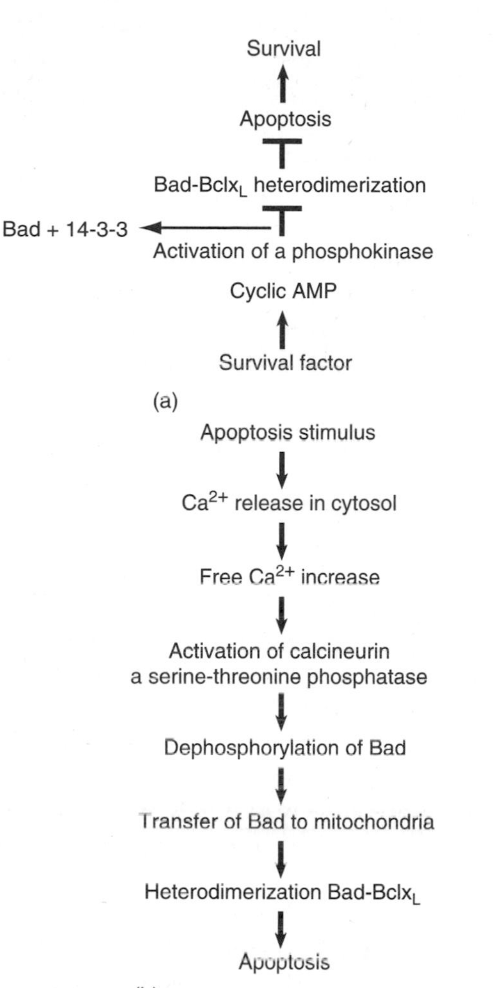


**FIGURE 5-6** (a) Bad phosphorylation and (b) the pathway to apoptosis, the dephosphorylation of Bad.

1. Translocation to the mitochondrial membrane where it binds to either the active or inactive Bcl-2 and thereby constructs a receptor that serves to trigger the apoptotic cascade.
2. tBid displaces Apaf 1 from Bcl-2 (see below).
3. tBid contributes to pore forming.

In 1999, Desagher et al., induced apoptosis in Hela cells with staurosporine. In the process, Bid translocates from cytosol to mitochondria. At the same time, a conformational change of Bax takes place in the mitochondria causing the unmasking of its NH2 terminal domain, the release of cytochrome c, and caspase activation. Bid added to cells devoid of Bax barely releases detectable amounts of cytochrome c. The Bax conformational changes are blocked by the antiapoptotic agents Bcl-2 and Bclx$_L$. The findings suggest that Bid is directly responsible for the conformational changes of Bax and that the interaction of Bid and Bax (two proapoptotic agents) is required to unleash, at least in certain cases, the sequence of steps that lead to apoptosis, in particular cytochrome c release and caspase activation (37).

The mode of action of Bid is revisited in the section on cytochrome c release in apoptosis.

## Blk

Blk is a proapoptotic member of the Bcl-2 family. The *blk* gene has been cloned in mouse and human. The protein is composed of 239 amino acids; it possesses a hydrophobic C terminal and three of the four BH domains: BH1, BH2, and BH3. It is an intermembrane protein with ion channel properties. Blk forms heterodimers with the antiapoptotic Bcl-2 and $Bclx_L$, the 19k kDa adenovirus E1B EBV and BHRF1. Blk is localized in the mitochondrial membrane. Mutations of the BH3 domain of the molecule prevent apoptosis. Fifty-two amino acids (43–94) encompassing the BH3 domain are sufficient for heterodimerization, and 18 amino acids encompassing the BH3 domain (57–74) form the core of the heterodimerization site. The proapoptotic properties of Blk are abolished by inhibitors of caspase 9, suggesting that Blk activates the cytochrome c-apaf-1-caspase 9 sequence (38).

## Bim

Another $BH_3$-only proapoptotic protein, Bim contains a conserved BH3 and a hydrophobic COOH terminal. The $BH_3$ domain is required for its proapoptotic activity. Three proapoptotic isoforms of Bim exist because of alternative splicing, namely: $Bim_S$, $Bim_L$, and $Bim_{EL}$. Bim is, however, the most potent of the three isoforms. The regulation of the proapoptotic activity of Bim is of interest. In healthy cells Bim is associated with the microtubule complex, more specifically with the LC8 cytoplasmic light chain of dynein that is part of the dynein-microtubules motor complex. In apoptotic cells Bim dissociates from dynein; once free, Bim binds to the mitochondrial Bcl-2 and thereby abrogates its antiapoptotic function. Caspases are required for transfer from dynein to Bcl-2. Therefore, it seems that the dissociation of Bim from LC8 may constitute an early step in the sequence of events signaling apoptosis (39). At present there are three known mechanisms by which a proapoptotic agent is sequestered in the cytoplasm: binding to 14-3-3 (Bax), binding to dynein (Bim), and binding to ced-2 (Egl-1).

## Bod

Bod is a BH3-only homologue of Bcl-2 that heterodimerizes with the antiapoptotic protein Mcl-1, $Bclx_L$. The protein is encoded by a gene conserved in various mammalian species.[4] Overexpression of Bod (Bcl-2-related ovarian death gene) in Chinese hamster ovary cells induces apoptosis that is blocked by caspase inhibitors (40).

### BNIPI

The protein Bnipi is encoded by a gene located on chromosome 5q33-p34 that has been cloned from human prostate. The gene encodes four splicing variants. BNIPIb contains a conserved $BH_3$ domain and interacts with Bcl-2 and $Bclx_L$ (or Bcl-2L1). The

[4]There are three isoforms of BOD, a long, medium, and short (198, 100, and 43 amino acids, respectively) in the open reading frame.

function of the protein is unknown, but is believed to be different from that of any other known $BH_3$-only members of the Bcl-2 family.

### Egl-1

Egl-1 is a $BH_3$-only proapoptotic protein found in *C. elegans.*

## RAD 9 (Human)

The *rad 9* gene[5] was isolated from a genomic DNA library. Rad 9 mutants are unable to delay cell cycle progress after exposure to radiation. The human Rad 9 functions as a $G_2$ checkpoint protein that binds to $Bclx_L$ and Bcl-2 through a $BH_3$ consensus sequence. The $BH_3$-like domain is necessary for binding to the antiapoptotic proteins. When Rad 9 is overexpressed in mammalian cells, in culture it induces apoptosis and colocalizes with Bcl-2. Removal of the $BH_3$-like domain prevents colocalization and apoptosis. Moreover, coexpression of Bcl-2 abrogates the proapoptotic properties of RAD 9.

## Noxa

A gene (*Noxa*), was identified and its cDNA encodes a 103 amino acid protein. The primary sequence of Noxa contains two putative $BH_3$ motifs. The sequence contains none of the other BH motifs. Noxa messenger RNA is constitutively expressed in small amounts in adult mouse brains, thymus, spleen, lung, and kidney. The promotor region of the *Noxa* gene contains a P53 responsive element. Introduction of an antisense Noxa nucleotide in a hematopoietic cell line in which apoptosis is induced by radiation and is therefore P53-dependent, blocks Noxa expression, and apoptosis. It appears that Noxa is a $BH_3$-only protein that could serve as a proxy for Bax in P53-dependent apoptosis (41).

**TABLE 5-1** lists some of the known members of the Bcl-2 family; some are proapoptotic and others are antiapoptotic. Although the existence of the BH1, BH2, and BH4 domains varies among the proapoptotic members of the family, the BH3 domain is consistently present.

**Table 5-1 Selected members of the Bcl-2 family**

| | Proapoptotic | |
|---|---|---|
| Antiapoptotic | | BH3 only |
| Bcl-2 | Bax | Bid |
| $Bcl_L$ | Bak | Bad |
| Boo | Bok | Bim |
| | Diva | Bik |
| | | Noxa |
| | | Puma |

[5]In yeast *Rad 9* is one of the six checkpoint genes that blocks the cell in the $G_2$ phase of the cell cycle in presence of incomplete DNA replication or DNA damage.

## Mechanism of Action of the Anti- and Proapoptotic Agents of the Bcl-2 Family

A long list of biological effects of Bcl-2 on intact cells has been assembled and is summarized by Kroemer (41). At a first glance, the multiplicity and the complexity of these effects makes it difficult to determine what function or functions need to be disrupted to start apoptosis or to signal the point of no return in apoptosis, since then a consensus has emerged, In some situations, the signal downstream the Bcl-2 inactivation involves mitochondria, whereas in others the signal transduction pathway may not include mitochondria (discussed in Chapters 6 and 7). The cytochrome c release will be discussed after considering the function of caspases (38–40).

## References and Recommended Readings


1. Hengartner, M.O., Ellis, R.E., and Horvitz, H.R. *Caenorhabditis elegans* gene ced-9 protects cells from programmed cell death. *Nature.* 1992;356:494–9.
2. Desjardin P., and Ledoux S. The role of apoptosis in neurodegenerative diseases. *Metab Brain Dis.* 1998;13:79–96.
3. Ameisen, J.C. The origin of programmed cell death. *Science.* 1996;272:1278–9.
4. Nakayama, K., Nakayama, K., Negishi, I., Kuida, K., Shinkai, Y., Louie, M.C., Fields, L.E., Lucas, P.J., Stewart, V., Alt, F.W., et al. Disappearance of the lymphoid system in Bcl-2 homozygous mutant chimeric mice. *Science.* 1993;261: 1584–8.
5. Kelekar, A., and Thompson, C.B. Bcl-2-family proteins: the role of the BH3 domain in apoptosis. *Trends Cell Biol.* 1998;8:324–30.
6. Oltvai, Z.N., and Korsmeyer, S.J. Checkpoints of dueling foil death wishes. *Cell.* 1994;79:189–92.
7. Tsujimoto, Y. Regulation of apoptosis by the Bcl-2 family of proteins. In Jacobson, M.D., and McCarthy, N. (eds). *The Molecular Biology of Programmed Cell Death.* 2002;136–60. New York: Oxford University Press.
8. Minn, A.J., Kettlun, C.S., Liang, H., Kelekar, A., Vander Heiden, M.G., Chang, B.S., Fesik, S.W., Fill, M., and Thompson, C.B. Bcl-xL regulates apoptosis by heterodimerization-dependent and -independent mechanisms. *EMBO J.* 1999;18: 632–43.
9. Muchmore, S.W., Sattler, M., Liang, H., Meadows, R.P., Harlan, J.E., Yoon, H.S., Nettesheim, D., Chang, B.S., Thompson, C.B., Wong, S.L., Ng, S.L., and Fesik, S.W. X-ray and NMR structure of human Bcl-xL, an inhibitor of programmed cell death. *Nature.* 1996;381:335–41.
10. Egan, B., Beilharz, T., George, R., Isenmann, S., Gratzer, S., Wattenberg, B., and Lithgow, T. Targeting of tail-anchored proteins to yeast mitochondria in vivo. *FEBS Lett.* 1999;451(3):243–8.
11. Hsu, S.Y., Lai, R.J., Finegold, M., and Hsueh, A.J. Targeted overexpression of Bcl-2 in ovaries of transgenic mice leads to decreased follicle apoptosis, enhanced folliculogenesis, and increased germ cell tumorigenesis. *Endocrinology.* 1996;137:4837–43.
12. Hsueh, A.J., Eisenhauer, K., Chun, S.Y., Hsu, S.Y., and Billig, H. Gonadal cell apoptosis. *Recent Prog Horm Res.* 1996;51:433–55.

13. Deng, X., Ito, T., Carr, B., Mumby, M., and May, W.S. Jr. Reversible phosphorylation of Bcl-2 following interleukin 3 or bryostatin 1 is mediated by direct interaction with protein phosphatase 2A. *J Biol Chem.* 1998;273:34157–63.
14. Ruvolo, P.P., Deng, X., Carr, B.K., and May, W.S. A functional role for mitochondrial protein kinase C alpha in Bcl-2 phosphorylation and suppression of apoptosis. *J Biol Chem.* 1998;273:25436–42.
15. Dimmeler, S., Breitschopf, K., Haendeler, J., and Zeiher, A.M. Dephosphorylation targets Bcl-2 for ubiquitin-dependent degradation: a link between the apoptosome and the proteasome pathway. *J Exp Med.* 1999;189:1815–22.
16. Breitschopf, K., Haendeler, J., Malchow, P., Zeiher, A.M., and Dimmeler, S. Posttranslational modification of Bcl-2 facilitates its proteasome-dependent degradation: molecular characterization of the involved signaling pathway. *Mol Cell Biol.* 2000;20:1886–96.
17. Cheng, E.H., Kirsch, D.G., Clem, R.J., Ravi, R., Kastan, M.B., Bedi, A., Ueno, K., and Hardwick, J.M. Conversion of Bcl-2 to a Bax-like death effector by caspases. *Science.* 1997;278:1966–8.
18. Simonen, M., Keller, H., and Heim, J. The BH3 domain of Bax is sufficient for interaction of Bax with itself and with other family members and it is required for induction of apoptosis. *Eur J Biochem.* 1997;249:85–91.
19. Knudson, C.M., Tung, K.S., Tourtellotte, W.G., Brown, G.A., and Korsmeyer, S.J. Bax-deficient mice with lymphoid hyperplasia and male germ cell death. *Science.* 1995;270:96–9.
20. Yin, X.M., Oltvai, Z.N., and Korsmeyer, S.J. BH1 and BH2 domains of Bcl-2 are required for inhibition of apoptosis and heterodimerization with Bax. *Nature.* 1994;369:321–3.
21. Cheng, E.H., Levine, B., Boise, L.H., Thompson, C.B., and Hardwick, J.M. Bax-independent inhibition of apoptosis by Bcl-XL. *Nature.* 1996;379:554–6.
22. Diaz, J.L., Oltersdorf, T., Horne, W., McConnell, M., Wilson, G., Weeks, S., Garcia, T., and Fritz, L.C. A common binding site mediates heterodimerization and homodimerization of Bcl-2 family members. *J Biol Chem.* 1997;272:11350–5.
23. Zha, H., and Reed, J.C. Heterodimerization-independent functions of cell death regulatory proteins Bax and Bcl-2 in yeast and mammalian cells. *J Biol Chem.* 1997;272:31482–8.
24. Wang, K., Gross, A., Waksmann, G., and Korsmeyer, S.J. Mutagenesis of the BH3 domain of BAX identifies residues critical for dimerization and killing. *Mol Cell Biol.* 1998;18:6083–9.
25. Jurgensmeier, J.M., Xie, Z., Deveraux, Q., Ellerby, L., Bredesen, D., and Reed, J.C. Bax directly induces release of cytochrome c from isolated mitochondria. *Proc Natl Acad Sci USA* 1998;95:4997–5002.
26. Frankowski, H., Missotten, M., Fernandez, P.A., Martinou, I., Michel, P., Sadoul, R., and Martinou, J.C. Function and expression of the Bcl-x gene in the developing and adult nervous system. *NeuroReport.* 1995;6:1917–21.
27. Gonzalez-Garcia, M., Garcia, I., Ding, L., O'Shea, S., Boise, L.H., Thompson, C.B., and Nunez, G. Bcl-x is expressed in embryonic and postnatal neural tissue and functions to prevent neuronal cell death. *Proc Natl Acad Sci USA.* 1995;92:4304–8.

28. Parsadanian, A.S., Cheng, Y., Keller-Peck, C.R., Holtzman, D.M., and Snider, W.D. Bcl-xL is an antiapoptotic regulator for postnatal CNS neurons. *J Neurosci.* 1998;18:1009–19.

29. Almeida, O.F., Conde, G.L., Crochemore, C., Demeneix, B.A., Fischer, D., Hassan, A.H., Meyer, M., Holsboer, F., and Michaelidis, T.M. Subtle shifts in the ratio between pro- and antiapoptotic molecules after activation of corticosteroid receptors decide neuronal fate. *FASEB J.* 2000;5:779–90.

30. Inohara, N., Gourley, T.S., Carrio, R., Muniz, M., Merino, J., Garcia, I., Koseki, T., Hu, Y., Chen, S., and Nunez, G. Diva, a Bcl-2 homologue that binds directly to Apaf-1 and induces BH3-independent cell death. *J Biol Chem.* 1998;273:32479–86.

31. Inohara, N., Koseki, T., del Peso, L., Hu, Y., Yee, C., Chen, S., Carrio, R., Merino, J., Liu, D., Ni, J., and Nunez, G. Nod 1, an Apaf-1-like activator of caspase-9 and nuclear factor-kappa B. *J Biol Chem.* 1999;274:14560–7.

32. Kelekar, A., Chang, B.S., Harlan, J.E., Fesik, S.W., and Thompson, C.B. Bad is a BH3 domain-containing protein that forms an inactivating dimer with Bcl-XL. *Mol Cell Biol.* 1997;17:7040–6.

33. Zha, J., Harada, H., Yang, E., Jockel, J., and Korsmeyer, S.J. Serine phosphorylation of death agonist BAD in response to survival factor results in binding to 14-3-3 not BCL-X(L). *Cell.* 1996;87:619–28.

34. Harada, H., Becknell, B., Wilm, M., Mann, M., Huang, L.J., Taylor, S.S., Scott, J.D., and Korsmeyer, S.J. Phosphorylation and inactivation of BAD by mitochondria-anchored protein kinase A. *Mol Cell.* 1999;3(4):413–22.

35. Wang, H.G., Pathan, N., Ethell, I.M., Krajewski, S., Yamaguchi, Y., Shibasaki, F., McKeon, F., Bobo, T., Franke, T.F., and Reed, J.C. $Ca^{2+}$ induced apoptosis through calcineurin dephosphorylation of BAD. *Science.* 1999;284:339–43.

36. Luo, X., Budihardjo, I., Zou, H., Slaughter, C., and Wang, X. Bid, a Bcl-2 interacting protein, mediates cytochrome c release from mitochondria in response to activation of cell surface death receptors. *Cell.* 1998;94:481–90.

37. Desagher, S., Osen-Sand, A., Nichols, A., Eskes, R., Montessuit, S., Lauper, S., Maundrell, K., Antonsson, B., and Martinou, J.C. Bid-induced conformational change of Bax is responsible for mitochondrial cytochrome c release during apoptosis. *J Cell Biol.* 1999;144:891–901.

38. Hegde, R., Srinivasula, S.M., Ahmad, M., Fernandes-Alnemri, T., and Alnemri, E.S. Blk, a BH3-containing mouse protein that interacts with Bcl-2 and Bcl-xL, is a potent death agonist. *J Biol Chem.* 1998;273(14):7783–6.

39. Puthalakath, H., Huang, D.C., O'Reilly, L.A., King, S.M., and Strasser, A. The proapoptotic activity of the Bcl-2 family member Bim is regulated by interaction with the dynein motor complex. *Mol Cell.* 1999;3:287–96.

40. Hsu, S.Y., Lin, P., and Hsueh, A.J. BOD (Bcl-2-related ovarian death gene) is an ovarian BH3 domain-containing proapoptotic Bcl-2 protein capable of dimerization with diverse antiapoptotic Bcl-2 members. *Mol Endocrinol.* 1998;12:1432–40.

41. Oda, E., Ohki, R., Murasawa, H., Nemoto, J., Shibue, T., Yamashita, T., Tokino, T., Taniguchi, T., and Tanaka, N. Noxa, a BH3-only member of the Bcl-2 family and candidate mediator of p53-induced apoptosis. *Science.* 2000;288:1053–8.

42. Kroemer, G. The proto-oncogene Bcl-2 and its role in regulation apoptosis. *Nat Med.* 1997;3(6):614–20.

CHAPTER

# INTERLEUKIN 1-β CONVERTING ENZYME (ICE) AND CASPASES

# 6

## ICE

Because DNA breakdown was long considered to be the hallmark of apoptosis, research focused on endonucleases as a potential determinant of the process. However, it soon became obvious that the executioners in apoptosis might not be in the nucleus, but in the cytoplasm. Cytoplasmic granules found in T lymphocytes and natural killer cells contain a pore forming protein, perforin, and a number of proteases. Granzyme B, one of these enzymes, cleaves proteins at the right side of aspartic residues (1). Exposing cells to this purified enzyme is enough to kill them.

The determination of the sequence of the *ced*-3 gene of *Caenorhabditis elegans* in Horvitz's laboratory led to the identification of the executioner in apoptosis (2). The sequence of CED-3 (the protein encoded by the *ced*-3 gene) revealed substantial homology with a mammalian cysteine protease, interleukin-1-converting enzyme (ICE).[1] The molecules share significant sequence identity: 29% over their entire lengths and 43% for a 115 amino acid sequence that includes a pentapeptide, Gln-Ala-Cs-Arg-Gly, which is essential to the active site (2).

Evidence supporting the notion that the ICE family of proteins contributes to the apoptotic sequence includes:

- Most mutations that affect the CED-3 activity involve sequences conserved in both CED-3 and ICE.
- Several investigators showed that ectopic expression of DNA's encoding members of the ICE protein family cause apoptosis.
- The use of an ICE inhibitor (Crm), a serine protease (serpin) encoded in the cow pox virus, provided evidence that ICE plays a critical role in apoptosis.
- Crm binds to ICE, inhibits it, and prevents apoptosis in cell culture (3). Moreover, overexpression of Crm prevents apoptosis triggered by the FAS ligand–FAS

[1]The functions of interleukin-1 are critical for the defense against infections. They include: secretion of acute phase proteins by hepatocytes, mobilization of polymorphonuclear leukocytes by the bone marrow, increase in body temperature, and induction of adaptive immune response. ICE cleaves the human prointerleukin at two sites: between asparagine 27 and glycine 28 and between asparagine 116 and leucine 117, and thereby activates it.

receptor engagement or that induced by tumor necrosis factor-alpha (TNF-α). Other inhibitors were also found, for example, the Baculovirus protein P35, which inhibits apoptosis in *Drosophila* as well as *C. elegans* (4; reviewed in 5).

These inhibitors bind to ICE through a tetrapeptide and to other cysteine proteases through tetrapeptides of different sequences, and block apoptosis. Discovery of a homology between ICE and the product of the *ced*-3 gene triggered the search for other cells' executioners in apoptosis. The observation that the homozygous destruction of the *ICE* gene in mice does not prevent apoptosis, suggested that proteases other than ICE can function in a similar way. An extensive family of proteases with functions similar to that of ICE was discovered, referred to as caspases (cysteinyl-aspartate-specific proteases).[2]

## Caspases

The caspase family is composed of highly specific proteases that cleave proteins only to the right of an aspartic residue or, more precisely, to the COOH side of an aspartic residue (Asp/X). Cleavage of the substrates requires that the aspartic residue be part of a conserved tetrapeptide sequence. For example, caspase 1 (ICE) recognizes the sequence Tyr-Val-Ala-Asp or YVAD in the prointerleukin-1β. Although similar tetrapeptides are critical for activity, their sequence varies considerably among the caspases (6). Caspases, like many proteolytic enzymes, are synthesized as zymogens or procaspases. Their molecular weight varies considerably, in part because of the size of the prodomain. The proenzyme contains three domains: the NH-2 terminal prodomain (molecular wt 32–55 kDa), a large subunit (close to 20 kDa, except for caspases 8, 9, and 10, molecular wts 43, 37, and 43 kDa, respectively) and a small subunit (molecular wt between 10 and 14 kDa). After cleavage of the prodomain, the two subunits form a heterohomodimer (a tetramer) and the enzyme is activated. Caspases are found in *C. elegans*, *Drosophila*, rodents, and humans. More than 20 caspases have been identified, among them 11 human enzymes. These proteins have been conserved through evolution and have several features in common:

- They are synthesized as procaspases, with a prodomain of various lengths, ranging from 23 amino acids for caspases 6 and 7 to 219 residues in caspase 10.
- The length of the prodomain determines whether a caspase triggers the execution, initiators (long prodomain) or is a member of the firing squad, effectors (shorter prodomains).
- The prodomain contains two structurally related motifs, the dead effector domain (DED) and the caspase recruitment motif (CARD). The contribution of these motifs to caspase's function is clarified later in this chapter. The effector caspases have a shorter prodomain but they also contain two other subunits, a large one (molecular wt mostly close to 20) and a short one (molecular wt 10 to 14).
- Caspases are capable of autoprocessing. The maturation of caspases occurs in two steps: (1) separation of the small from the large subunits; (2) separation of the prodomain from the two subunits. Both cleavages take place at an Asp/X peptide

[2]Early after the discovery of ICE, a large number of proteases with caspase activity were discovered in several laboratories. Each was referred to by a different acronym. Depending on who discovered them, similar or identical enzymes were given different names. To bring order to this complex and often confusing nomenclature, a new word was coined to refer to all members of the ICE family: caspases. The C stands for the catalytic cysteine and the A for the special specificity of the caspases that cleave peptide bonds only to the right of an aspartic residue; S means specific.

bond; after cleavage a heterohomodimer is formed. The cleavage is most likely autocatalytic (**FIGURE 6-1**) (6).[3]

- The cleavage site of the substrates involves an aspartic acid residue located upstream of the COOH terminal Asp/X.

The tripeptide preceding the aspartic residue (e.g., YVAD/X for ICE or caspase 1) is critical in the determination of the cleavage site, although this sequence varies from one caspase to another. The aspartic residue at the COOH terminal site is referred to as P1 and the three amino acid residues upstream are P2, P3, and P4. P4 contributes most effectively and P2 least effectively to the specificity of the caspase.[4]

There are many common denominators in the quaternary structure of caspases. An active caspase is formed by the association of two identical proenzymes. The autocatalytic process occurs in two steps:

1. Separation of the large subunit from the small subunit; both subunits are required for activity.
2. Cleavage of the prodomain—the latter is not required for catalytic activity. These four subunits (2 small and 2 large) associate to form a tetramer, an active enzyme with two catalytic sites. In the tetramer, the small and large subunits form six central β sheets trapped between several α helices, a unique quaternary structure.

The active dimer has two catalytic sites.

- The catalytic function of the molecule involves three conserved amino acids: cysteine 285, histidine 237, and glycine 228. The proximity of these amino acids led Wilson and associates (9) to propose a tentative mechanism for the activation of the catalytic site.
- The aspartates of the substrate drops into a deep pocket where several amino acids are conserved in human caspases (Arg 179, Gly 283, Arg 341, Ser 47). Both the large and small units of the tetramer contribute to the formation of the catalytic site.
- A multigene family that encodes for ICE-associated proteins was discovered. The genes encode more than 20 different caspases. Some are primarily involved in inflammation that cleaves procytokinines (caspases: 1 [ICE], 4, 5, 11, 12, 13) and others are involved in programmed cell death (caspases: 2, 3, 6, 7, 8, 9, 10). The latter group of caspases, on the basis of their role in apoptosis, has been subdivided into two groups: initiator caspases (caspases 2, 8, 9, 10) and effector caspases (caspases 3, 6, 7; Figure 6-1) (10).

## Initiator and Effector Caspases and Pathways for Apoptosis

The initiator caspases unleash a caspase cascade that involves the effector caspases. There are at least two pathways for the consummation of apoptosis: one independent of cytochrome C release and one that depends on cytochrome C release. The former was discussed earlier.

[3]Because the sites of cleavage in the zymogen are so far removed from the catalytic site, it is suggested that each zymogen activates the other through the process of transcatalysis (6).

[4]Determination of the P1–P4 sequence at the cleavage site of each caspase has facilitated: (a) the synthesis of short substrates composed of a four amino acid recognition sequence linked through its carboxyterminal aspartate to a chromogenic or fluorogenic amine; and (b) the synthesis of tetrapeptidic inhibitors (e.g., Ac-YVDA-CHO for caspase 1 and Ac-DEVD-CHO for caspases 3 and 7, respectively).

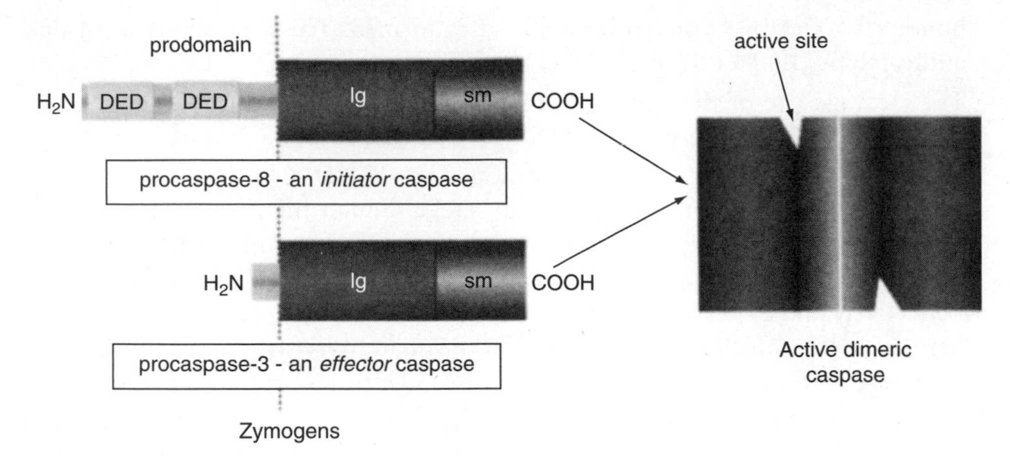


**FIGURE 6-1** Diagram (left) showing the structure of zymogens for caspase 8, an initiator caspase, and caspase 3, an effector caspase. The prodomsin of caspase 8 contains two dead effector domains (DED) (7). Other initiation caspases may have a caspase recruitment domain (CARD) (8) instead. The heteromeric enzyme (right) results from proteolytic activation. (Reproduced with permission of Earnshaw, W.C., Martins, L.M., and Kaufman, S.H. Structure, activation, substrates, and functions during apoptosis. *Annu Rev Biochem.* 1999; 68:383–424.)

Consider the pathway triggered by TNF-α. TNF-α binds to its trimeric receptor, TNFR1, and its cytoplasmic domain engages several adapter molecules (TRADD, RIP, RAIDD) and procaspase 8. Procaspase 8 is special; it possess a long prodomain that contains two similar death effector domains. During activation of caspase 8, its prodomain is cleaved. The catalytic unit then separates from the large molecular complex formed by TNF-α and TNFR1, the adapter proteins, and is released in the cytosol where it activates the effector caspases (caspases 3, 6, 7). Similarly, the activation of the FAS receptor by the FAS ligand results in the binding of caspase 8 through adapter molecules (FADD/MORT1), followed by activation of the initiator (caspase 8). Release of the active initiator caspase into the cytosol triggers activation of effector caspases (11, 12). Observations show that the inhibition of caspase 8 blocks the manifestations of a sequence of proteolytic events and support the caspase-cascade concept (**FIGURES 6-1** and **6-2**) (13).

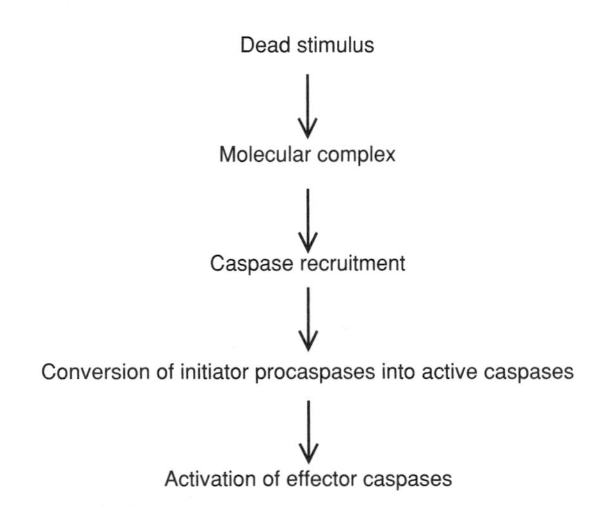


**FIGURE 6-2** Effector caspase activation.

The activation of initiator and effectors is thus modulated differently. Dead stimuli trigger the activation of initiator caspases. The formation of special molecular complexes, referred to as scaffolds, is required for caspase recruitment and activation (**FIGURE 6-2**). Thus, activated FAS links to the adapter FADD, which in turn binds to the prodomain of caspase 8. Three specialized domains are involved in the formation of the scaffold: the death domain, the death effector domain, and the caspase recruitment domain. Each of these three domains comprises six antiparallel α helices that fold into a similar three-dimensional structure and are linked by short loops. All three domains can be involved in homophilic reactions. FAS and FADD are linked by their DD domains. The latter contain clusters of charged amino acids in their α2 and α3 helices. Therefore, it appears that the FAS and FADD are linked by electrostatic bonds, as is the case for the bonding of TNF-α and TNFR1.

In contrast, FAD and caspase 8 interact through their death effector domains (DED) and the latter has a three-dimensional structure similar to that of the DD; however, they differ from the DD by the length of both the α helices and their linking loops as well as by the direction of the α helices. Moreover, the motif contains two hydrophobic patches. Thus, the adapter protein FADD interacts upstream through its dead domain with that of FAS and downstream with the prodomain of caspase 8 through their respective DED domains (**FIGURE 6-3**) (14–16). In conclusion, in the cytochrome C-independent pathway, the dead stimulus involves receptors of the TNF receptors (TNFr) family in the cell membrane, in particular TNFR1 and FAS. TNF-α or the FAS ligands are the dead stimuli. These stimuli engage the trimerized receptors α (TNFr or FAS), thereby bringing the dead domains of the cytosolic component of each of the monomers in close contact. The trimerization of the receptor engages the binding of receptor associated proteins, also called adapter proteins (TRADD, RIP, and RAIDD or FADD/MORT 1). These proteins contain dead domains in their COOH terminal sequence that bind to a dead domain in the NH terminal sequence of the molecule located upstream in the molecular complex (an illustration of the scaffold metaphor). The NH terminal sequence of the adapter molecules reacts with the dead effector domains found in the prodomain of caspase 8 and thereby triggers the autocatalytic activation of caspase 8.

## Caspase Substrates in Apoptosis and the Significance of Their Proteolysis

A comparison of the two dimensions of electrophoresis patterns of cells before or during apoptosis show few changes, yet a long list of proteins (albeit few in comparison of the total protein population of the cell) cleaved by caspases in cell-free systems has been compiled by Earnshaw, Martins, and Kauffman (6). The list includes:

- Some proteins involved in regulating apoptosis
- Some proteins critical to the cell membrane structure

FAS — DD
|
DD — FADD
|
DED
|
DED
|
Caspase 8

**FIGURE 6-3** The role of death domains and death effector domains in the activation of caspase 8 by FAS.

- Nuclear lamins
- DNA repair enzymes
- Cell cycle regulatory proteins
- Numerous protein kinases
- Several proteins involved in transduction pathways

Several proteins that are both caspase substrates and critical components for the consumption of apoptosis have been discussed previously. The degradation of Bcl-2 and $Bclx_L$ leads to the formation of inactive polypeptides or proapoptotic fragments, processes that may render apoptosis irreversible, and the cleavage of ICAD that leads to activation of CAD, the endodeoxyribonuclease that is likely to be responsible (partly or wholly) for chromatin's orderly fragmentation. Clearly, the proteolysis of these or similar functional substrates are likely to be critical to the progression of apoptosis. However, it is still difficult to make sense of the long list of caspase substrates reported without being selective and, therefore, possibly ignore the critical role that other caspase substrates, known or unknown, may play in the process of apoptosis. Moreover, the function of ubiquitin and the proteosome in the degradation of proteins functional in apoptosis cannot be ignored.

Consider the potential role in apoptosis of proteins associated with the cell membrane. Fodrin is a large component of the cell membrane and it was discovered in the red cell. Red cells have only one membrane; they move from large vessels where they can assume a relaxed oval shape, to small capillaries whose diameter is smaller than that of the erythrocytes. Their passage from the large vessels to the capillaries requires that the cell changes its shape from the relaxed oval shape to a compressed elongated shape. The switch in shape requires a flexible membrane, and this elasticity (most likely limited) is secured by a flexible cytoskeleton located at the cytoplasmic side of the cell membrane. The most abundant protein in the fibrous network is spectrin. Spectrin fibers are antiparallel heterodimers containing two chains, an alpha and a beta chain. Each chain is comprised of repeated 106 amino acids domains joined by a flexible interlink domain, and these two chains intertwine to form a spectrin fiber. At the head of the fiber, the COOH terminal of one chain self-associates with the COOH terminal at the second chain to form a tetramer. At the tail end of the fiber, the chain that carries the COOH terminal is phosphorylated. Four to five tetramers are lined by 13 actin monomers to other cytoskeletal proteins to form junctional complexes. Among the cytoskeletal protein population are: bands 4, 1, and 3, and ankerin, tropomyosin, and adducin. A nonerythroid spectrin named fodrin as well as actin tropomyosin and ankerin, and the intramembrane protein band 3 are found in the plasma membrane of most, if not of all, cells (**FIGURE 5-2**).

Clearly, the partial or complete proteolytic collapse of anyone of these proteins could have dreadful consequences for the cell structure and perhaps its chance of survival. Fodrin, the most abundant of these cytoplasmic membrane proteins, has been the target of investigators for its potential role in apoptosis. Like spectrin, it is composed of an $\alpha$ and a $\beta$ chain.

Caspase 3 activation induced by TNF-$\alpha$ or the FAS ligands, in association with their receptors, in Jurkat and Hela cells causes the cleavage of the 240 kDa, $\alpha$ chain of fodrin to break down into two segments (150 and 120 kDa) (17). Fodrin is cleaved in Burkitt lymphoma cells exposed to radiation, but this cleavage is blocked by calpain inhibitors. While calpain activation and fodrin cleavage are concomitant and occur 15 minutes after irradiation, the activation of caspase 3 takes place only two hours later. Thus, it seems that the cleavage of fodrin by calpain occurs upstream of the activation of caspase 3 (18).

The role of the migration of phosphatidylserine in the phagocytosis of the apoptotic bodies has already been discussed. Using FAS ligation to the cytoplasmic membrane of Jurkat cytoplasts, Martin et al. showed that the activation of caspase 3 not only causes the cleavage of fodrin, but also the externalization of phosphatidylserine (19). These and other studies also suggest that some apoptotic events are entirely controlled by cytoplasmic cellular components and may not require the presence of a nucleus.

## Nuclear Substrates of Caspases

Even if nuclear events associated with apoptosis are sometimes or always caused by primary cytoplasmic events, they cannot be ignored, because they may be critical to the sequence of steps in apoptosis.[5]

According to Kerr and associates, the morphology of apoptosis is quite complex (20). The first morphologic event is clearly condensation and segregation of chromatin, which is often associated with convolution of the nuclear outline. Under the electron microscope, these events precede cytoplasmic membrane changes and formation of the apoptotic bodies. Are the nuclear events that are primary or secondary to discrete cytoplasmic changes not detectable by electron microscopy? The breakdown of the nuclear chromatin and the digestion of ICAD by caspases suggest that, at least in some if not in all cases, chromatin fragmentation might be secondary to cytoplasmic events. Whatever the sequence, nuclear changes are part of apoptosis and, many nuclear proteins are cleaved by caspases. Among the nuclear caspase substrates are lamins and enzymes involved in DNA metabolism and DNA repair.

### Lamins

The nucleus is separated from the cytoplasm by two concentric membranes. The inner membrane rests on a filamentous meshwork called the nuclear lamina. Among the protein families associated with the lamina are the laminal proteins and the lamins (divided in three categories: lamins A, B, and C); both families are part of what is sometimes referred to as the nuclear matrix. Lamins are fibrous proteins that provide a link between the cytoskeleton and the molecules of the interior of the nucleus. These important connections involve interactions between lamins and many proteins, among which are those present in the nuclear membrane and the chromatin. The functions of the lamins in the cell's life are linked to these molecular interactions. Thus, lamins play a critical role in the cell cycle, cell differentiation, and possibly cell death. The importance of lamins to the maintenance of the cells' homeostasis is sadly illustrated by the many heritable conditions associated with the mutations of these molecules, including: muscular dystrophies, lipodystrophies, cardiomyopathies, and some form of progeria (21). In this section, our concern is only with the fate of lamins in apoptosis. Mammalian lamins are substrates for some caspases, and their breakdown is suspected to contribute to the nuclear events associated with apoptosis (22, 23). However lamins break down does not seem to be essential to apoptosis in *C. elegans* (24).

[5]The absence of a nucleus, however, might be a step in the ultimate demise of the cell, as in the case of the red cell.

## Nuclear Enzymes

Among the nuclear enzymes that have been shown to be substrates of caspases, polyadenosyl ribose polymerase (PARP) may be critical.

***PARP*** PARP catalyzes the adenosine 5′-diphosphate (ADP) ribosylation of the acceptor substrate (protein or DNA) in the presence of nicotinamide dinucleotide (oxidized form; $NAD^+$). In this process, it transfers the adenosine-ribose of $NAD^+$ to the acceptor to yield a monoadenosine-ribosylated or a polyadenoside ribolsylate product. Thus, $NAD^+$ is the substrate of PARP (**FIGURE 6-4**).

PARP is a ubiquitous housekeeping enzyme found from the embryo through adulthood, with the highest levels in the thymus and lowest levels in the liver and kidney, two organs with low rates of cell proliferation. PARP, like α polymerase, is located at the periphery of the nucleus. The enzyme acts on multiple acceptors, which include histone 1 topoisomerase, DNA polymerase α and β, and RNA polymerase. The enzyme is activated by single and double strand breaks. It plays critical roles in DNA replication, DNA repair, and genetic recombination.

In humans the gene is located on chromosome q41-42, and two pseudogenes are located on chromosome 13. The human gene is large: 43 kb long with 23 exons (25). The promoter site contains a main initiation site located –160 bp upstream of the transcription initiation site. The latter includes CCAAT and TATA boxes located too far from the SPI binding site to permit transcription. In any event, the PARP mRNA is estimated to contain 3.7 kb with an open reading frame of 3042 bp, which transcribes into a protein made of 1014 amino acids (26). In contrast, the *Drosophila* gene is a single copy composed of six exons, and the organization of the exons seems to correspond with the functional domains of the enzyme (27). The *Drosophila* gene contains TATA and CAAT boxes in the promoter region. The protein is divided in three domains:

a. A DNA binding domain that contains two zinc finger motifs

b. An automodification domain

c. A catalytic domain.[6]

The entire molecule weighs 113 kDa ($M_r$). After DNA damage caspase 3 cleaves it to yield a 24 kDa and an 89 kDa product. The 24 kDa fragment remains in the nucleus and contains the DNA binding domain, but is unable to repair DNA; the 89 kDa fragment is translocated to the cytoplasm where it interacts with PARP and prevents isomerization (28).

PARP is activated in presence of DNA damage and $NAD^+$ is hydrolyzed. Thus, PARP also controls the $NAD^+$ concentration in the cell. After exposure of organisms or cells to UV or ionizing radiation and monoalkylating agents, PARP is the enzyme most directly responsible for the depletion of $NAD^+$. The enzyme polyadenylates the strand breaks and thereby facilitates DNA repair. When lymphocytes are placed in a medium

$$[(\text{ADP-D-ribosyl})n + \text{acceptor}] + \xrightarrow[\text{PARP}]{} [(\text{ADP-D-ribosyl})n + 1\} + \text{acceptor}]$$

**FIGURE 6-4** Mode of action of PARP.

[6]In *Drosophila,* (a) is coded for by exons 1, 2, 3, and 4; (b) is by exon 5; and (c) is by exon 6.

deprived of $NAD^+$, the activity of PARP is decreased and DNA repair inhibited. Similarly, inhibition of PARP activity increases the cell's radiosensitivity and potential lethal and sublethal damage. In contrast, preventive or postirradiation administration of nicotinamide increases cell survival (**FIGURE 6-5**).

Nicotinamide adenosine dinucleotide (NAD) is a major electron acceptor. Oxidations of energy generating molecules require electrons, and the conversion of NAD into its reduced form (NADH) generates the major source of electrons for the synthesis of ATP.

The sequence of events leading to the exhaustion of $NAD^+$ after DNA damage is as follows: Ionizing radiation (~15 Gy) causes single and double strand breaks. PARP is activated and triggers the polyribose adenylation of the broken ends of the strands, but the PARP reaction uses and depletes the cell $NAD^+$ pool. This depletion of the pool leads to an impairment of the Krebs cycle, glycolysis, and ATP depletion. Ultimately, cell death follows by apoptosis or necrosis.

Radiation induced apoptosis was shown to lead to a decrease in PARP activity. Moreover inhibitors of PARP cause apoptosis in cultured mammalian cells. He and associates reported that apoptosis induced by photodynamic agents activates caspase 3 and causes the cleavage of PARP, and that PARP breakdown takes place at the same time as the DNA fragmentation (29). The question is, how do the cells die? The answer is: by apoptosis or necrosis. There seems to be little doubt that the cleavage of PARP

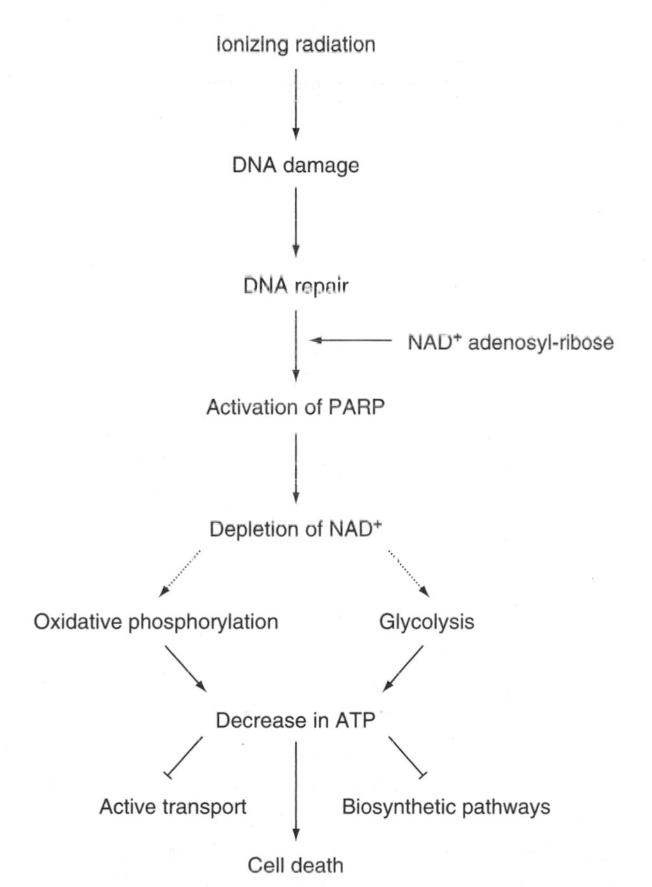


**FIGURE 6-5** Mode of action of PARP after exposure to ionizing radiation (---indicates a substantial decrease in activity).

only occurs early in apoptosis, possibly after a period of activation of the enzyme. However, the significance of the proteolytic breakdown is not clear.

Studies with knockout mice have helped our understanding of the significance of PARP fragmentation in apoptosis. PARP –/– mice present no phenotypic anomalies except for epidermal hyperplasia and obesity as they age. In culture, their fibroblasts grow more slowly, have normal DNA excision repair, and undergo apoptosis in vitro. Still, they are not more sensitive to UV radiation or alkylating agents (30).

Thus, PARP is not required for apoptosis, but it is needed to maintain genomic integrity. Leist and colleagues compared the sensitivity to apoptosis of hepatocytes, thymic cells, and neurons derived from wild-type and PARP –/– mice (31). Apoptosis was induced by FAS-ligand or TNF-α in hepatocytes, by ceramide, dexamethasone, or etoposide in thymic cells, and by colchicine, potassium withdrawal, peroxynitrate, and the neurotoxin MPP in neurons. The sensitivity to apoptosis was—in all cases—the same in the cells derived from the wild-type and the null mice. Leist et al. interpreted their findings as proof that, although PARP is inactivated by apoptotic stimuli XX, the latter cleavage could not be the cause of apoptosis (31). As mentioned earlier, mice that are PARP-deficient show no development or phenotypic anomalies and DNA excision repair is normal. However, when fibroblasts derived from these mice are grown slowly, they show high levels of sister chromatid exchange, indicating high levels of recombination. Recombination is a frequent contribution to cancer in mammals.

If the activity of PARP is rate limiting for some critical reactions, for example, DNA repair, its overexpression can be expected to increase survival after exposure to γ radiation, administration of N-methyl-N′-nitroguanidine (NNNG) or cisplatin (30). This is not the case, however, and it seems likely that the activities of normal levels of nuclear enzyme are sufficient to deal with critical housekeeping functions. Whether that is true after massive DNA damage is not clear. Moreover, whenever the function(s) of an enzyme are examined, there is a need to exclude the possibility of alternative pathways. Therefore, it is of particular interest that the existence of such a pathway(s) has been discovered by Shieh and associates (33). Embryonic cells derived from PARP –/– cells exposed to NNNG synthesize homopolymers of ADP ribose in the presence of $NAD^+$. The polymers could not be distinguished from those synthesized by PARP +/+ cells (33). A nuclear PARP (PRAPII) different, but with marked homology in its catalytic domain, from that of the traditional PARP (PARPI) was also discovered (34).

Yet it would seem that the alternative pathway is not the major one. Indeed, even if the PARP –/– develop normally and are phenotypically normal, they are more sensitive to γ radiation and N-methyl-N-nitrosourea (MNU). Trucco et al. exposed primary and immortalized mouse fibroblasts derived from PARP –/– mice to methylmethane sulfonate. The monofunctional alkylating agent caused a delay in the repair of DNA strand breaks, $G_2$/M arrest, and cell death (35). These findings further suggest that even in the presence of an alternative pathway for the synthesis of polyadenine-ribose polymers, the traditional PARP is still required to secure DNA repair in the presence of extensive damage.

The confusing, if not necessarily conflicting findings reported for PARP activation versus cleavage, at least in part, were reconciled by the studies of Simbulan-Rosenthal and colleagues (36). Their study resorted to a human osteosarcoma cell line that undergoes spontaneous apoptosis. In these cells, the apoptosis is not only reproducible, but takes place slowly, the entire process takes over a week. The study used immunofluorescent antibodies to three critical compounds: Poly(ADP)ribose (PAR), PARP, and the cleavage fragment of PARP containing the DNA binding domain (including the two zinc fingers). The fate of these compounds was followed at two, four, and six days. The levels of expression of PARP and PAR, synthesized from $NAD^+$, increases early (at day 2) before internucleosomal cleavage takes place and before the cell is committed to apoptosis. This stage is therefore reversible; indeed, cells repleted prior to the internucleosomal cleavage are viable. It is logical to conclude that during the early hours after exposure to γ radiation and MNU, the enzyme (PARP) is activated, polyadenine ribose (PAR) is synthesized, proteins and DNA strand breaks are polyadenylated, and excision repair takes place. When the genetic integrity is restored, the cells divide and survive. At four days the levels of PARP and that of PAR decrease. At the sixth day in culture, the DNA binding fragments of PARP are detected in both cytoplasm and nucleus by both immunochemical and Western blot analysis. This event coincides with the appearance of DNA strand breaks. Thus, in this second stage PARP is cleaved, PAR is no longer synthesized and DNA strand breaks can no longer be repaired.

Simbulan-Rosenthal et al. compared the early stages of apoptosis in PARP antisense cells and PARP null cells. In controls, a burst of PARP occurs early in apoptosis and precedes the internucleosomal cleavage and the commitment to apoptosis. This is followed by a PARP decrease coincident with caspase 3 activation and the cleavage of PARP. A burst did not take place when PARP antisense was used with PARP –/– cells, again suggesting that PARP is required to prevent apoptosis when the cell is threatened to die (36). The sensitivity of PARP –/– cells and PARP +/+ cells to γ radiation and NNNG is the same in growth arrest. However, during exponential growth, PARP null cells are more sensitive to both γ rays and NNNG (37). Chatterjee et al. explain these differences by postulating that PARP maintains the genome integrity at least in part by protecting against recombination, thereby preventing an increase in sister chromatid exchange.

In conclusion, PARP plays a role in protection against DNA damage and maintaining the integrity of the genome probably by facilitating excision repair and by preventing recombination. Addition of PARP and its substrates to a cell-free DNA excision repair system suggests that PARP is directly involved in the process (38). PARP null mice have normal DNA repair. When the cellular insults cause strands to break, the activation of PARP is likely to precede apoptosis. Whether $NAD^+$ exhaustion or PARP degradation causes apoptosis may depend on the extent of the damage or on special circumstances.

Degradation of PARP certainly occurs in some forms of apoptosis, but it is not clear whether it is an early or a late event. Experiments using cells that undergo apoptosis slowly suggest that PARP activation is an early event and PARP degradation a late event. The late event coincides with internucleosomal degradation. Such coincidences are no direct proof that PARP causes the internucleosomal breakdown, however.

Experiments with knockout mice suggest that PARP is not necessary for apoptosis. Moreover, overexpression of PARP does not increase protection against apoptosis.

The significance of the existence of PARP isozymes or of alternative pathways at present, despite the important role of PARP, do not help to clarify the conflicting results reported for the role of PARP in apoptosis.

Then, what is the significance of PARP degradation by caspase 3 and what is the role of PARP in apoptosis? Could PARP be one of the molecules that operate at the crossroad where signal pathways that lead to apoptosis and necrosis separate? All DNA damage is not equal. Sustainable DNA damage might lead to PARP activation, DNA repair, and resumption of the cell cycle. More severe damage might be associated with active DNA repair, accelerated PARP activation, and rising depletion of NAD. These metabolic changes may at a certain point lead the cell to recognize that it is not redeemable or that its survival would endanger other cells and, therefore, it must be sacrificed discretely through apoptosis. Massive damage overwhelms DNA repair, activates PARP, causes sudden NAD depletion, and the cell dies by necrosis (39, 40).

***DNA-Dependent Protein Kinase and Topoisomerase*** DNA-dependent protein kinase is a member of the phosphoinositol 3 kinases family. It is composed of a catalytic (460 kDa) and a regulator (70/86 kDa heterodimer) domain. The regulator domain is also referred to as the KU fraction, and it modulates the binding of the catalytic unit of DNA-PK to DNA. The enzyme is activated by single and double strand breaks (41). Like PARP, it is down-regulated during cell senescence and possibly in apoptosis.

Other nuclear proteins have also been shown to be caspases substrates, for example, components of the splicesosome (40) and the DNA replication complex C (43). Still other nuclear proteins like topoisomerase II are degraded through the ubiquitin-proteosomal pathway (44).

The above discussion of selected substrates of caspases illustrates the importance of identifying some of the problems encountered in determining which substrates are cleaved in vivo in the process of apoptosis. The problems are technical or inherent to the specificity of the caspases and their distribution in cells.

- Most studies have been done in cells in culture, and even then there is considerable heterogeneity in substratés cleavage.
- Many studies involve the use of inhibitors. Penetration of the inhibitor through the cell membrane and its effectiveness inside the cell likely varies from cell type to cell type if not from cell to cells In any event, in most cases the amount of inhibitors used largely exceeds those needed to inhibit purified enzymes.
- All substrates are not cleaved in all cells during apoptosis. PARP is cleaved in hepatoma cells but not in hepatocytes. Actin is cleaved in neurons and thymocytes, but not in some other cells.
- The site of cleavage may vary from cell to cell (45).
- All caspases may not have been identified in each cell type or in a family of cell types.
- Other proteolytic enzymes may be involved in apoptosis, for example, caspase B, caspase J, and calpain.
- All caspases are not activated at the same time. This is discussed in the section on the caspase cascade.

## Apoptosis in Caspase-Deficient Mice

With the aid of genetic engineering, a gene that encodes a specific protein can be expressed in the proper vector, and thereby the function of that protein can be investigated in the normal phenotype either in cells in culture or in the entire organism. Knockout mice are unable to express both the maternal and paternal copy of the selected gene.

Mice deficient of caspases 1, 2, 3, 9, and 11 have been generated. Consider the potential phenotypical manifestations in the mouse when genes encoding a given caspase are knocked out.

- There may be no changes at all. In that case, since it is known that many cells die during development (for example, in the brain), the caspase cannot be critical to programmed cell death during development.
- During development there may be changes that interfere with apoptosis in some organs, for example, the brain where some cells are normally programmed to die. Such a finding suggests that the knockout caspase plays a role in apoptosis.
- If there is only one molecular pathway for apoptosis, it can be anticipated that apoptosis will be interfered with in all cells of the mice carrying the knocked out caspase.
- In contrast, if apoptosis can be induced in some cell types of the knocked out mice and not in others, it can be assumed that the pathway for apoptosis must vary with the cell type and/or with the inducer.

The most severe changes in the phenotype are observed in caspase 3 –/– and caspase 9 –/– mice. The development of the brain is most affected and the anomalies are explainable by defects in apoptosis. The defects are more severe for caspase 9 than caspase 3. The differences between the two types of caspase knockout mice concerns development of the embryo, apoptosis among different cell types, and response to different inducers of apoptosis.

Caspase 3 –/– mice show abnormal brain development that can be explained by a defect in apoptosis. In contrast, development of thymus lung, liver, kidney, spleen, and testis is normal. There are also differences in the capacity of various organs to respond to apoptosis. While neutrophils and activated T cells are defective in apoptosis, the thymus responds normally to a variety of inducers of apoptosis: dexamethasone, FAS ligation, ceramide, staurosporine, and $\gamma$ radiation. Depending on the stimulus, embryonic stem cells respond differently. While UV radiation and osmotic shock are inefficient inducers of apoptosis, apoptosis is normal after $\gamma$ radiation and heat shock. In view of these findings, in caspase 3 –/– mice, one must conclude that either the regulatory mechanisms or the individual steps of the pathway to apoptosis, or both, are likely to vary with the organ, the cell type, and the mode of induction of apoptosis.

The findings in caspase 9 –/– mice confirmed those made with caspase 3, but extended the interpretation of the results to the relative roles of the two caspases in the pathway. The damage to the brain is more severe in caspase 9 –/– mice than in caspase 3 –/– mice. They show extensive malformation of the cerebellum that can be explained by defective apoptosis and most of the offspring dies either before or soon after birth. Apoptosis in thymocytes is normal in caspases 9 –/–. Therefore, it would appear that caspase 9 has no role in the important thymocyte selection process. Thymocytes are resistant to the induction of apoptosis by some stimuli (dexamethasone, $\gamma$ radiation, and etoposide), but they remain sensitive to the activation of the FAS receptors by its ligand (an observation in keeping with the already-described pathway for apoptosis involving the FAS receptor, which recruits caspase 8 through the FADD/MORT adapter molecules).

As discussed in the next section, cytochrome c interacts with Apaf-1 and caspase 9, and ultimately caspase 3 is activated. This activation of caspase 3 does not take place in caspase 9 –/– embryonic tissue lysates. Moreover, the reconstitution of the deleted gene restores caspase 3 activation. The findings indicate that caspase 9 is essential for activation of the cytochrome c pathway for apoptosis (46–48).

In conclusion, mice knocked out for caspase 3 and caspase 9 demonstrate that the caspases are necessary for development, primarily of the brain. Studies of these mice also suggest that more than one pathway for apoptosis exists (according to Hakem et al., there are 4 pathways) (45), and that the pathway selected may differ with the type of tissue and the nature of the inducer. Moreover, caspase 9 plays a critical role in the cytochrome c dependent pathway for apoptosis, where it acts upstream of caspase 3. In contrast, it is unlikely to play a role in the pathway triggered by the FAS receptor activation, a finding in agreement with the observation made on caspase 8 –/– mice (47).

## References and Recommended Reading


1. Shi, L., Kraut, R.P., Aebersold, R., and Greenberg, A.H. A natural killer cell granule protein that induces DNA fragmentation and apoptosis. *J Exp Med.* 1992;175:553–66.
2. Yuan, J., Shaham, S., Ledoux, S., Ellis, H.M., and Horvitz, H.R. The *C. elegans* cell death gene ced-3 encodes a protein similar to mammalian interleukin-1 beta-converting enzyme. *Cell.* 1993;19:641–52.
3. Wang, X.M., Helaszek, C.T., Winter, L.A., Lirette, R.P., Dixon, D.C., Ciccarelli, R.B., Kelley, M.M., Malinowski, J.J., Simmons, S.J., Huston, E., et al. Production of active human interleukin-1 beta-converting enzyme in a baculovirus expression system. *Gene.* 1994;5:273–7.
4. Juo, P., Kuo, C.J., Yuan, J., and Blenis, J. Essential requirement for caspase-8/FLICE in the initiation of the Fas-induced apoptotic cascade. *Curr Biol.* 1998;8:1001–8.
5. Clem, R.J., and Miller, L.K. Control of programmed cell death by the baculovirus genes p35 and iap. *Mol Cell Biol.* 1994;5212–22.
6. Earnshaw, W.C., Martins, L.M., and Kauffman, S.H. Mammalian caspases: structure, activation, substrates, and functions during apoptosis. *Annu Rev Biochem.* 1999;68:383–424.
7. Chinnaiyan, A.M., O'Rourke, K., Tewari, M., and Dixit, V.M. FADD, a novel death domain-containing protein, interacts with the death domain of Fas and initiates apoptosis. *Cell.* 1995;81:505–12.
8. Hofmann, K., Bucher, P., and Tschopp, J. The CARD domain: a new apoptotic signalling motif. *Trends Biochem Sci.* 1997;22:155–6.
9. Wilson, K.P., Black, J.A., Thomson, J.A., Kim, E.E., Griffith, J.P., Navia, M.A., Murcko, M.A., Chambers, S.P., Aldape, R.A., Raybuck, S.A., et al. Structure and mechanism of interleukin-1 beta converting enzyme. *Nature.* 1994;370:270–5.
10. Wolf, B.B., and Green, D.R. Suicidal tendencies: apoptotic cell death by caspase family proteinases. *J Biol Chem.* 1999;274:20049–52.
11. Duan, H., Orth, K., Chinnaiyan, A.M., Poirier, G.G., Froelich, C.J., He, W.W., and Dixit, V.M. ICE-LAP6, a novel member of the ICE/Ced-3 gene family, is activated by the cytotoxic T cell protease granzymes B. *J Biol Chem.* 1996;271:16720–4.
12. Boldin, M.P., Goncharov, T.M., Goltsev, Y.V., and Wallch, D. Involvement of MACH, a novel MORT1/FADD-interacting protease, in Fas/APO-1- and TNF receptor-induced cell death. *Cell.* 1996;85:803–25.

13. Orth, K., O'Rourke, K., Salvesen, G.S., and Dixit, V.M. Molecular ordering of apoptotic mammalian CED-3/ICE-like proteases. *J Biol Chem.* 1996;271:20977–80.
14. Eberstadt, M., Huang, B., Chen, Z., Meadows, R.P., Ng, S.C., Zheng, L., Lenardo, M.J., and Fesik, S.W. NMR structure and mutagenesis of the FADD (Mort 1) death-effector domain. *Nature.* 1998;392:941–5.
15. Huang, B., Eberstadt, M., Olejniczak, E.T., Meadows, R.P., and Fesik, S.W. NMR structure and mutagenesis of the Fas (APO-1/CD95) death domain. *Nature.* 1996;384:638–41.
16. Chou, J.J., Matsuo, H., Duan, H., and Wagner, G. Solution structure of the RAIDD CARD and model for CARD/CARD interaction in caspase-2 and caspase-9 recruitment. *Cell.* 1998;94:171–80.
17. Cryns, V.L., Bergeron, L., Zhu, H., Li, H., and Yuan, J. Specific cleavage of alpha-fodrin during Fas- and tumor necrosis factor-induced apoptosis is mediated by an interleukin-1beta-converting enzyme/Ced-3 protease distinct from the polyp (ADP-ribose) polymerase protease. *J Biol Chem.* 1996;271:31277–82.
18. Waterhouse, N.J., Finucane, D.M., Green, D.R., Elce, J.S., Kumar, S., Alnemri, E.S., Litwack, G., Khann, K., Lavin, M.F., and Watters, D.J. Calpain activation is upstream of caspases in radiation-induced apoptosis. *Cell Death Differ.* 1998; 5:1051–61.
19. Martin, N.J., Finucane, D.M., Green, D.R., Elce, J.S., Kumar, S., Alnemri, E.S., Litwack, G., Khann, K., Lavin, M.F., and Watters, D.J. Phosphatidylserine externalization during CD95-induced apoptosis of cells and cytoplasts requires ICE/CED-3 protease activity. *J Biol Chem.* 1996;271:28753–6.
20. Kerr, J.F., Gobe, G.C., Winterford, C.M., and Harmon, B.V. Anatomical methods in cell death. (Review) *Methods Cell Biol.* 1995;46:1–27.
21. Gruenbaum, Y., Goldman, R.D., Myuhas, R., Mills, E., Margalit, A., Fridkin, A., Dayani, Y., Prokocimer, M., and Enosh, A. The nuclear lamina and its functions in the nucleus. (Review) *Int Rev Cytol.* 2003;226:1–62.
22. McConkey, D.J. Calcium-dependent, interleukin 1-converting enzyme inhibitor-insensitive degradation of lamin B1 and DNA fragmentation in isolated thymocyte nuclei. *J Biol Chem.* 1996;271:22398–406.
23. Rao, L., Perez, D., and White, E. Lamin proteolysis facilitates nuclear events during apoptosis. *J Cell Biol.* 1996;135:1441–55.
24. Tzur, Y.B., Hersh, B.M., Horvitz, H.R., and Gruenbaum, Y. Fate of the nuclear lamina curing *Caenorhabditis elegans* apoptosis. *J Struct Biol.* 2002;137:146–53.
25. Auer, B., Nagl, U., Herzog, H., Schneider, R., and Schweiger, M. Human nuclear NAD+ ADP-ribosyltransferase (polymerizing): organization of the gene. *DNA.* 1989;8:575–80.
26. Oei, S.L., Herzog, H., Hirsch-Kauffman, M., Schneider, R., Auer, B., and Schweiger, M. Transcriptional regulation and autoregulation of the human gene for ADP-ribosyltransferase. *Mol Cell Biochem.* 1994;138:99–104.
27. Hanai, S., Uchida, M., Kobayashi, S., Miwa, M., and Uchida, K. Genomic organization of Drosophila polyp(ADP-ribose) polymerase and distribution of its mRNA during development. *J Biol Chem.* 1998;273:11881–6.

28. Kim, J.W., Kim, K., Kang, K., and Joe, C.O. Inhibition of homodimerization of poly(ADP-ribose) polymerase by its C-terminal cleavage products produced during apoptosis. *J Biol Chem.* 2000;275:8121–5.

29. He, J., Whitacre, C.M., Xue, L.Y., Berger, N.A., and Oleinick, N.L. Protease activation and cleavage of poly(ADP-ribose) polymerase: an integral part of apoptosis in response to photodynamic treatment. *Cancer Res.* 1998;58:940–6.

30. Wang, Z.Q., Auer, B., Sting, L., Berghammer, H., Haidacher, D., Schweiger, M., and Wagner, E.F. Mice lacking ADPRT and poly(ADP-ribosyl)ation develop normally but are susceptible to skin disease. *Genes Dev.* 1995;9:509–20.

31. Leist, M., Single, B., Kunstle, G., Volbracht, C., Hentze, H., and Nicotera, P. Apoptosis in the absence of poly-(ADP-ribose) polymerase. *Biochem Biophys Res Commun.* 1997;233:518–22.

32. Bernges, F., Burkle, A., Kupper, J.H., and Zeller, W.J. Functional overexpression of human poly(ADP-ribose) polymerase in transfected rat tumor cells. *Carcinogenesis.* 1997;18:663–8.

33. Shieh, W.M., Ame, J.C., Wilson, W.V., Wang, Z.Q., Koh, D.W., Jacobson, M.K., and Jacogson, E.L. Poly(ADP-ribose) polymerase null mouse cells synthesize ADP-ribose polymers. *J Biol Chem.* 1998;273:30069–72.

34. Ame, J.C., Rolli, V., Schreiber, V., Niedergang, C., Apiou, F., Decker, P., Muller, S., Hoger, T., Menissier-de Murcia, J., and de Murcia, G. PARP-2, A novel mammalian DNA damage-dependent poly(ADP-ribose) polymerase. *J Biol Chem.* 1999;274:17860–8.

35. Trucco, C., Oliver, F.J., de Murcia, G., and Menissier-de Murcia, J. DNA repair defect in poly(ADP-ribose) polymerase-deficient cell lines. *Nucl Acids Res.* 1998;26:2644–9.

36. Simbulan-Rosenthal, C.M., Rosenthal, D.S., Iyer, S., Boulares, H., and Smulson, M.E. Involvement of PARP and poly(ADP-ribosyl)ation in the early stages of apoptosis and DNA replication. *Mol Cell Biochem.* 1999;193:137–48.

37. Chatterjee, S., Berger, S.J., and Berger, N.A. Poly(ADP-ribose) polymerase: a guardian of the genome that facilitates DNA repair by protecting against DNA recombination. *Mol Cell Biochem.* 1999;193:23–30.

38. Dantzer, F., Schreiber, V., Niedergang, C., Trucco, C., Flatter, E., De La Rubia, G., Oliver, J., Rolli, V., Menissier-de Murcia, J., and de Murcia, G. Involvement of poly(ADP-ribose) polymerase in base excision repair. *Biochimie.* 1999;81:69–75.

39. Soldani, C., and Scovassi, A.I. Poly(ADP-ribose) polymerase-1 cleavage during apoptosis: an update. (Review) *Apoptosis.* 2002;4:321–8.

40. Chatterjee, S., and Berger, N. Poly ADP-ribose polymerase in response to DNA damage. In: Nickoloff, J.L., and Hoekstra, M., eds. *DNA Damage and Repair.* Totowa, NY: Humana Press; 1998;2:577–638.

41. Weinfled, M., Chaudhry, M.A., D'Amours, D., Pelletier, J.D., Poirier, G.G., Povirk, L.F., and Lees-Miller, S.P. Interaction of DNA-dependent protein kinase and poly(ADP-ribose) polymerase with radiation-induced DNA strand breaks. *Radiat Res.* 1997;148:22–8.

42. Waterhouse, N., Kumar, S., Song, Q., Strike, P., Sparrow, L., Dreyfuss, G., Alnemri, E.S., Litwack, G., Lavin, M., and Watters, D. Heteronuclear ribonucleo-

proteins C1 and C2, components of the splicesosome, are specific targets of interleukin 1 beta-converting enzyme-like proteases in apoptosis. *J Biol Chem.* 1996; 271:29335–41.

43. Ubeda, M., and Habener, J.F. The large subunit of the DNA replication complex C (DSEB/RF-C140) cleaved and inactivated by caspase-3 (CPP32/YAMA) during Fas-induced apoptosis. *J Biol Chem.* 1997;272:19562–8.

44. Nakajima, T. Degradation of topoisomerase II alpha precedes nuclei degeneration during adenovirus E1A-induced apoptosis and is mediated by the activation of the ubiquitin dependent proteolysis system. *Nippon Rinsho.* 1996;54:1828–35.

45. Samejima, K., Svingen, P.A., Basi, G.S., Kottke, T., Mesner, P.W., Stewart, L., Durrieu, F., Poirier, G.G., Alnemri, E.S., Champoux, J.J., Kaufmann, S.H., and Earnshaw, W.C. Caspase-mediated cleavage of DNA topoisomerase I at unconventional sites during apoptosis. *J Biol Chem.* 1999;274:4335–40.

46. Varfolomeev, E.E., Schuchmann, M., Luria, V., Chiannilkulchai, N., Beckmann, J.S., Mett, I.L., Rebrikov, D., Brodianski, V.M., Kemper, O.C., Kollet, O., Lapidot, T., Soffer, D., Sobe, T., Avraham, K.B., Goncharov, T., Holtmann, H., Lonai, P., and Wallach, D. Targeted disruption of the mouse caspase 8 gene ablates cell death induction by the TNF receptors, Fas/Apo1, and DR3 and is lethal prenatally. *Immunity.* 1998;2:267–76.

47. Hakem, R., Hakem, A., Duncan, G.S., Henderson, J.T., Woo, M., Soengas, M.S., Elia, A., de la Pompa, J.L., Kagi, D., Khoo, W., Potter, J., Yoshida, R., Kaufman, S.A., Lowe, S.W., Penninger, J.M., and Mak, T.W. Differential requirement for caspase 9 in apoptotic pathways in vivo. *Cell.* 1998;94:339–52.

48. Kuida, K., Haydar, T.F., Kuan, C.Y., Gu, Y., Taya, C., Karasuyama, H., Su, M.S., Rakic, P., and Flavell, R.A. Reduced apoptosis and cytochrome c-mediated caspase activation in mice lacking caspase 9. *Cell.* 1998;94:325–37.

49. Juo, P., Kuo, C.J., Yuan, J., and Blenis, J. Essential requirement for caspase-8/FLICE in the initiation of the Fas-induced apoptotic cascade. *Curr Biol.* 1998;8:1001–8.

CHAPTER

# 7 THE CYTOCHROME C–DEPENDENT PATHWAY TO APOPTOSIS

## The Pathway

The pathway for apoptosis described in Chapter 6 began with the activation of specific death receptors: the FAS and the tumor necrosis factor (TNF) receptors. A glance at the list of the triggers of apoptosis is a reminder that all programmed or induced control cell death is not caused by direct activation of death receptors. Some triggers are hormones and others are toxins or xenobiotic agents including reactive oxygen species, some of which are endogenously produced. In the latter cases, the death signal is different from that which activates the classical death receptors. Do these different early events unleash identical or different molecular sequences leading to programmed cell death?

The discovery that a critical molecule involved in the second pathway was cytochrome c surprised many investigators in the field. Few suspected that this small (104 amino acids) but highly respected molecule, which faithfully and assiduously shuttles electrons from complex III (ubiquinol-cytochrome c reductase) to complex IV (cytochrome oxidase), a molecule comfortably tucked in the outer surface of the inner mitochondrial membrane, could be an accomplice in cell death. The evidence for the unsuspected function of cytochrome c involvement is compelling. In 1966, Liu and associates working in Xiaodong Wang's laboratory prepared a cell-free system derived from Hela cells grown in culture, and discovered that cytochrome c activates caspase 3 (1). This seminal observation was soon followed by the discovery of the molecules that contribute to the elaboration of the molecular scaffold that leads to apoptosis in the second pathway. A critical step was the discovery of the mammalian homologue[1] of the product of the *ced-4* gene in *Caenorhabditis elegans,* the apoptosis-activating factor 1 (Apaf-1). Cytochrome c was discovered to form a molecular complex with Apaf-1 and caspase 9.

The components and the sequence of events that generate the molecular complex that causes the initiation of apoptosis in the cytochrome c–dependent pathway are

[1]The mammalian homologue has 21% identity and 53% similarity with *Ced-4.*

described in this section. The components of the complex are: cytochrome c, a nucleoside triphosphate (ATP or dATP) (2),[2] and Apaf-1 and caspase 9.

- Cytochrome c is released from its membrane support by a mechanism to be discussed later in this chapter.
- Present evidence demonstrates that caspase 9 is the initiator caspase in the cytochrome c–dependent pathway. Caspase 9 is the homologue of Ced-3 and some of it is located in the mitochondrial inner membrane. Caspase 9 possesses a CARD motif, a caspase recruitment domain in its N terminal. After ultraviolet (UV) irradiation of embryonic fibroblasts derived from caspase 9 –/– mice, cytochrome c[3] is released but caspase 3 is not activated.
- Apaf-1 is a 130-kDa protein found in the cytosol. The molecule shares a substantial homology with Ced-4 mainly in its N terminal segment. The NH terminal (85 amino acids) is flanked by a caspase recruitment domain or CARD and Wheeler motifs. The C terminal region (320 amino acids) lacks homology with Ced-4 and includes 12 WD40 repeats that mediate protein-protein interactions(3). Bcl-x binds to the C terminal sequence of Apaf-1. The C terminal is the portion of the molecule that contains the WD40 repeats and blocks the maturation of caspase 9 mediated by Apaf-1 (**FIGURE 7-1**) (4).

Apaf-1 knocked out mice die approximately two weeks after conception with very severe brain damage; elimination of the interdigital webbing is delayed and fusion of the palate fails. No anomalies were detected in other organs (5, 6).

Some of the findings made in embryonic fibroblasts derived from Apaf-1 –/– mice support what was already learned about Apaf-1, but others are intriguing. As expected, the cells are resistant to many triggers of apoptosis, and in those cases the activation of caspase 9 and secondarily of caspase 3 are markedly decreased. However, engaging of FAS by the FAS ligand activates caspase 8. Moreover, the cells undergo normal apoptosis when exposed to staurosporine, which is expected to trigger the cytochrome c pathway, a finding that can be interpreted to mean that molecules other than Apaf-1 play a critical role in activating initiator caspases or that in some cases apoptosis may choose a different pathway. Indeed, homologues of Apaf-1 have been described (7) but their exact role in apoptosis is still unclear.

An elegant and seminal paper that originated in Wang's laboratory demonstrated the sequence of events that take place in initiating apoptosis in the cytochrome c–dependent pathway. Zou et al. assembled the components described above in a purified cell-free system (3). They used purified cytochrome c recombinant Apaf-1 and recombinant caspase 9 to demonstrate that the activation of caspase 9 occurs in three steps and leads to the formation of the apoptosome:

- dATP or ATP binds to Apaf-1 through its consensus nucleoside-triphosphate domain, and dATP or ATP are respectively hydrolyzed to dADP and ADP. Mutations in the ATP binding site inactivate Apaf-1.
- However, when the released cytochrome c binds to Apaf-1, it promotes the multimerization of the bimolecular complex and dATP or ATP hydrolysis is required for

[2]The nucleoside triphosphate is 1 μM of ATP or 1 mM of dATP, a considerable difference. Moreover, the triphosphates of 2 chloro-2′-deoxyadenosine and fludarabine can substitute for either adenosine 5′-triphosphate (ATP) or deoxyadenosine 5′-triphosphate (dATP).

[3]The newly synthesized apocytochrome is unable to activate the initiator caspase; only cytochrome c released from mitochondria can secure the activation. Still, cytochrome c release from mitochondria devoid of nuclei can activate caspase 9.

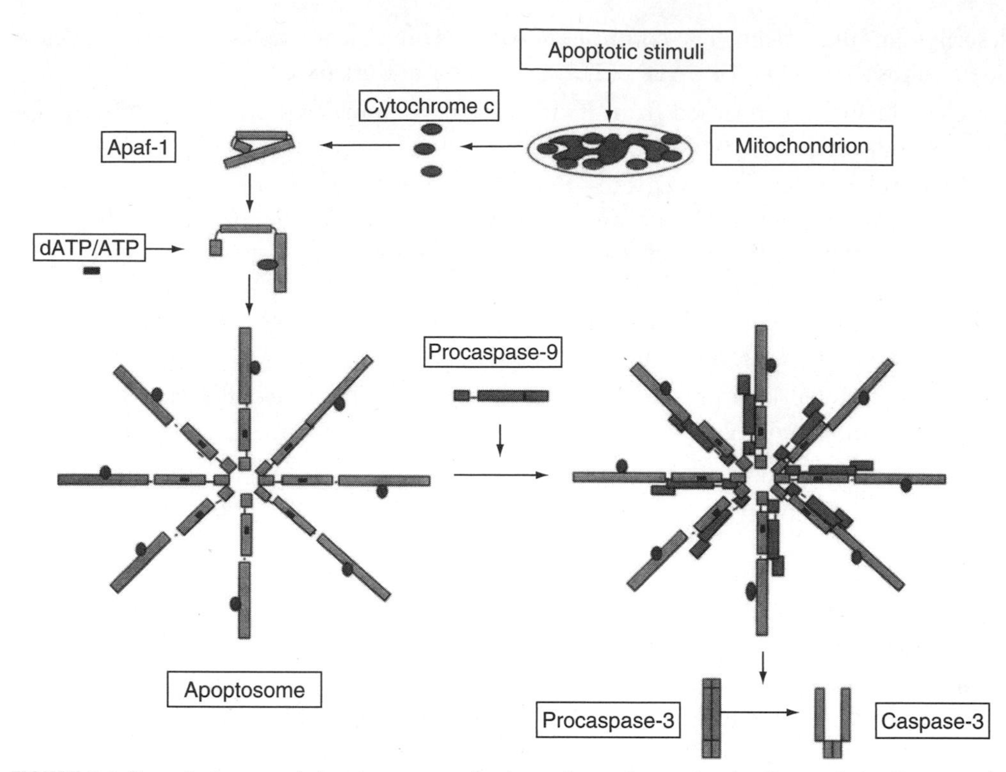


**FIGURE 7-1** The cytochrome c–induced caspase activation pathway. Apoptotic stimuli exert their effects on mitochondria to cause the release of cytochrome c. Cytochrome c in turn binds to Apaf-1, a cytosolic protein that normally exists as an inactive monomer. The binding of cytochrome c induces a conformational change in Apaf-1, allowing it to bind the nucleotide dATP or ATP. The nucleotide binding to the Apaf-1–cytochrome c complex triggers its oligomerization to form the apoptosome, which recruits procaspase-9. The binding of procaspase-9 to the apoptosome forms the caspase-9 holoenzyme that cleaves and activates the downstream caspases, such as caspase-3. (From Wang, X. The expanding role of mitochondria in apoptosis, *Genes Dev.* 2001; 15:2922–2933, with permission.)

the process. Eight molecules of Apaf-1 join each molecule of cytochrome c. The function of cytochrome c is to facilitate the multimerization of Apaf-1 molecules in the presence of dATP and not to regulate nucleoside-ribose binding to Apaf-1 or its hydrolysis.

- The multimeric complex then recruits procaspase 9 through respective CARD domains in a 1:1 molar ratio of Apaf-1 to procaspase 9. The formation of a multimeric complex is likely to facilitate the transcatalytic activation of caspase 9. Indeed, the multimeric complex no longer requires cytochrome c and ATP to activate procaspase 9.
- Finally, caspase 9 is released from the molecular complex and triggers the caspase cascade by proteolytic cleavage. The large molecular complex (>1.3 million Da) has been named the apoptosome. Thus, the formation of the apoptosome requires dATP/ATP hydrolysis and, once formed, the complex activates the initiator caspase 9, which in turn triggers the caspase cascade (3).

In summary, the WD40 repeats of Apaf-1 bind to holocytochrome c; this association is followed by the binding of ATP or dATP to CARD and multimerization and binding of caspase 9. The result is an enormous complex of cytochrome c–Apaf-1 and caspase 9 (~10 M Da; **FIGURE 7-1**) (8–11).

Thus, currently two molecular constructions are known to be formed at the onset of apoptosis: one after the interaction of the ligand with the FAS or TNF in trimerized receptor, recruitment of adapter molecules and of caspase 8; the other, the cytochrome c–dependent pathway with the formation of the apoptosome. However, it cannot be excluded that other pathways for apoptosis, which do not require death receptor activation or cytochrome c release, may exist.

## Homophilic Interaction Domains and Apoptosis

Hoffman and Bucher, among the molecules that constitute the two scaffolds that activate the initiator caspases, identified three homophilic-interaction domains (12): the death domain (DD); the dead effector domain (DED); and the caspase recruitment domain (CARD). Each of these domains transmits death signals by protein-protein interactions. The DD is found in the C terminal sequence of FAS and TNF receptors. They signal the oligomerization of the receptors to the downstream DD found in adapter molecules such as TRADD, FADD, and MORT. The death effector domain found in the N terminal, of for example, FADD, signals caspase activation by homophilic-interaction with the DED in the caspase, for example caspase 8.

The caspase recruitment domain found in RAIDD, an adapter protein similar to FADD and Apaf-1, is also found in the N terminal prodomain of caspases: ICE, CED-3, caspase 2, caspase 4, caspase 9, and caspase 10. The N terminal CARD of RAIDD interacts with the CARD found in the N terminal of CED-3 and caspase 2, but not with other mammalian caspases. Thus, the death signal communicated by the death effector molecule (RAIDD) to the caspase is very specific. Chou et al. used magnetic nuclear resonance (MNR) spectroscopy to solve the structure of the N terminal portion of RAIDD, which contains the CARD domain. The three dimensional structure of RAIDD-CARD is a six helices bundle. Three of the helices are vertically stacked on one side and three are stacked on the other side of the bundle. There is one acidic patch and one basic patch on each opposite side. The observation made by MNR spectrophotometry coupled with mutational studies allowed Chou et al. to propose a mechanism of interaction between the CARD domains of RAIDD and caspase 2 (13). The positively charged helices 1, 3, and 4 of the RAIDD CARD interact electrostatically with the negatively charged helices 2, 5, and 6 of the caspase CARD. Moreover, modeling of the CARDs of Apaf-1 and caspase 9 revealed that the basic and acidic surfaces are highly conserved, suggesting that the mode of interaction of the RAIDD-CARD with the CARD of caspase 2 can be extended to the homophilic interaction between caspases and Apaf-1, and most likely generalized to most CARD/CARD interactions (13).

### Caspase Cascade

The events associated with apoptosis are unleashed by the sequential activity of several caspases. The sequence is referred to as the caspase cascade. The cascade is put in gear by caspases containing long $NH_2$ terminals. Two different pathways for apoptosis are known: the cytochrome c–independent and the cytochrome-dependent pathways. The first is triggered by engagement of the ligand TNF-$\alpha$ or the FAS ligand with the death receptor (TNFR1, TNFR2, or FAS), and a caspase, for example, caspase 8, is recruited to the cell surface by the dead induced signal complex (DISC) where it interacts with the FADD-MORT complex (refer to Chapters 3 and 6). The caspases form multimolecular complexes and are activated autocatalytically. The cytochrome c–dependent pathway does not involve the death receptor. Among the agents known to activate this pathway are: growth factor deprivation, excessive DNA damage, and chemotherapy.

Thus, a physiologic or a pathologic message reaches the cell, which in turn signals the release of cytochrome c. Then in the presence of dATP/ATP, caspase 9 forms a multimolecular complex with Apaf-1 and cytochrome c, and is activated autocatalytically. It is possible that the point of no return in the path of cell death may be at the level of activation of the first caspase, for example, caspase 8 or caspase 9. In the next step(s), caspases 8 and 9 activate caspases devoid of a long $NH_2$ terminal namely, caspases 3, 6, and 7. Caspases 3, 6, and 7 in turn cleave a population of substrates that in sequence or simultaneously dismantle the cell.[4]

In vitro reconstruction of the caspase cascade (14) suggests that the cascade occurs in three steps. Prior to activation of apoptosis the initiator (procaspase 9) and the effector procaspases (procaspase 3) exist in separate dimeric complexes (60–90 kDa). During activation of the caspases a much larger complex is formed ($M_r$ 700,000 kDa) through the redistribution and activation of the caspases. This large multiheterodimeric complex is called the aposome,[5] which includes Apaf-1 and caspases 9, 3, and 7. In the last step, a smaller particle is released, ($M_r$ 200,000 kDa), which contains little Apaf-1, caspase 3, and caspase 7 (the latter does not have detectable caspase activity).

## Mechanisms of Cytochrome c Release in Apoptosis

The exact mechanism or mechanisms of cytochrome c release are debatable. The existing data and proposed hypothesis will be covered briefly after reviewing some basic properties of mitochondria.

There are four major components to mitochondria: the outer membrane, the inner membrane, and the two membranes that separate two intramitochondrial regions: the large internal space, the matrix, and a small intermembrane space. The inner membrane repeatedly projects into the matrix forming the cristae, a process that markedly increases the membrane surface and facilitates contact between the inner membrane and the matrix.

The mitochondria, first seen by microscopists in the late 1800s, kept their secrets for a long time. It took several decades to unravel the mechanism of generation of ATP in the presence of oxygen, by far the major source of cellular energy. This process was discovered to be dangerous when vitiated because reactive oxygen species (ROS) are formed; it wasn't until the late 1990s when it was suspected that mitochondria participated in a conspiracy to kill the cell. When one of the key members of the electron transport chain, cytochrome c, was suspected and later proved to be a double agent and a member of this conspiracy, it generated surprise if not skepticism, but the evidence soon became overwhelming (1). Cytochrome c, in contrast to many other important members of the electron transport chain, is a soluble mitochondrial matrix protein (104 amino acids, 11.60 kDa in humans). When released in the cytosol it becomes part of the machinery that causes cell death. The molecular composition of mitochondria is listed in **TABLE 7-1** and **FIGURE 7-2**.

Cytochrome c is encoded by a nuclear gene and after transcription and translation the protein appears in the cytosol in the form of an apoprotein. The latter is trans-

[4]If the point of no return in the apoptotic process, as suggested by Cain and his colleagues, is located at the site of activation of the initiator caspases upstream of the activation of the effector caspases (14), it suggests that identification of the enzymes and their substrates in the following steps would be of little interest.

[5]This is not to be confused with the apoptosome, which is much smaller ($10^6$ Da) composed of Apaf-1, cytochrome c, and caspase 9. Of course, it cannot be excluded that the formation of the apoptosome precedes that of the aposome.

**Table 7-1 Some active content of the mitochondrial components**

| Location | Molecular Composition of Mitochondria |
|---|---|
| Outer membrane | Porins allow passage of (VDAC) molecules 5,000–10,000 Da or less |
| | Mitochondrial lipid synthetic machinery |
| Inner membrane | Mitochondrial oxidative phosphorylation electron transport chain |
| | ATP synthase |
| | ADP-ATP translocator (ANT) |
| Intermembrane space | Phosphorylation of nucleotides |
| matrix | Enzymic machinery of the |
| | • Krebs cycle enzymes |
| | • Pyruvic acid[a] oxidation |
| | • Fatty acid[a] oxidation |
| | Gene expression machinery |
| | • Mitochondrial DNA |
| | • Mitochondrial ribosomes |
| | • Mitochondrial tRNA |
| | • Enzymes involved in DNA replication and protein synthesis |

[a]Transported from the cytosol.

ported to the mitochondrial inner membrane where a haem ligase binds the apoprotein (104 amino acids, 11.60 kDa in humans) to haem. The holoenzyme is localized in the inner membrane where it is safely secured to be released only when the membrane barrier becomes dysfunctional.

## Mitochondrial Permeability, Transition Pore, and Apoptosis

Most if not all cell membranes contain segments made of protein complexes whose construction permits the passage of molecules from one site of the membrane to another. These passages are called channels or pores. The passages are selective and their penetration depends in part on the size and the charge of the molecule. This passage may be "gated," meaning that the amount and the timing of the flow of the molecules that pass through the channel are carefully regulated, depending on the cell type or the membrane organelle (e.g., mitochondria) requirements. Like diffusion, the molecules or ions flow only down a concentration gradient that is from the highest concentration on one side of the membrane toward the lowest concentration on the

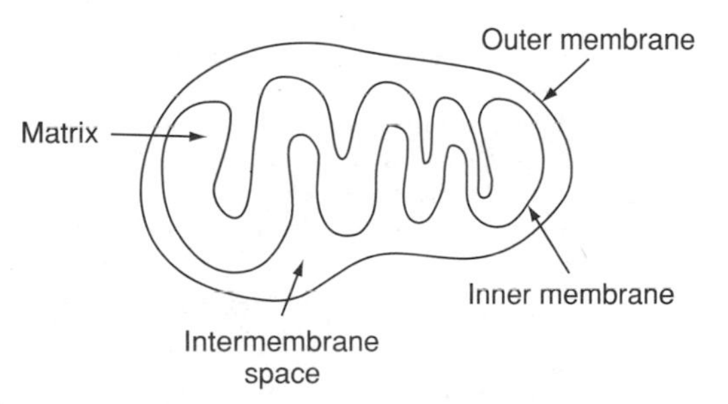


**FIGURE 7-2** Fundamental mitochondrial components.

other side. When the flow is gated, various mechanisms control the gate: voltage dependence (ions $Na^+$, $K^+$, and $Ca^{2+}$), chemical or transmitter dependence (acetylcholine), or second messenger dependence (cyclic AMP).

One mechanism(s) by which cytochrome c, an inner membrane protein, is regulated by Bcl-2, an outer membrane protein, and by Bax, may involve mitochondrial transition pores. Consider the structure and function(s) of the pore and its role in apoptosis.

Permeability transition consists of a permeability increase of the inner mitochondrial membrane. This permeability is mediated by a regulated channel whose formation involves proteins derived from both the inner and outer membranes (the mitochondrial transition pore [MTP]). The fundamental structure of MTP is composed of the outer membrane voltage-dependent anion (VDAC) and the inner membrane adenine nucleotide translocase (ANT) and cyclophilin D (CYP-D) (**FIGURE 7-3**). MTP thus forms a multiple protein complex that establishes contact between the inner and outer membranes. VDAC is a 30-kDa transmembrane protein forming a hydrophilic trimeric hollow β barrel pore.[6] The complex is abundant in the outer mitochondrial membrane. The porin secures permeability for polar molecules and ions, including ATP and metabolites that flow from cytoplasm to mitochondria.[7]

It is not clear what portion of the molecular complex forms the pore. But it is generally believed that VDAC permits low molecular weight compounds to reach transport systems found in the inner membrane. The latter, in time, controls the entry of specific molecules in the inner membrane or the matrix. ANT[8] secures the ADP ⇔ ATP exchange between compartment and provides the chemical energy needed for mitochondrial functions involving molecular imports and exports. Although the role of CYP-D is not clear, it is believed to be involved in protein folding. Investigators have reconstructed the VDAC-ANT-CYP-D complex in vitro and proved that the components of the complex assemble readily. This led to the assumption that the MTP may exist and function under physiological conditions in vivo.

It is known that the contact sites between inner and outer membrane are essential in energy transduction and specific metabolic functions. The specificity of the event is in

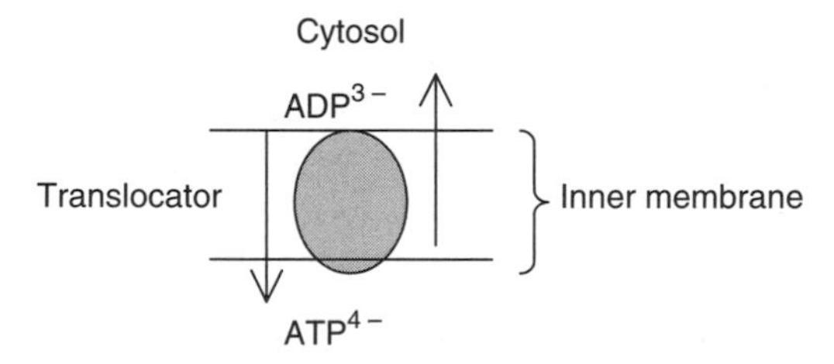


**FIGURE 7-3** Schematic representation of mitochondrial ADP-ATP translocation in the inner membrane.

[6]The barrel sizes are 35 Å and 45 Å, respectively; a single β sheet is believed to contain 16 strands (250 amino acids) disposed at an inclination of 45°.

[7]The porin also serves as a binding site for some cytosolic enzymes (hexokinase, glucokinase, glycerol kinase, and creatine kinase), thereby giving these enzymes preferential access to mitochondrial ATP. Multiple forms of VDAC exist in multicellular organisms. Close to twelve such channels have been cloned, and they have only low sequence homology with each other and none with bacterial porins. Still, mitochondrial and bacterial porins have a similar quaternary structure: the β barrel structure.

[8]Oxidative phosphorylation is associated with a one-to-one exchange of cytosolic ADP with mitochondrial ATP. At physiological pH, ATP carries four and ADP three negative charges. Consequently, the one-to-one exchange between cytosol and mitochondria could lead to an imbalance unless it is associated with proton movement or the movement of another charged molecule. The latter is achieved by the adenine nucleotide translocator, a 60,000 kDa homodimer that is inhibited by atractylate. It forms a multispanned protein inserted in the inner membrane of the mitochondrion.

part secured by the attraction of special molecules to the VDAC-ANT-CYP-D complex. Indeed, VDAC attracts molecules other than ANT and CYP-D, namely, enzymes (glycerol kinase, hexokinase, creatine kinase) and at least one receptor (benzodiazepine).

Thus, VDAC recruits proteins to the VDAC-ANT-CYP-D complex to assemble a communication path between the outer membrane (whose permeability is not strictly specific) to specific enzymes or receptors, and thereby it contributes to specific mitochondrial functions. The transmembrane electric potential difference and matrix pH regulate the channel. The first favors the pores opening and the second their closure. The opening of the channel is inhibited by cyclosporin A (reviewed in 15, 16).

The physiologic functions of the channel include:

- Energy transduction: the association of hexokinase, glycerol kinase and the VDAC-ANT-CYP-D complex provides a mode of transport of the ATP that is generated by the Krebs cycle and oxidative phosphorylation to the kinases.
- Steroidogenesis: cells that contribute to steroid anabolism need to transfer cholesterol to the inner membrane of the mitochondrion where it is converted to pregnenolone. The association of VDAC-ANT to the benzodiazepine receptor is believed to secure the transfer of cholesterol.
- Some pathologic conditions such as ischemia and appearance of oxygen reactive species (e.g., reperfusion) cause a loss of permeability. The formation of the megachannel leads to mitochondrial permeability transition with the following consequences: diffusion of the mitochondrial membrane potential; uncoupling of electron transport and oxidative phosphorylation; overproduction of superoxide anions; outflow of matrix $Ca^{2+}$ and glutathione; and release of intermembrane proteins. Such an event could be critical to the onset of apoptosis, if it is proved that dissipation of the mitochondrial transmembrane potential ($\Delta\Psi$) precedes the apoptotic events.

## Cytochrome c Release

Three mechanisms for cytochrome c release by mitochondria have been invoked: (1) anti- and proapoptotic molecules form channels in the mitochondrial membrane, the latter when disturbed or activated release cytochrome c; (2) anti- and proapoptotic molecules activate pre-existing mitochondrial channels that facilitate the release of cytochrome c; and (3) cytochrome is released by caspase 8.

### Pro- and Anti-Apoptotic Molecules Form Channels

The first model was inspired by the similarity of the $Bclx_L$ structure[9] and that of pore forming proteins, among them bacterial toxins such as diphtheria toxin[10] and colicin (17). The proapoptotic $Bclx_L$ could form an ion channel whose function buffers osmotic stress inside the mitochondria and prevents cytochrome c release (18–20).

Several observations indirectly support this model. Mitochondria are required for the manifestations of apoptosis to take place. Bcl-2 is an integral part of the mitochondrial membrane (21, 22). Overexpression of Bcl-2 in mitochondria blocks apoptosis and cytochrome c release most likely by raising the Bcl-2/Bax ratio (23, 24). These findings

[9]The $Bclx_L$ structure has two central hydropholic core helices surrounded by four amphipathic helices.
[10]Diphtheria is composed of two polypeptidic chains, A and B. The A chain is an enzyme that inhibits the EF-2 elongation factor, and the B chain a pore in the plasma membrane that facilitates the passage of the A chain.

suggest that in the chain of events that end in apoptosis, members of the Bcl-2 family constitute the first link and determine whether the cell will die or live by interaction at the level of the mitochondrial membrane. Is it significant, therefore, that death antagonists (Bcl-2, $Bclx_L$) are an integral part of the mitochondrial membrane and that death agonists (Bax and Bid) are often translocated to mitochondria to exert these effects?

However, it has been suggested that Bax and Bid may also form membrane channels. The enforced dimerization of Bax leads to its translocation to the mitochondrial membrane and ultimately causes apoptosis (25). Inducible expression of Bax is associated with alterations of the permeability of the mitochondrial membrane, formation of reactive oxygen species, and cytochrome c release. These events take place independently of transition pore activation (discussed below) (26). The cytochrome c release and the caspase activation that follow can be inhibited by $Bclx_L$ (27).

The structure of the proapoptotic Bid strongly suggest that it is capable of forming pores. Bid's structures resemble that of $Bclx_L$: two central hydrophobic $\alpha$ helices surrounded by four $\alpha$ amphipathic helices and the $BH_3$ domain. The similarity between the two structures led to the suggestion that Bid, like $Bclx_L$, forms intramembrane channels. Moreover, Bid does not have a large hydrophobic pocket and its $BH_3$ is buried, and therefore is not available to react with $Bclx_L$ (28). In conclusion, it is conceivable that proapoptotic agents (Bcl-2 or $Bclx_L$) form channels that prevent cytochrome c release from mitochondria and that proapoptotic agents (Bax and Bid) may also facilitate cytochrome c's release through channel formation.

### Cytochrome c and the Transition Pore

The second mechanism for release of cytochrome c postulates that instead of modulating the function of specific antiapoptotic and proapoptotic channels ($Bclx_L$, Bax, and Bid), the antiapoptotic Bcl-2 prevents the swelling of mitochondria by modulating the permeability of the transition pore. In contrast, proapoptotic agents activate the mitochondrial permeability transition pore (MTP) (refer to the section above).

Some experimental data support the hypothesis that MTP pores reconstituted in liposomes function like the natural pores of intact mitochondria. The antiapoptotic agents Bcl-2 and $Bclx_L$ prevent the opening of the purified megachannel after their insertion in liposomes (29). Mutated Bcl-2 proteins with low antiapoptotic activity lose their ability to maintain the MTP closed; in contrast, caspases 1, 2, 3, 4, and 6 cause the opening of MTP. Thus, changes in Bcl-2 activity and activation of caspases are events associated with MTP opening.

As for the antiapoptotic member of the Bcl-2 family, Bax, it dissipates the inner transmembrane potential ($\Delta\Psi m$) and releases cytochrome c through the outer membrane, manifestations seen with agents that open the permeability transition pore. Intracellular microinjection of the adenine nucleotide translocator, a component of the transition pore (ANT or ATP carrier; Figure 7-3) causes dissipation of $\Delta\Psi m$ followed by apoptosis. When Bax is overexpressed in yeast, it leads to apoptosis of the wild-type yeast but not in the ANT-deficient mutant. These experiments and others led Marzo and associates to conclude that the antiapoptotic Bax and mitochondrial adenine nucleotide translocator act together within the permeability pore complex to increase the mitochondrial membrane permeability, which leads to cytochrome c release, caspase activation, and apoptosis (29).

In conclusion, it appears that a death signal can in certain cells trigger the following sequence of events. Increased free $Ca^{2+}$ in the cytosol activates the phosphatase calcineurin, which dephosphorylates Bax in the cytosol. The proapoptotic agent moves to the mitochondria and forms heterodimers with one or more of the anti-apoptotic members of the Bcl-2 family (Bcl-2-Bclx$_L$). Cytochrome c is released in cytosol through selective (Bax or Bid) or unselective channels (MTC), and caspase 9 is activated leading to the opening of the caspase cascade (caspase 2, 6, 7, and 10; **FIGURE 7-4**).

Although it is certain that cytochrome c is released in some pathways to apoptosis, the molecular mechanism of cytochrome c release is not always understood and it may vary with the circumstances.

## Caspase 8 and Cytochrome c Release

The third proposed mechanism for the release of cytochrome c is based on the following reasoning: (1) FAS-FAS ligand interaction rapidly activates caspase 8; (2) FAS activation rapidly interrupts electron transport and releases cytochrome c; and (3) these events are blocked by a broad range of caspase inhibitors. The findings suggest that caspase 8 functions upstream of cytochrome c and is responsible for its release.

Cells respond differently to the engagement of FAS and its ligand. Type I cells (e.g., SKW cells) after formation of DISC recruit caspase 8 in substantial amounts, the initiator caspase activates the effector caspases 3, and the critical caspase substrates are degraded leading to the manifestations of apoptosis (DNA degradation, PARP degradation). This pathway is not inhibited by Bcl-2. Type II cells (e.g., Furkat cells) have slow and low DISC formation and only small amounts of caspase 8 are recruited. Inac-

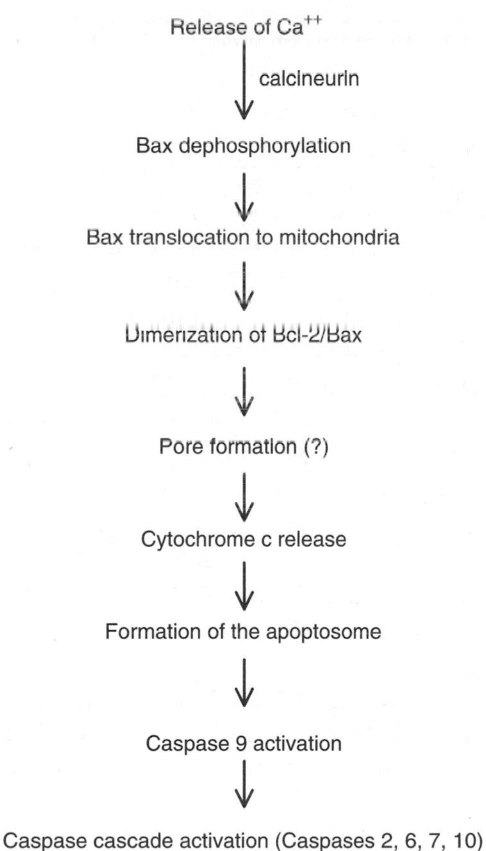


**FIGURE 7-4** Upstream and downstream steps to apoptosis in the cytochrome c release pathway.

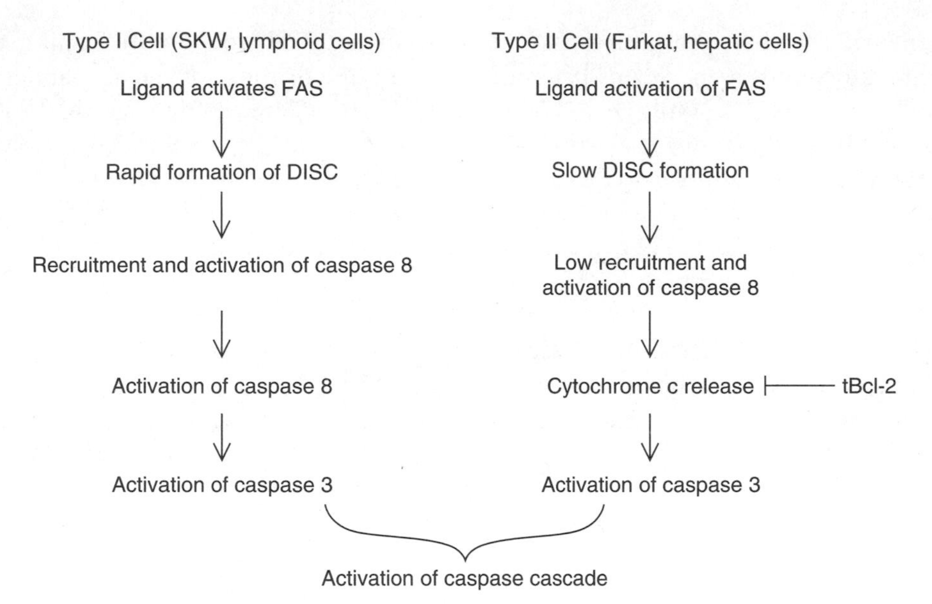

**FIGURE 7-5** Different pathways of caspase cascade activation depending on the circumstances.

tivation of the Bcl-2 leads to cytochrome c release that binds to Apaf-1 and recruits caspase 9 or another initiator caspase, and thereby amplifies the activation of caspase 3 and the unleashing of the caspase cascade (30). Strasser and colleagues show that Bcl-2 provides almost no protection to B-lymphocytes and thymocytes after engagement of the FAS receptor, and suggest that Bcl-2 may regulate apoptotic pathways different from those obtained in lymphocyte apoptosis (31). In contrast, the in vivo administration of anti-FAS antibody to mice kills them because of massive apoptosis of the hepatic cells. However, if the hepatocytes carry a Bcl-2 transgene in vivo, they largely resist FAS-mediated apoptosis (32–34). Therefore, lymphoid cells, including T cells, belong to the type I cells while the hepatocyte is likely to be a type II cell (**FIGURE 7-5**).

Luo and colleagues investigated the mechanism of cytochrome c release by the Bcl-2 by Bid, a 22 kDa cytosolic compound, and demonstrated that Bid acts as an intermediate between the release of caspase 8 and mitochondria (35). After caspase 8 release, Bid transduces the signal to the mitochondria.

Bid, a member of the "BH3 domain-only" subfamily, exists in an inactive form in the cytosol. It is activated by caspase 8 at aspartic acid 59. After losing its COOH terminal, the truncated Bid (tBid) translocates from cytosol to mitochondria. The translocation is sufficient for cytochrome c release. Truncated Bid is much more efficient than Bax in securing cytochrome c release.[11] Moreover, tBid causes cytochrome c release without the outer membrane bursting and matrix swelling, events that have been postulated to cause cytochrome c release and, thus, Bid is not likely to cause a loss of mitochondrial potential.

The BH3 domain is required for tBid activity; it may be that the truncated Bid heterodimerizes with Bcl-2 or Bax. However, mutants of tBid that bind to mitochondria (protein-protein or protein-lipid binding) do not induce cytochrome c release,

[11]Bax releases 20% of cytochrome c even at high concentrations. Truncated Bid releases 100% of cytochrome c at concentrations 500 times lower.

suggesting that the action of tBid requires binding of the BH3 domain in mitochondria (34, 35).

Yin and associates constructed Bid null mice (Bid –/–). They found that:

- Activation of FAS does not kill Bid –/– mice but wild-type mice die quickly after FAS activation.
- There is massive apoptosis of the hepatocytes in the wild-type mice, but little apoptosis of the hepatocytes in the Bid –/– mice.
- There is no difference between wild-type and Bid –/– in the FAS-mediated level of apoptosis in embryonal fibroblast and thymocytes.

These studies confirm the existence of two types of cells with respect to their capacity to respond to FAS ligand activation, and that thymocytes (and embryonal fibroblasts) are type I cells while hepatocytes fall in the type II category (36). This pathway can be constructed as shown in **FIGURE 7-6**.

One cannot fail to notice the analogy between the type II cell pathway and the traditional pathway in *C. elegans*, except of course for the involvement of cytochrome c.

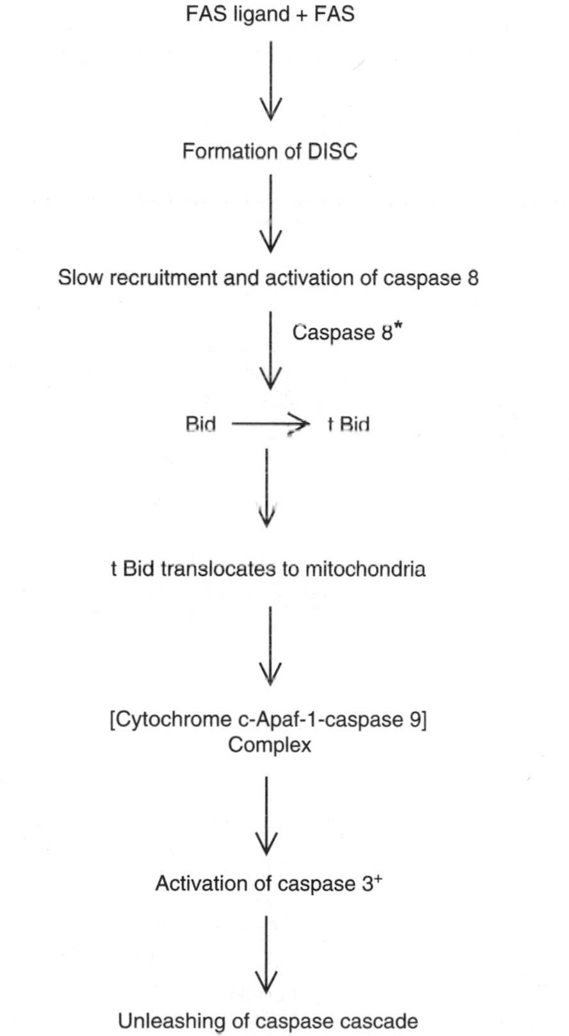


**FIGURE 7-6** The role of Bid in cytochrome c release. (*initiator caspase; +effector caspase)

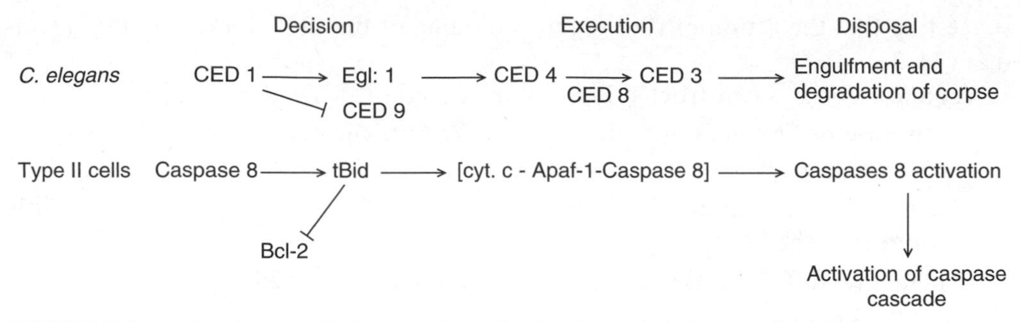


**FIGURE 7-7** Comparison between *C. elegans* and type II cell pathways for programmed cell death.

In *C. elegans* at least 15 genes have been identified that are involved in programmed cell death. The products of some of these genes are involved in deciding whether the cell is to die while the product of other genes determine the execution of the cells (**FIGURE 7-7**) (reviewed in 37).

In conclusion, there are two major death pathways one primarily controlled by death receptors the other by the release of cytochrome c. The initiator caspase is caspase 8 or 10 in the first and caspase 9 in the second. The two pathways are interconnected in type II cells where the activation of FAS causes activation of caspase 8, which in turn activates the release of cytochrome c, and the formation of the cytochrome c–Apaf-1–caspase 9 complex after cleaving its substrate Bid to tBid (**FIGURE 7-8**).

Tang et al. showed that caspase 8 activation is not absolutely required for securing cytochrome c release and formation of the [cytochrome c–Apaf-1–caspase 9] multimeric

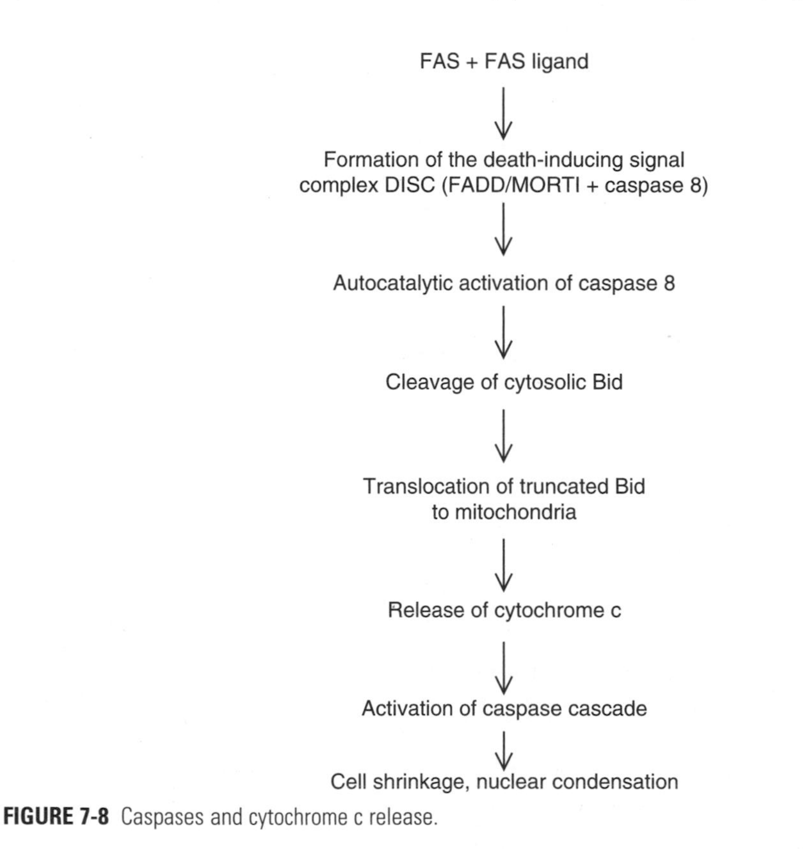


**FIGURE 7-8** Caspases and cytochrome c release.

complex (38). However, when caspase 8 is available and its substrate Bid is cleaved, the apoptotic pathway is amplified and the cell's demise is greatly accelerated (38). Thus, there is clear evidence for crosstalk between different mammalian pathways for apoptosis and in some cases the pathway may secure a quick rather than a slow cell death.

Cohen and associates show that cytochrome c release does not require a reduction of the inner membrane potential and the release precedes the mitochondrial morphologic events associated with apoptosis. Their study design used a combination of microscopic examination, flow cytometric measurements of $\Delta\Psi m$, and Western blot detection of cytochrome c in human monocytic cells in culture (THP.1 cells) (8).

Vander Heyden had already shown that the release of cytochrome c is associated with the appearance of discontinuities in the outer membrane of the mitochondria, but not with a loss of $\Delta\Psi m$ (20). Such discontinuities may provide a route for cytochrome c redistribution, a route that is not necessarily associated with mitochondrial swelling (20).

Mitochondrial hypercondensation is a common finding during cell injury. It is transient; after the return to a normal appearance the mitochondria swell in the THP.1 cell. Hypercondensation is the earliest morphologic change in apoptosis of THP.1 cells; it is not transient, is always associated with a reduction in $\Delta\Psi m$, and is preceded by the release of cytochrome c. The sequence was demonstrated with the use of a caspase inhibitor (Z-VAD.Ink) that prevents the morphological changes but not the cytochrome c redistribution. The morphologic changes in the mitochondria are followed by changes in the cytoskeleton, in particular, the formation of cytoplasmic inclusions containing actin and alterations of the Golgi apparatus. These changes are also prevented by caspase inhibitors (39). Dinsdale et al. conclude that release of cytochrome c is not caused by a change in mitochondrial permeability transition, loss of inner membrane potential, or rupture of the mitochondrial membrane, because the release of cytochrome occurs upstream of these events in the apoptotic sequence (39).

## Role of Glutathione in Cytochrome c Release

Ghibelli and associates argue that glutathione may play a role in cytochrome c release in at least two of the mechanisms proposed (40):

- The MPTP opening, because it is controlled by the mitochondrial redox potential.
- The translocation of proapoptotic proteins Bid or Bax, because they require homodimerization dependent on the oxidation of SH groups to yield disulfide bonds.

The cycling between reduced and oxidized glutathione is a most important cellular antioxidant mechanism. Apoptosis is associated with a decrease of the levels of reduced glutathione in many types of cells and overexpression of Bcl-2 prevents both reduced glutathione decrease and apoptosis (41–43). Zoratti and Szabo showed that: (1) a decrease in reduced glutathione either by inhibiting neosynthesis with puromycin or by causing its extrusion with BSO always leads to the release of cytochrome c; and (2) cytochrome c release can be transient and not associated with apoptosis in cells in which reduced glutathione is depleted (44). These findings coupled with the observation that cytochrome c can be released without causing apoptosis led to the conclusion that redox imbalances are responsible for the cytochrome c release. The latter may be transient and not associated with apoptosis, but it is associated with apoptosis when the apoptotic stimuli are overwhelming. The findings imply that the cytochrome c release is not the point of no return in the process of apoptosis. However, it should be noted that the role of glutathione in apoptosis is complex. Indeed, activation of caspases requires high levels and a high total glutathione/reduced glutathione ratio for activation (45).

### Role of HSP 27 in the Cytochrome c Pathway

Chaperones protect against many forms of cellular stress including oxidative stress. Sustained oxidative stress causes apoptosis. Investigators have focused their studies on the role of the small HSP 27 chaperones in apoptosis caused by the engagement of the Fas ligand, staurosporine, or the epipodophyllotoxin etoposide (VP 16). Overexpression of HSP 27 inhibits the activation of caspase 9 and consequently that of the caspase cascade and apoptosis. However, in the sequence of events leading to apoptosis induced by etoposide, while overexpression of Bcl-2 present in the mitochondria prevents the release of cytochrome c, that of HSP 27 does not. In contrast, using a cell-free system, Garrido and colleagues showed that, while the addition of Bcl-2 to the system was unable to prevent the activation of caspase 9 and that of caspase 3, HSP 27[12] inhibited the maturation of caspase 9, although it did not prevent the release of cytochrome c from the mitochondria (46). The exact molecular mechanism by which HSP 27 prevents the maturation of caspase 9 and/or the formation of the apoptosome is not yet clear, however, these findings may be the reason for the poor prognosis of tumors overexpressing HSP 27 and the role of HSP 27 in drug resistance (46).

In conclusion, if our knowledge of the mechanism of release of cytochrome c is still incomplete and sometimes confusing, there is no doubt that cytochrome c is released in some forms of apoptosis namely in the cytochrome c–dependent pathway. Cytochrome c release is blocked by overexpression of antiapoptotic and stimulated by proapoptotic proteins. The cytochrome c–dependent pathway can also be modulated by different mechanisms and at different sites by other molecules known (e.g., glutathione and HSP 27) or unknown. Its engagement in the formation of the apoptosome (multimeric [Apaf-1, caspase 9 and cytochrome c] complex) may well constitute the point of no return in apoptosis. If so, it is critical to understand the mechanism of cytochrome c release. True, the release of cytochrome c is an early event in the cytochrome c–dependent pathway to apoptosis. However, transient cytochrome c release occurs in the absence of apoptosis. Also, the release of cytochrome c can be a secondary event in other forms of apoptosis. At least two membrane mechanisms have been proposed for primary cytochrome c release: the activation of existing channels and the formation of new channels. In a third pathway, caspase 8 acts upstream of cytochrome c. It is crucial to find out by which molecular gymnastics cytochrome c contributes to cell survival by participating in the electron transport chain, and to cell death by formation of the apoptosome (47, 48).

## The Apoptosome

Genetic investigations on programmed cell death in *C. elegans* reveal that at least three proteins—CED-3, CED-4, and CED-9—are needed to effect apoptosis. Multimerization of the complex accelerates the process. The notion that a multimeric structure that includes an inhibitor of apoptosis, an adapter protein, and a procaspase, functions to kill the cell led to the concept of a specialized cellular structure responsible for the process: the apoptosome. It cannot be ruled out, however, that in nematodes the apop-

[12]Hsp is a low molecular weight chaperone. Chaperones are known to protect against many forms of cellular stress including oxidative stress.

tosome harbors proteins other than CED-3, -4, and -9, and it has been suggested that Egl-1, a proapoptotic $BH_3$-only protein, may be a participant (49).

Each molecule involved in programmed cell death in *C. elegans* has a mammalian homologue. Moreover, in the cytochrome c–dependent pathway, a trimeric complex (cytochrome c–Apaf-1-procaspase 9) is formed. Zou and colleagues reconstructed de novo the caspase 9 activation pathway, using recombinant Apaf-1 and procaspase 9 and demonstrated that the caspase 9 took place in three steps (3):

- dATP binds to Apaf-1 through a consensus nucleotide-binding domain.
- Cytochrome c released in the cytosol binds to Apaf-1 in an unstable manner in absence of dATP, but Apaf-1 multimerization is mediated by dATP hydrolysis.
- The multimeric complex recruits procaspase 9, which is activated by autocatalysis and transcatalysis. Neither cytochrome c nor dATP are required in this step.

Thus, oligomerization of Apaf-1 in the presence of cytochrome c requires dATP hydrolysis. But the recruitment and activation of procaspase 9 that follows oligomerization do not require dATP hydrolysis.

Several studies appropriately argue that the multimeric complex is a functional apoptosome for at least two reasons (3, 9)

1. The assemblage of the complex requires dATP hydrolysis.
2. The formation of the complex, by recruiting multiple procaspase 9 molecules and bringing them in close proximity, facilitates their catalysis.

Cohen and collaborators have shown that two apoptosomes of different sizes are formed (one of ~1.4 M Da and one of ~700 kDa) in cell lysates activated by dATP. The smaller apoptosome (~700 kDa) does not appear in reconstruction studies with recombinant Apaf-1 and caspase 9 in the presence of cytochrome c. Moreover, in lysates the smaller apoptosome is formed rapidly and it efficiently processes caspase 9. In contrast, the ~14 M Da complex is practically inactive in cell lysates. Cain et al. concluded that the ~700 kDa is most likely the functionally active apoptosome in the dying cell (**FIGURE 7-9**) (8).

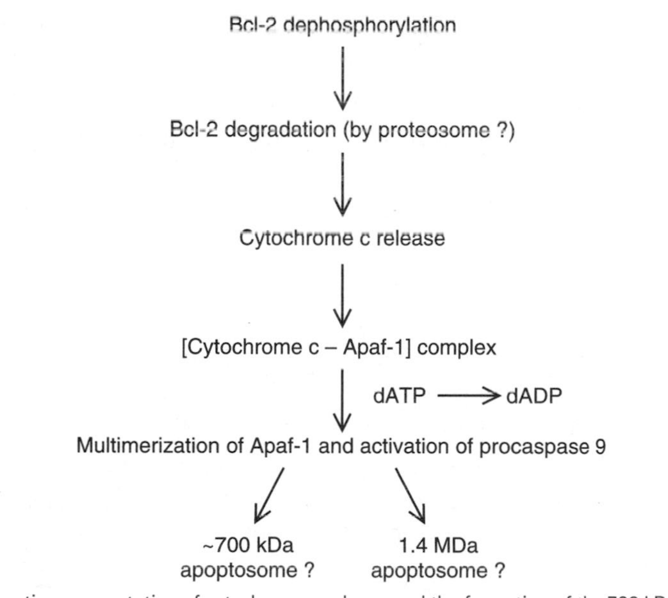


**FIGURE 7-9** Schematic representation of cytochrome c release and the formation of the 700 kDa and 1.4 MDa complexes.

Although Bcl-2 and $Bclx_L$ can prevent the formation of an active apoptosome by securing the retention of cytochrome c in the mitochondrial membrane, they cannot inhibit procaspase 9 activation once the apoptosome is formed (47). Bcl-2 and $Bclx_L$ are membrane proteins, while Apaf-1 is cytosolic. Because the intramembrane and the cytosolic proteins cannot colocalize, the cytochrome c released by the death stimulus is free to complex with Apaf-1 to generate a multimeric apoptosome (48). Thus, Bcl-2 and $Bclx_L$ act in preventing apoptosis upstream of cytochrome c release and the formation of the apoptosome mostly by facilitating the release of cytochrome c rather than by unleashing Apaf-1 from sequestration. After stimulation of human endothelial cells by TNF-α, Bcl-2 is dephosphorylated and degraded by the ubiquitin proteosomal pathway, thereby leaving proapoptotic agents to trigger the emergence of the active apoptosome. Thus, there may be a critical link between the function of the proteosome and that of the apoptosome (50). Refer to the recent review of Jiang and Wang, and the references therein, for further information about the role of cytochrome c in apoptosis (51).

## References and Recommended Reading


1. Liu, X., Kim, C.N., Yang, J., Jemmerson, R., and Wang, X. Induction of apoptotic program in cell-free extracts: requirement for dATP and cytochrome C. *Cell.* 1996;86:147–57.
2. Leoni, L.M., Chao, Q., Cottam, H.B., Genini, D., Rosenbach, M., Carrera, C.J., Budihardjo, I., Wang, X., and Carson, D.A. Induction of an apoptotic program in cell-free extracts by 2-chloro-2′-deoxyadenosine 5′-triphosphate and cytochrome c. *Proc Natl Acad Sci USA.* 1998;95:9567–71.
3. Zou, H., Henzel, W.J., Liu, X., Lutschg, A., and Wang, X. Apaf-1, a human protein homologue to *C. elegans* CED-4, participates in cytochrome c–dependent activation of caspase-3. *Cell.* 1997;90:405–13.
4. Hu, Y., Ding, L., Spencer, D.M., and Nunez, G. WD-40 repeat region regulates Apaf-1 self-association and procaspase-9 activation. *J Biol Chem.* 1998;273:33489–94.
5. Cecconi, F., Alvarez-Bolado, G., Meyer, B.I., Roth, K.A., and Gruss, P. Apaf 1 (CED-4 homolog) regulates programmed cell death in mammalian development. *Cell.* 1998;94:727–37.
6. Yoshida, H., Kong, Y.Y., Yoshida, R., Elia, A.J., Hakem, A., Hakem, R., Penninger, J.M., and Mak, T.W. Apaf 1 is required for mitochondrial pathways of apoptosis and brain development. *Cell.* 1998;94:739–50.
7. Bertin, J., Nir, W.J., Fischer, C.M., Tayber, O.V., Errada, P.R., Grant, J.R., Keilty, J.J., Gosselin, M.L., Robison, K.E., Wong, G.H., Glucksmann, M.A., and DiStefano, P.S. Human CARD4 protein is a novel CED-4/Apaf-1 cell death family member that activates NF-kappaB. *J Biol Chem.* 1999;274:12955–8.
8. Cain, K., Bratton, S.B., Langlais, C., Walker, G., Brown, D.G., Sun, X.M., and Cohen, G.M. Apaf-1 oligomerizes into biologically active approximately 700-kDa and inactive approximately 1.4-Mda apoptosome complexes. *J Biol Chem.* 2000;275:6067–70.

9. Saleh, A., Srinivasula, S.M., Acharya, S., Fishel, R., and Alnemri, E.S. Cytochrome c and dATP-mediated oligomerization of Apaf-1 is a prerequisite for procaspase-9 activation. *J Biol Chem.* 1999;274:17941–5.
10. Jiang, X., and Wang, X. Cytochrome c promotes caspase-9 activation by inducing nucleotide binding to Apaf-1. *J Biol Chem.* 2000;275:31199–203.
11. Wang, X. The expanding role of mitochondria in apoptosis. (Review). *Genes Dev.* 2001;15:2922–33.
12. Hofmann, K., Bucher, P., and Tschopp, J. The CARD domain: a new apoptotic signaling motif. *Trends Biochem Sci.* 1997;22:155–6.
13. Chou, J.J., Matsuo, H., Duan, H., and Wagner, G. Solution structure of the RAIDD CARD and model for CARD/CARD interaction in caspase-2 and caspase-9 recruitment. *Cell.* 1998;94:171–80.
14. Cain, K., Brown, D.G., Langlais, C., and Cohen, G.M. Caspase activation involves the formation of the aposome, a large (approximately 700 kDa) caspase-activating complex. *J Biol Chem.* 1999;274:22686–92.
15. Bernardi, P., Colonna, R., Costantini, P., Eriksson, O., Fontaine, E., Ichas, F., Massari, S., Nicolli, A., Petronelli, V., and Scorrano, L. The mitochondrial permeability transition. (Review). *Biofactors.* 1998;8:273–81.
16. Crompton, M. The mitochondrial permeability transition pore and its role in cell death. *Biochem J.* 1999;341:233–49.
17. Muchmore, S.W., Sattler, M., Liang, H., Meadows, R.P., Harlan, J.E., Yoon, H.S., Nettesheim, D., Chang, B.S., Thompson, C.B., Wong, S.L., Ng, S.L., and Fesik, S.W. X-ray and NMR structure of human Bcl-xL, an inhibitor of programmed cell death. *Nature.* 1996;381:335–41.
18. Minn, A.J., Velez, P., Schendel, S.L., Liang, H., Muchmore, S.W., Fesik, S.W., Fill, M., and Thompson, C.B. Bcl-x(L) forms an ion channel in synthetic lipid membranes. *Nature.* 1997;385:353–7.
19. Schendel, S.L., Xie, Z., Montal, M.O., Matsuyama, S., Montal, M., and Reed, J.C. Channel formation by antiapoptotic protein Bcl-2. *Proc Natl Acad Sci USA.* 1997;94:5113–8.
20. Vander Heiden, M.G., Chandel, N.S., Williamson, E.K., Schumacker, P.T., and Thompson, C.B. Bcl-xL regulates the membrane potential and volume homeostasis of mitochondria. *Cell.* 1997;91:627–37.
21. Hockenbery, D., Nunez, G., Milliman, C., Schreiber, R.D., and Korsmeyer, S.J. Bcl-2 is an inner mitochondrial membrane protein that blocks programmed cell death. *Nature.* 1990;348:334–6.
22. Green, D., and Kroemer, G. The central executioners of apoptosis: caspases or mitochondria? *Trends Cell Biol.* 1998;8:267–71.
23. Oltvai, Z.N., Milliman, C.L., and Korsmeyer, S.J. Bcl-2 heterodimerizes in vivo with a conserved homolog, Bax, that accelerates programmed cell death. *Cell.* 1993;74:609–19.
24. Xiang, J., Chao, D.T., and Korsmeyer, S.J. BAX-induced cell death may not require interleukin 1 beta-converting enzyme-like proteases. *Proc Natl Acad Sci USA.* 1996;93:14559–63.

25. Gross, A., Jocket, J., Wei, M.C., and Korsmeyer, S.J. Enforced dimerization of BAX results in its translocation, mitochondrial dysfunction and apoptosis. *EMBO J.* 1998;17:3878–85.

26. Eskes, R., Antonsson, B., Osen-Sand, A., Montessuit, S., Richter, C., Sadoul, R., Mazzei, G., Nichols, A., and Martinou, J.C. Bax-induced cytochrome c release from mitochondria is independent of the permeability transition pore but highly dependent on $Mg^{2+}$ ions. *J Cell Biol.* 1998;143:217–24.

27. Finucane, D., Bossy-Wetzel, E., Waterhouse, N.J., Cotter, T.G., and Green, D.R. Bax-induced caspase activation and apoptosis via cytochrome c release from mitochondria is inhibitable by Bcl-xL. *J Biol Chem.* 1999;274:2225–33.

28. McDonnell, J.M., Fushman, D., Milliman, C.L., Korsmeyer, S.J., and Cowburn, D. Solution structure of the proapoptotic molecule BID: a structural basis for apoptotic agonists and antagonists. *Cell.* 1999;96:625–34.

29. Marzo, I., Brenner, C., Zamzami, N., Susin, S.A., Beutner, G., Brdicska, D., Remy, R., Xie, Z.H., Reed, J.C., and Kroemer, G. The permeability transition pore complex: a target for apoptosis regulation by caspases and bcl-2-related proteins. *J Exp Med.* 1998;187:1261–71.

30. Scaffidi, C., Fulda, S., Srinivasan, A., Friesen, C., Li, F., Tomaselli, K.J., Debatin, K.M., Krammer, P.H., and Peter, M.E. Two CD95 (APO-1/Fas) signaling pathways. *EMBO J.* 1998;17:1675–87.

31. Strasser, A., Harris, A.W., Huang, D.C., Krammer, P.H., and Cory, S. Bcl-2 and Fas/APO-1 regulate distinct pathways to lymphocyte apoptosis. *EMBO J.* 1995;14: 6136–47.

32. Rodriguez, I., Matsuura, K., Khatib, K., Reed, J.C., Nagata, S., and Vassalli, P. A bcl-2 transgene expressed in hepatocytes protects mice from fulminant liver destruction but not from rapid death induced by anti-Fas antibody injection. *J Exp Med.* 1996;183:1031–6.

33. Lacronique, V., Mignon, A., Fabre, M., Viollet, B., Rouquet, N., Molina, T., Porteu, A., Henrion, A., Bouscary, D., Varlet, P., Joulin, V., and Kahn, A. Bcl-2 protects from lethal hepatic apoptosis induced by an anti-Fas antibody in mice. *Nat Med.* 1996;2:80–6.

34. Li, H., Zhu, H., Xu, C.J., and Yuan, J. Cleavage of BID by caspase 8 mediates the mitochondrial damage in the Fas pathway of apoptosis. *Cell.* 1998;94:491–501.

35. Luo, X., Budihardjo, I., Zou, H., Slaughter, C., and Wang, X. Bid, a Bcl2 interacting protein, mediates cytochrome c release from mitochondria in response to activation of cell surface death receptors. *Cell.* 1998;94:481–90.

36. Yin, X.M., Wang, K., Gross, A., Zhao, Y., Zinkel, S, Klocke, B., Roth, K.A., and Korsmeyer, S.J. Bid-deficient mice are resistant to Fas-induced hepatocellular apoptosis. *Nature.* 1999;400:886–91.

37. Liu, Q.A., and Hengartner, M.O. The molecular mechanism of programmed cell death in *C. elegans. Ann NY Acad Sci.* 1999;887:92–104.

38. Tang, D., Lahti, J., and Kidd, V. Caspase-8 activation and bid cleavage contribute to MCF7 cellular execution in a caspase-3-dependent manner during staurosporine-mediated apoptosis. *J Biol Chem.* 2000;275:9303–7.

39. Dinsdale, D., Lee, J.C., Dewson, G., Cohen, G.M., and Peter, M.E. Intermediate filaments control the intracellular distribution of caspases during apoptosis. *Am J Pathol.* 2004;164:395–407.

40. Ghibelli, L., Coppola, S., Fanelli, C., Rotilio, G., Civitareale, P., Scovassi, A.I., and Ciriolo, M.R. Glutathione depletion causes cytochrome c release even in the absence of cell commitment to apoptosis. *FASEB J.* 1999;13:2031–6.

41. Merad-Boudia, M., Nicole, A., Santiard-Baron, D., Saille, C., and Caballos-Picot, I. Mitochondrial impairment as an early event in the process of apoptosis induced by glutathione depletion in neuronal cells: relevance to Parkinson's disease. *Biochem Pharmacol.* 1998;56:645–55.

42. Armstrong, J.S., Steinauer, K.K., Hornung, B., Irish, J.M., Lecane, P., Birrell, G.W., Peehl, D.M., and Knox, S.J. Role of glutathione depletion and reactive oxygen species generation in apoptotic signaling in a human B lymphoma cell line. *Cell Death Differ.* 2002;9:252–63.

43. Lee, Y.J., Chen, J.C., Amoscato, A.A., Bennouna, J., Spitz, D.R., Suntharalingam, M., and Rhee, J.G. Protective role of Bcl2 in oxidative stress-induced cell death. *J Cell Sci.* 2001;114:677–84.

44. Zoratti, M., and Szabo, I. The mitochondrial permeability transition. *Biochim Biophys Acta.* 1995;1241:139–76.

45. Musallam, L., Ethier, C., Haddad, P.S., Denizeau, F., and Bilodeau, M. Resistance to Fas-induced apoptosis in hepatocytes: role of GSH depletion by cell isolation and culture. *Am J Physiol Gastrointest Liver Physiol.* 2002;283:G709–18.

46. Garrido, C., Bruey, J.M., Fromentin, A., Hammann, A., Arrigo, A.P., and Solary, E. HSP27 inhibits cytochrome c–dependent activation of procaspase-9. *FASEB J.* 1999;13:2061–70.

47. Green, D.R., and Reed, J.C. Mitochondria and apoptosis. *Science.* 1998;281: 1309–12.

48. Newmeyer, D.D., Bossy-Wetzel, E., Kluck, R.M., Wolf, B.B., Beere, H.M., and Green, D.R. Bcl-xL does not inhibit the function of Apaf-1. *Cell Death Differ.* 2000;7:402–7.

49. Hausmann, G., O'Reilly, L.A., Van Driel, R., Beaumont, J.G., Strasser, A., Adams, J.M., and Huang, D.C. Pro-apoptotic apoptosis protease-activating factor 1 (Apaf-1) has a cytoplasmic localization distinct from Bcl-2 or Bcl-x(L). *J Cell Biol.* 2000;149:623–34.

50. Dimmeler, S., Breitschopf, K., Haendeler, J., and Zeiher, A.M. Dephosphorylation targets Bcl-2 for ubiquitin-dependent degradation: a link between the apoptosome and the proteosome pathway. *J Exp Med.* 1999;189:1815–22.

51. Jiang, X., and Wang, X. Cytochrome C-mediated apoptosis. (Review). *Annu Rev Biochem.* 2004;73:87–106.

CHAPTER

# 8 INHIBITOR OF APOPTOSIS PROTEINS (IAPs)

Programmed cell death is used as a defense mechanism against viral invasion (1). Perhaps this is why insect and mammal viruses (such as, cowpox) produce inhibitor of apoptosis proteins (IAPs) in self defense (2). IAPs constitute a large family of highly conserved proteins that suppress apoptosis. They were first discovered in Baculoviridae, which generates two classes of apoptosis inhibiting proteins:

1. P35 is encoded in the genome of *Autographa californica*, a general inhibitor of caspase (3, 4).
2. P35 is cleaved by caspases, and a stable fragment complexes with the enzyme and blocks its activity.
3. An apoptosis inhibitory protein (AIP) able to rescue mutant virus that lost P35 activity was first discovered in baculovirus (5). Such IAPs block apoptosis in insect cells infected with *Autographa alifornica* nuclear polyhydrosis virus. The still expanding populations of IAPs were later found in baculovirus, insects, birds, and mammals (mouse, pig, and human). The acronyms of some species are listed in **TABLE 8-1**.

Although IAP's structure varies, each member of the family is characterized by:

1. Two to three tandems of ~70 amino acid conserved repeats referred to as baculovirus IAP repeats (BIRs). BIRs are located in the $NH_2$ terminus and the central region of the molecule.
2. A ring finger domain in the COOH terminus.
3. Hoffman and colleagues have identified a caspase recruitment domain (CARD) in c-IAP1 and c-IAP2 (6).

The domain most unique to IAPs is the conserved BIR domain. The number of BIRs per IAP molecule may vary. The presence of this conserved sequence raises a number of questions. Is BIR central to the inhibition of apoptosis? Is more than one BIR necessary to inhibit apoptosis? Are the BIR domains needed for inhibition of apoptosis by IAPs? Do BIRs have other functions in the cell's life and death cycle? This chapter addresses these questions, but not necessarily separately.

**Table 8-1 Partial list of inhibitor of apoptosis proteins (IAPs) acronyms**

| Species | Acronym |
|---|---|
| Baculovirus | CP-IAP |
| | Op-IAP |
| *Drosophila* | DAIP 1 |
| | DAIP 2 |
| Chicken | ITA |
| Mouse | MINA |
| Human | c-IAP 1 |
| | c-IAP2 |
| | NIAP |
| | XIAP (hILP) |
| | Apollon |
| | Survivin |

## P35 and IAPs in Baculovirus

Caspases exist in the form of proenzymes that require cleavage for activity. P35 possibly prevents caspase activation in vivo downstream of the procaspase cleavage step by interfering with caspase maturation (7). After infection of insect (*Sporoptera frugiperda*) cells with the *Autographica alifornica* multiple nuclear polyhydrosis virus, all of the cells die. However, viral mutants defective in P35 fail to kill all of the cells, and these surviving cells can be cloned. These cells continue to harbor virus. Moreover, persistent infection can be established by transfecting the *P35* gene in the host genome or by inserting the *IAP* gene in the viral genome. These findings establish the existence of two classes of antiapoptotic genes (*P35* and *IAP* genes) in the baculovirus genome (8).

Most baculoviruses, among them the *Cydia pomonella* granulovirus and the *Oryga pseudotsugata* multinucleocapsid nucleopolyhydrovirus, elaborate IAPs (e.g., Cp-IAP and Op-IAP) that possess two baculovirus IAP repeat motifs (BIRs) and a ring finger. They are infective in cell lines derived from insects, such as the SF21 cells, and are able to rescue P35-deficient virus (*Autographa alifornica* multinucleocapsid nucleopolyhidrovirus or AC-MNPV). However, the *Trichoplusia ni* granulovirus IAP is unable to rescue P35-deficient AC-MNPV despite the fact that the protein (~35 kDa) also contains two BIR motifs and a ring finger, and shares 21% to 27% similarity and 28% to 53% identity in its amino acid sequence with that of several other baculovirus IAPs, including Cp-IAP and OP-IAP (9). In conclusion, although most baculovirus IAPs can rescue P35-deficient mutant, the *Trichoplusia ni* granulovirus fails to do so. It may be that the effectiveness of IAPs in rescuing the defective virus depends on the cell type.

## Human and Other Mammalian IAPs

The first discovered human IAP is the neuronal apoptosis inhibitor protein (NAIP), which plays a role in spinal muscular atrophy (SMA). SMA is an inherited autosomal-recessive disease with variable severity. It is the most frequent fatal neuromuscular disease in infants and children. This disease causes muscle weakness affecting the legs more than the arms and the muscles of the trunk, and the upper motor and sensory neurons are spared (10). In all patients, the disease is associated with a defect in the chromosome 5q11.3-13.1 region. This region includes two genes that are suspected to be at the source of SMA: the survival motor neuron (*SMN*) gene and the *NAIP* gene. This chapter considers the *NAIP* gene only; it is a 70 kB gene encoding a 1232 amino acid protein. Two segments of the gene (exons 5-7 and 4-11) include sequences that encode polypeptides with amino acid sequences with substantial homology (33%) over 180 or 189 amino acids with the IAPs of the *Cydia pomonella* granulosis virus and the *Oryga pseudotsugata* nuclear polyhydrosis virus.

The role of the *NAIP* gene in the pathogenesis of SMA remains unsettled. However, the first two codons of the gene that encodes NAIP are deleted in 70% of patients with the most severe form of SMA (Werdnig-Hoffman disease). There is no doubt that apoptosis plays a critical role in neurogenesis of the embryo and, therefore, it cannot be excluded that SMA may result from the persistence or reoccurrence of apoptosis in the affected neurons (11).

Since NAIP inhibits caspases 3 and 7, it has been suggested that IAPs could help to treat SMA and possibly other neurological conditions resulting from apoptosis (12). Also, attempts to overexpress NAIP using an adenoviral vector for the gene might assist the recovery of hippocampal neurons of the forebrain ischemia. Thus, antiapoptotic therapy might be of assistance someday in acute and chronic degenerative diseases of the nervous system.

Since the discovery of IAPs in Baculoviridae and of NAIP in humans, several other IAPs have been identified and their mode of action investigated.[1] In humans, they include c-IAP1, c-IAP2, and XIAP. All three proteins have two to three BIR domains, a ring finger and, in the case of c-IAP1 and c-IAP2, a CARD motif. Sun and associates investigated the MNR structure of a region incorporating the BIR-2 of XIAP (also hILP) (13). The region includes three-stranded antiparallel β sheets and four α helices, a structure that resembles the shape of a classical ring finger. Also of significance is the sequence that links BIR-1 to BIR-2, the region essential for the inhibition of caspase 3 (13). All three IAPs are potent inhibitors of apoptosis. When the process is put in gear by triggers that activate caspase 9 or 8, which then activate the effector, caspase 3, unleashing the caspase cascade, each of the IAPs can block apoptosis either upstream of caspase 3 or by direct interaction with caspase 3. CIAP1, c-IAP2, and XIAP block the release of cytochrome c after inactivation of the initiator caspase (caspase 9) and thereby prevent activation of caspase 3. In the case of caspase 8-induced apoptosis, the IAPs inhibit caspase 3 directly and thereby prevent the unleashing of the caspase cascade (**FIGURE 8-1**) (14).

Duckett and associates compared human IAPs to $Bclx_L$ with respect to their ability to intervene with one of the several steps that lead to apoptosis (15). Overexpression of the IAPs did not prevent the release of cytochrome c triggered by a variety of stimuli. In contrast, apoptosis directly induced by microinjection of cytochrome c was inhibited by IAPs and not by $Bclx_L$. Their study concluded that, while Bclx protects against

[1]CIAP was identified because the protein is recruited by TNFR2 through TRAF2 in a linkage that involves the BIR domain.

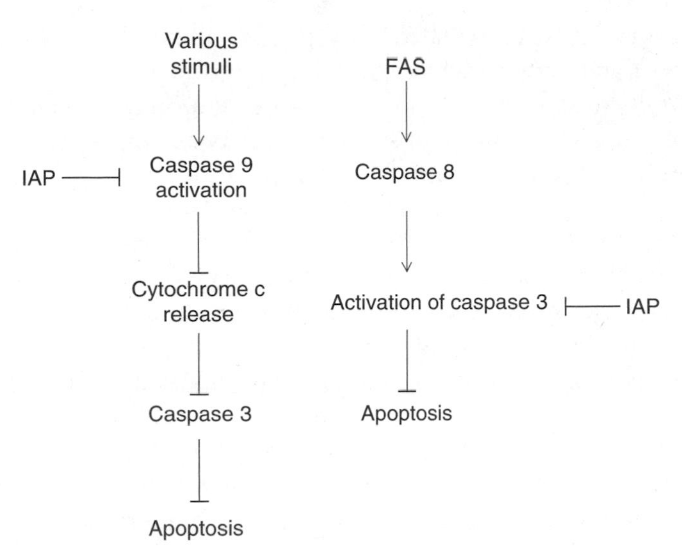

**FIGURE 8-1** Sites of action of inhibitor of apoptosis proteins (IAPs) in the apoptotic pathway.

cytochrome c's release, IAPs function downstream of cytochrome c (possibly by inhibiting caspase 3; **TABLE 8-2**).

## Other IAPs: Apollon and Survivin

*Apollon* is a human IAP gene that encodes a large protein (530 kDa) that contains only one BIR domain and a ubiquitin-conjugating enzyme domain. Apollon was found to be expressed in brain and ovarian carcinoma where it causes resistance to chemotherapy. The brain cancer cell lines SMB 78, which express high levels of the IAP, are resistant to chemotherapeutic agents because of inhibition of apoptosis, thereby facilitating tumor growth (16).

Survivin, the smallest member of the IAP family, was discovered by Ambrosini and associates through hybridization of a human genomic library (17). While most mammalian IAPs have two to three BIR domains and a ring finger domain, survivin contains only a single BIR, no ring finger domain, and a coiled-coil domain in the COOH terminal sequence. In contrast to c-AIP1 and c-IAP2, it possesses no caspase recruitment domain (CARD), which is required for interaction with TRAFs. Like other IAPs, despite having only a single BIR, survivin is an inhibitor of apoptosis; it inhibits the processing of caspases 3 and 7 and actively binds to them (18). However, survivin is less effective an inhibitor of apoptosis than other human IAPs, for example, XIAP. While cotransfection of cell lines with survivin protected only 65% of the cells from apoptosis induced by FAS, cotransfection with XIAP leads to 93% of survival. Protection of

**Table 8-2 Comparison of the effects of $Bclx_L$ and hIAPs on molecules involved in the apoptotic pathway**

| | Cytochrome c Release | Caspase Activation | Site of Control |
|---|---|---|---|
| $Bclx_L$ | Blocked | 0 | Mitochondrial integrity |
| HIAPs | No effect | Blocked | Caspase 3 (?) |

apoptosis with survivin was associated with specific inhibition of, and binding to, the effector caspases 3 and 7, thus excluding inhibition of the initiator caspase 8.

The mode of expression of survivin sheds some light on its function as a regulator of apoptosis. In mitosis, expression of survivin persists during the entire process starting at the beginning of prophase and disappearing at the end of telophase. Survivin is associated to the microtubules of the mitotic spindle. Thus, survivin is expressed during mitosis in a cell cycle-dependent manner (19). In mice survivin is up-regulated in $G_2$/M phase (20). These observations suggest that survivin may contribute to the regulation of the cell cycle as well as to that of apoptosis, and raise concern about the exact role of proteins containing a single BIR domain.

*C. elegans* contains a single BIR domain protein that is not involved in apoptosis. Yeast, the enzymic repertoire of which contains no caspases, also harbors a gene encoding a single BIR protein. If these proteins do not contribute to inhibiting apoptosis, what is their role in cell metabolism and how is that role relevant to the function of survivin? Disruption of the interaction of survivin with microtubules causes a loss of survivin antiapoptotic function and leads to activation of caspase 3 followed by cell death during mitosis. Overexpression of survivin in cancer cells allows the cells to bypass the disruption and avoid apoptosis (19). Indeed, survivin is up-regulated in cancer. Because inhibition of apoptosis is critical to cancer prognosis, it is not surprising that the worst clinical stages in cancer patients are associated with high levels of expression of Bcl-2 and survivin (21).

Conway and associates characterized a mouse survivin gene and its complementary DNA (22). The gene is located on chromosome 11E2 and, like the human gene, it contains four exons. The gene encodes three different survivins with various lengths: survivin 140, survivin 121, and survivin 40. Survivin 140 resembles the human survivin and contains a BIR domain and a COOH terminal coiled-coil domain. Survivin 121 lacks the coiled-coil domain and survivin 40 lacks both the BIR and the COOH coiled-coil domain. Only survivin 140 blocks apoptosis. The significance of this study resides in the existence of several isoforms of survivins (23). Is it possible that one or more play different roles in the life of the cell or are the survivins 121 and 40 simply artifacts? The pattern of expressions of survivin during the cell cycle clearly suggests that, in addition to causing inhibition of apoptosis, it also participates in the regulation of mitosis.

It was hoped that investigations of the mode of action of BIR domains in *C. elegans* and yeast might help to better understand the function(s) of survivin. Fraser and colleagues identified and characterized the only two BIR-containing proteins in *C. elegans*: BIR-1 and BIR-2 (23).

While BIR-2 is only detectable in embryos and adults, BIR-1 is highly expressed during embryogenesis. Overexpressed BIR-1 cannot arrest apoptosis in developing embryos. Moreover, the incidence of cell death is not increased in embryos in which BIR's expression is suppressed. However, cells of the embryos are unable to complete cytokinesis, but cytokinesis, at least in part, can be preserved by transgenic expression of survivin in embryos in which BIRs expression is suppressed. These findings led Fraser et al. to suggest that in addition to regulating apoptosis, BIRs regulate cytoskeletal changes, particularly during cytokinesis (23).

*Saccharomyces cerevisiae* and *Schizosaccharmyces pombe* each encode a single protein containing two BIR domains, referred to as BIR-1 and bir 1, respectively. The mitotic cycle in mutants of BIR-1 and bir 1 is blocked in the metaphase/anaphase transition because the mitotic spindle is unable to elongate (24). Li and associates found that the sequence of the *S. cerevisiae* genome contains a single open reading frame that encodes a protein 95 amino acids long with two BIR domains (spanning residues 12 to 116 and 145-240, respectively), named *Bir-1* and *Bir-2* (25). Their study found that:

- The protein presents substantial homology with *C. elegans* (BIR-2 and BIR-1) and with human survivin.
- Inactivation of BIR-1 causes dysregulation of the cell's ploidy, morphology, and growth rate, suggesting that BIR-1 contributes to the control of cell division.
- To determine which segment of the sequence of the *C. elegans* BIR-1 contributed to or was necessary for function, the human homologue survivin was chosen. Yeast cells were transformed with three forms of survivin: the full length (wild-type) and two truncated forms: the C survivin devoid of BIR domain and the N survivin containing the BIR domain. Only N survivin, which contains the BIR domain, markedly increased the time required for the cell population to double (248 minutes for the N survivin compared to 163 minutes for the parent vector), a finding indicating that the survivin BIR domain inhibits mitosis in a dominant negative manner.

In summary, this study concluded that ancient IAPs containing a single BIR domain function mainly to control several aspects of cell division rather than preventing apoptosis. Survivin is likely to have evolved (from yeast to mammals) into a protein that links both functions: the control of mitosis and that of apoptosis (25).

Anticipating that the discovery of the presence of survivin homologues in mice would help the understanding of the role of the molecule in vivo, Li and Altieri searched a mouse library by hybridization screening (26). They were able to characterize the mouse survivin locus. The gene contains four axons and three introns spanning over 50 kb on the telomeric site of chromosome 11E2. It generates an mRNA of 0.85 kb (compared to 1.9 kb in humans). The transcript is made of 140 amino acids (~16.00 Da). Its sequence is 84% identical to that of human survivin, with a BIR domain and a COOH coiled-coil domain. The sequence of the gene itself is of particular interest. The promotor region, the site for initiation of transcription located 5′ to gene sequences, does not contain a TATA box, but does contain:

- A CpG island of 1 to 2 kb sequences containing a high density of CpG dinucleotides frequently found in animal tissues upstream of the promotor region.
- Several sites for a transcription factor that has a DNA binding domain and contains three zinc fingers (Sp 1).
- Two cell cycle-dependent elements (CDEs; a group of DNA sequences found in centromeres).
- One cycle gene homology region (CHR), which functions as a periodic repressor of gene *cdc-25* that encodes a phosphatase, is itself activated by phosphorylation. CHR is often found in $G_2M$ genes.
- Three transcription start sites located at positions –32, –36, and –40 from the initiating trinucleotide, ATG.
- A minimal promoter region 174 bp upstream from the first ATG.

The transcription activity of the gene was reduced to almost half, or less than half, by mutations of the CED, CHR, and SP1 sites alone or in combination in cells growing asynchronously. Moreover, the synchrony of $G_2$/M synchronized cells was disturbed when the mutated genes were expressed. The study concluded that the coordinate transcription of CED elements, the cell cycle homology region, and Sp7 are needed for survivin expression during the cell cycle. It also suggests that these components may serve as targets for suppression of survivin overexpression in cancer (26).

Otaki and associates discovered another murine homologue of human survivin named TIAP (27). The protein molecule contains one BIR domain, binds with processed caspases 3, and inhibits apoptosis. TIAP is mainly found in the embryo and three growing tissues in the adult: the thymus, testis, and intestine. It is expressed in the S and $G_2$/M phase in a cell cycle-regulated manner. During the $G_2$/M phase, it is associated with the microtubules of the mitotic spindle. Here again is a protein that contains a single BIR domain that blocks apoptosis and is expressed in a cell cycle-regulated manner. The authors also investigated the 5′ flanking region of the *TIAP* gene, finding one CHR and nine CDEs upstream of the ATG codon of the mouse *TIAP* survivin gene. Again, mutations of CED and the CHR as well as that of enhancer elements caused dysregulated expression of the survivor gene (27). These findings further emphasize the importance of understanding the expression of the various components of the 5′ flanking region of the survivin gene and, in particular, that of the CED/CH3 element.

A close relationship exists between the gene believed to encode the receptor of the Xa blood coagulation factor and the gene that encodes the effector cell protease receptor (EPR-1). A gene cluster on chromosome 17q25 includes two distinct sequences topographically separated by 75 to 130 kb, namely that of EPR-1 and that of survivin. The cluster generates two mRNAs: (1) a 1.3kb mRNA for EPR-1 found in fetal tissues (kidney and liver) and many adult tissues (skeletal muscle, heart, pancreas, spleen, lymph nodes, and hematopoietic cells); and (2) that for survivin (1.9 kb) not found in adult tissue. Induction of EPR-1 suppresses endogenous expression of survivin (determined by immunoblotting with an antisurvivin antibody) and causes massive apoptosis as well as inhibition of cell growth in HeLa cells. Clarification of EPR-1/survivin expression is needed to better understand the mode of regulation of survivin and its contribution to the pathogenesis of neurodystrophies and cancer.

## Mitosis Process

Before discussing the mode of action of survivin during mitosis further, in particular chromosomal segregation, a brief review of some of the pertinent elements of the process is in order. After completion of mitosis, the cell may either enter the $G_1$ cell cycle or exit for a time in $G_0$ to later return to $G_1$ or differentiate and die. During this interphase, chromosomes cannot be identified by light or electron microscopy. This is not because they are entirely disorganized, as shown by in situ hybridization using several different probes for specific sequence separated by known distances. However, the distance separating the clones identified on the in situ hybridization preparation does not reflect the distance that separates these clones identified on linear DNA; indeed, after in situ hybridization some of the clones appear to be closer to each other while others are separated. The differences between measurements made in situ and those made on linear DNA can best be explained by assuming that, during interphase, the 30-nm chromatin fiber forms on either side of a scaffold to which loops of variable sizes (one to four million bases) are attached. During the S phase, each DNA strand is replicated, each chromosome is duplicated, and each of the daughter strands of a new chromosome is a chromatid. Thus, in diploid cells approximately at the end of $G_2$, the number of chromosomes is doubled (4N). During metaphase the equatorial plate is formed. Each sister chromatid is attached at the center of the spindle, which is a specialized chromosomal region called the centromere that is itself connected to the spindle through a multiprotein structure called the kinetochore, which is assembled to the centrosome. During anaphase, the sister chromatids separate and one of each pair moves at the opposite pole prior to cell division. This separation and migration of sister chromatids must be tightly regulated to secure equal distribution of chromosomes in the daughter cells.

The anaphase promoting complex (APC) is a protein complex that functions as an E3 enzyme (ubiquitin ligase). The APC causes several proteins involved in the regulation of mitosis to be ubiquinated.

A multiple protein complex includes cohesins and holds sister chromatids together. Cohesin(s) are located at the site of the centromere and at several other sites on the chromosome. The function of cohesin is regulated by the anaphase inhibitor, which is a target for APC ubiquination and proteosomal proteolysis. After chromosomal duplication, the anaphase inhibitor and another protein cooperate to secure the association of cohesin with the daughter chromatid at the appropriate sites, thus maintaining mother and daughter chromatids together in the equatorial plate. At the onset of anaphase, APC degrades the anaphase inhibitor and each chromatid moves at the opposite pole through the impulse of the combined forces of the kinetochore and the spindle.

The chromosomal segregation ultimately secures the distribution of the information stored in its genome. Speliotes and associates investigated the role of BIR-1 and an aurora serine tyrosine kinase in dividing cells (28). Antibodies were prepared against the BIR-1 of *C. elegans*, and their association with BIR-1 was detected by Western blot analysis. Using RNA-mediated gene interference, a band approximately 8 kDa was found that disappeared when BIR-1 was deleted from *C. elegans* embryos. This band was clearly associated with dividing cells (highest in embryos and lowest in larvae). BIR-1 was further localized using a combination of appropriate antibodies aimed at the spindle midzone, microtubules, BIR-1, and DNA. Little BIR-1 was found in differentiating cells, but BIR-1 was clearly detectable in dividing cells where BIR-1 staining overlapped with DNA rather than the kinetochore during prophase and anaphase. In contrast, BIR-1 staining was associated with both chromosomes and spindle midzone during anaphase, and disappeared from DNA, but was still found in the spindle midzone during telophase. The BIR-1 localization on chromosomes and the spindle midzone suggests that BIR-1 functions in some way during chromosome segregation in mitosis and meiosis.

Speliotes et al. also showed that in the absence of BIR-1, the morphology of meiosis in germ cells and diploid cells was abnormal (28). Chromosomal behavior and spindle organizations were disrupted. Since the movement of the spindle, and the function of the kinetochore and that of other structures or molecules involved in chromosomal segregation and cytokinesis require multiple phosphorylations and dephosphorylations, the investigators sought to identify a potential kinase involved in the process. Their study found that embryos and fertilized oocytes defective in AIR-2 (a member of the aurora family of serine threonine kinases) show defects similar to embryos and fertilized oocytes defective in BIR-1. Moreover, in the absence of BIR-1, AIR-2 is not detected in chromosomes. Because both histone H3 and the kinetochore are required for chromosomal condensation, alignment on the spindle, and bipolar segregation, appropriate stains were used to determine the state of histone 3 phosphorylation and that of the kinetochore in BIR-1– and AIR-2–defective fertilized oocytes. Histone H3 and the marker for the kinetochore stained poorly or not at all in BIR-1– and

AIR-2–defective fertilized oocytes, a finding that again suggests a close functional association between BIR-1 and AIR-2. The similarities between BIR survivin functions and BIR-1 were also studied; transgenic BIR-1 defective embryos were constructed and it was shown that survivin partially rescues the BIR-1 defective embryos.

The study thus establishes that BIR-1 plays a critical role in chromosomal function (alignment segregation and condensation in the latter phases of the cell cycle and cytokinesis) (28). Is it possible, as proposed by the authors, that BIR-1's function is to localize AIR-2 at the appropriate sites during these processes? In any event, the rescue of BIR-1– and AIR-2–defective embryos indicates that the function of BIR-1 and survivin are conserved (28) and, therefore, it cannot be excluded that mutations of survivin might lead to disturbances in chromosomal segregation and cytokinesis, as they are often seen in cancer cells. Such notions are in keeping with the demonstration that survivin is associated with several elements involved in chromosomal alignment, segregation (microtubules, centrosome), and cytokinesis by connecting to molecules that join the daughter cells at the end of telophase.

Survivin thus appears to have two distinct functions: one that involves the spindle (chromosomal segregation, cytokinesis, etc.) and another that regulates apoptosis. Since survivin inhibits apoptosis in the $G_0$ cell cycle, the first of these functions must be separated from the latter (29, 30). Indeed, the first of these functions but not the latter exist in *C. elegans*, and both functions exist in mice and humans. Thus, the spindle-associated functions of BIR proteins are conserved through evolution.

## Structure of Survivin

The structure of survivin is significant because of the many intermolecular contacts the molecule needs to make in performing its functions. These include: inhibit caspases, bind to tubules, and bind to IAP inhibitors. The first of these are clearly relevant to the maintenance of homeostasis and genomic integrity. Despite the diligent work ongoing in many laboratories, the connections between the structure and functions of survivin are still incomplete.

Some of this work is briefly discussed here. Chantalat and coworkers used a human survivin (31) and Muchmore et al. used a mouse survivin (32). Some of their findings are related to:

- The N terminal sequence and the BIR domains.
- The C terminal α helix outside the BIR domain.
- Survivin dimerization.

Mouse survivin contains an N terminal sequence that includes a single BIR domain (residues 10–80) and an extended COOH terminal forming a α helix. The overall amino acid sequence of mouse survivin is 84% identical to that of human survivin. The sequences that differ are dispersed throughout the mouse and human molecules, but remain located at conserved sites in each of the survivins.

The crystal structure reveals the BIR motif that includes a zinc finger fold containing three antiparallel β sheets surrounded by four short α helices. Four highly conserved amino acids, three cysteines (Cys 57, Cys 60, and Cys 84), and one histidine (Hin 79) chelate a single Zinc motif. The integrity of the $Zn^{2+}$ coordination sphere is vital to the BIR's inhibition of apoptosis, as revealed by the replacement of cysteine 84 by alanine, which increases caspase 3 activity in $G_2$/M synchronized cultures (31). The side chain of arginine 18 forms hydrogen bonds with alanine 39 and leucine 12, thereby stabilizing helices 1 and 2 in the structure. A tight turn occurs between the end of helix 2 and the start of the first β strand; lycine 42 is located at the junction.

The carboxy terminal that follows the BIR domain (residues Leu 87 to Lys 120) constitutes a helix with mixed amphiphatic, hydrophobic, and hydrophylic nature. The C terminal helix interacts with the rest of the survivin molecule by hydrophobic interactions and hydrogen bonding.

In addition to the hydrophobic bonds that interact between the C terminal α chain outside the BIR domain, the side chain of arginine 108 forms a hydrogen bond with phenylalanine 58. The remaining segment of the C terminal amino acid sequence is hydrophylic and was shown to be needed for the binding of survivin to microtubules. Truncated human survivin mutants (minus the last 42 amino acids) are unable to localize to microtubules (31).

## Functional Survivin Is a Dimer

A dimer structure is supported by both studies in solution and crystal analysis. Both the human and the mouse survivin monomers possess a globular N terminal zinc finger domain and a coiled coil C terminal united by a short sequence, the linker region. At first it was assumed that the dimerization interphasing involved the coiled coil C terminal. Survivin lacking part of the C terminal α helices still dimerize, indicating that it is the N terminal segment of the molecule that is involved in the dimer's interphase. In human survivin, three components of the monomer are involved in dimerization: the N terminal zinc finger–methionine 1–serine 88, the linker v 89-T 97, and the N terminus of helix α 1. The long C terminal helix extends from lysine 98 to aspartic 142 (31). Seventy percent of the amino acids located at the monomer's interface are hydrophobic and monopolar. Phenylalanine 93, leucine 98, and phenylalanine 101 are at the core of the hydrophobic contact with their counterpart in the opposite monomer. In addition, the two monomers are held together by hydrogen bonds and van der Waals interactions (31). In the mouse, a second $Zn^{2+}$ ion is found that is most likely involved in the interface between the monomers; it forms symmetric covalent bonds between glutamine 76 and histidine 80. Again, hydrogen bonds and hydrophobic reactions are also part of the interface connection (32).

Thus, the functional survivin is a dimer. Dimerization occurs mainly through the N terminal and the linker region. The core of the dimer extends into two C terminal α helical regions not involved in the dimerization. The dimer assumes a unique shape with an extended loop (sometimes referred to as a bow tie structure) that may favor docking properties on tubules or other molecular complexes, and thereby contribute to the dual role of survivin: inhibition of apoptosis and regulation of cytokinesis.

In conclusion, survivin, a single BIR motif-containing protein, functions as an antiapoptotic agent in HeLa cells stimulated to die by various agents. Targeting of survivin genes causes apoptosis in proliferating cells (19) and, thus, antisense survivin promotes caspase activation and cell death.

Survivin is up-regulated in the $G_2$/M phase of the cell cycle (33). Destruction of microtubules abrogates the ability of survivin to cause apoptosis (34, 35).

At least in *C. elegans* and yeast, interference with the BIR domain causes distortion in embryogenesis and interferes with chromatid separation and cytokinesis. Similarly, survivin may be the prototype of special proteins that affect, possibly at different times and under different circumstances, both apoptosis and cell division. The unique structure of survivin supports its ability to interact with several multiprotein complexes as a docking protein. Finally, overexpression of survivin could play a dual role in cancer by: (1) by blocking apoptosis and thereby maintaining the capacity of the cells to grow; and (2) by accelerating cell division and thereby contributing to the expansion of the cancer cell population.

## Inhibitors of IAPs

For the process of apoptosis to succeed, IAPs, which are the major natural inhibitors of caspases, must be negatively regulated. This can be achieved in various ways, such as reduction of transcription, interference with translation, or specific protein inhibition. The molecular mechanism of IAPs' regulation has been investigated most often in *C. elegans* and *Drosophila*, and to a much lesser extent in mammalian cells. Although regulation of transcription and translation is not excluded, degradation by ubiquitination seems to play the key role.

In the absence of an apoptotic stimulus in *Drosophila*, DIAP ubiquitinates a protein (DRONC) that blocks caspase's activities. In the presence of an apoptotic stimulus, levels of DIAP are controlled by decreasing transcription and DIAP ubiquitination followed by proteolysis. In mammalian cells the most investigated negative regulation focused on that of XIAP and functions by suppression of translation and degradation. At the initiation of the apoptotic signal, translation is rapidly suppressed. Initiation of translation is blocked by modification of the translational initiator (36).

Proteosomal inhibition prevents apoptosis in thymocytes and neurons (37, 38); XIAP contains a ring finger and an E3 ubiquitin ligase activity that permits autoubiquitination, and thereby leads to self-degradation. Of course, it should be pointed out that self-degradation does not exclude the IAPs from other proteins' degradations (e.g., by caspases) (39, 40).

Direct inhibition of IAPs in *Drosophila* has been known. Three different molecules were incriminated: Grim, Reaper, and HID. In 2000, Du et al. reported the discovery of a mitochondrial protein released concomitantly with cytochrome c that activates caspases by IAP inhibition (41). The protein is referred to as Smac, for "Second Mitochondrial-derived Activator of Caspases." Independently but simultaneously, a second group of investigators discovered a mammalian protein that binds and inhibits IAPS, named DIABLO, for direct IAP binding protein with low pI (42). The Smac/DIABLO precursor N terminal possesses an N terminal mitochondrial signal sequence that secures mitochondrial incorporation where it is located in healthy cells. After apoptotic stimulus caused by cellular stress, the protein, now free of its mitochondrial signal sequence, is released in the cytosol where it competes with the effector caspase (caspase 9) as well as with other caspases (caspases 3 and 7) for binding with the IAPs causing their inhibition. The binding involves the BIR repeats with variable affinities.

Chai and colleagues investigated the structural and biochemical basis of apoptotic activation by Smac/DIABLO (43). They studied the crystal structure of Smac/DIABLO at 2.2 Å resolution and found that the molecule forms a 21 kDa homodimer through a large hydrophobic interface. Missense mutations in the sequence that forms the hydrophobic interface interfere with the dimerization. Their study showed that Smac/DIABLO interacts with both BIR-2 and BIR-3 of XIAP, but not with BIR-1, despite the similarities in BIR's structures. Yet, Smac/DIABLO must recognize the same structural features in BIR-2 and BIR-3 because they exclude each other when exposed to Smac/DIABLO. The replacement of the N terminal alanine residue by the more bulky methionine residue disables the interaction with XIAP through structural hindrance. Therefore, it is postulated that the N terminus of Smac/DIABLO fits tightly within the surface groove of the BIR. Chai et al. also identified an N terminal seven amino acid sequence that activates procaspase 3 (NH2-AVPIAQK-COOH). On the basis of this elegant research, Chai et al. (43) and Verhagen and Vaux (reviewed in 44) propose a tentative model for the interaction of Smac/DIABLO and BIR-2. Li and colleagues have further shown that Smac/DIABLO is unable to inhibit mature caspase 9; inhibition occurs only after maturation (41). The interaction takes place between the N terminal tetrapeptide of Smac/DIABLO (AVPI) and the p12 subunit of human caspase 9 (ATPP). Apparently, caspase 9 uses the same conserved tetrapeptide to react with XIAP and Smac/DIABLO (45).

## Mitochondrial Inhibitors of IAPs

HtrA2 is a member of conserved serine proteases found in bacteria, humans, and other mammals (46). The bacterial enzyme HtrA has been investigated extensively. HtrA has a dual role: that of a chaperone at normal temperature, and that of a serine protease at high temperatures. The enzyme is required for bacterial thermotolerance (47). The mammalian protein, a homologue of the bacterial DegP/HtrA, was first found in the nucleus and the endoplasmic reticulum. Later it was obvious that the newly synthesized protein ultimately locates in the mitochondrion, where it is found in the intermembrane space, where it is released upon apoptotic stimuli (48). When synthesized, HtrA2 has a 49 kDa protein that matures in the mitochondrion to yield a 37 kDa product. The mechanism of the proteolysis is still unknown; autocatalysis is not excluded. The molecule is made of 458 amino acids; it possesses a transmembrane segment, a trypsinlike catalytic domain, and a PDZ domain located in the carboxy terminal region. The N terminal starts with a tetrapeptide (AVPS), the composition of which is reminiscent of that of four first amino acids of the N terminal of Smac/DIABLO (AVPI) (44). Such a finding suggests that, like Smac/DIABLO, HrtA2 might bind IAPs. This observation would explain why overexpression of HrtA2 in culture cells leads them to die and that the cause is associated with increased caspase activity. Yang and associates investigated the interaction between IAPs and HtrA2 (49). It was known that the active molecular complex is a homotrimer with PDZ domains covering the catalytic domain of the molecule (49), but Yang and colleagues demonstrated that HtrA2 binds c-IAP, cleaves it, inhibits caspase ubiquitination, and causes apoptosis. The cleavage is restricted to those IAPs that are potent caspase inhibitors (e.g., c IAP 1 and 2, XIAP, and DIAP). Survivin is not cleaved by HrtA2 probably because the protein's main function concerns cytokinesis rather than apoptosis. In contrast to the binding of Smac/DIABLO with IAP, which is stoichiometric and therefore can be reversed by excess IAPs, the cleavage of c-IAP by HrtA2 is irreversible.

In conclusion, two inhibitors of IAPs have been discovered and investigated: Smac/DIABLO and Omi/HrtA2. Both are mitochondrial proteins released under cellular stress at the same time as cytochrome c and both bind IAPS. But there are important differences with respect to organic distribution function and control of transcription. Smac/DIABLO is found only in heart, liver, kidney, and testis; in contrast, HrtA2 is ubiquitous. Thus, unless other antiAIP agents are discovered, HrtA2 is the only generalized AIP inhibitor available. The two inhibitors may function simultaneously or in sequence in the organs that contain Smac/DIABLO. The interaction between

Smac/DIABLO and the IAPs is stoichiometric and reversible by excess IAPs; the c-IAP cleavage by HrtA2 is irreversible. It has been suggested that HrtA2 transcription is under the control of P53. If confirmed, this would provide a mechanism for p53-induced apoptosis (reviewed in 50, 51).

Independent of its importance in understanding the control of apoptosis, the significance of this research resides in the therapeutic potential of IAP inhibitors, especially for cancer cells that are apoptosis resistant.

## References


1. Shen, Y., and Shenk, T.E. Viruses and apoptosis. *Curr Opin Genet Dev.* 1995;5:105–11.
2. Deveraux, Q.L., Stennicke, H.R., Salvesen, G.S., and Reed, J.C. Endogenous inhibitors of caspases. *J Clin Immunol.* 1999;19:388–98.
3. Bump, N.J., Hackett, M., Hugunin, M., Seshagiri, S., Brady, K., Chen, P., Ferenz, C., Franklin, S., Ghayur, T., Li, P., et al. Inhibition of ICE family proteases by baculovirus antiapoptotic protein p35. *Science.* 1995;269:1885–8.
4. Salvesen, G.S., and Dixit, V.M. Caspases: intracellular signaling by proteolysis. (Review). *Cell.* 1997;91:443–6.
5. Crook, N.E., Clem, R.J., and Miller, L.K. An apoptosis-inhibiting baculovirus gene with a zinc finger-like motif. *J Virol.* 1993;67:2168–74.
6. Hofmann, K., Bucher, P., and Tschopp, J. The CARD domain: a new apoptotic signaling motif. *Trends Biochem Sci.* 1997;22:155–6.
7. LaCount, D.J., Hanson, S.F., Schneider, C.L., and Friesen, P.D. Caspase inhibitor P35 and inhibitor of apoptosis Op-IAP block in vivo proteolytic activation of an effector caspase at different steps. *J Biol Chem.* 2000;275:15657–64.
8. Lee, J.C., Chen, H.H., and Chao, Y.C. Persistent baculovirus infection results from deletion of the apoptotic suppressor gene p35. *J Virol.* 1998;72:9157–65.
9. Bideshi, D.K., Anwar, A.T., and Federici, B.A. A baculovirus anti-apoptosis gene homolog of the *Trichoplusia ni* granulovirus. *Virus Genes.* 1999;19:95–101.
10. Panozzo, C., Frugier, T., Cifuentes-Diaz, C., and Melki, J. Spinal muscular atrophy. In: Scriver, C.R., Beaudet, A.L., Sly, W.S., Valle, D., Childs, B., Kinzler, K.W., and Vogelstein, B. *The Metabolic and Molecular Bases of Inherited Disease,* 8th ed., vol 14. New York, NY: McGraw-Hill; 2001; 5834–43.
11. Roy, N., Mahadevan, M.S., McLean, M., Shutler, G., Yaraghi, Z., Farahani, R., Baird, S., Besner-Johnston, A., Lefebvre, C., Kang, X., et al. The gene for neuronal apoptosis inhibitory protein is partially deleted in individuals with spinal muscular atrophy. *Cell.* 1995;80:167–78.
12. Robertson, G.S., Crocker, S.J., Nicholson, D.W., and Schulz, J.B. Neuroprotection by the inhibition of apoptosis. *Brain Pathol.* 2000;10:283–92.
13. Sun, C., Cai, M., Gunasekera, A.H., Meadows, R.P., Wang, H., Chen, J., Zhang, H., Wu, W., Xu, N., Ng, S.C., and Fesik, S.W. NMR structure and mutagenesis of the inhibitor-of-apoptosis protein XIAP. *Nature.* 1999;401:818–22.
14. Deveraux, Q.L., Roy, N., Stennicke, H.R., Van Arsdale, T., Zhou, Q., Srinivasula, S.M., Alnemri, E.S., Salvesen, G.S., and Reed, J.C. IAPs block apoptotic events

induced by caspase-8 and cytochrome c by direct inhibition of distinct caspases. *EMBO J.* 1998;17:2215–23.

15. Duckett C.S., Li, F., Wang, Y., Tomaselli, K.J., Thompson, C.B., and Armstrong, R.C., Human IAP-like protein regulates programmed cell death downstream of Bcl-xL and cytochrome c. *Mol Cell Biol.* 1998;18:608–15.
16. Chen, Z., Naito, M., Hori, S., Mashima, T., Yamori, T., and Tsuruo, T. A human IAP-family gene, apollon, expressed in human brain cancer cells. *Biochem Biophys Res Commun.* 1999;264:847–54.
17. Ambrosini, G., Adida, C., and Altieri, D.C. A novel anti-apoptosis gene, survivin, expressed in cancer and lymphoma. *Nat Med.* 1997;3:917–21.
18. Tamm, I., Wang, Y., Sausville, E., Scudiero, D.A., Vigna, N., Oltersdorf, T., and Reed, J.C. IAP-family protein survivin inhibits caspase activity and apoptosis induced by Fas (CD95), Bax, caspases, and anticancer drugs. *Cancer Res.* 1998;58:5315–20.
19. Li, F., Ambrosini, G., Chu, E.Y., Plescia, J., Tognin, S., Marchisio, P.C., and Altieri, D.C. Control of apoptosis and mitotic spindle checkpoint by survivin. *Nature.* 1998;396:580–4.
20. Kobayashi, K., Hatano, M., Otaki, M., Ogasawara, T., and Tokuhisa, T. Expression of a murine homologue of the inhibitor of apoptosis protein is related to cell proliferation. *Proc Natl Acad Sci USA.* 1999;96:1457–62.
21. Tanaka, K., Iwamoto, S., Gon, G., Nohara, T., Iwamoto, M., and Tanigawa, N. Expression of survivin and its relationship to loss of apoptosis in breast carcinomas. *Clin Cancer Res.* 2000;6:127–34.
22. Conway, E.M., Pollefeyt, S., Cornelissen, J., DeBaere, I., Steiner-Mosonyi, M., Ong, K., Baens, M., Collen, D., and Schuh, A.C. Three differentially expressed surviving cDNA variants encode proteins with distinct antiapoptotic functions. *Blood.* 2000;95:1435–42.
23. Fraser, A.G., James, C., Evan, G.I., and Hengartner, M.O. *Caenorhabditis elegans* inhibitor of apoptosis protein (IAP) homologue BIR-1 plays a conserved role in cytokinesis. *Curr Biol.* 1999;9:292–301.
24. Uren, A.G., Beilharz, T., O'Connell, M.J., Bugg, S.J., van Driel, R., Vaux, D.L., and Lithgow, T. Role for yeast inhibitor of apoptosis (IAP)-like proteins in cell division. *Proc Natl Acad Sci USA* 1999;96:10170–5.
25. Li, F., Flanary, P.L., Altieri, D.C., and Dohlman, H.G. Cell division regulation by BIR1, a member of the inhibitor of apoptosis family in yeast. *J Biol Chem.* 2000;275:6707–11.
26. Li, F., and Altieri, D.C. The cancer antiapoptosis mouse surviving genes: characterization of locus and transcriptional requirements of basal and cell cycle-dependent expression. *Cancer Res.* 1999;59:3143–51.
27. Otaki, M., Ogasawara, T., Kuriyama, T., and Tokuhisa, T. Cell cycle-dependent regulation of TIAP/m-survivin expression. *Biochem Biophys Acta.* 2000;1493:188–94.
28. Speliotes, E.K., Uren, A., Vaux, D., and Horvitz, H.R. The survivin-like *C. elegans* BIR-1 protein acts with the aurora-like kinase AIR-2 to affect chromosomes and the spindle midzone. *Mol Cell.* 2000;6:211–23.
29. Deveraux, Q.L., and Reed, J.C. IAP family proteins—suppressors of apoptosis. *Genes Dev.* 1999;13:239–52.

30. Li, F., Ambrosini, G., Chu, E.Y., Plescia, J., Tognin, S., Marchisio, P.C., and Altieri, D.C. Control of apoptosis and mitotic spindle checkpoint by survivin. *Nature.* 1998;396:580–4.

31. Chantalat, L., Skoufias, D.A., Kleman, J.P., Jung, B., Dideberg, O., and Margolis, R.L. Crystal structure of human survivin reveals a bow tie-shaped dimer with two unusual α-helical extensions. *Mol Cell.* 2000;6:183–189.

32. Muchmore, S.W., Chen, J., Jakob, C., Zakula, D., Matayoshi, E.D., Wu, W., Zhang, H., Li, F., Ng, S-C., and Altieri, D.C. Crystal structure and mutagenic analysis of the inhibitor-of-apoptosis protein survivin. *Mol Cell.* 2000;6:173–82.

33. Beardmore, V.A., Ahonen, L.J., Gorbsky, G.J., and Kallio, M.J.. Survivin dynamics increases at centromeres during G2/M phase transition and is regulated by microtubule-attachment and Aurora B kinase activity. *J Cell Sci.* 2004;117(Pt 18):4033–42.

34. Carvalho, A., Carmena, M., Sambade, C., Earnshaw, W.C., and Wheatley, S.P. Survivin is required for stable checkpoint activation in taxol-treated HeLa cells. *J Cell Sci.* 2003;116(Pt 14):2987–98.

35. Mollinedo, F., and Gagate, C. Microtubules, microtubule-interfering agents and apoptosis. (Review). *Apoptosis.* 2003;8:413–50.

36. Clemens, M.J., Bushell, M., Jeffrey, I.W., Pain, V.W., and Morley, S.J. Translation initiation factor modifications and the regulation of protein synthesis in apoptotic cells. (Review). *Cell Death Differ.* 2000;7:603–15.

37. Sadoul, R., Fernandez, P.A., Ouiquerez, A.L., Martinou, I., Maki, M., Schroter, M., Becherer, J.D., Irmler, M., Tschopp, J., and Martinou, J.C. Involvement of the proteosome in the programmed cell death of NGF-deprived sympathetic neurons. *EMBO J.* 1996;15:3845–52.

38. Yang, Y., Fang, S., Jensen, J.P., Weissman, A.M., and Ashwell, J.D. Ubiquitin protein ligase activity of IAP's and their degradation in proteasomes in response to apoptotic stimuli. *Science.* 2000;288:874–7.

39. Li, X., Yang, Y., and Ashwell, J.D. TNF-RII and c-IAP1 mediate ubiquination and degradation of TRAF2. *Nature.* 2002;416:345–7.

40. Huang, H., Joazeiro, C.A., Bonfoco, E., Kamada, S., Leverson, J.D., and Hunter, T. The inhibitor of apoptosis, cIAP2, functions as a ubiquitin-protein ligase and promotes in vitro monoubiquitination of caspases 3 and 7. *J Biol Chem.* 2000; 275:26661–4.

41. Du, C., Fang, M., Li, Y., Li, L., and Wang, X. Smac, a mitochondrial protein that promotes cytochrome c-dependent caspase activation by eliminating IAP inhibition. *Cell.* 2000;102:33–42.

42. Verhagen, A.M., Ekert, P.G., Pakusch, M., Silke, J., Connolly, L.M., Reid, G.E., Moritz, R.L., Simpson, R.J., and Vaux, D.L. Identification of DIABLO, a mammalian protein that promotes apoptosis by binding to and antagonizing IAP proteins. *Cell.* 2000;102:43–53.

43. Chai, J., Du, C., Wu, J.W., Kyin, S., Wang, X., and Shi, Y. Structural and biochemical basis of apoptotic activation by Smac/DIABLO. *Nature.* 2000;406:855–62.

44. Verhagen, A.M., and Vaux, D.L. Cell death regulation by the mammalian IAP antagonist Diablo/Smac. (Review). *Apoptosis.* 2002;7:163–6.

45. Li, W., Srinivasula, S.M., Chai, J., Li, P., Wu, J.W., Zhang, Z., Alnemri, E.S., and Shi, Y. Structural insights into the proapoptotic function of mitochondrial serine protease HtrA2/Omi. *Nat Struct Biol.* 2002;9:436–41.
46. Hu, S.I., Carozza, M., Klein, M., Nantermet, P., Luk, D., and Crowl, R.M. Human HtrA, an evolutionarily conserved serine protease identified as a differentially expressed gene product in osteoarthritic cartilage. *J Biol Chem.* 1998;273: 34406–12.
47. Spiess, C., Beil, A., and Ehrmann, M. A temperature-dependent switch from chaperone to protease in a widely conserved heat shock protein. *Cell.* 1999;97:339–47.
48. Suzuki, Y., Imai, Y., Nakayama, H., Takahashi, K., Takio, K., and Takahashi, R. A serine protease, HtrA2, is released from the mitochondria and interacts with XIAP, inducing cell death. *Mol Cell.* 2001;3:613 21.
49. Yang, Q.H., Church-Hajduk, R., Ren, J., Newton, M.L., and Du, C. Omi/HtrA2 catalytic cleavage of inhibitor of apoptosis (IAP) irreversibly inactivates IAPs and facilitates caspase activity in apoptosis. *Genes Dev.* 2003;17:1487–96.
50. Vaux, D.L., and Silke, J. Mammalian mitochondrial IAP binding proteins. *Biochem Biophys Res Commun.* 2003;304:499–504.
51. Van Gurp, M., Festijens, N., van Loo, G., Saelens, X., and Vandenabeele, P. Mitochondrial intermembrane proteins in cell death. *Biochem Biophys Res Commun.* 2003;304:487–97.

PART

II

# Integrity of the Genome

CHAPTER

# 9 THE *P53* GENE

Several genes contribute to the maintenance of genomic integrity; some are involved in carcinogenesis (oncogenes or suppressor genes) and others are implicated in apoptosis. In discussing their roles, it seems logical to start with the gene that encodes the wild-type P53 protein, sometimes referred to as the guardian of the genome. However, as is often the case, the function of a single molecule cannot be discussed in isolation. Other molecules also serve to guard the genome, among them the other members of the P53 family, a number of kinases (ATM kinase, ATR), and other proteins (such as the BRCA 1 and 2, products of the genes responsible for the Nijmegen, Bloom, and Fanconi syndromes).

Mutations of the *p53* gene (missense mutations and, less frequently, deletions or truncations) are present in half of human cancers including the most devastating, such as cancers of the breast, lung, and colon. Also, a large percentage of the remaining cancers harbor cellular or viral oncogenes that inhibit functions of wild-type *p53* (1). At first it was believed that cancer was caused by overexpression of the *p53* gene encoding a dominant phenotype associated with a gain in function, as with an oncogene. Thus, it was not apoptosis, but cell proliferation and cancer that generated interest in the P53 protein

Originally, P53 was nothing more than another tumor antigen; the large T antigen of the simian virus (SV40) was found to bind to a protein with a molecular weight of 53 kDa, referred to as the P53 protein. In 1989, Finlay et al. and Friedman et al. demonstrated that wild-type *p53* prevents the transformation induced by oncogenes in cell cultures (2, 3) and arrests tumor cell growth in vitro. Mutant P53 did not have such effects on cells in culture transfected with P53 cDNA derived from tumors. These cells were immortalized and it was later discovered that all transforming P53s were mutants (4). This finding suggests that in cancers associated with overexpression of a mutant *p53* gene, the expression of the wild-type *p53* allele is inhibited. Thus, the mutated gene causes a phenotype to emerge that is similar to the one observed when both alleles are deleted and in which the function of the wild-type *p53* is lost in the phenotype[1] (5, 6). In summary, *p53* is a tumor suppressor gene, a gene which suppresses growth, and not a proto-oncogene that facilitates growth.

[1]The high frequency of *p53* missense mutations in cancer results from their dominant negative effect. The mutant interferes with the function of the wild-type *p53* allele and causes decreased stability of the gene with loss of heterozygosity as well as P53 function.

**Table 9-1 Exons of P53 in the evolutionary conserved domains**

| Domain | Amino Acid | Exon |
|---|---|---|
| 1 | All AA encoded | 1 |
| 2 | 129-146 | 4 |
| 3 | 171-179 | 5 |
| 4 | 234-260 | 7 |
| 5 | 270-287 | 8 |

The *p53* gene maps to chromosome 17 and mutations of the second *p53* allele were found in all cases in which the gene causes cancer. The role of chromosome alterations in colon cancer helps to understand the mode of action of the *p53* gene. Heterozygosity is common in colorectal cancer:

- The most frequent chromosomal deletions (~80%) occur on 18q and 17p.
- The smallest deletion on 17p was found to map to the locus of the *p53* gene.
- In tumors, the non-deleted *p53* allele is often mutated (7).
- Missense mutations are concentrated in four hot spots conserved from amphibians to mammals.
- The offspring of parents who carry a germ line mutation of one *p53* allele, such as in the Li-Fraumeni syndrome, are predisposed to cancers.

These findings led to the conclusion that the wild-type *p53* gene is indeed a tumor suppressor gene.

The human *p53* gene is located on the short arm of chromosome 17 band 13 (17p13); it is 20 kb long and generates a 2.8 kb mRNA transcript. The gene encodes a 53 kDa nuclear phosphoprotein composed of 393 amino acids containing five evolutionary conserved domains from *Xenopus* to humans,[2] which are believed to be essential for normal function of the wild-type P53 protein (**TABLE 9-1**).

## P53 Protein

P53, a nuclear protein with a short half-life, is prominent among those proteins that protect the integrity of the genome. It is scarce under normal conditions, but is stabilized by posttranslational modifications (phosphorylation and acetylation) after exposure to various sources of genotoxic stress. P53 functions as a multifunctional transcription transactivator that modulates the transcription of over 30 genes that regulate the cell cycle. Among them are:

- P21, a cyclin kinase inhibitor.
- MDM2, a regulator of *p53* transcription transactivation.
- GADD45, a protein induced by DNA damage that binds to PCNA and is involved in DNA repair.
- Bax, a member of the Bcl-2 family.
- The insulin-like growth factor binding protein (IGF-BP3) that may inhibit the growth factor's signal for the initiation of mitosis.

[2] P53 is not detected in invertebrates.

There are also complex interactions between the P53 protein and several other gene products, including Rb protein, cyclin kinase inhibitor, ink4a, and others. After briefly discussing the structure of P53, the functions of the molecule are considered.

## P53's Structure

P53 is a tetramer. The human oligomer is composed of 393 amino acids and contains four domains: the transactivation, the core, the regulatory, and the tetraisomerization domains. The tetraisomerization domains also include a nuclear export domain.

### Transcription Transactivation Domain

The transcription transactivation domain (residues 1–73) is located in the acidic N terminal portion of the molecule (residues 1–99). Sequences in this domain bind with TATA associated members (among them TAF II 70 and TAF II 3). The sequence of residues 13-23 of human P53 is conserved and identical in a number of different species. The amino acids F19, L22, and W23 are required for transcription activation and negative regulation (8, 9). The amino terminal sequence also contains several serines and one threonine that can be phosphorylated.

### Core Domain

The central core sequence is a specific DNA binding domain (residues 100–293). Because of the manner in which the molecule is folded, the sequence is protease-resistant. The domain contains a zinc atom that is indispensable for DNA binding (10). Some of the missense mutations are located in the DNA binding domain, and these fall into two categories: those that cause defective contacts (e.g., mutations of R248 and R273) and those that alter the conformation of P53. Forty percent of the latter occur at the site of amino acids R175, G245, R228, R273, and R282 (10, 11). Some of the properties of the core domain have been identified by protein digestion by Pavletich and associates (10). Their findings show that the sequence extending from amino acid 102 to 292 forms a compact structural domain separated from the regulatory C-terminal sequence 311–364. The core domain binds to specific DNA sequences in vitro and contains a zinc atom required for binding and function.

In the mammalian genome, the DNA sequences targeted by P53 are numerous (~300) albeit somewhat variable. The consensus sequence contains two decamers joined head to tail and each decamer is made of two pentameric inverted repeat sequences (**FIGURE 9-1**) (12).

Cho et al. crystallized a segment of P53 (amino acid 94–312) complexed to a 21 base sequence of the *p53* gene containing the pentamer duplex (10). The cocrystal includes one DNA duplex and three core units, only one of which is specifically DNA-bound.

Decamer

[P-Pu-Pu-C(T/a)_(A/t) G-Py-Py-Py]

1 2 3 4 5 5' 4' 3' 2' 1'

Pentamer Pentamer'

ATAATT G GG CA AGTCTA GGAA

ATTAA C CC GT TCAGAT CCTT

**FIGURE 9-1** The consensus sequence contains two decamers joined head to tail, and each decamer is made of two pentameric inverted repeat sequences. (Adapted from Cho, Y., et al. 1994 [10].)

The binding does not seem to alter the overall structure of the core. The core forms two antiparallel β-sheets containing four and five β strands arranged to form a β sandwich. This large β structure is not directly involved in DNA binding, but serves as a scaffold for the structural components that operate at the DNA protein interface. These include three different elements: a loop-sheet helix motif (LSH) and two large loops (L2 and L3). The LSH binds to the major groove of the DNA and the sequence extending from residue 278 to residue 286 forms an α helix that fits into the groove (**TABLE 9-2**). The two large loops interact. The interaction is stabilized by a zinc atom held in place through metal binding ligands that involve three cysteines and one histidine residue (C196, H174, C238, and C292). The interaction of the two large loops (L2 and L3) is such that a critical arginine (R248) is presented to the minor DNA groove at the site of an A/T rich region. The contact between that portion of the P53 core sequence and the DNA minor groove shrinks the latter, and thereby provides a tight pocket for the R248 residue. The interaction between R248 and the minor grove is, however, delicate enough that any alteration in L2 or L3 disturbs it. Arginine 175 and a zinc atom (that is not part of a classical zinc finger motif, but is essential to the P53 structure) contribute to stabilize the interaction between L2 and L3 (see Figure 6A and B of Cho et al. [10]).

The significance of this study resides in part in the location of the mutations associated with cancer that occur in the core domain of P53. The most frequent residues are found near or at the site of the P53-DNA interface, and the less frequent or never mutated residues are generally located far away from that segment of the DNA (Table 9-1).

On the basis of these observations, Cho and associates proposed two mechanisms for the inactivation of *p53* by core mutations: mutations that prevent DNA contact between the P53 binding surface, and the DNA target sequence and mutations that destabilize the folding of the core antigen (10, 13, 14).

Investigation of the P53 structure not only serves to identify the mechanism of action between DNA and P53 and the importance of the mutation sites in the development of cancer but may also prove useful for the design of drugs that would restore P53 conformation and increase its thermodynamic stability. Because the P53 core domain is most highly conserved between species, Wong and colleagues used magnetic nuclear resonance (MNR) spectroscopy to investigate, in solution, the difference in conformations of five common P53 mutants: V143A, G245S, R248Q, and R249S (15). The conformation of the core domains of the mutants were compared to that of the crystallized wild-type P53 core domains and found to be similar. Earlier it was mentioned that arginine 248 is the amino acid that makes critical contact with the minor

**Table 9-2 Relative degrees of contact of the three elements of the core domain with the DNA binding site**

| Core Domain | Degree of Contact |
|---|---|
| L3 | 30% minor groove contact |
| LSH | 25% major groove contact |
| L2 | 167% no direct contact with DNA, but extensive contact with cancer (in %) |

Both LSH and L3 make critical contacts with phosphate groups. The location of P53 mutations is associated with cancer (in %). Data are from Cho et al. (10).

groove of the DNA (10). MNR spectroscopy investigations of the core in solution confirm that arginine 248 is important for the thermodynamic stability of the core domain. The stability is challenged by replacing the arginine by glutamine or serine, mutations that alter the L2-L3 interactions. Thus, Arg 248, in addition to being a contact mutant, also acts as a stabilizer of L2-L3 (15).

Can antibodies detect the so-called "mutation conformations?" Cho and associates considered the binding of P(Ab)240 to P53 mutants. P(Ab)240 binds to mutated P53 but not to wild-type P53. The wild-type P53 epitope (residues 212–217) is located on the β strand S7, which is inaccessible to the antibody because it dwells in the hydrophobic core of the β sandwich; therefore, P(Ab)240 does not bind to wild-type P53. The finding suggests that mutants may not only have undergone a conformation change, but also have become partly unfolded (or denatured) (10). Wong and associates show that the MNR spectra of G245S and R249S mutants in solution are not different from that of wild-type P53. They propose that it should be possible to restore proper folding and DNA contacts using drugs that increase the thermodynamics of the mutant and thus restore the proper conformation of the DNA binding region (15).

These views are supported by experiments of Brachman and associates, who tested the function of P53 in yeast and in BKW cells (16). In yeast they followed P53 function by measuring the expression of a reporter gene[3] (UAS53::URA3). Expression depends on P53 binding upstream of URA 3. This construct not only allows yeast carrying the reporter gene to survive without uracyl, but also sensitizes the cells to 5-fluoroorotic acid. Consequently, cells that harbor a wild-type P53 will be resistant to 5-fluoroorotic acid. In BKH cells, the function of wild-type P53 and mutants were tested by observing morphological manifestations of apoptosis after second site mutations were introduced in the core domain of P53 mutants. The percentage of apoptotic cells in the system is approximately 30% with wild-type P53, but it is less than 10% with the V143A, G245S, and the G249S mutants. However, the level of apoptosis is partly restored by a second site mutation (20% with V143A + N239Y; G245S + T123P; G245S + N239Y). In summary, second site mutations can partially restore the function of certain P53 mutants (among them: V143A, G245S, and 249S). Therefore, it is believed that clarification of the structural changes imposed on the P53 core domain mutant by these second site mutations might help identify small molecules able to restore P53 function in cells harboring P53 mutants.

The work of Joerger and associates (17) provides an elegant demonstration of the molecular performance of an intragenic suppression mutation. The authors solved the crystal structure of the oncogenic P53 mutants R273H, R249S, and H168R, and investigated their effects on the structure and function of the P53 molecule.

The R273H mutation deprives the P53 DNA binding site from an essential arginine. Therefore, it is a classical contact mutant, unable to bind to its normal contact molecule (DNA).

The R249S mutation destabilizes the L3 loop and, thereby, prevents the arginine 248 from making DNA contact. R249S is a classical structural mutant.

The H168R mutation is an intragenic suppressor mutant that rescues the R249S mutation when both mutations occur at the same time by offering arginine as a substitute for that of the R249S mutation in binding to DNA.

[3]A reporter gene is a coding sequence inserted in a vector whose product can easily be assayed, for example, an antibiotic (chloramphenicol). When such a sequence is connected to a promoter, its expression measures the promoter's function.

All three of these mutations differ significantly from the global suppressor rescue P53 mutations that restore P53 thermodynamic stability without affecting its molecular structure.

Wild-type P53 also possesses 3′ to 5′ exonuclease activity (18) associated with the core domain. The exonuclease is part of a multicatalytic function that includes phosphatase, a 3′ to 5′ exonuclease, and an endonuclease. The structure of the catalytic unit resembles that of a similar enzyme described in *Escherichia coli* (19).

## Carboxy Terminal Domain

The basic carboxy terminal domain of P53 extends from amino acids 293 to 393. In addition to regulating the transcription transactivation of the acidic N terminal sequence, it also includes three functional motives whose sequences partially overlap: an oligomerization domain, a nuclear export signaling (NES) motif, and an exonuclease regulatory motif.

The structure of the P53 molecule must accommodate multiple functions: modulation of transcription transactivity, direct participation in DNA repair, nuclear export signaling, and regulation of the core domain. The P53 oligomers tetramerize most likely to secure optimal activity. The tetramer is made of two dimers, each of which is formed by two antiparallel β sheets and two antiparallel α helices (20). The tetramer is thus a dimer of dimers. Relatively small amino acid sequences (326-355) determine the ability of the monomers to dimerize and for the dimers to tetramerize (21). The molecules fold as they tetramerize. Each monomer contains nine hydrophobic residues that are critical for dimerization followed by tetramerization. These residues are conserved in human P53, but not in P73. There is also evidence that tetramerization is influenced by the N terminal domain of P53 (22).

The tetramerization domain also contains a leucine-rich nuclear export sequence (NES) that spans over residues 340-351. Stommel and associates demonstrated that mutations of L344, L348, and L350 interfere with tetramerization (23). The failure to tetramerize is associated not only with reduced transcription activation and interference with arrest of the cell cycle but also with failure of P53 retention in the nucleus. Stommel et al. propose that the NES is structurally shielded in wild-type tetramerized P3, thereby preventing P53 export from nucleus to cytoplasm. In contrast, mutations of L344 and L348, which are leucine residues located within the NES site, prevent tetramerization and facilitate export (23).

The C-terminal regulatory domain of P53 also regulates P53 transcription transactivity. Posttranslation events that affect the C terminal domain all modulate the regulatory function of the C terminal.[4] The mechanism of the allosteric regulatory function of the C terminal domain of P53 results from its binding to a different monomer of the P53 tetramer, thereby preventing the access of the core binding site to DNA. Small peptides with sequences identical to that of residues 369 thru 383 of the C terminal activate DNA binding probably by modulating conformational changes associated with transcription activation (24). Also, a 22-mer peptide sequence (GSRAHSSHLK-SKKGQSTSRHKK) extending from amino acid 361 to 382 in the C terminal domain, and reactivates mutated P53 by a molecular mechanism that as yet is not entirely clarified. In any event, the above and similar studies may help in the design of small pep-

[4]These posttranslation events include: phosphorylation acetylation, binding of the calcium binding protein S100 B, and the Pab antibody that attaches to a C terminal epitope.

tides that restore function to defective P53 mutants (24). Mutations of the C terminus domain are rare, but when they occur, they cause distortions of the P53 function(s). Zhou and associates studied the effect of several P53 mutations that occur in the C terminal domain, among them point mutations and deletions. Although the mutants bind to DNA and stimulate the transcription of P21, Bax, and IGF-BP3, they fail to adequately stimulate apoptosis, which proved to be significantly reduced. Because the mutants were tumor derived, one is led to conclude that their transcription activation of P53 remains associated with growth, but not with the ability to induce apoptosis (25).

We mentioned earlier that the complexity and variety of P53 functions requires that the active molecules be tetramerized in the nucleus. Tetramerization selectively and specifically activates several transcription sites, possibly because the tetrameric structure provides several DNA binding sites. The presence of four P53 binding domains markedly increases the degree of DNA binding induced by interaction between tetramer and DNA: (32° to 36° with 4 subunits and 51° to 57° with the wild-type tetramer). Tetramerization of wild-type P53 further over twists the DNA response element. The modifications in binding and twisting allow the DNA to assume multiple conformations adapted to specific interactions with chromatin. Moreover, the observation suggests that the N and C terminal domains that flank the core-binding domain also may contribute to the interaction between P53 and DNA. The molecular details of the influence of the latter remain speculative, however. For example, the partly exposed N terminus might be accessible to other proteins after the tetramer is bound to DNA and thereby may play a role in downstream signal transduction, or the C terminus could exert an allosteric inhibitory effect on DNA binding.

The C-terminal regulatory domain of P53 also regulates the activity of the P53 core domain exonuclease. Thus, truncation of the P53 C-terminal regulatory domain by 30 amino acids enhances the exonuclease activity (~10-fold) without interfering with the ability of the P53 core domain to bind DNA (26). However, it is still unclear whether the 5′ exonuclease functions in vivo.

The agents listed in **TABLE 9-3** cause base alterations, single or double strand breaks, and crosslinks that, once recognized, need repair because if they remain unrepaired, damaged DNA can be transferred from one generation of cells to the next. Indeed, a variety of genetic alterations—point mutations, deletions, translocations, gene amplifications and chromosomal aneuploidy—are associated with the loss of P53 function (26–30).

**Table 9-3 Amino acids that are hotspots for P53 mutations**

| Amino Acid | Mutations, % | Role in Structure |
|---|---|---|
| Arg 248 | 9.6 | DNA contact |
| Arg 273 | 8.8 | DNA contact |
| Arg 175 | 6.1 | No contact with DNA, but critical role in stabilizing |
| Gly 245 | 6.0 | the P53 binding surface by participating in |
| Arg 249 | 5.6 | electrostatic, Van der Walls, and hydrogen bonding |
| Arg 286 | 4.0 | with side chains or carbonyl groups. |

Wild-type P53 secures the integrity of the genome in two ways, by arresting the cell cycle or triggering apoptosis. The mechanism by which P53 blocks the cell cycle is reasonably well known. P53 transactivates genes that encode cyclin inhibitors (*p21* or *gDD 45*) that cause cell cycle arrest at the $G_1$/S or the $G_2$/M check points. In contrast, by transactivating the *mdm2* gene, P53 unleashes a negative feedback pathway that inhibits its transcription activation capacity. Most likely the arrest of the cycles gives the cell time to repair the DNA damage. However, if repair fails, the cell is killed by apoptosis.

## Summary

Functional P53 is a tetramer, a dimer of dimers. The monomer is composed of four classical domains. The acidic N terminal DNA transactivates transcription. The core DNA binding site is made of a β sandwich that supports a loop-helix-sheet motif and two other loops (L2-L3) brought together by a zinc atom. The LHS and the loops bind to DNA and their sequence harbors most of the cancer-associated mutations. This core sequence also includes a 5′exonuclease. The C-terminal regulatory domain modulates the DNA binding of the core domain and thereby exerts negative feedback on the effect of the transcription factor (by association of expressed MDM2). It also includes the tetramerization domain, a nuclear export sequence, and an exonuclease regulatory function.

## References and Recommended Reading

1. Scheffner, M., Werness, B.A., Huibregtse, J.M., Levine, A.J., and Howley, P.M. The E6 oncoprotein encoded by human papillomavirus types 16 and 18 promotes the degradation of p53. *Cell.* 1990;63:1129–36.
2. Finlay, C.A. Hinds, P.W., and Levine, A.J. The p53 proto-oncogene can act as a suppressor of transformation. *Cell.* 1989;57:1083–93.
3. Friedman, S.L., Shaulian, E., Littlewood, T., Resnitzky, D., and Oren, M. Resistance to p53-mediated growth arrest and apoptosis in Hep 3B hepatoma cells. *Oncogene.* 1997;15:63–70.
4. Chen, P.L., Chen, Y. M., Bookstein, R., and Lee, W.H. Genetic mechanisms of tumor suppression by the human p53 gene. *Science.* 1990;250:1576–80.
5. Michalovitz, D., Halevy, O., and Oren, M. p53 mutations: gains or losses? *J Cell Biochem.* 1991;45:22–9.
6. Hann, B.C., and Lane, D.P. The dominating effect of mutant p53. *Nat Genet.* 1995;9:221–2.
7. Baker, S.J., Fearon, E.R., Nigro, J.M., Hamilton, S.R., Preisinger, A.C., Jessup, J.M., vanTuinen, P., Ledbetter, D.H., Barker, D.F., Nakamura, Y., et al. Chromosome 17 deletions and p53 gene mutations in colorectal carcinomas. *Science.* 1989; 244:217–21.
8. Lin, J., Wu, X., Chen, J., Chang, A., and Levine, A.J. Functions of the p53 protein in growth regulation and tumor suppression. *Cold Spring Harb Symp Quant Biol.* 1994;59:215–23.
9. Thut, C.J., Chen, J.L., Klemm, R., and Tjian, R. p53 transcriptional activation mediated by coactivators TAFII40 and TAFII60. *Science.* 1995;267:100–4.

10. Cho, Y., Gorina, S., Jeffrey, P.D., and Pavletich, N.P. Crystal structure of a p53 tumor suppressor-DNA complex: understanding tumorigenic mutations. *Science.* 1994;265:346–55.
11. Hollstein, M., Marion, M.J., Lehman, T., Welsh, J., Harris, C.C., Martel-Planche, G., Kusters, I., and Montesano, R. p53 mutations at A:T base pairs in angiosarcomas of vinyl chloride-exposed factory workers. *Carcinogenesis.* 1994;15:1–3.
12. el-Deiry, W.S., Kern, S.E., Pietenpol, J.A., Kinzler, K.W., and Vogelstein, B. Definition of a consensus binding site for p53. *Nat Genet.* 1992;1:45–9.
13. Prives, C. How loops, beta sheets, and alpha helices help us to understand p53. *Cell.* 1994;78:543–6.
14. Levine, A.J. p53, the cellular gatekeeper for growth and division. *Cell.* 1997;88: 323–31.
15. Wong, K.B., DeDecker, B.S., Freund, S.M., Proctor, M.R., Bycroft, M., and Fersht, A.R. Hot-spot mutants of p53 core domain evince characteristic local structural changes. *Proc Natl Acad Sci USA.* 1999;96:8438–42.
16. Brachmann, R.K., Vidal, M., and Boeke, J.D. Dominant-negative p53 mutations selected in yeast hit cancer spots. *Proc Natl Acad Sci USA.* 1996;93:4091–5.
17. Joerger, A.C., Ang, H.C., Veprintsev, D.B., Blair, C.M., and Fersht, A.R. Structures of p53 cancer mutants and mechanism of rescue by second-site suppressor mutations. *J Biol Chem* 2005;280:16030–7.
18. Mummenbrauer, T., Janus, F., Muller, B., Wiesmuller, L., Deppert, W., and Grosse, F. p53 Protein exhibits 3′-to-5′ exonuclease activity. *Cell.* 1996;85:1089–99.
19. Mol, C.D., Kuo, C.F., Thayer, M.M., Cunningham, R.P., and Tainer, J.A. Structure and function of the multifunctional DNA-repair enzyme exonuclease III. *Nature.* 1995;374:381–6.
20. Clore, G.M., Ernst, J., Clubb, R., Omichinski, J.G., Kennedy, W.M., Sakaguchi, K., Appella, E., and Gronenborn, A.M. Refined solution structure of the oligomerization domain of the tumour suppressor p53. *Nat Struct Biol.* 1995;2: 321–33.
21. Mateu, M.G., and Fersht, A.R. Mutually compensatory mutations during evolution of the tetramerization domain of tumor suppressor p53 lead to impaired hetero-oligomerization. *Proc Natl Acad Sci USA.* 1999;96:3595–9.
22. Chene, P. Influence of the N-terminal region on the oligomerisation between human and Xenopus laevis p53. *J Mol Biol.* 1999;288:883–90.
23. Stommel, J.M., Marchenko, N.D., Jimenez, G.S., Moll, U.M., Hope, T.J., and Wahl, G.M. A leucine-rich nuclear export signal in the p53 tetramerization domain: regulation of subcellular localization and p53 activity by NES masking. *EMBO J.* 1999;18:1660–72.
24. Selivanova, G., Ryabchenko, L., Jansson, E., Iotsova, V., and Wiman, K.G. Reactivation of mutant p53 through interaction of a C-terminal peptide with the core domain. *Mol Cell Biol.* 1999;19:3395–402.
25. Zhou, X., Wang, X.W., Xu, L., Hagiwara, K., Nagashima, M., Wolkowicz, R., Zurer, I., Rotter, V., and Harris, C.C. COOH-terminal domain of p53 modulates p53-mediated transcriptional transactivation, cell growth, and apoptosis. *Cancer Res.* 1999;59:843–8.

26. Dudenhoffer, C., Rohaly, G., Will, K., Deppert, W., and Wiesmuller, L. Specific mismatch recognition in heteroduplex intermediates by p53 suggests a role in fidelity control of homologous recombination. *Mol Cell Biol.* 1998;18:5332–42.
27. Kuerbitz, S.J., Plunkett, B.S., Walsh, W.V., and Kastan, M.B. Wild-type p53 is a cell cycle checkpoint determinant following irradiation. *Proc Natl Acad Sci USA.* 1992;89:7491–5.
28. Livingstone, L.R., White, A., Sprouse, J., Livanos, E., Jacks, T., and Tlsty, T.D. Altered cell cycle arrest and gene amplification potential accompany loss of wild-type p53. *Cell.* 1992;70:923–35.
29. Yin, Y., Tainsky, M.A., Bischoff, F.Z., Strong, L.C., and Wahl, G.M. Wild-type p53 restores cell cycle control and inhibits gene amplification in cells with mutant p53 alleles. *Cell.* 1992;70:937–48.
30. Kastan, M.B., Zhan, Q., el-Deiry, W.S., Carrier, F., Jacks, T., Walsh, W.V., Plunkett, B.S., Vogelstein, B., and Fornace, A.J. Jr. A mammalian cell cycle checkpoint pathway utilizing p53 and GADD45 is defective in ataxia-telangiectasia. *Cell.* 1992;71:587–97.

CHAPTER

# 10 INTEGRITY OF THE GENOME AND PROGRAMMED CELL DEATH

## P53 and DNA Repair

The principal role of wild-type P53 is to maintain the integrity of the genome. P53 is stabilized or neoexpressed in response to a variety of DNA insults (**TABLE 10-1**). P53 is activated by DNA damage. If repair is successful, the cell reenters the cycle and, if repair fails, the cell dies. However, if the cell persists to enter the cell cycle without adequate DNA repair, then cancer may follow (**FIGURE 10-1**).

## P53 Activation

Several mechanisms for signaling P53 activation in response to DNA damage have been proposed, but the nature of the ultimate DNA insult (which include bulky adducts, single or double strand breaks, and crosslinks) is likely to be the common denominator that links them. Latent wild-type P53 is scarce in undamaged cells, mostly because it is present in an inactive form, but also because of the short lifespan (a matter of minutes) of the active molecule. Therefore, in order to secure P53 function (e.g., $G_1$/S arrest or apoptosis) activation requires either changes in P53 conformation, de novo transcription, a block of P53 degradation, or all of the above. The P53 activating agents are believed to function upstream of P53.

### P53 Conformation and Activation

Changes of P53 conformation can result from binding of P53 to DNA or from post-translational alterations (phosphorylation or acetylation). In vitro evidence suggests that DNA strand breaks cause P53 activation (1, 2) possibly by stimulating sequence specific binding of the P53 C-terminal domain (3, 4). Thus, it cannot be ruled out that wild-type P53 recognizes DNA damage and thereby triggers DNA repair. Zotchev and associates investigated the binding ability of the P53 C-terminal domain to several kinds of double stranded oligonucleotides constructed in the laboratory (5). The nucleotides contain lesions mimicking those seen after exposure to DNA damaging

**Table 10-1 Agents that cause DNA damage**

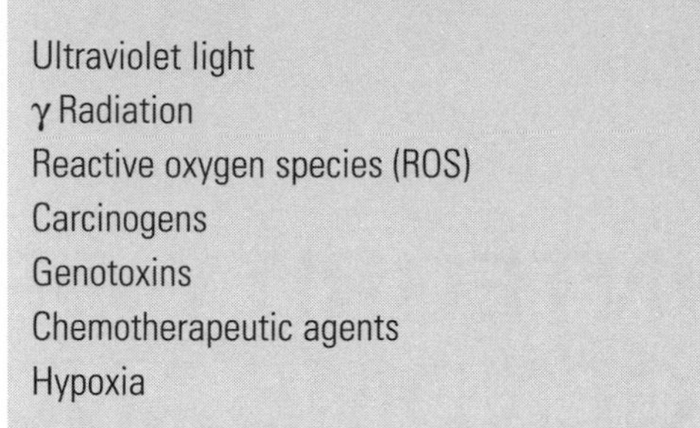

| Agents that cause DNA damage |
|---|
| Ultraviolet light |
| $\gamma$ Radiation |
| Reactive oxygen species (ROS) |
| Carcinogens |
| Genotoxins |
| Chemotherapeutic agents |
| Hypoxia |

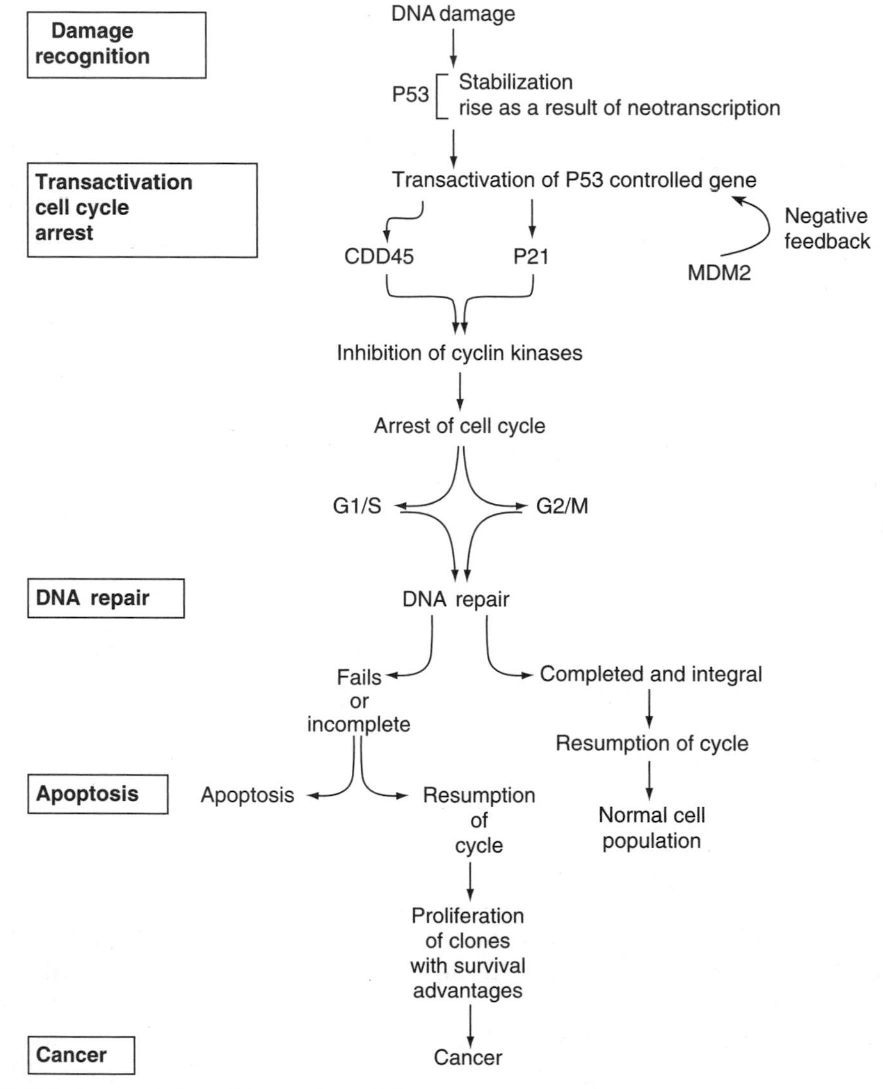

**FIGURE 10-1** An overall view of the functions of P53.

agents. There are many of those lesions, among them a single strand break that protrudes from their double strand construct and single strand gaps (for example, caused by nucleotide excision repair). The response of P53 to either type of damage is highly sensitive. The C terminal binds with both strand ends and internal gaps, but there is a difference. Binding of the C terminal with strand ends is associated with P53 core binding; in contrast, P53 C terminal gaps' binding prevents that of the P53 core domain. This indicates that wild-type P53 binding to DNA is essential for its functions (cell cycle arrest and possibly apoptosis), and the conformation of P53 bound to single strand ends differs from that of the P53 bound to a gap. Zotchev et al. further suggest that the P53 C terminus binds to single and double strand breaks in vivo (5).

The bifunctional antimetabolite cisplatin forms DNA adducts as a result of intra strand crosslinks. The latter interfere with the binding affinity of tetrameric wild-type P53 to consensus DNA response elements probably because of the steric hindrance generated by the adducts. There is, however, significant binding of P53 to portions of the damaged DNA devoid of consensus DNA response elements or to some chromosomal proteins (e.g., the high mobility group [HMG-1]). Thus, cisplatin and probably other bifunctional adducts interfere with wild-type P53 function in two ways: by preventing wild-type P53 from binding at functional sites and by trapping P53 on DNA at nonfunctional sites or by binding it to nuclear proteins (6).

Although the above findings confirm that the binding of P53 to damaged DNA is critical and one of the earliest events that trigger the various wild-type P53 functions, they do not prove that DNA damage serves as an immediate and sufficient trigger for wild-type P53 activation and do not exclude the existence of signals operating upstream of P53. After briefly reviewing the various mechanisms of DNA repair, to emphasize the significance of the role of this genome in apoptosis and cancer, this chapter will return to the subject of repair. Because the function of P53 is linked to DNA repair, apoptosis, and cancer, a brief overview of the more traditional forms of DNA repair are presented, which include homologous recombination, nonhomologous recombination, and excision repair.

## DNA Repair

At a first glance DNA is a long but monotonous molecule compared to protein. Four nucleotides are each made of a different base, but contain the same sugar deoxyribose and phosphorus. The sum of purines equals that of pyrimidines in the molecule, and the number of thymines equals the number of adenines, and that of cystosine that of guanines. The studies of Avery, MacLeod, and McCarty published in 1944 (7) left scientists with little doubt that DNA stored genetic information. Based on what was known at the time, two among many questions were raised: Could the molecule duplicate without making faults? How did it give rise to the great variety of proteins? Watson and Crick discovered the DNA structure (8) and this led to an understanding of its mode of replication. The discovery of DNA repair took more than another decade. Yet before then it was clear that evolution, sexual reproduction, crossing over, and physiologic and immunologic recombination at some point require double strand breaks and recombination. This implied that DNA fragments involved in these processes needed to dig into the genome for a homologous sequence tucked within a chromosome inside the nucleus. In summary, recombination guarantees two important genetic capabilities (reviewed in 9, 10):

- Increased genetic variation, for example, by homologous recombination between chromosomes during meiosis.
- Genomic stability in the process of recombination repair.

## Recombination Repair

Various genotoxic agents, including ionizing radiation, cause double strand breaks. Unless the discontinuity of the DNA double strand is restored error-free, a defective genome is transferred from one cell generation to the next. The cell can repair double strand breaks in at least two ways:

- Error-free restoration of the double strand break (dsb) by homologous recombination (11).[1]
- More simple end-to-end repair (non-homologous recombination).

End-to-end repair is often associated with deletions, translocation, and chromosomal rearrangement that may result in loss or gain of genetic information and contribute to a loss of genomic integrity. Non-repaired strands are potentially lethal. Repair requires P53 activation and, therefore, the function of the latter and that of DNA repair must in some way be linked.

## Model for Homologous Recombination

The model presented here is based on the descriptions by Petukhova, Lee, and Sung (10). The hallmarks of homologous combination include:

- The necessity to identify a homologous sequence in the genome.
- Interaction between two DNA molecules that involves information exchange.

Most of the information about recombination is derived from investigations on bacteria or in yeast. A number of proteins have been shown to be involved in the process, including: RAD50, RAD51, RAD52, RAD54, RAD55, RAD57, RAD59, MER11, and others yet to be discovered. Human homologues to RADs 51, 52, 54, and 50 have been shown to exist and to participate to homologous recombination. However, details of the molecular mechanism of homogenous recombination remain to be uncovered. The following is a highly simplified description of the process, which is divided into four steps:

- The first step involves an attack by a 5′ exonuclease that generates a long 3′ single strand overhang. The latter recruits several proteins (or recombination factors) and forms a novel nucleoprotein complex at the end or close to the end of the strand.
- In the second step, the nucleoprotein complex searches for a homologous DNA sequence on another chromosome or a sister chromatid. Once the latter is identified, the nucleoprotein complex invades the homologous region.
- In the third step, the invading complex triggers DNA synthesis by a polymerase using the homologous sequence as a template. This generates an extended DNA sequence that includes the sequence lost at the site of the double strand break.
- In the fourth step, the extended DNA sequence re-anneals to the homologous sequence in the broken chromosome. The sequences that are nonhomologous to the missing sequences are appropriately trimmed and the re-annealed segment is ligated.

[1]Homologous recombination is to an extent linked to the stage of the cell cycle during which the breaks take place; the first predominates in the $G_1$ and the second during the S and $G_2$ phases of the cell cycle.

In summary, the association of proteins to the extended DNA single strand generates a nucleoprotein filament that sneaks its way through the genome to find a homologous sequence inside the duplex, and it uses the latter as a template to restore the lost segment. Restoring the chromosomes involves a number of different proteins including a polymerase, a ligase, endonucleases, and several cofactors. In mammalian cells, 30 to 50% of double strand breaks are likely to be repaired by homologous recombination, at least in vitro (**FIGURE 10-2**) (13–16).

## Non-Homologous End Joining

The non-homologous end joining (NHEJ) requires little or no homology between the recombining strands (17), but causes a distortion of the normal DNA sequence at the point of contact through either deletions or insertions. Still, NHEJ is important to

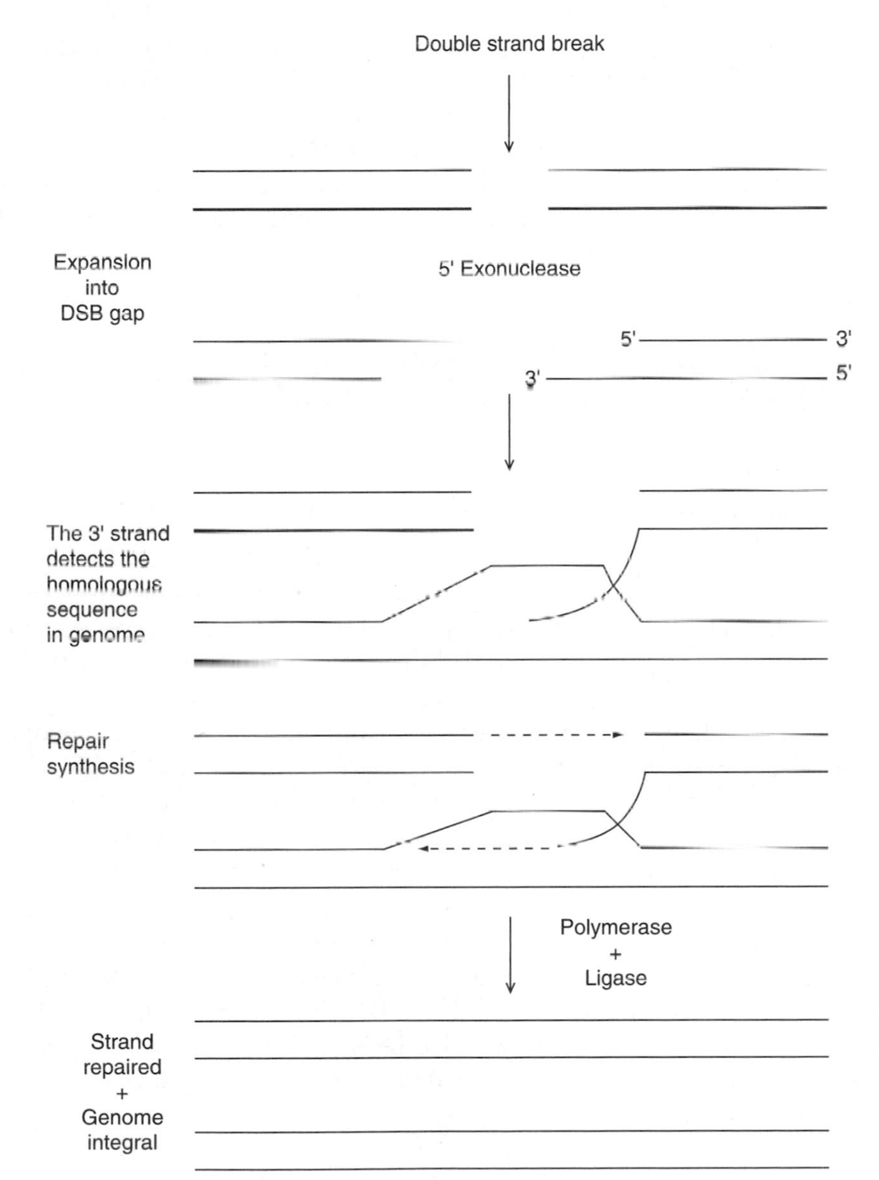


**FIGURE 10-2** Model for homologous recombination. (See Singleton and Jeggo (1999) [15] and Jasin (2001) [16].)

**Table 10-2 Genes involved in DNA repair in yeast after ultraviolet exposure**

| ERCC Gene | Gene in XP Complementation Group |
|---|---|
| *ERCC 1* | No XP equivalent |
| *ERCC 2* | *XPD* |
| *ERCC 3* | *XPB* |
| *ERCC 4* | *XPF* |
| *ERCC 11* | *XPF* |
| *ERCC 5* | *XPF* |

mammalian integrity. A number of proteins are also involved in nonhomologous end joining. They include the DNA-dependent protein kinase (see below), which is absolutely required, and an assemblage of proteins, namely, DNA ligase IV, XRCC4 (x-ray repair cross complementing gene IV), and the MRE11–RAD50 complex (16, 18). Note that the latter functions in both homologous and nonhomologous double strand repair (see Nymegen breakage syndrome in Chapter 23).

XRCC4[2] (19) is a 334 amino acid protein that is an effective substrate for DNA-PK in vitro and complexes with DNA ligase IV. The exact molecular function of the MER 11/Rad 50/Xrs2 is still not entirely clarified.

**TABLE 10-2** is adapted from a review by Hanspeter Naegeli (20); for further information about these molecules, refer to reference (21).

## PCNA and Recombination Repair: Error-Free and Error-Prone Pathways

### Role of Ubiquitination

When DNA lesions are present during DNA replication, they arrest the process of replication at the fork. Pathways, sometimes referred to as damage tolerance pathways, have evolved that allow cell replication to proceed with the lesion in place. Two such pathways exist: one is error-free (also called the damage avoidance pathway), and the other is error-prone (or the translesion pathway).

The error-free pathway uses an undamaged copy of the sister strand as a template to bypass the block caused by the damaged sequence. The process was first discovered in yeast but also exists in humans (22). The mechanism that gears the cell in either the error-prone or the error-free pathway is not completely known; however, new studies have, at least in part, revealed the molecular components that contribute to switching gears in the direction of error-free postreplication repair (discussed below). In addition to proliferating cell nuclear antigen (PCNA), a number of specialized molecules (among them, polymerase[s] or molecular complexes) involved in DNA replication and postreplication repair are required. All the DNA repair molecules probably have not yet been identified, but much progress has been made in understanding postreplication repair. A few of the molecules required in the process are considered: RAD6, ubc 3–MMS2, RAD5, and Rad18.

[2]XRCC4 is a substrate for DNA-PK; however, phosphorylation is not required for function and, therefore, it may not be physiologically relevant.

## RAD6

The *Rad6* gene is a member of a group of genes associated with DNA repair. The encoded protein is a UV inducible 20 kDa protein that functions as a ubiquitin-conjugating $E_2$ enzyme (14).

## UBC13-MMS2

UBC13 is an $E_2$ enzyme that catalyzes the formation of ubiquitin chains linked at the Lys63 of that molecule. It is involved in postreplication repair and is believed to be conserved among eukaryotes (23). UBC13 forms a complex with MMS2 and this complex is required for postreplication repair in *Saccharomyces cerevisiae* (24) as well as in human cells (14, 25). The UBC13-MMS2 heterodimer links ubiquitin units into a polyubiquitin chain by forming an isopeptide bond between the C terminal of one unit and the lysine 63 of another. The *MMS2* gene encodes a ubiquitin conjugating variant. These molecules have a sequence similar to that of traditional $E_2$ enzymes (ubiquitin-conjugating enzymes). McKenna and colleagues suggested that the function of MMS2 is to secure the proper orientation of the ubiquitin toward its target and eventually prevent any molecular interference with the extension of the ubiquitin chain (26). The crystal structure of hMMS2 has been reported (27).

## RAD18

RAD18 is a 55 kDa protein inducible by ultraviolet light and ionizing radiation. It is a member of the structural maintenance chromosome family of proteins (SMC6) and is involved in eukaryotic DNA replication and DNA repair (28). It possesses a ring finger protein (28) that binds DNA, and is believed to be involved in structural alterations of chromatin required for DNA repair (29).

## RAD5

RAD5 is a DNA-dependent helicase zinc finger motif that is required for postreplication repair of yeast exposed to ultraviolet radiation (24, 25).

## PCNA

Proliferating cell nuclear antigen (PCNA) is involved in cell replication and is recruited to DNA by replication factor C (RFC). RFC recognizes the junction of primer and template, binds it to the site, and recruits PCNA and the polymerase. PCNA is a trimeric ring that encircles the double DNA strand and serves as a sliding clamp along the strand transporting those enzymes contributing to DNA repair, in particular the appropriate polymerases (30).

PCNA contributes to many aspects of DNA replication, such as:

- Biosynthesis of both the leading and the lagging strands.
- S-$G_2$ transition arrest by interacting with P21.
- Almost all forms of DNA repair (31).

Mechanisms of postreplication repair are complex and require strict regulation. For example, as mentioned earlier, PCNA loading and unloading on the DNA template is modulated by a regulator: RFC. Moreover, PCNA interacts with inhibitors of the cell cycle, in particular P21. Little is currently known about the mechanisms that regulate PCNA functions, including those that are involved in various forms of DNA repair. However, it is clear that the initiation of DNA synthesis occurs in at least three well-established steps:

1. DNA replication starts with an RNA polymerase chain (polymerase α).
2. This is followed by a short installment by the polymerase of a 3′ to 5′ DNA sequences.
3. Only after these steps does PCNA and polymerase δ bind to DNA and form the traditional sliding clamp.

This initiation process may vary somewhat depending upon the type of DNA repair replication, but the central function of PCNA is to serve as a moving clamp for several polymerases in various forms of DNA replication, including postreplication repair. In postreplication repair the lesion is ignored; DNA synthesis takes place after replication and fills gaps by using undamaged strands as a template. The repair can be error-free, but if that pathway fails, the repair becomes error-prone. Again, the question is, what directs the process into either an error-free or an error-prone pathway (25)? The discovery that ubiquination and sumoylation are associated with postreplication repair helped to clarify, at least in part, the molecular regulatory mechanisms of PCNA, and postreplication repair itself (31). The involvement of RAD6 in that form of repair gave a clue as to the possible role of ubiquitin in the process. Indeed, RAD6 is a ubiquitin conjugating enzyme. However, direct evidence for ubiquitination, particularly that of PCNA, was not available until it was discovered by Hoege and associates in 2002 (32).

In addition to being polyubiquitinated, PCNA can also be sumoylated. PCNA was found to be the target of monoubiquitination, polyubiquitination, and sumoylation of lysine sites in the PCNA structure. In yeast, SUMO binds to DNA bound PCNA during the S phase, preferentially on lysine 164 and much less on lysine 127, by the SUMO ligase S121. Sumoylation inhibits PCNA's function in repair.[3]

In summary, exposure to DNA damage is associated with complex interactions between DNA and proteins that lead to monoubiquitination, polyubiquitination, and sumoylation of DNA bound PCNA.

- DNA damage is associated with the appearance of RAD18 to chromatin, a ring finger protein that recruits RAD6 (the $E_2$ ubiquitin conjugating enzyme). This combination leads to monoubiquitination of lysine 164 of the DNA-bound PCNA ring.
- This step is followed by translocation of the heteromeric UBC13-MMS2 (another ubiquitin conjugating enzyme) to the nucleus, where it interacts with chromatin bound RAD5.
- DNA damage is further associated with the appearance of RAD18 to chromatin (another ring finger protein) that recruits RAD6, the $E_2$ ubiquitin-conjugating enzyme.
- The translocation of the heteromeric UBC13-MMS2 (another ubiquitin conjugating enzyme) to the nucleus follows, where it interacts with chromatin bound RAD5.
- RAD6 recruits PCNA to the chromatin, and the UBC13-MMS2 heterodimer is brought to RAD6 on the chromatin by the two ring finger chromatin-binding proteins, RAD18 and RAD5.
- This brings together two ubiquitin-conjugating enzymes and two ubiquitin ligases, which provide the machinery needed to extend the ubiquitin of the monoubiquitinated PCNA into a polyubiquitinated chain through the K63-C terminal bonds.

---

[3]SUMO (small ubiquitin modifier) refers to a family of proteins that bind covalently to many proteins and thereby modify their cellular functions. Sumoylation, for example, may modulate the interactions of its targeted proteins with DNA and other proteins, or it may prevent the binding of ubiquitin to the sumyolated protein. The biochemical pathway for sumoylation is reminiscent of that of ubiquitination, as enzymes with similar specificities function in both cases (33).

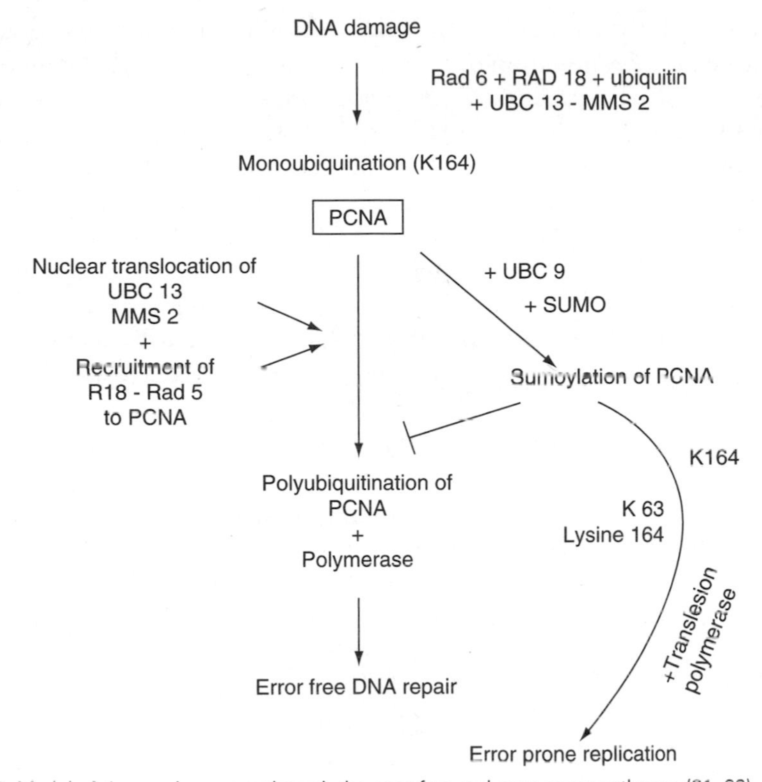


**FIGURE 10-3** Model of the repair process through the error-free and error prone pathways (31, 32).

In conclusion, polyubiquitination of PCNA is required for error free DNA repair. Polyubiquitinated PCNA is needed to engage replication beyond the block imposed by DNA damage at the fork and secure error free postreplication repair (**FIGURE 10-3**).

## Excision Repair

Excision repair involves excision of a segment of the DNA that includes a damaged base or nucleotide sequence. Although a panoply of different enzymes and pathways for excision repair exists and varies with the type of damage, it follows the same general sequence of basic steps:

- Incision of the DNA sequence at the site or close to the site of the DNA damage.
- Excision of a stretch of DNA that includes the damaged site.
- Restoration of the damage strand through DNA polymerization using the sequence of the complementary strand to restore the correct sequence of the excised stretch.
- Ligation of the repaired stretch within the DNA molecule.

### Base Excision Repair

The base excision repair pathway is activated when a base is lost in the DNA sequence of one strand. The latter may result from the attack of the bond between a damaged base and its adjacent deoxyribose sugar by glycosylases, by spontaneous breaks or because of a break caused by exposure to environmental agents (e.g., UV or x-radiation).

Glycosylases hydrolyze the N-C bond that links a damage base and a sugar. Bases may be damaged by (deamination of cytosine to yield uracyl, methylation of adenine, formation of trihydroxyperoxides [after irradiation, etc.]). Some glycosylases remove the damaged base without further attacking the DNA sequence, but cleavage of the DNA backbone follows the split of the N-C bond caused by others.

In the first case, after a single nick by the AP endonuclease, the new complementary base is inserted by polymerase β, an enzyme that also excises the downstream deoxyribose phosphate residue. In the last step, the new nucleotide is sealed in the DNA sequence by ligase. Because the process involves only a single nucleotide, it is called short patch repair. The abasic site generated by spontaneous hydrolysis is also repaired by the short patch pathway.

In other cases, the attack on the N-C bond by a glycosylase is associated with the excision of several nucleotides (2 to 6 or more). Similarly, several nucleotides may be removed after single strand breaks caused by ionizing radiation, and in those cases the pathway diverges from that of the short patch repair. The gap is recognized by poly-ADP-ribose polymerase and maybe DNAPK, an oligonucleotide sequence (dNTP) is dissociated from the complementary strand forming a flap, polymerase β or polymerase δ/ε construct the new complementary sequence, the flap is removed by the flap endonuclease (FEN I), and the restored sequence is sealed by ligase I. Note that in all cases of BER after a strand break has appeared, either spontaneously after a glycosylase attack or after exposure to ionizing radiation, the repair pathway is started by an apurinic/apyrimidinic endonuclease (AP endonuclease) that generates a 3′ OH, a requirement for the polymerases. The pathway may diverge or overlap from there on. What determines the selection of a specific pathway is not always known (detailed and referenced in 31).

## Nucleotide Excision Repair

The environment of all living organisms, including humans, is a rich source of DNA insults. Unprotected skin bathes in the sun much of the time, and our foods contain additives (nitrites) and contaminants (acetylaminofluorene). Cooking generates carcinogenic byproducts (benzopyrene). Our air is polluted with smoke and toxic particles and our homes may be the target of radon. Ultraviolet rays, ionizing radiation, and reactive chemicals attack the DNA of our skin or penetrate deep in our bodies, causing the formation of multiple and varied DNA adducts that need to be repaired for cell survival. This is possible only if the cells of our organism possess either a large arsenal of enzymic machineries that deal with each specific adduct, or possess one or more enzymic complexes able to clear a multitude of adducts at any time, including under special circumstances, such as when DNA is transcribed. Nucleotide excision repair (NER) is such a versatile mechanism. This universal and permanent surveyor of the genome is divided into two pathways, each involving an integrated panoply of enzymes and cofactors that may to an extent overlap in both pathways. One NER system detects any distortion anywhere in the genome, the global genome NER (GGNER). Another more specialized repair process surveys and clears the genome of distortions at the time of transcription: the transcription coupled repair (TCR). It is no surprise that these survival tools appear early and persist through evolution. In eukaryotes the DNA repair pathways were investigated in yeast, rodents, and humans (reviewed in 10, 31, 34). NER uses multiprotein complexes instead of single enzymes to do the job(s); in essence it operates in six steps:

1. Damage recognition.
2. DNA unwinding around the lesion.
3. Double incision of the distorted oligonucleotide (~25–30 nucleotides).
4. Excision.
5. Repair synthesis of the lost sequence.
6. Ligation.

Fusion between cells obtained from different patients with Xeroderma pigmentosum uncovered seven complementation groups. The complementation cell lines are labeled XP-A through XP-G, and the genes defective in each case are *XPA* through *XPG*. The "excision repair cross complementation" gene (*ERCC*) from *XPB* to *XPF* was identified by complementation of NER-defective hamster cell lines using human genomic libraries (**TABLE 10-3**). *XPA* and *XPG* were cloned directly by complementation of repair defect in XP cell lines.

Once the gene for complementation was identified, it was overexpressed and the encoded protein was purified and characterized. With the availability of these tools, investigators succeeded in reconstructing NER in vitro (35, 36).

A set of interacting proteins that is necessary and sufficient for NER, including TCR, was identified. These proteins are: XPA, RPA, XPC-hH Rad23B, TFIIH, and XPF-ERCC1 XPG. The exclusion of XPA, RPA, XPC-hHRAD23 B, and TFIIH from the system abolishes NER and that of XPG prevents bimodal incision, whereas that of XPF-ERCC1 uncouples excision on the 3′ side.

Note that instead of using single enzymes functioning in sequence to perform NER, the cell uses protein complexes. Because of the need for such complexes, the notion of an NER repair organelle emerged in which the molecules necessary for NER are assembled in an organized structure. The alternative to the organized attack of the damaged DNA site by an organelle is a rigidly regulated cascade of events in which the molecules act in sequence.

As mentioned earlier, there are two NER pathways: a permanent surveyor of the entire genome, that is, the global genome NER (GGNER) and a transcription-activated NER pathway (transition coupled repair; TCR). The difference between the interactions of each of the two pathways at the site of the DNA damage consists in the presence of RNA polymerase II, which is absent in the first (GGNER) and needs to be removed in the second (TCR). The other steps of the pathways' recognition, double incision and excision of the damaged strands, and restoration of the partially amputated strand, are similar if not identical. The initial step involves identification of the site of damage followed by association of the multiprotein complex with the site. Thus, the XPA, RPA, and XPC-hHR23B, TFIIH protein complex contribute to the recognition, unwinding, and stabilization of the site of DNA damage (Table 10-2).

In TCR, the stalled Pol II must be displaced prior to the 5′→3′ and the 3′→5′ incisions. This step requires two TCR-specific factors, CSB and CSA. During unwinding, an oligonucleotide chain containing the bulky adduct (15–30 NTP long) is separated from the healthy complementary strand. This step most likely requires helicase activity associated with TFIIH. The nucleotide sequence containing the adduct is next incised at both ends: at the 5′→3′ end by the XPF-ERCC1 heterodimer, and at the 3′→5′ end by the single polypeptide XPG.

The ~30-nucleotide single stranded gap caused by the double incision is restored by the PCNA-dependent polymerase in the presence of RPA and RFC factors, and is ultimately ligated by ligase I (see Table 10-3).

**Table 10-3 Some proteins involved in DNA repair**

| Protein | Subunits | Properties | |
|---|---|---|---|
| XPA | Single polypeptide | 273 AA, 31 kDa | Necessary for NER |
| | | Zn finger motif[a] | |
| | | 4 domains for protein interactions with subunits | |
| | | ERCC1 AA 72-84 | Location of DNA damage |
| | | TFIIH AA 226-273 | |
| RPA | 3 Polypeptides | Replication protein (A) | |
| | 14 kDa | Necessary for NER and DNA | Replication accessory |
| | 34 kDa | replication | Precedes and prepares DNA |
| | 70 kDa | Binds to single stranded DNA | incision |
| | | 34 and 70 subunits bind XPA | Targets nuclease subunits of |
| | | Zinc finger near C terminal | NER (XPF-FRCC1α XPG) |
| | | Interaction domains for XPA, XPG, and P53 | |
| XPC-hHR23B | 2 Polypeptides | Not essential for incision | |
| | homolog of Rad 23 | May stimulate NER | |
| | Yeast NER protein | Binds single and double stranded DNA | Chaperone help fold Protein complex |
| | | No catalytic activity | Interaction with chromatin? |
| | | Ubiquitin-like N terminal | Contributes to degradation of |
| | | Interacts with TFIIH | NER factors? |
| TFIIH | 6–9 Polypeptides | Multiple catalytic activities | Transcript or initiator required for Pol II |
| | P89/XB/ERCC3 | DNA-dependent protein kinase | |
| | P80/XPD/ERCC2 | DNA helicase activities (5′-3′, 3′-5′) | |
| | P62/TFB1 | Protein kinase | Unwinds DNA duplex through |
| | P52/TFB2 | XPB | helicase 5′ → 3′, 3′ → 5′ ? |
| | P44/SSL1 | XPD | |
| | P38/cdk7/M015 | P62 | |
| | P37/cyclin H | Forms several distinct complexes harboring different quantities of the various polypeptides | Secures a link between transcription – NER and cell cycle? |
| | P36/MAT 1 | | Phosphorylates P53 (through |
| | P34 | | protein kinase) |
| XPF-ERCC1 | Heterodimer | Acts preferentially on single stranded DNA | 5′ → 3′ incision |
| | | RPA interact with XPF-ERCC1 and stops single strand degradation | |

*Continued*

**Table 10-3 Some proteins involved in DNA repair—*cont'd***

| Protein | Subunits | Properties | |
|---|---|---|---|
| XPG | Single polypeptide | Member of family of eukaryotic nucleases that includes the Flap endonuclease (FEN-1)<br>Interacts with RPA, TFIIH, PCNA | $3' \rightarrow 5'$ incision |

[a]Zinc finger with the following motif (Cyc 105– $(X)_2$ – Cys 108 $(X)_{17}$ – Cys 126 – $(X)_2$ – Cys 129). The cysteines are required for function. Replacement of cystenie by serine abolishes function.
*Source:* After H. Naegeli (20) (DNA recombination and repair. Smith & Jones 1999: p 99–137).

In conclusion, NER is the most frequent and versatile form of DNA repair in eukaryotes. It functions as a global surveyor of the genome at all times and as a remover of stalled RNA polymerase II during transcription. The process takes place in a sequence of six steps: DNA damage recognition; unwinding; 5′→3′ incision; 3′→5′ incision of the damage strand (~30 nucleotides); restoration of the gap by PCNA dependent polymerase; and ligation.

The last four steps of the GGNER and the TCR pathways are essentially the same. The fundamental difference between the two resides in recognition and the need for displacement of RNA polymerase II in TRC. They are:

- Recognition: Two molecules cooperate to secure recognition of the DNA distortion, Xeroderma pigmentation complementation group C (XPC) and the human homologue of yeast RAD 23 B (hHR23B). The molecules survey the DNA for distortion and identify the altered base.[4]
- Unwinding: Again, a molecular complex participates in the unwinding; in GGNER it is associated with the transcription factor II H (TFIIH) and the Xeroderma pigmentosum complementation group C (XPC).

At least three molecules are known to be involved in TCRNER:

- CSA, the human homologue of yeast RAD 28.
- TFIIH, the transcription factor II H.
- XPG, the Xeroderma pigmentosum complementation group G.

Finally, it cannot be ruled out that other molecules could be involved in GGNER and TCR.

## Mismatch Repair

Mismatch repair is a third form of DNA excision repair. The last steps of the pathway are shared with other excision repair pathways. The differences between mismatch repair and BER and NER reside in the nature of the DNA distortion and the enzymatic battery put in gear to correct the distortion.

In contrast to BER and NER, the target of mismatch repair is not a damaged base, but an undamaged one that is misplaced in the DNA sequence and thereby causes a

[4]Since DNA is not a simple double helix, it is essential that the recognition mechanism distinguishes natural contortions of the molecule from true lesions.

distortion in the complementary strand. Such misplacement or mismatch occurs because of:

- Defective proofreading of the nucleotide sequence of the replicating strand.
- Defective homology after recombination.
- DNA polymerase slippage.

When normal bases—adenine, guanine, cytosine or thymine—are mispaired, the defective strand must be corrected. This requires that the pathway involved with elimination of the mispair starts by recognizing:

- The existence of a mispair; and
- The strand that carries the wrongly inserted base.

The latter step is critical because only if the wrongly inserted base is removed and replaced by that which is complementary to the wild-type base sequence can the mutation caused by the mismatch be corrected. Otherwise the mutation will be perpetuated.[5] Mismatch repair was discovered in bacteria (e.g., *Escherichia coli*), but soon similar pathways were discovered in yeast and humans. In *E. coli* the adenines of the GATC and the cTAG complementary sequences are both methylated.

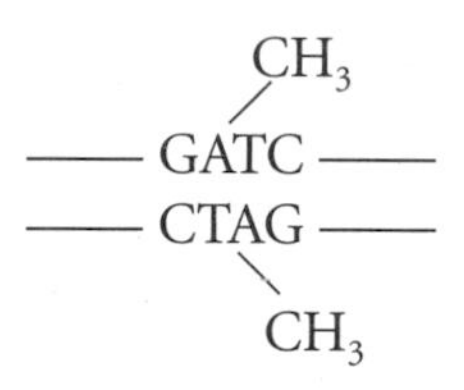


But in the newly replicated strand (-----) CTAG is not immediately methylated.

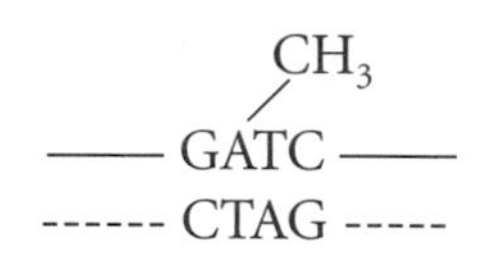


Thus, the newly replicated strand can be distinguished from the parental strand during the period that separates completion of polymerization and ligation[6] (0.5–13.0 minutes). Since the mispair is always in the unmethylated strand, the enzyme knows which strand is the newly replicated one. Three specialized proteins are involved in *E. coli* mismatched repair: Mut S; Mut L; and Mut H. Mut S (95 kDa) binds to the mismatch DNA or close to the mismatch, where it homodimerizes. Mut L (a homodimer of a 68 kDa subunit) is recruited at the site. Then Mut H joins the two other dedicated mismatch repair proteins. Mut H is an endonuclease activated by Mut L, which incises the mismatched DNA sequence. The site of the cut, located at a distance from the mismatch, is dictated by a specific sequence (GATC). Mut S and Mut L steps are both ATP-dependent; Mut S is also a weak ATPase.

Homologues of the Mut system have been found in yeast and humans. In humans, at least three dedicated proteins are required for mismatch recognition: hMSH2, hMSH3, and hMSH6 (105, 127, and 160 kDa, respectively). As is the case in bacteria, the process is ATP-dependent, but the exact role of ATP remains unclear. The mispaired site is

[5]Consider the case of deamination of 5-methyl cytosine to thymine, generating a GT sequence rather than GC. In the wild-type double strand, the GC is complementary to CG. In the mutated double strand, GT is paired to GC instead of CA. To avoid perpetuating the mutation, the enzyme must recognize that GT is the mutant and not CG.
[6]The adenine is methylated in position 6 by, a methylase encoded by the *dam*$^+$ gene.

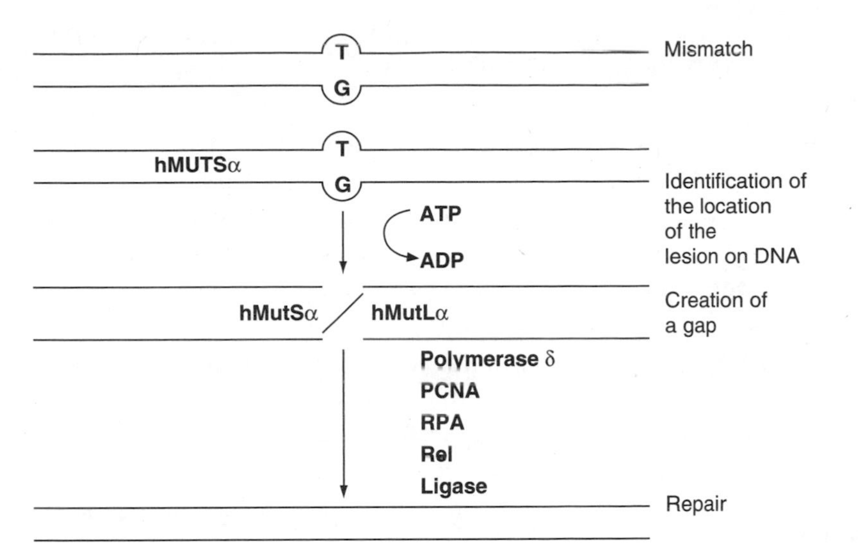


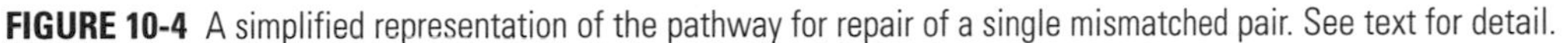

**FIGURE 10-4** A simplified representation of the pathway for repair of a single mismatched pair. See text for detail.

recognized by the heterodimers hMut S α (hMSH2 and hMSH6) and hMut S β (hMSH2 and hMSH3). The exact mechanism of recognition of the newly synthesized strand is not known; there is no adenine methylase in humans. Methylcytosine or a free single-strand DNA end may signal strand discrimination.

HMut S α and hMut S β are joined by another heterodimer hMut L α composed of two subunits homologous to Mut L: hMLH1 and HPMS2. The function of Mut L α and the hMut S β-DNA complexes is not clear. In any event, the association of hMut S α and hMut S β with hML H α is followed by excision of the mismatch strand by exonuclease. Polymerase δ or ε using the PCNA clamp secures the repair synthesis. The single strand replication protein RPA is also required (reviewed in 34, 35).

The repair of a single mismatched base pair can be divided into three phases:

1. The mismatched base pair generates a distortion of the double strand (a loop) hMUTSα, the heterodimer hMSH6/hMSH2, identifies the site of the mismatch on the DNA.
2. HMUTSα is joined by hMUTLα, the heterodimer hMSH2/hMLH1, which in the presence of ATP and other factors, identifies the strands that carry the mismatched base. The same heterodimer contributes to the creation of an endonuclecytic gap that extends on both sides of the mismatched pair.
3. Polymerase δ, in the presence of PCNA, RPA and RCL, inserts the correct complementary sequence and ligase seals it in the repaired strand (**FIGURE 10-4**).

The pathway of mismatch repair is also discussed in Chapter 24.

## References and Recommended Reading


1. Nelson, W.G., and Kastan, M.B. DNA strand breaks: the DNA template alterations that trigger p53-dependent DNA damage response pathways. *Mol Cell Biol.* 1994;14:1815–23.
2. Huang, L.C.; Clarkin, K.C., and Wahl, G.M. Sensitivity and selectivity of the DNA damage sensor responsible for activating p53-dependent G1 arrest. *Proc Natl Acad Sci USA.* 1996;93:4827–32.

3. Jayaraman, J., and Prives, C. Activation of p53 sequence-specific DNA binding by short single strands of DNA requires the p53 C-terminus. *Cell.* 1995; 81:1021–9.
4. Selivanova, G., Iotsova, V., Kiseleva, E., Strom, M., Bakalkin, G., Grafstrom, R.C., and Wiman, K.G. The single-stranded DNA end binding site of p53 coincides with the C-terminal regulatory region. *Nucleic Acids Res.* 1996;24:3560–7.
5. Zotchev, S.B., Protopopova, M., and Selivanova, G. p53 C-terminal interaction with DNA ends and gaps has opposing effect on specific DNA binding by the core. *Nucleic Acids Res.* 2000;28:4005–12.
6. Kasparkova, J., Pospisilova, S., and Brabec, V. Different recognition of DNA modified by antitumor cisplatin and its clinically ineffective trans isomer by tumor suppressor protein p53. *J Biol Chem.* 2001;276:16064–9.
7. Avery, O.T., MacLeod, C.M., and McCarty, M. Studies on the chemical nature of the substance inducing transformation of pneumococcal types. *J Exp Med.* 1944;79:137–158.
8. Watson, J.D., Crick, F.H.C. Genetical implications of the structure of deoxyribonucleic acid. *Nature (Lond).* 1953;171:964–7.
9. Alberts, B. DNA replication and recombination. *Nature.* 2003;421:431–5.
10. Friedberg, E.C. DNA damage and repair. *Nature.* 2003;421:436–40.
11. Takata, M., Sasaki, M.S., Sonoda, E., Morrison, C., Hashimoto, M., Utsumi, H., Yamaguchi-Iwai, Y., Shinohara, A., and Takeda, S. Homologous recombination and non-homologous end-joining pathways of DNA double-strand break repair have overlapping roles in the maintenance of chromosomal integrity in vertebrate cells. *EMBO J.* 1998;17:5497–508.
12. Petukhova, G., Lee, E.Y.H.P., and Sung, P. DNA end processing and heteroduplex DNA formation during recombinational repair of DNA double-strand breaks. In: Nickoloff, J. and Hoekstra, M., eds. *DNA Damage and Repair*, vol III. Totowa, NJ: Humana Press; 2001:125–46.
13. Liang, F., Han, M., Romanienko, P. J., and Jasin, M. Homology-directed repair is a major double-strand break repair pathway in mammalian cells. *Proc Natl Acad Sci USA.* 1998;95:5172–7.
14. Nickoloff, J.A., and Hoekstra, M.F. Double-strand break and recombinational repair in *Saccharomyces cerevisiae.* In: Nickoloff, J., and Hoekstra, M., eds. *DNA Damage and Repair.* Totowa, NJ: Humana Press; 1998:335.
15. Singleton, B.K., and Jeggo, P.A. Double strand break repair and V(D)J recombination. In: Smith, P.J., and Jones, C.J., eds. *DNA Recombination and Repair.* New York: Oxford University Press; 1996:16–37.
16. Jasin, M. Double-strand break repair and homologous recombination in mammalian cells. In: Nickoloff, J., and Hoekstra, M., eds. *DNA Damage and Repair*, vol III. Totowa, NJ: Human Press; 2001:207–35.
17. Taccioli, G.E., Gottlieb, T.M., Blunt, T., Priestley, A., Demengeot, J., Mizuta, R., Lehmann, A.R., Alt, F.W., Jackson, S.P., and Jeggo, P.A. Ku80: product of the *XRCC5* gene and its role in DNA repair and V(D)J recombination. *Science.* 1994;265: 1442–5.

18. Petrini, J.H.J., Maser, R.S., and Bressan, D.A. The MREII-RAD50 complex: Diverse functions in the cellular DNA damage response. In: Nickoloff, J., and Hoekstra, M., eds. *DNA Damage and Repair*, vol III. Totowa, NJ: Humana Press; 2001:147–72.
19. Guidos, C.J., Williams, C.J., Grandal, I., Knowles, G., Huang, M.T., and Danska, J.S. V(D)J recombination activates a p53-dependent DNA damage checkpoint in scid lymphocyte precursors. *Genes Dev.* 1996;10:2038–54.
20. Naegali, H. Enzymology of human nucleotide excision repair. In: Smith, P.J., and Jones, C.J. *DNA Recombination Repair.* Series. *Frontiers in Molecular Biology.* Hames, B.D., and Glover, D.M., Series ed. Oxford, UK: Oxford University Press; 1999.99–137.
21. Colin, A.B., and Nickoloff, J.A. Ultraviolet light-induced and spontaneous recombination in eukaryotes: roles of DNA damage and DNA repair proteins. In: Nickoloff, J., and Hoekstra, M., eds. *DNA Damage and Repair*, vol III. Totowa, NJ: Humana Press; 2001:329–58.
22. Li, Z., Xiao, W., McCormick, J.J., and Maher, V.M. Identification of a protein essential for a major pathway used by human cells to avoid UV-induced DNA damage. *Proc Natl Acad Sci USA.* 2002;99:4459–64.
23. Ashley, C., Pastushok, L., McKenna, S., Ellison, M.J., and Xiao, W. Roles of mouse UBC13 in DNA postreplication repair and Lys63-linked ubiquitination. *Gene.* 2002;285:183–91.
24. Torres-Ramos, C.A., Prakash, S., and Prakash, L. Requirement of RAD5 and MMS2 for postreplication repair of UV-damaged DNA in Saccharomyces cerevisiae. *Mol Cell Biol.* 2002;22:2419–26.
25. Ulrich, H.D. Natural substrates of the proteasome and their recognition by the ubiquitin system. *Curr Top Microbiol Immunol.* 2002;268:137–74.
26. McKenna, S., Spyracopoulos, L., Moraes, T., Pastushok, L., Ptak, C., Xiao, W., and Ellison, M. J. Noncovalent interaction between ubiquitin and the human DNA repair protein Mms2 is required for Ubc13-mediated polyubiquitination. *J Biol Chem.* 2001;276:40120–6.
27. Moraes, T.F., Edwards, R.A., McKenna, S., Pastushok, L., Xiao, W., Glover, J.N., and Ellison, M.J. Crystal structure of the human ubiquitin conjugating enzyme complex, hMms2-hUbc13. *Nat Struct Biol.* 2001;8:669–73.
28. Fujioka, Y., Kimata, Y., Nomaguchi, K., Watanabe, K., and Kohno, K. Identification of a novel non-structural maintenance of chromosomes (SMC) component of the SMC5-SMC6 complex involved in DNA repair. *J Biol Chem.* 2002;277: 21585–91.
29. Verkade, H.M., Teli, T., Laursen, L.V., Murray, J.M., and O'Connell, M.J.A homologue of the Rad18 postreplication repair gene is required for DNA damage responses throughout the fission yeast cell cycle. *Mol Genet Genomics.* 2001; 265:993–1003.
30. Kelman, Z. PCNA: structure, functions and interactions. *Oncogene.* 1997;14: 629–40.
31. Hoeijmakers, J.H. Genome maintenance mechanisms for preventing cancer. *Nature.* 2001;411:366–74.

**32.** Hoege, C., Pfander, B., Moldovan, G.L., Pyrowolakis, G., and Jentsch, S. RAD6-dependent DNA repair is linked to modification of PCNA by ubiquitin and SUMO. *Nature.* 2002;419:135–41.

**33.** Johnson, E.N. Protein modification by SUMO. (Review). *Annu Rev Biochem.* 2004;73:355–82.

**34.** Wood, R.D. DNA repair in eukaryotes. *Annu Rev Biochem.* 1996;65:135–67.

**35.** Mu, D., Park, C.H., Matsunaga, T., Hsu, D.S., Reardon, J.T., and Sancar, A. Reconstitution of human DNA repair excision nuclease in a highly defined system. *J Biol Chem.* 1995;270:2415–8.

**36.** Aboussekhra, A., Biggerstaff, M., Shivji, M.K., Vilpo, J.A, Moncollin, V., Podust, V.N., Protic, M., Hubscher, U., Egly, J.M., and Wood, R.D. Mammalian DNA nucleotide excision repair reconstituted with purified protein components. *Cell.* 1995;80:859–68.

**37.** Karran, P., and Bignami, M. DNA damage tolerance, mismatch repair and genome instability. *Bioessays.* 1994;16:833–9.

**38.** Boland, R. Hereditary nonpolyposis colorectal cancer (HNPCC). In: Sciver, C., Beaudet, A., Sly, W., Valle, D., Childs, B., Kinzler, K., and Vogelstein, B., eds. *The Metabolic and Molecular Bases of Inherited Disease.* New York, NY: McGraw-Hill Medical Publishing Division; 2000.

CHAPTER

# ENZYMES POTENTIALLY INVOLVED IN P53 ACTIVATION

# 11

## Indirect Activation of P53 and DNA Repair

P53 activation most likely involves molecules that sense the presence of DNA damage and then phosphorylate P53. Kinases that sense DNA damage include DNA-PK and the product of the wild-type gene that causes ataxia telangiectasia (*ATM* gene), the product of the ataxia telangiectasia-related gene (*ATR*) and *brca-1*.

## DNA-PK

DNA-PK is a member of a family of serine threonine kinases, a relative of phosphatidylinositol-3-kinase (1) that acts as one of the surveyors and protectors of the DNA. When the DNA's integrity is violated DNA-PK facilitates its repair. DNA-PK is made of three components: a large one, DNA-PKcs (460 kDa), harbors the kinase activity, and two smaller ones (referred to as Ku 80 and Ku 70) form a heterodimer. Although not found in all living cells (it is not found in yeast), DNA-PK homologs are present in *Xenopus*, *Drosophila*, mice, hamsters, horses, and humans. Homologues derived from mouse, horse, and humans have been cloned. The human locus encoding DNA-PK is located on chromosome 8 band q11 (2).

### DNA-PK and DNA Repair

Because the A-P endonuclease is phosphorylated by several kinases, it has been proposed that DNA-PK is involved in base excision repair. A-P endonuclease is an enzyme involved in the incision of DNA stretches devoid of purine bases (apurinic or abasic sites). The restoration of these sites (base excision repair) is essential for maintenance of both nuclear and mitochondrial DNA (reviewed in 3). Although A-P endonuclease is not a good substrate for DNA-PK, there is evidence for direct P53 involvement in base excision repair (4). Moreover, DNA-PK-Ku70/80 complex is one of several protein[1] complexes involved in the repair of double strand breaks (5). Double strand breaks result from interruption of the genome (e.g., ionizing radiation or metabolic

[1]These include DNA PK, Ku 70/80, DNA ligase IV, XRCC4, and the Mre II/Rad 50 complex.

cleavage (e.g., meiotic or V[D]J recombination). Repair takes place by homologous recombination or non-homologous joining. Failure to repair double strand breaks by either of these pathways threatens genomic stability (6). The Ku heterodimer has been directly implicated in DNA repair in part because it protects DNA ends against unwarranted nucleolytic degradation and brings the two ends together (7, 8). DNA damage activates the DNA-PKcs. The catalytic segment of the latter binds to DNA; however, the kinase activity is minimal in absence of Ku[2] (9). Ku itself shows high affinity for all forms of double stranded DNA. The binding of Ku to DNA has a triple effect: it stabilizes the binding of DNA-PKcs to DNA, triggers the sliding of Ku on the DNA toward the break sites, and increases the activity of DNA-PKcs (~ 8 times) (10, 11).

These findings led to the development of the following model for the interaction of DNA-PKcs and Ku at the site of DNA double strand breaks: Ku detects the damaged DNA, binds to it, moves to the site of the double strand breaks without the need of adenosine 5′-triphosphate (ATP) for translocation, recruits DNA-PKcs, and activates it (11). Of course, at some point in the sequence other components also known to be required for double strand breaks repair must be recruited (DNA ligase IV, XRCC4, and MRE11-RAD51-NBS-1 complex) (**FIGURE 11-1**).

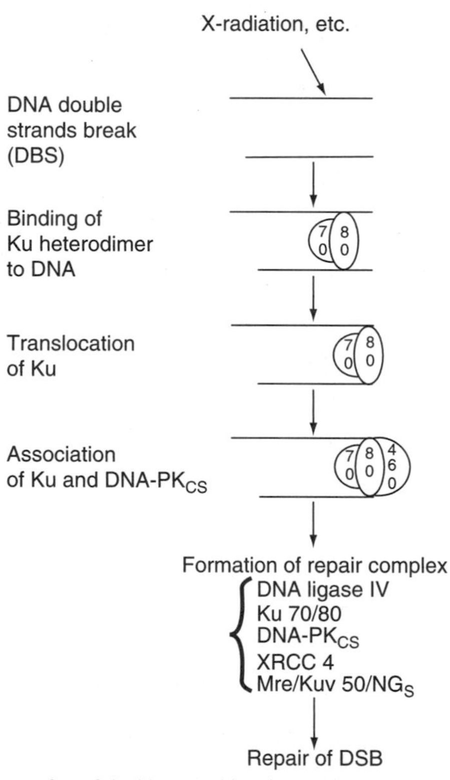


**FIGURE 11-1** Schemetic representation of double strand breaks repair.

[2]Ku was discovered by Mimori and associates in 1981 in sera of patients with polymyositis/scleroderma. It is a high molecular weight acidic nuclear protein that acts as an autoantibody. *XRCC6* encodes the 70 kDa monomer and the 86 kDa monomer is encoded by the *XRCC5* gene.

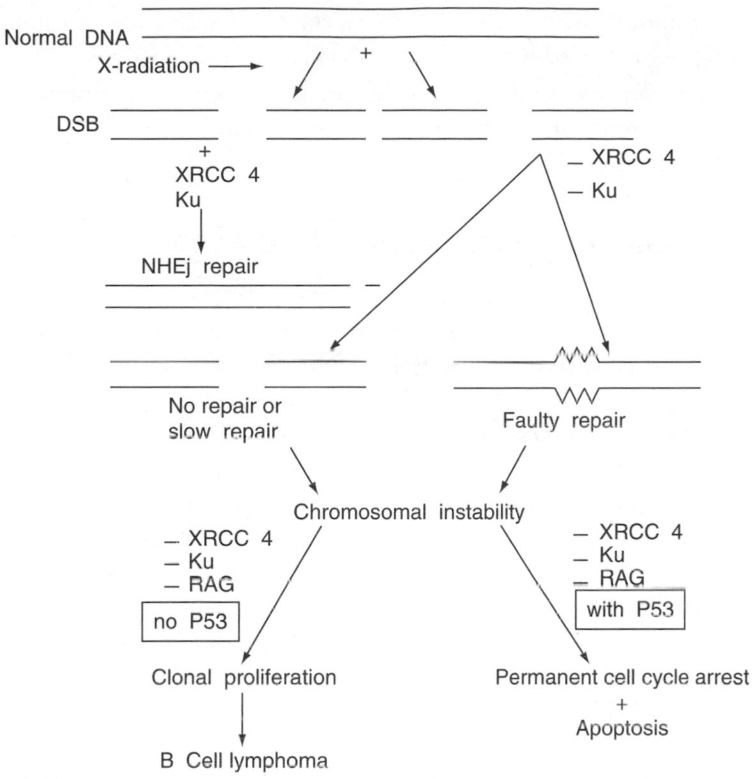

**FIGURE 11-2** Schematic representation of NHEj showing the repair process and the paths to cancer or apoptosis.

**FIGURE 11-2** illustrates the path to cancer in absence of repair of double strand breaks. DiFilippantonio et al. (12) and Gao et al. (13) showed that mice defective in XRCC4 and Ku are not only unable to either correctly repair or simply reconnect DNA double strand breaks. In either case, the defects also led to chromosomal instability, which in the absence of P53 was associated with clonal proliferation and the appearance of B cell lymphomas at a young age (11, 12).

- Ku 80 deficient mice live, but are afflicted with severe immunodeficiency because defective joining of the breaks made during V(D)J recombination prevents rearrangement of antigen receptor genes (14). The Ku 30 deficient mice also show increased levels of apoptosis.
- Mice deficient in XRCC4 are immunodeficient, develop severe apoptosis of the neurons, and die during embryogenesis, but these embryos can be saved by P53 deletion only to have the young mice develop B cell lymphomas resembling Burkitt lymphomas (13). The chromosomal changes in these lymphomas include specific translocations.
- Triple mutants were obtained by mating K80- and P53-deficient mice with mice deficient in the *RAG* gene. As anticipated, the triple mutants were not able to produce DNA end joining by V(D)J recombination, since they lack the *RAG* gene whose product activates recombination. Moreover, since these mice also lack a protein (Ku 80) required for non-homologous end joining, they are unable to repair these ends and, therefore, are unable to generate translocation and gene amplification. Consequently, they remain lymphoma-free (14) probably because of permanent arrest of

the cell cycle or apoptosis. In any event, these elaborate experiments establish that XRCC4 and Ku are required for NHEj repair[3] (Figure 11-2) (16).

In conclusion, the DNA recombination pathways are essential to maintenance of the integrity of the genome and, when defective, they lead to cancer in absence of P53 and to apoptosis in the presence of P53.

## *ATM* Gene, P53, and the Response to DNA Damage

When mutated, the wild-type gene that causes ataxia telangiectasia (*ATM*) includes a domain that encodes a protein kinase. The ATM protein has been shown to function upstream of P53, and it is believed to recognize DNA damage and phosphorylate P53. The details of the gene's function remain unknown, and yet sequence comparison with homologous genes in mice, *Saccharomyces cerevisiae*, and *Drosophila* suggests that the *ATM* gene is a member of a family of genes that regulate the cell cycle (discussed later in Chapter 23).

Ataxia telangiectasia is an autosomal-recessive hereditary disease with a highly pleiotropic phenotype. The mutated gene maps on chromosome 11q23.1 and spans over 184 kb genomic DNA. It gives rise to a messenger RNA (13 Kb) with an open reading frame of 9168 nucleotides. The AT protein (350 kDa) is made of 3056 amino acids (17).

Ataxia telangiectasia is caused by mutation of the *ATM* gene. The mutated phenotype includes:

- Progressive cerebellar ataxia
- Telangiectasias of the eyes and skin
- Immunodeficiencies
- Premature aging
- High levels of α fetoprotein in serum
- Increased sensitivity to ionizing radiation but not ultraviolet light
- Defective transduction of signals triggered by ionizing radiations
- Marked chromosomal instability
- Predisposition to cancer (T cell polymorphocytic leukemia and B cell lymphocytic leukemia)

This heavy and varied panoply of manifestations led researchers to suspect that the nonmutated gene must function in some way by protecting the integrity of the genome (18). This suspicion was soon supported by the discovery of the phosphoinositol-like kinase activity associated with the 350 amino acids of the COOH terminal region of the large AT protein, a function related to DNA damage repair (17). The discovery that the kinase phosphorylates P53 further supported this notion. However, P53 is not the only target of the AT protein phosphorylating activity. CAbl, c-Jun ctIP (a polypeptide that complexes with BRCA-1) (19), and other molecules are also phosphorylated (20).

[3]Because cells defective in DNA-PK do not prevent P53 cell cycle arrest, it has been suggested that the kinase is unlikely to recognize DNA damage in an event that occurs upstream of the cell cycle arrest. However, it cannot be excluded that one or more other molecules function upstream of P53 and contribute to the recognition of DNA damage and that DNA-PK is only one of them. Even if DNA-PK is not involved in the recognition of DNA damage, it operates upstream of P53 and is required for P53 function.

DNA is a dynamic structure modulated throughout the entire lifespan of the cell. It must be integrally replicated anytime a cell divides, correctly transcribed anytime synthesis of new or replacement of proteins is required, appropriately repaired after DNA damage, and fully degraded when the cell dies. The pleiotrophic manifestations of ataxia telangiectasia and the large size of the AT molecule imply that the AT molecule plays multiple functions, and phosphorylation of molecules involved in maintaining the integrity of the genome may be only one of them. A role in signal transduction has been suggested. For example, lymphocytes of AT patients fail to proliferate in response to various membrane stimuli, among the application of phorbol myristate acetate (21, 22), because of impairment of cytoplasmic-nuclear signals. It also has been shown that the *ATM* gene may contribute to apoptosis (23).

Cells obtained from ataxia telangiectasia heterozygote patients are more sensitive to ionizing radiation and patients afflicted with ataxia telangiectasia are also oversensitive to ionizing radiation. Moreover, P53 activation is slow in ATM cells exposed to $\gamma$ radiation (24, 25). These findings suggest a connection between P53 activation and the increased radiosensitivity of AT patients.[4]

The *AT* gene also contributes to the transient cell cycle arrest that follows DNA damage, thus giving the cell time to repair its DNA prior to its replication, and thereby preventing the DNA lesion from being transferred to the daughter cells (26, 27).

Before any protein can contribute to the repair of any kind of DNA or chromosomal lesions the damage must be recognized. Does the AT protein operate alone or in cooperation with other molecules in sensing DNA damage, and thereby contribute to the regulation of the cell cycle checkpoints? To try to understand the potential role of this protein, first consider what was learned from yeast. In *Saccharomyces pombe*, at least six proteins have been identified that contribute to the detection of alterations of the DNA structure and the regulation of different checkpoints in the cell cycle (28).[5] They are: RAD1, RAD3, RAD9, RAD17, RAD26, and HusI. Each of these molecules contributes to the regulation of several cell cycle checkpoints (**TABLE 11-1**) (29).

The RAD3-RAD26 complex is believed to be the "first sensor" of DNA damage (30). Mammalian homologues of each of these genes and regulatory proteins except RAD-

**Table 11-1 Molecules that contribute to cell cycle regulation**

| Molecule | Action |
|---|---|
| RAD1 | A 3′-exonuclease that nicks single strand DNA tails; the enzyme complexes |
| RAD3 | A 3′-5′ helicase that unwinds the DNA at the site of the lesion |
| RAD9 | Controls the $G_1$/S and $G_2$/M checkpoints and may contribute to the S checkpoint; the protein has exonuclease activity. |
| RAD17 | Controls the $G_1$/S, $G_2$/M, and meisosis checkpoints, and contributes to the control of the S checkpoint. |
| RAD26 | Interacts with RAD9 |
| Hus1 | Complexes with RAD1 |

[4]However, P53 activity is normal in *ATM* null cells exposed to adriamycin, a finding in conflict with a potential role for ATM in DNA damage recognition, at least after adriamycin exposure.

[5]The operation of checkpoints is discussed in a later section.

26 have been identified. In human hRAD9, hRAD1, and hRAD17 complements, the function of their homologues in yeast mutants and the hRAD1, hRAD9-hHus complex associates with chromatin after DNA damage (31, 32).

The product of the gene mutated in ataxia telangiectasia (*ATM* gene) and in ataxia telangiectasia-related gene (*ATR* gene) show structural and functional similarities with the *S. pombe* RAD3 and cells obtained from patients with ataxia telangiectasia have cell cycle checkpoint defects in $G_1$ and $G_2$/M (22) reminiscent of those caused by defects of RAD9 and RAD17.

What is the contribution of the product of the *ATM* gene to sensing DNA damage after ionizing radiation? Constitutive hRAD9 is a phosphorylated protein. After exposure to radiation in vitro or in vivo, hRAD9 is hyperphosphorylated on $Ser^{272}$ by ATM. The hyperphosphorylated hRAD9 complexes with hRAD1 and hHus1. The latter complex can then join a hRAD17 complex. The ultimate multiprotein complex identifies the DNA break sites, binds to them, and activates the cell cycle checkpoint (**FIGURE 11-3**).

The *ATR*[6] gene may be similarly activated after exposure to ultraviolet radiation or hydroxyuridine (31).

In conclusion, after exposure to clinically significant doses of ionizing radiation, the phosphokinase activity of the AT protein is activated. Several proteins critical to recognition of DNA damage and control of cell cycle checkpoints are turned on, thus allowing DNA repair to take place.

What is the relationship between the *ATM* gene and P53? After exposure to ionizing radiation ATM phosphorylates the serine 15 of P53 (33). Phosphorylation is critical for P53 function; after phosphorylation MDM2 is released from its attachment to P53 and the transcription activation of P53 is put in gear. P21 and GDD45 activate the cycle checkpoints, Bax contributes to the induction of apoptosis, and MDM2 exerts its negative feedback on P53 transcription, all in due time (see below) (**FIGURE 11-4**).

Brown and associates discovered that Ku70 is up-regulated in human fibroblasts nuclei after exposure to ionizing radiation. Both ATM and P53 are required for up-regulation. The significance of the Ku70 up-regulation remains unresolved. What is certain is that it is not required for DNA-PK activation (34).

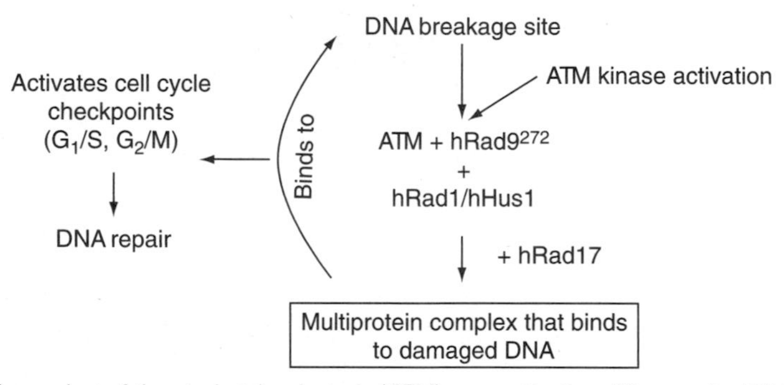


**FIGURE 11-3** The product of the ataxia telangiectasia (*ATM*) gene activation. After sensing DNA damage because of ionizing radiation, the *ATM* gene product binds to $hRAD9^{272}$ and complexes with hRAD1 and hHus1. The resulting multiprotein then identifies the DNA strand break sites and binds to them, thus activating the cell cycle checkpoints at $G_1$/S and $G_2$/M. Note: 272 refers to the serine 272 (S272) that is phosphorylated.

[6] ATR, is like ATM, a large nucleoprotein that harbors protein kinase activity. It activates the $G_1$/S and $G_2$/M checkpoints when DNA is damaged and thereby blocks DNA replication. This type of function is in keeping with the overexpression of ATR-kia kinase, an inactive protein that causes an increased sensitivity to ultraviolet (UV) and ionizing radiations, and defective cell cycle arrest.

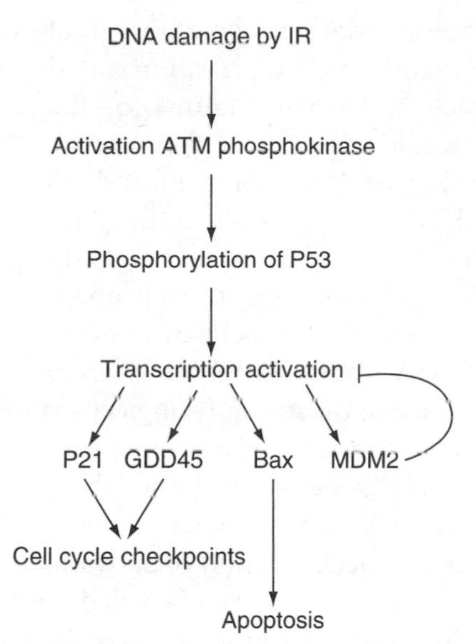


**FIGURE 11-4** Relationship between *ATM* and P53. After exposure to ionizing radiation (IR), ATM phosphorylates the serine 15 of P53, which causes MDM2 to be released, thus activating P53-targeted transcription. P21 and GDD45 activate the cell cycle checkpoints for repair, Bax contributes to apoptosis and MDM2 exerts negative feedback on P53 transcription.

## *BRCA1* and *BRCA2* Genes

Two other nuclear proteins critical to genome integrity are implicated in sensing DNA damage: BRCA1 and BRCA2. Both are phosphorylated by the ATM kinase and mutations of the genes that encode them are associated with some cancers of the breasts and ovaries. A panoply of molecular functions is attributed to both these molecules, but it remains unclear when and how much these functions contribute to cellular adjustments before or after DNA damage. The BRCA1 gene (*brca-1*) is located on chromosome 17 (17q21).

The BRCA2 gene (*brca-2*), located on chromosome 13 (13q12/q13), is composed of 25 exons and generates several splice variants. There are no sequence similarities with *brca-1*; however, exon 3 shares some sequence similarities with c-*jun*. Again, the molecule has several binding sites and RAD 51 is a major protein found to be associated with BRCA-2.

The products of those two genes, BRCA-1 and BRCA-2, contribute to the stability of chromosomes. Chromosomal anomalies accumulate in *brca-1* and *brca-2* mutant cells probably as a result of defective DNA repair. Carriers of *brca-1* and *brca-2* germ line mutations have hereditary predispositions to breast cancer and show chromosomal rearrangements more frequently than patients with sporadic breast cancer. The functions of BRCA1 and BRCA2 are discussed further in Chapter 25.

## P53 Activation, Phosphorylation, and Acetylation

Molecules operating upstream of P53 must recognize the presence of DNA damage and activate P53 functions by stabilizing the molecule. The half-life of P53 is measured in minutes and, thus, activation of the molecules demands that it be immediately stabilized. Fast activation (or inactivation) of a protein is often achieved by phosphorylation or dephosphorylation. There is no doubt that P53 can be phosphorylated in vitro. Evidence that P53 is also phosphorylated in vivo is accumulating (35, 36). Faced with these facts and what is known of the structure and the catalytic capacity of DNA-PK, it would appear that the latter qualifies as a good model for both functions: recognition of DNA damage and phosphorylation of P53. The attraction of the DNA-PKcs by Ku to the site of the damage and the activation of the kinase by the same, coupled to the finding that DNA-PK phosphorylates P53 (most likely on the right serine residues: serine 15 in humans, serine 18 in mice) suggest that DNA-PK is uniquely suited for the activation of P53.

However, DNA-PK is not without competition. Other members of the phosphoinositide-3-kinase family, that is, the ataxia telangiectasia mutated (ATM) kinase, the ataxia telangiectasia related kinase (ATR), and most likely others, also fit some of the criteria for DNA damage recognition and P53 phosphorylation. Add to these findings the fact that, at least in vitro, P53 can be phosphorylated at multiple sites and it is no surprise that at first glance the mechanism(s) of activation of P53 through phosphorylation appear confusing. Of course, it cannot be ruled out that the phosphoinositide-3-kinase family may be involved with other kinases, in a sequence that starts with recognition of the damage and continues with P53 activation by other kinases in a cascade of phosphorylations.

The existence of many sites of phosphorylation (at least 13 sites; 8 in the N terminus and 5 in the C terminus) is also in keeping with the dynamics of the P53 molecule. Between the moment of P53 activation (by phosphorylation?) and that of its inactivation by degradation, the molecular functions include:

- Translocation to the nucleus
- Tetramerization
- DNA specific binding
- Transactivation of several genes that:
  - Command arrest of the cell cycle (e.g., through *p21* and *Gdd4*)
  - Trigger P53 inactivation by causing P3 degradation (*mdm2*)
  - Cause apoptosis (*Bax*)
  - Activate DNA repair P21 (*GDD4*)
  - Trigger resumption of the cell cycle

Such multiple functions might explain why P53 is phosphorylated at different sites and it may be that other posttranscription events (such as acetylation) further contribute to modulate the functions of the molecule (**TABLE 11-2**) (37–39).

Attempts to correlate the sites of P53 phosphorylation with its molecular functions have yielded fruit. The correlations are suggested from available data, but remain inconclusive in many cases. Indeed, many different kinases have been shown to phosphorylate P53 and the correlation between kinase and site of phosphorylation is not always consistent. This could result from differences in species, cell types, and modes of DNA damage (e.g., UV versus ionizing radiations). Note that the pattern of phosphorylation in general does not vary with exposure to UV or ionizing radiations, but the enzymes involved may vary (e.g., DNA-PK vs. ATM) (**TABLE 11-3**).

There are also indications that, all other variables being constant, P53 may be phosphorylated at more than one site either at the same time or in sequence. Phosphorylation of serine 15 precedes that of threonine 18 (40, 41).

**Table 11-2 P53 activity**

| Assumed Function Activation | Amino Acid Phosphorylated | Enzyme Involved |
|---|---|---|
| Nuclear localization | Serine 315 | Cyclin/cdk2 |
| Tetramerization | Serine 392 | Casein kinase |
| Inhibition of MDM2 binding and | Serine 15 | ATM/ATR, ERK 12 |
| P53 stabilization | Serine 20 | CHK1, CHK2 |
| | Serine 37 | DNA-PK |
| DNA sequence specific binding | Serine 37 | Protein kinase C |
| | Serine 315 | Cyclin A/cdk2 |
| | Serine 392 | |
| P53 transactivation | Serine 37 | DNA-PK |
| | Serine 392 | P38 |
| Apoptosis | Serine 46 | P38 kinase |
| Inhibition of 14-3-3 binding | Serine 376 | Protein kinase C |
| | Serine 392 | |
| Enhancement of 14-3-3 binding | Serine 378 | Protein kinase C |
| | Serine 392 | Double stranded RNA activated protein kinase |
| Unknown | Serine 6 | Casein kinase 1 |
| | Serine 9 | Casein kinase 1 |
| Second phosphorylation following that of Ser 15 | Threonine 18 | Casein kinase 1 |

Correlation between site of phosphorylation and potential sites of P53 activity (see Zoe Stewart and Jennifer Pietenpol, 38). Stoichiometric phosphorylation at serine 315 has been associated with P53 dependent transcription (Blaydes et al., 39).

In summary, phosphorylation of P53 is likely required for function. However, the site of phosphorylation on the serine residues in the P53 sequence may vary with the cell type, the animal used in the experiment, the nature of the DNA damage inflicted, or the functional mode of P53. Activation of the molecular function(s) of P53 is likely to involve a cascade of phosphorylations, the exact sequence of which needs to be determined in many cases.

An enzyme (P300 histone acetyltransferase) at first believed to acetylate only histones, was found to acetylate P53 as well (**TABLE 11-4**) (42–44). Later, it was shown that the P300 histone transacetylase was not the only acetylase that uses P53 as a substrate; P/CAF does it, too. Moreover, acetylation of P53 occurs in vivo after genotoxic stress, but at different sites in the P53 molecule depending whether the P300 histone acetyltransferase or P/CAF is involved (37).

It was at first assumed that P53 acetylation facilitates P53 binding to DNA, but this was disproved (45). Although the exact role of P53 acetylation is still unclear, Prives and others have suggested a plausible model (46–48): Acetylation of P53 coupled to

**Table 11-3 Enzymes involved in P53 phosphorylation**

| Kinase | Sites of P53 Phosphorylation |
|---|---|
| DNA-PK | Ser 15–Ser 37 |
| ATM | Ser 15 |
| ATR | Ser 36 |
| Casein kinases | |
| I | Ser 6, Ser 9, Ser 37 |
| II | Ser 292 |
| Protein kinase C | Ser 376, Ser 378 |
| Cyclin-dependent kinases | |
| Cyclin A/cdk 2 | Ser 315 |
| Cyclin B1/cdk 2 | Ser 315 |
| Double strand RNA-activated protein kinase | Ser 392 |
| Chk 1 | Ser 15, Ser 18, Ser 20 |
| Chk 2 | Ser 20 |

**Table 11-4 Some P53 sites of acetylation**

| | P300/Histone Acetyl Transferase | P/CAF |
|---|---|---|
| Site of acetylation in P53 | COOH terminus | Links region between the DNA binding and the tetramerization domains |
| Lysines acetylated | K 372<br>K 373<br>K 381<br>K 382 | K 320 |

that of histones contributes in changing chromatin structure, thereby facilitating the recruitment of the transcription machinery, including transcription factors, modulators, and RNA polymerase.

In summary, the operation is postulated to occur in the following steps:

- P53 is bound to DNA and the binding is independent of its acetylation.
- Phosphorylation of P53 frees it from MDM2 and activates it.
- Acetylases (P300/histone acetyltransferase and P/CAF acetylate P53 and histones.
- Histones and P53 acetylation facilitate the recruitment of the transcription machinery:
  a. Transcription factor
  b. Mediators
  c. RNA polymerase II and transcription activation
- Deacylation of P53 and histones allows cell cycle reactivation

The current hypothesis has the advantage of providing a schematic vision for P53 transcription activation. P53 transactivates the transcription of P21, which in turn arrests the cell cycle and allows for DNA repair; when the repair has been completed, the cycle can resume.

## References and Suggested Reading


1. Poltoratsky, V.P., Shi, X., York, J.D., Lieber, M.R., and Carter, T.H. Human DNA-activated protein kinase (DNA-PK) is homologous to phosphatidylinositol kinases. *J Immunol.* 1995;155:4529–33.
2. Connelly, M.A., Zhang, H., Kieleczawa, J., and Anderson, C.W. The promoters for human DNA-PKcs (PRKDC) and MCM4: divergently transcribed genes located at chromosome 8 band q11. *Genomics.* 1998;47:71–83.
3. Strauss, P.R., and O'Regan, N.E. Structure and function of major human AP endonuclease HAP1/ref12001. In: Nickoloff, J.A., and Hoekstra, M.F, eds. *DNA Damage and Repair. Volume III. Advances from Phage to Humans.* Totowa, NJ: Humana Press; 2001:43–86.
4. Offer, H., Wolkowicz, R., Matas, D., Blumenstein, S., Livneh, Z., and Rotter, V. Direct involvement of p53 in the base excision repair pathway of the DNA repair machinery. *FEBS Lett.* 1999;450:197–204.
5. Paull, T.T., and Gellert, M. Nbs1 potentiates ATP-driven DNA unwinding and endonuclease cleavage by the Mre11/Rad50 complex. *Genes Dev.* 1999;13:1276–88.
6. Ferguson, D.O., Sekiguchi, J.M., Chang, S., Frank, K.M., Gao, Y., DePinho, R.A., and Alt, F.W. The nonhomologous end-joining pathway of DNA repair is required for genomic stability and the suppression of translocations. *Proc Natl Acad Sci USA.* 2000;97:6630–3.
7. Cary, R.B., Peterson, S.R., Wang, J., Bear, D.G., Bradbury, E.M., and Chen, D.J. DNA looping by Ku and the DNA-dependent protein kinase. *Proc Natl Acad Sci USA.* 1997;94:4267–72.
8. Liang, F., and Jasin, M. Studies on the influence of cytosine methylation on DNA recombination and end joining in mammalian cells. *J Biol Chem.* 1995; 270:23838–44.
9. Featherstone, C., and Jackson, S.P. Ku, a DNA repair protein with multiple cellular functions? *Mutat Res.* 1999;434:3–15.
10. Yaneva, M., Kowalewski, T., and Lieber, M.R. Interaction of DNA-dependent protein kinase with DNA and with Ku: biochemical and atomic-force microscopy studies. *EMBO J.* 1997;16:5098–112.
11. Hammarsten, O., and Chu, G. DNA-dependent protein kinase: DNA binding and activation in the absence of Ku. *Proc Natl Acad Sci USA.* 1998;95:525–30.
12. DiFilippantonio, M.J., Zhu, J., Chen, H.T., Meffre, E., Nussenzweig, M.C., Max, E.E., Ried, T., and Nussenzweig, A. DNA repair protein Ku80 suppresses chromosomal aberrations and malignant transformation. *Nature.* 2000;404:510–4.
13. Gao, Y., Ferguson, D.O., Xie, W., Manis, J.P., Sekiguchi, J., Frank, K.M., Chaudhuri, J., Horner, J., DePinho, R.A., and Alt, F.W. Interplay of p53 and DNA-repair protein XRCC4 in tumorigenesis, genomic stability and development. *Nature.* 2000;404:897–900.

14. Zhu, C., Bogue, M.A., Lim, D.S., Hasty, P., and Roth, D.B. Ku86-deficient mice exhibit severe combined immunodeficiency and defective processing of V(D)J recombination intermediates. *Cell.* 1996;86:379–89.
15. Roth, D.B., and Gellert, M. New guardians of the genome. *Nature.* 2000;404:823–5.
16. Woo, R.A., McLure, K.G., Lees-Miller, S.P., Rancourt, D.E., and Lee, P.W. DNA-dependent protein kinase acts upstream of p53 in response to DNA damage. *Nature.* 1998;394:700–4.
17. Savitsky, K., Sfez, S., Tagle, D.A., Ziv, Y., Sartiel, A., Collins, F.S., Shiloh, Y., and Rotman, G. The complete sequence of the coding region of the ATM gene reveals similarity to cell cycle regulators in different species. *Hum Mol Genet.* 1995;4:2025–32.
18. Gatti, R. Ataxia telangiectasia. In: Scriver, CR, Beaudet, AL, Sly WA, Valle, D, eds. *The Metabolic Basis of Inherited Disease,* 8th ed. New York, NY: McGraw-Hill; 2000:705–32.
19. Yu, X., Wu, L.C., Bowcock, A.M., Aronheim, A., and Baer, R. The C-terminal (BRCT) domains of BRCA1 interact in vivo with CtIP, a protein implicated in the CtBP pathway of transcriptional repression. *J Biol Chem.* 1998;273:25388–92.
20. Shafman, T., Khanna, K.K., Kedar, P., Spring, K., Kozlov, S., Yen, T., Hobson, K., Gatei, M., Zhang, N., Watters, D., Egerton, M., Shiloh, Y., Kharbanda, S., Kufe, D., and Lavin, M.F. Interaction between ATM protein and c-Abl in response to DNA damage. *Nature.* 1997;387:520–3.
21. Garcia-Perez, M.A., Allende, L.M., Corell, A., Varela, P., Moreno, A.A., Sotoca, A., Moreno, A., Paz-Artal, E., Barreiro, E., and Arnaiz-Villena, A. Novel mutations and defective protein kinase C activation of T-lymphocytes in ataxia telangiectasia. *Clin Exp Immunol.* 2001;123:472–80.
22. Lavin, M.F., and Shiloh, Y. The genetic defect in ataxia-telangiectasia. *Annu Rev Immunol.* 1997;15:177–202.
23. Hotti, A., Jarvinen, K., Siivola, P., and Holtta, E. Caspases and mitochondria in c-Myc-induced apoptosis: identification of ATM as a new target of caspases. *Oncogene.* 2000;19:2354–62.
24. Kastan, M.B., Zhan, Q., el-Deiry, W.S., Carrier, F., Jacks, T., Walsh, W.V., Plunkett, B.S., Vogelstein, B., and Fornace, A.J Jr. A mammalian cell cycle checkpoint pathway utilizing p53 and GADD45 is defective in ataxia-telangiectasia. *Cell.* 1992; 71:587–97.
25. Lu, X., and Lane, D.P. Differential induction of transcriptionally active p53 following UV or ionizing radiation: defects in chromosome instability syndromes? *Cell.* 1993;75:765–78.
26. Weinert, T. DNA damage and checkpoint pathways: molecular anatomy and interactions with repair. *Cell.* 1998;94:555–8.
27. Wang, J.Y. Cellular responses to DNA damage. *Curr Opin Cell Biol.* 1998; 10:240–7.
28. Hawley, R.S., and Friend, S.H. Strange bedfellows in even stranger places: the role of ATM in meiotic cells, lymphocytes, tumors, and its functional links to p53. *Genes Dev.* 1996;10:2383–8.
29. Rhind, N., and Russell, P. Mitotic DNA damage and replication checkpoints in yeast. *Curr Opin Cell Biol.* 1998;10:749–58.

30. Edwards, R.J., Bentley, N.J., and Carr, A.M. A Rad3-Rad26 complex responds to DNA damage independently of other checkpoint proteins. *Nat Cell Biol.* 1999;1:393–8.

31. Chen, M.J., Lin, Y.T., Lieberman, H.B., Chen, G., and Lee, E.Y. ATM-dependent phosphorylation of human Rad9 is required for ionizing radiation-induced checkpoint activation. *J Biol Chem.* 2001;276:16580–6.

32. Burtelow, M.A., Kaufmann, S.H., and Karnitz, L.M. Retention of the human Rad9 checkpoint complex in extraction-resistant nuclear complexes after DNA damage. *J Biol Chem.* 2000;275:26343–8.

33. Li, L., Story, M., and Legerski, R.J. Cellular responses to ionizing radiation damage. *Int J Radiat Oncol Biol Phys.* 2001;49:1157–62.

34. Brown, K.D., Lataxes, T.A., Shangary, S., Mannino, J.L., Giardina, J.F., Chen, J., and Baskaran, R. Ionizing radiation exposure results in up-regulation of Ku70 via a p53/ataxia-telangiectasia-mutated protein-dependent mechanism. *J Biol Chem.* 2000;275:6651–6.

35. Jimenez, G.S., Nister, M., Stommel, J.M., Beeche, M., Barcarse, E.A., Zhang, X.Q., O'Gorman, S., and Wahl, G.M. A transactivation-deficient mouse model provides insights into Trp53 regulation and function. *Nat Genet.* 2000;26:37–43.

36. Chao, C., Saito, S., Anderson, C.W., Appella, E., and Xu, Y. Phosphorylation of murine p53 at ser-18 regulates the p53 responses to DNA damage. *Proc Natl Acad Sci USA.* 2000;97:11936–41.

37. Sakaguchi, K., Saito, S., Higashimoto, Y., Roy, S., Anderson, C.W., and Appella, E. Damage-mediated phosphorylation of human p53 threonine 18 through a cascade mediated by a casein 1-like kinase. Effect on Mdm2 binding. *J Biol Chem.* 2000;275:9278–83.

38. Stewart, Z.A., and Pietenpol, J.A. p53 Signaling and cell cycle checkpoints. *Chem Res Toxicol.* 2001;14:243–63.

39. Blaydes, J.P., Luciani, M.G., Pospisilova, S., Ball, H.M., Vojtesek, B., and Hupp, T.R. Stoichiometric phosphorylation of human p53 at Ser315 stimulates p53-dependent transcription. *J Biol Chem.* 2001;276:4699–708.

40. Dumaz, N., and Meek, D.W. Serine15 phosphorylation stimulates p53 transactivation but does not directly influence interaction with HDM2. *EMBO J.* 1999; 18:7002–10.

41. Gu, W., and Roeder, R.G. Activation of p53 sequence-specific DNA binding by acetylation of the p53 C-terminal domain. *Cell.* 1997;90:595–606.

42. Sakaguchi, K., Herrera, J.E., Saito, S., Miki, T., Bustin, M., Vassilev, A., Anderson, C.W., and Appella, E. DNA damage activates p53 through a phosphorylation-acetylation cascade. *Genes Dev.* 1998;12:2831–41.

43. Herrera, J.E., Saito, S., Miki, T., Bustin, M., Vassilev, A., Anderson, C.W., and Appella, E. DNA damage activates p53 through a phosphorylation-acetylation cascade. *Genes Dev.* 1998;12:2831–41.

44. Pearson, M., Carbone, R., Sebastiani, C., Cioce, M., Fagioli, M., Saito, S., Higashimoto, Y., Appella, E., Minucci, S., Pandolfi, P.P., and Pelicci, P.G. PML regulates p53 acetylation and premature senescence induced by oncogenic Ras. *Nature.* 2000;406:207–10.

45. Barlev, N.A., Liu, L., Chehab, N.H., Mansfield, K., Harris, K.G., Halazonetis, T.D., and Berger, S.L. Acetylation of p53 activates transcription through recruitment of coactivators/histone acetyltransferases. *Mol Cell.* 2001;8:1243–54.

46. Prives, C., and Manley, J.L. Why is p53 acetylated? *Cell.* 2001;107:815–8.

47. Abraham, R.T. PI 3-kinase related kinases: 'big' players in stress-induced signaling pathways. *DNA Repair (Amst).* 2004;3:883–7.

48. Iliakis, G., Wang, H., Perrault, A.R., Boecker, W., Rosidi, B., Windhofer, F., Wu, W., Guan, J., Terzoudi, G., and Pantelias, G. Mechanisms of DNA double strand break repair and chromosome aberration formation. (Review). *Cytogenet Genome Res.* 2004;104:14–20.

# P53 AND THE CELL CYCLE

# 12

P53 must arrest the cell cycle at least at the $G_1/S$ checkpoint to allow DNA repair to take place. To facilitate an understanding of this important function of P53, an overview of the cell cycle regulation is briefly presented.

## Overview of Cell Cycle Regulation

Quiescent cells enter the cell cycle after stimulation with growth factors. In the absence of a $G_0$ phase, as is the case for cells in culture, the cells are released into the $G_1$ phase soon after mitosis. Over 30 years ago, Temin demonstrated that the growth of chicken cells switched from growth factor dependence to growth factor independence prior to entering the S phase (1). The $G_1$ phase prepares the cell's entrance into other phases of the cycle. In $G_1$, proteins and RNAs are synthesized, but DNA is not. Pardee showed that in quiescent cells the $G_1$ can be divided in two stages (2): the first is growth factor dependent and the second is growth factor independent (**FIGURE 12-1**). The two stages are separated by the restriction point (R). At the R point, the cell makes the commitment to replicate its chromosomes or to differentiate (and sometimes in the presence of the proper stimuli, to enter apoptosis). DNA synthesis, a critical event in the replication of the chromosome, takes place during the S phase. Before that phase, during $G_1$, the cell makes an inventory of its biochemical resources and determines whether its substrates, enzymes, adenosine 5′-triphosphate (ATP), etc., are adequate to secure the orderly passage from $G_1$ to S, and thereby ensure the integral replication of the DNA of the mother cell. If the DNA cannot be integrally replicated because of DNA damage, the cell must first repair the damage. If repair fails, the cell must die rather than generate a progeny of mutants that may cause inborn errors of metabolism (when derived from germ cells) or cancer (when derived from somatic cells). P53 plays a critical role in determining whether the cell should arrest in $G_1$ or enter apoptosis.

Passage from one phase of the cycle to another is regulated in part by cyclins and cyclin-dependent kinases. Cyclins are members of a population of proteins that rise when the cell enters a new phase of the cycle and drop when the phase is completed. The cyclins activate cyclin-dependent kinases. There are four different types of cyclins in mammalian cells (cyclins A, B, D, E); the D type (D1, D2, D3) binds to two distinct, but homologous cyclin-dependent kinases (cdk4 and cdk6). The cdks are only active when complexed with cyclins. In these complexes, the cdks are the catalytic units and the cyclins are the necessary cofactors. These complexes regulates the cell's passage from the $G_1$ to S phase as follows:

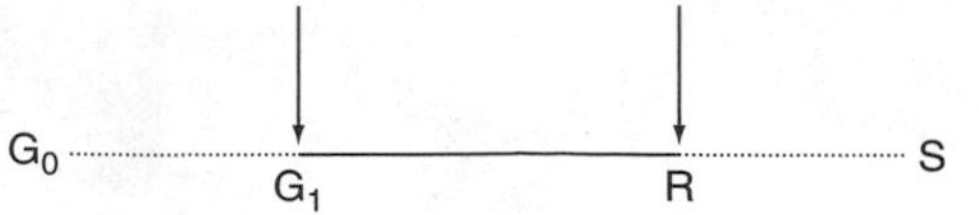


**FIGURE 12-1** The two phases of $G_1$: One, growth-factor dependent (—); the other, growth factor independent.

- The D type cyclins complex with cdk4 and cdk6, and are involved in the $G_1$/S transition phase from $G_1$ to S.
- The E cyclins bind to cdk2 and also contribute to the regulation of the passage from $G_1$ to S.
- The A cyclins bind to cdk2 and regulate the S phase.
- The B cyclins bind to cdk1 and regulate the passage from $G_2$ to M.
- The cdks are themselves inhibited by distinct cdk inhibitors that include proteins of the P21 family[1] and the INK4 family.

Although the role of P53 and P21 in the control of the cell cycle is not limited to the passage of $G_1$ to S, the latter phase is the most investigated. After Rb phosphorylation, E2F is released and the progress of the cycle is blocked at the restriction point (R), a few hours before the cycle enters the S phase (Figure 12-1). Passage through R requires that the D cyclins and the cdk4/cdk6 form complexes. The cdks are constitutively expressed and the rise in the levels of cyclin D requires growth factors. In contrast to cyclin D, cyclin E is present in quiescent cells and complexes with cdk2, but requires activation for function. Cyclins can be activated by decreasing the levels of cdk inhibitors (P21 or P27, P57, and the INK4 family), by activating the inhibitor's degradation, for example, in the proteosome or by modulation of the activity of the growth factor(s). In a next step preceding the entry of the cycle in the S phase, cyclin A interacts with cdk2.

In absence of stress, activation of cdk4/cdk6 phosphorylates the Rb protein (discussed further below). In quiescent cells the Rb protein is bound to transcription factors of the E2F family. The Rb protein when bound to E2F functions as a repressor of transcription. Phosphorylation of Rb releases E2F transcription factor(s), leading to the transcription of several genes that encode, among other proteins, cdk2 and cyclins A and E (**FIGURE 12-2**), and the cycle continues.

Under genotoxic stress, activated P53 activates the transcription of several new genes. The *p21* gene is the one of immediate concern. The induction of the P21 protein, an unspecific and potent kinase inhibitor that blocks the activities of cdk4 and cdk2, thus prevents RB phosphorylation. The *Rb* suppressor gene remains bound to E2F and activation of the genes that engage the progress of the cycle does not take place. The cycle is arrested in $G_1$/S (**FIGURE 12-3**) (3). P21 is a cyclin kinase with broad specificity. Therefore, it is not surprising that it also interferes with the $G_2$/M and the spindle checkpoints.

[1]The activities of the cyclin kinases are in turn modulated by two groups of cyclin kinase inhibitors, the P21 family that also includes P27 and P57, and the P16 and $P19^{Arf}$ family. This chapter focuses on the role of the P21 family in the regulation of the cell cycle and then returns to the P16/$P19^{Arf}$ family.

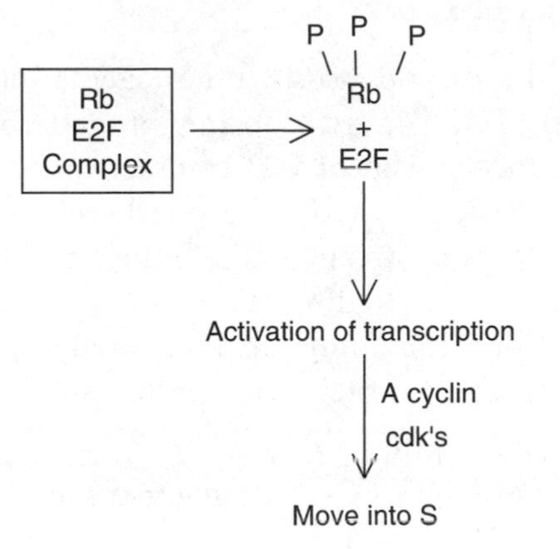


**FIGURE 12-2** Rb phosphorylation.

The p21 (WAF1, CIP1, Sdi1) (4) or the *cDKNIA* gene encodes the P21 protein (CDKNIA), a nuclear protein that inhibits cyclin-dependent kinases. The kinase inhibitor participates in regulation of: cell growth, some forms of terminal differentiation, apoptosis, and senescence. The transcription of the *p21* gene is activated by P53.

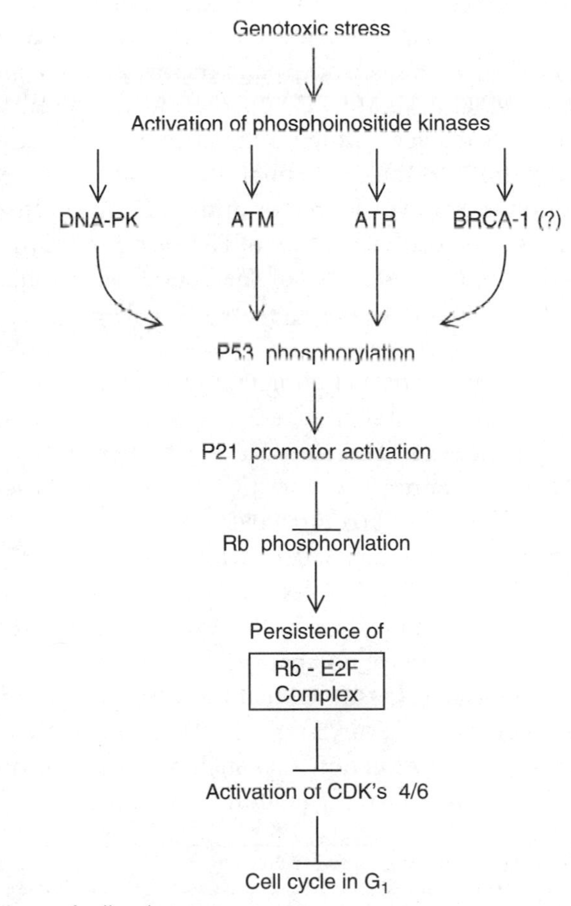


**FIGURE 12-3** The pathway of cell cycle arrest.

## Gene Transactivation by P53

The *p21* gene is induced by P3 and also by other agents such as retinoic acid, transforming growth factor-β (TGF-β), cyclosporine, progesterone, interferon-γ (IFN-γ), and phorbol esters. The mechanism of P53 transactivation of its target genes is far from completely clear, but fragments of the puzzle are beginning to fall together. Because P53 induction occurs upstream of the induction of its targets (among them P21, MDM2, and Bax) it is logical to discuss it here. The transcription transactivation domain of P53 is located in a 150 amino acid terminal sequence. To effectuate transcription, P53 recruits a group of proteins needed for the process, including:

- The TATA box binding protein (TBP)
- The TAF II component of TFII D, a molecule that binds to amino terminal of P53 (5,6).

The CREB binding protein (7, 8), the transcription coactivation protein P300, the P300 (9) cofactors (1Mγ and PCAF), and AMF1 (a modulator of P53 transactivation) are included among the proteins that contribute to transcription transactivation. The protein associates with P53 in vivo and in vitro and enhances the P53 response (e.g., for $G_1$ arrest and apoptosis) by increasing P53-dependent transactivation (of, e.g., P21) (10). Unfortunately, the precise interaction of these molecules among each other or with others, or their sequence of action in the process of P53 genes' transactivation is still not completely known.

There are two DNA motifs involved in the *p21* gene expression: a distal motif that extends from the nucleotide in position –2300 to that in position –210, and a proximal motif located between the –210 and +1 base pairs relative to the transcription initiation site. P53, retinoic acid, and vitamin $D_3$ operate on the distal whereas phorbol esters, TGF-β, and progesterone act on the proximal motifs. The latter contains GC-rich DNA sequences that bind to the Sp1 family of ubiquitous transcription factors. c-jun is, after dimerization, a positive modulator of the P21 promotor via Zip domains located in the 122–331 regions. Transactivation of P21 by c-jun is Sp1-dependent (11). Several molecules modulate the expression of the gene that encodes P21; some up-regulate it (e.g., c-jun, thioredoxin) and others down-regulate it (e.g., thymidylic synthetase).

P53 is activated by genotoxic stress including that generated by ultraviolet (UV) and X-radiations. The cell contains a number of reducing agents, and major among them are thioredoxine and glutathione. Therefore, it is of interest that thioredoxin[2] is induced by oxidative stress and appears to be critical to the regulation of DNA binding by P53, P21 induction, and cell cycle arrest (**FIGURE 12-4**).

Down-regulation of the *p21* gene is associated with a decrease in P21 mRNA and the P21 protein and impairment of cell cycle arrest, and causes resistance to chemotherapy and radiotherapy. The mechanism involved in the down-regulation of P21 expression remains unknown. However, there are no changes in P53 activity associated with the down-regulation and, therefore, it is assumed that the latter is P53-independent. Some cancer cells do not up-regulate the expression of P21 triggered by interferon. IFN-γ activates a pathway of signal transduction that includes the transducer activator of transcription: the STAT I protein. In some rhabdomyosarcomas or rhabdomyosarcoma

[2]Thioredoxin functions as a dithiol hydrogen donor for several proteins, including the ribonucleotide reductase(s). The active site includes highly conserved sequences (e.g., two cysteine residues, Cys 32 and Cys 35). Thioredoxin catalyzes the reversible oxidation-reduction of the cysteines in the presence of NADPH.

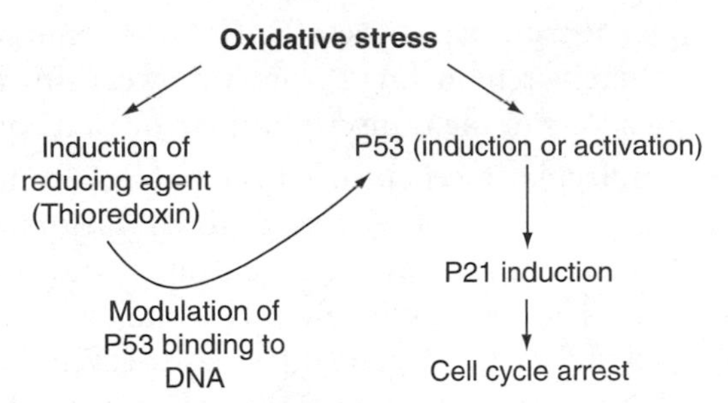


**FIGURE 12-4** Potential mechanism of induction of P53 and P21 by oxidative stress.

cell lines, P21 expression by INF-γ does not take place. Chen and associates have shown that:

- In normal cells, a STAT-1 responsible element, the *sis* inducible element (SIE-1), contains a "CpG sequence" (located upstream of the P21 CpG island) that is incompletely methylated.
- In contrast, in cancer cells the CpG sequence is completely methylated.
- Demethylation of SIE-1 allows expression of P21.

In conclusion, modulation of the methylation of the P21 promotor region prevents binding of Stat I to SIE-1 and blocks P21 expression in at least some cells (12).

## The Rb Protein

The Rb protein is the product of a tumor suppressor gene that is a member of a large class of genes whose products normally regulate cellular proliferation. When mutated, these genes become the primary cause of some cancers (retinoblastoma and osteosarcoma) and a frequent participant for many other cancer types. However, to cause cancer, both alleles need to be mutated. Therefore, these genes are referred to as recessive oncogenes,[3] and since many are involved in regulation of the cell cycle, they are also called suppressor genes.

The wild-type gene (responsible when mutated for the eye cancer retinoblastoma) encodes a 928 amino-acid phosphoprotein that plays a critical role (the Rb protein) in cell cycle regulation by shuttling between phosphorylation and dephosphorylation. The hypophosphorylated Rb forms a complex with transcription factors (the E2F family of transcription factors), which activate genes encoding, among others, proteins involved in the S phase of the cycle (13, 14). The gene that encodes the Rb protein is located on chromosome 13; it harbors a 200 kb locus (13q14) that encodes a 110 kDa protein mostly found in the nucleus, a suppressor gene expressed in all human and rat tissues.[4] In humans and some animals, several mutations of the Rb protein cause retinoblastoma by a mechanism predicted by the two-hit hypothesis, which is discussed in the introduction to the section of the life and death of the cancer cell.

During the cell cycle, the Rb protein inhibits the E2F transcription factors by two different mechanisms:

1. It binds to the E2F transactivation domain and thereby blocks its transcriptional activation.

[3]They are called recessive oncogenes in contrast to oncogenes. In the case of oncogenes, only one allele needs to be mutated to cause the malignant phenotype.

[4]The Rb protein forms complexes with over 50 proteins, but prominent among them are the E2F proteins.

2. The Rb/E2F complex recruits enzymes (HDACs) that remove acetyl groups from histone octamers, thereby remodeling chromatin by causing nucleolar condensation and impairing access of the transcription factor to its promoter.

Investigations of the molecular mechanism of the passage from the first (requiring growth factors) to the second stage (not requiring growth factors) of $G_1$ and the completion of the interval that separates $G_1$ from the entrance into the S phase have been clarified to a large extent. The process involves two major groups of activators of the cycle: the cyclins and the cyclin-dependent kinases (cyclinD/Cdk4-Cdk6, cyclin E/Cdk-2, cyclin A/Cdk-2) as well as inhibitors of the cycle among them inhibitors of the cyclin D4 kinase (the P21 family of unspecific and the INK-4 family of specific cyclin kinase inhibitors). The concerted action of these molecules leads to the separation of the transcription activator E2F from the Rb protein. Thus, the passage from one phase of the cycle to another is regulated, in part, by cyclins and cyclin-dependent kinases.

Cyclins are members of a population of proteins that rise when the cell enters a new phase of the cycle and drop when the phase is completed. Cyclins activate cyclin dependent kinases. The cyclin-dependent kinases are only active when complexed with their respective cyclins. There are at least seven different types of cyclins (from A to H)[5] in mammalian cells (15), but this chapter focuses on cyclins A, B, D, and E. The D type (D1, D2, D3) binds to two distinct but homologous cyclin-dependent kinases (Cdk4 or Cdk6). In the complex, the Cdks constitute the catalytic units and the cyclin is the necessary coactivator. The complex regulates the passage from $G_1$ to S, that is from the initiation of the cycle to DNA synthesis. Cdk2 is associated with several cyclins: cyclin E for the passage from $G_1$ to S, and cyclin A during the S phase and the passage of $G_2$ to M. The B cyclins bind to Cdk2 and regulate mitosis. The Cdk's themselves are inhibited by distinct cdk inhibitors and proteins of the P21 and the INK4 families.

E2F/DP heterodimers complex with Rb (P107 and P130). The E2F/DP family can function as transcription activators and transcription repressors. In the case of transcription activation, the sequence of events is as follows:

- In quiescent cells, E2F (mainly E2F4) is bound to Rb, which represses the E2F transcription factor and expression of target genes.
- The passage from $G_0$ to $G_1$ requires dissociation of the complex triggered by hyperphosphorylation of the repressor: the Rb protein. The phosphorylation of Rb occurs at two steps: phosphorylation by cyclin D kinase, followed by phosphorylation by cyclin E or A kinase.
- It has been proposed that dissociation of the E2F/RB complex may involve E2F phosphorylation as well.
- Relocation of most of the E2F from cytoplasm to the nucleus in dividing cells, by mechanisms still unknown.
- Promotor activation and gene transcription by either free E2F or E2F complexed with activator molecules.
- *De novo* synthesis of E2F may further contribute to transcription transactivation.

[5]For optimal activity, cdks require additional phosphorylation (thyronine 160 in human Cdk-2). The kinases responsible are the cdk activating kinases (CAK). A CAK has been found in budding yeast and another in fission yeast, *Drosophilas*, and humans; in humans this is a complex of three proteins (cyclin H, Cdk-7, and an assembly factor [MnATI]). P53 can arrest the cycle in $G_1$ by interfering with CAK rather than activating Cdk inhibitors. The human genes for cyclin H and MNAT1 have been mapped to 5q13.3-14 and 14q23, respectively. The complex cyclin H–Cdk-7–MNAT1 also has been found to bind to the TFIIH protein that is involved in transcription and DNA repair.

- As the cells exit the cycle or proceed into differentiation, E2F is inactivated by three potential mechanisms: shifts in phosphorylation, degradation by the ubiquitin proteosomal system, and appearance of repressors (e.g., E2F6) of E2F transcription of target genes.

## Function of Rb

The Rb protein is primarily a phosphorylated nuclear protein that exists in a hyperphosphorylated form in dividing cells and in a hypophosphorylated form in quiescent and differentiated cells. The Rb molecules contain several functional motifs in the A and B conserved domains (the pocket), in the C terminal domain, and in the N terminal domain.

The "pocket" is an important functional component of the molecule. The motif includes two highly conserved domains (A and B) and directly participates in transcriptional repression. The pocket constitutes a binding site for a LXCXE motif, and the Rb pocket bound to the LXCXE peptide has been crystalized (16). Both the A and B domains are required for the formation of a functional pocket. Although the pocket's LXCXE binding site is best characterized, its role needs further clarification. It is known that it binds deacetylases (HDAC 1 and 2), some ATPases, and components of the nucleosome remodeling complex. There is evidence that histone deacetylases are probably involved in the function of Rbs (e.g., the repression of transcription by E2F) (17, 18).

As pointed out by Harbour and Dean, these findings are in keeping with the function of Rb (19). Indeed, regulation of gene transcription requires remodeling of chromatin (19).At least three critical molecules bind the Rb carboxy terminal functional domain:

1. E2F, the transcriptor factor, binds to sites located in the pocket (other than the LXCXE binding sites and the C terminal domain).
2. MDM2 (discussed later in this chapter) contributes to the degradation of P53. Hsieh and associates have shown the P53 activity is inhibited by the formation of a Rb-MDM2-P53 complex (20).
3. c-abl tyrosine kinase: the binding of the kinase to Rb is likely to be critical to the arrest or the slowdown of the cell cycle (21).

The aminoterminal domain includes at least two functional domains:

1. Consensus Cdk phosphorylation sites.
2. Binding sites for several proteins whose role in the function of the Rb protein is still unclear.

However, there is little doubt that the aminoterminal is required for full functional activity of the Rb protein. For details, see the excellent review of Harbour and Dean (19).

## The Family of E2F Proteins

The E2F family constitutes a second group of proteins that participate in the control of the $G_1$ (and S) checkpoint(s). E2F is bound to Rb in quiescent cells and released in dividing cells.[6] E2Fs are heterodimers composed of one polypeptide encoded by one of the six E2F genes and another encoded by one of the two DP genes. Two domains are

---

[6]Dryson has emphasized the complexity of the functions of the RB and the E2F families of protein (13). Indeed, related proteins may perform different activities, including conflicting ones.

highly conserved in each of the monomers (E2F and DP): a DNA binding domain and a dimerization domain. A transactivation domain is found in E2F1-E2F5, but not in E2F6[7] and DP. DP activates E2F transcription-dependent activity indirectly.

### Conclusion

There are two segments in $G_1$. The first half of the $G_1$ phase starts either after mitosis or at the end of $G_0$ and stops at the restriction point R, a few hours before the cycle enters the S phase. Passage through R first requires that the D cyclins and Cdk4/Cdk6 to form complexes. The Cdk's are constitutively expressed, but the addition of growth factors to the medium increases the level of cyclins D. In the second stage of $G_1$, cyclin E, which in contrast to cyclins D is present in quiescent cells, complexes with Cdk2. This complex needs to be activated, and this can be achieved by decreasing the levels of P21 or other cyclin inhibitors through the ubiquitin proteosomal pathway. It cannot be ruled out that the levels of growth factors influence this step.

In the next step, cyclin A forms a complex with Cdk2, an event that coincides with entry in the S phase. The Rb protein is hypophosphorylated during the mitotic phase and remains in that state during quiescence. Hypophosphorylated Rb binds to several proteins, among them members of the E2F family (the activating members) and the DP1 transcription factors. The Rb molecule pocket domain is required for binding. The complex acts as a repressor for transcription of several genes, including some that encode proteins required for DNA synthesis.

In the course of $G_1$, at the restriction point, the Rb protein is phosphorylated by the Cdk4/Cdk6/cyclins D1, D2, D3 complexes. This causes the release of members of the E2F family of proteins that act either as transcription factors or as factors repressing target genes that trigger cell proliferation or apoptosis (**FIGURE 12-5**) (13, 22).

## Cyclin Kinases

The interaction between cyclins and Cdk's is fundamental to the evolution of the cell cycle, and is reviewed in this section. Cdk's are (Ser/Thr) protein kinases; they bring ATP and the protein substrate together for the purpose of phosphorylating the protein at a specific site and thereby activating its function. The process of phosphorylation is highly specific; it links the γ-phosphate of ATP to the oxygen atom of an hydroxyl residue of a threonine residue in the protein. Such delicate interaction requires:

- Binding of both ATP and the protein substrate to the kinase.
- Conformational adjustments of the kinase that facilitate and secure the specificity of the interaction between t phosphorus and oxygen.

In view of these requirements, it is not surprising that structural conformities are present within the Cdk family. They are:

- The length of the molecules is approximately 300 amino acids long.
- The molecules are highly conserved (35% to 65% identical to the prototypes cdc2 and cdc28).
- The conserved sequence generates similar conformations (to be discussed further).
- The structure includes several critical components located:

[7]E2F6 represses E2F dependent transcription.

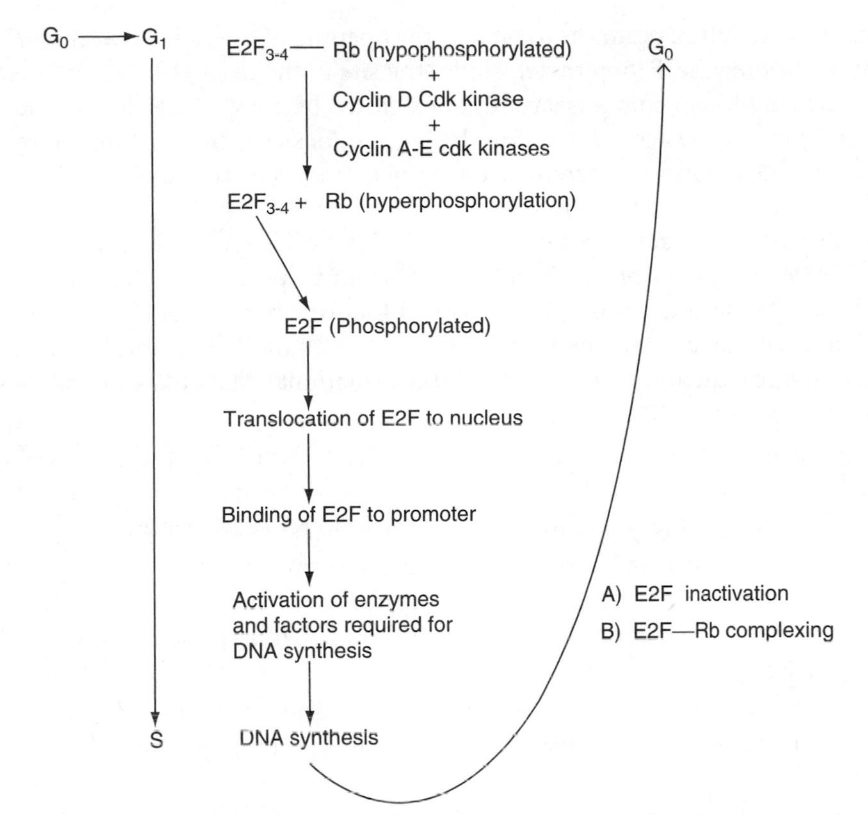


**FIGURE 12-5** Schematic illustration of the role of Rb and E2F in the passage of $G_1$ to S.

1. At the ATP binding site;
2. In the PSTAIRE helix that is involved in the association between the cdk and the cyclin;
3. In the T loop that protects the active site.

## Structure of the Cyclin A Complex

Russo and associates (23) studied the crystal structure of Cdk2 bound to a truncated cyclin A ($A_t$).[8] The structure is divided in three components, a central compact domain composed of five helices surrounded by the extended helices in the N&C terminal. The cyclin box is composed of the five helices ensconced in the core and two of those helices (3 and 5) contribute to the binding of the cyclin to Cdk2.

Cyclin's structures are heterogeneous compared to that of cdk. They vary in size (35–90 kDa) with little sequence homogeneity, except for a 100 amino acid sequence referred to at the cyclin core. Cdk's are only active when bound to cyclins.

The structure of cyclin A is essentially unchanged after binding to Cdk2. The helices 5 and 3 of the cyclin box form hydrogen bonds with amino acids found in the PSTAIRE helix in Cdk2. In addition, there are links between the non-conserved N terminal of cyclin A and the C lobe of the Cdk2. In contrast, the association of Cdk2 to cyclin A induces significant conformational changes of the kinase and the latter are required for activation of Cdk2 (24).

[8] $A_t$ is cyclin A minus 172 N terminal amino acids.

In essence, activation occurs as a result of the opening of the cleft between the N and C lobes, by flipping the T loop that occludes the site in the inactive Cdk2. It is as if the 300 amino acid protein contains a central box closed by a flap (the T loop), and when Cdk2 comes in contact with the cyclin, the loop unlocks the box by flipping by a 20° angle) in the open position, thereby allowing penetration of the substrate. Details are in (24) and the references therein.

Morris et al. investigated the kinetics of the Cdk2/cyclin A complex formation. As mentioned, displacement of the T loop by 20° from its position in the inactive Cdk2 opens the active site and exposes $Thr^{160}$. This is followed by phosphorylation of $Thr^{160}$ by CAK and additional conformational changes in both the T loop and the C terminal lobe that stabilize the binding site. These conformational changes, rapid as they are, take place in two steps (**FIGURE 12-6**):

1. Interaction between the PSTAIRE of the Cdk2 and the a3 and a5 helices of cyclin A causes a rotation of the PSTAIRE and conformational changes in the ATP binding site, yet the threonine 160 of the T loop remains inaccessible.
2. Isomerization of Cdk2 and A into a mature form. The latter involves contacts between the C lobe and the T loop that results in the exposure of the T loop, which opens the substrate binding site and confers full kinase activity. Threonine 160 makes it accessible to CAK.

The first step is rapid. The second is relatively slow and rate limiting for the reaction of Cdk, and is likely to be responsible for the cyclin selectivity by the Cdk (e.g., Cdk2 and cyclin A) (25).

## Cyclin Kinase Inhibitors: P21 and INK

In discussing cell cycle arrest, this chapter has focused on P21, which is a member of the non-specific inhibitors family of the cyclin kinases. However, P21 is not the only inhibitor of the Cdk's. Cdk inhibitors fall in two families: the CIP family that includes P21, P27, and P57; and the INK4 family that include P15, P16, P18, and P19. The chromosomal loci for these proteins are listed in **TABLE 12-1**.

The INK4's are specific inhibitors of cyclin D/Cdk4 and Cdk6. They inhibit the complex in two ways by binding to cyclin D at a site near the N terminus and by directly inhibiting the cyclin D/Cdk4 complex (**FIGURE 12-7**) (27).

The function of P16 (and the other members of the INK4 family) is to impede Rb phosphorylation and thereby arrest the cycle in $G_1$ (12). *p16* is a tumor suppressor gene. Mice harboring a *p16* defective gene are prone to develop spontaneous sarcomas

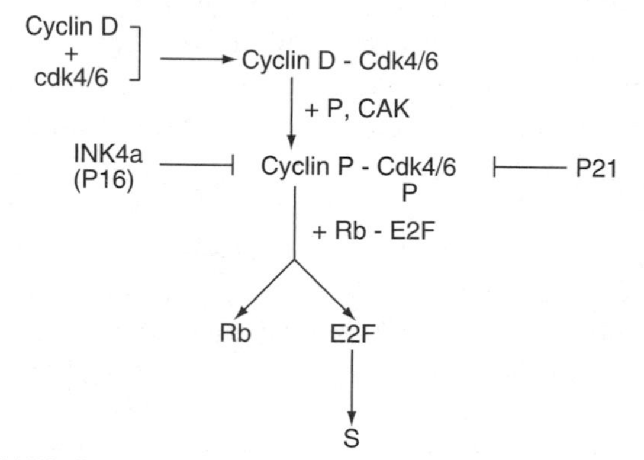


**FIGURE 12-6** The role of CAK in S-phase activation.

**Table 12-1 Chromosomal loci of cyclin kinase inhibitor**

| Cyclin Inhibitors | Chromosomal Locus |
|---|---|
| P15 | 9p21 |
| P16 | 9p21 |
| P18 | 1p32 |
| P19 | 19p13 |
| P21 | 6p21 |
| P27 | 12p12-13 |
| P57 | 11p15.5 |

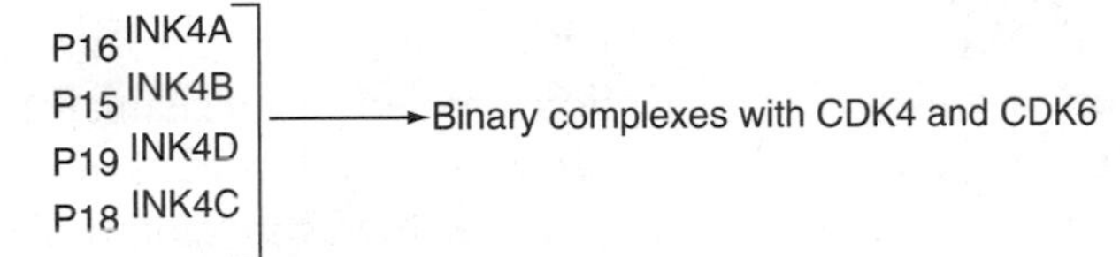


**FIGURE 12-7** Inhibitors of the $G_1$-S phase transition.

and induced cancers. Human cancers (e.g., melanomas) are also often associated with P16 mutations. In contrast, P16 overexpression causes persistent inhibition of phosphorylation of Rb, cell cycle arrest, and senescence (28, 29).

Distortion of the relationship between INK4a/s (P16) and cyclins D 4/6 (21) may block arrest of the cycle in the $G_1$ phase and cause cancer (30). These observations led several laboratories to investigate the structural relationship between INK4a and Cdk's, and their modification imposed upon them by site mutations in the INK4a. Russo and colleagues investigated the structural parameters of the binding INK4a and Cdk6 (31). In brief, Cdk6 is composed of two domains (referred to as lobes) the N (aa1–103) rich in β sheets and a C terminal domain (amino acids 104–301). The catalytic cleft is found at the junction of the lobes. P16[9] (31) binds to one site on the left (the site opposite that of the cyclin binding site and next to the ATP binding site) and interacts with residues of both the N and C terminal domains. The INK4a causes conformational changes that interfere with ATP binding and deform the catalytic site. The binding of P16 to the catalytic cleft is secured by a network of hydrogen bonds. Several of the mutations leading to cancers involve residues found to contribute to the hydrogen bonding.

Brotherton and colleagues investigated the three dimensional structure of the P19 $^{INK4d}$/Cdk6 by x-ray diffraction, and thereby identified the critical residues involved in the interaction between inhibitor and cyclin D–dependent kinase (32). The Cdk6, like other Cdk's, is composed of an $NH_2$ (5–10 amino acids) and a longer COOH (amino acids 101–309), terminal domains that ensconce the catalytic cleft P19, binds to the cleft, and the binding induces conformational changes in the Cdk that prevent effective ATP binding and convert an active into an inactive kinase. By identifying the critical residues

[9]P16 contains four ankyrin repeats that form a concave surface, one to the convex phase of the N domain of the Cdk and the tips of the first three ankyrins bind to the C domain; and the last two bindings involve residues contributing to the catalytic cleft.

involved in the interaction between Cdk and P19, it also became possible to explain their mode of action and the significance of the mutations (32; reviewed in 25 and 33).

## Conclusion

In brief, the findings include the following:

- Cdk6, like other Cdks, is composed of an $NH_2$ (5–10 amino acids) and a longer COOH (amino acids 101–309), terminal domains that ensconce the catalytic cleft, and P19 binds to it.
- The P19 binding induces conformational changes in the Cdk that prevent effective ATP binding and convert an active into an inactive kinase.
- By identifying the critical residues involved in the interaction between Cdk and P19, it also became possible to explain the significance of the P19 mutations.

Activation of the proteosomal proteolytic system secures degradation of cyclin kinases. Transcription and posttranscription regulation mechanisms govern expression of P21. The latter is likely to include proteosomal degradation of P21. Inhibition of the 20S proteosome by lactocyclin up-regulates the expression of P21 by increasing its half-life and activates its cytostatic effect (34). These findings suggest a close interaction between P21 and the proteosomes during the cell cycle. Regulation of P21 activity at the level of transcription cannot be excluded. In the absence of DNA damage caused by external agents, P53 regulates the constitutive levels of P21 by transcriptional inactivation after sequence specific binding to DNA.

Cyclin inhibitors must disappear before the S phase to allow DNA synthesis to take place, a notion supported by findings in mammalian cells (35, 36).

P53 acts as a transcription factor that transactivates transcription of a large number of proteins, among them, P21. P21 is a universal cdk inhibitor and a major contributor to cell cycle arrest (albeit not the only one). Cell cycle arrest is required from the end of mitosis to quiescence ($G_0$) and for differentiation and senescence. In stressed cells it also precedes DNA repair and apoptosis. Thus, the harmonious development of the organism sometimes demands delay or arrest of the cycle. Terminal differentiation (e.g., in grown skin or neurons) and senescence are associated with complete arrest; a temporary arrest may precede DNA repair, followed by resumption of the cycle, and prolonged arrest of the cycle also may be linked to apoptosis. In summary, P53 and P21 are likely to intervene at one or more of the cell cycle checkpoints, prior to DNA repair, differentiation, and senescence, which may lead to apoptosis. Therefore, the roles of P53 and P21 in the regulation of each of these phases of the cell's lifespan are examined in Chapter 13.

## References and Recommended Reading


1. Temin H.M. Stimulation by serum of multiplication of stationary chicken cells. *J Cell Physiol.* 1971;78:161–70.
2. Pardee, A.B. $G_1$ events and regulation of cell proliferation. *Science.* 1989;246:603–8.
3. el-Deiry, W.S., Kern, S.E., Pietenpol, J.A., Kinzler, K.W., and Vogelstein, B. Definition of a consensus binding site for p53. *Nat Genet.* 1992;1:45–9.
4. Marx, J. How p53 suppresses cell growth. *Science.* 1993;262:1644–5.

5. Thut, C.J., Chen, J.L., Klemm, R., and Tjian, R. p53 transcriptional activation mediated by coactivators TAFII40 and TAFII60. *Science.* 1995;267:100–4.
6. Lu, H., and Levine, A.J. Human TAFII31 protein is a transcriptional coactivator of the p53 protein. *Proc Natl Acad Sci USA.* 1995;92:5154–8.
7. Gu, W., Shi, X.L., and Roeder, R.G. Synergistic activation of transcription by CBP and p53. *Nature.* 1997;387:819–23.
8. Avantaggiati, M.L., Ogryzko, V., Gardner, K., Giordano, A., Levine, A.S., and Kelly, K. Recruitment of p300/CBP in p53-dependent signal pathways. *Cell.* 1997;89:1175–84.
9. Lu, H., and Levine, A.J. Human TAFII31 protein is a transcriptional coactivator of the p53 protein. *Proc Natl Acad Sci USA.* 1995;92:5154–8.
10. Peng, Y.C., Kuo, F., Breiding, D.E., Wang, Y.F., Mansur, C.P., and Androphy, E.J. AMF1 (GPS2) modulates p53 transactivation. *Mol Cell Biol.* 2001;21:5913–24.
11. Kardassis, D., Papakosta, P., Pardali, K., and Moustakas, A. c-Jun transactivates the promoter of the human p21(WAF1/Cip1) gene by acting as a superactivator of the ubiquitous transcription factor Sp1. *J Biol Chem.* 1999;274:29572–81.
12. Chen B., He L., Savell V.H., Jenkins J.J., Parham D.M. Inhibition of the Interferon-gamma/signal transducers and activators of transcription (STAT) pathway by hypermethylation of a STAT-binding site in the p21 WAF1promoter region. *Cancer Res.* 2000;60:3290–8.
13. Dryson, N. The regulation of E2F by pRB-family proteins. *Genes Dev.* 1998: 12:2245–62.
14. Weinberg, R.A. The retinoblastoma protein and cell cycle control. *Cell* 1955; 81:323–30.
15. Eki, T., Okumura, K., Abe, M., Kagotani, K., Taguchi, H., Murakami, Y., Pan, Z.Q., and Hanaoka, F. Mapping of the human genes encoding cyclin H (CCNH) and the CDK-activating kinase (CAK) assembly factor MAT1 (MNAT1) to chromosome bands 5q13.3-q14 and 14q23, respectively. *Genomics.* 1998;47:115–20.
16. Lee, J.O.; Russo, A.A., and Pavletich, N.P. Structure of the retinoblastoma tumour suppressor pocket domain bound to a peptide from HPV E7. *Nature.* 1998; 391:859–65.
17. Brehm, A., Miska, E.A., McCance, D.J., Reid, J.L., Bannister, A.J., and Kouzarides, T. Retinoblastoma protein recruits histone deacetylase to repress transcription. *Nature.* 1998;391:597–601.
18. Magnaghi-Jaulin, L., Groisman, R., Naguibneva, I., Robin, P., Lorain, S., Le Villain, J.P., Troalen, F., Trouche, D., and Harel-Bellan, A. Retinoblastoma protein represses transcription by recruiting a histone deacetylase. *Nature.* 1998;391:601–5.
19. Harbour, J.W., and Dean, D.C. Chromatin remodeling and Rb activity. *Curr Opin Cell Biol.* 2000;12:685–9.
20. Hsieh, J.K., Chan, F.S., O'Connor, D.J., Mittnacht, S., Zhong, S., and Lu, X. RB regulates the stability and the apoptotic function of p53 via MDM2. *Mol Cell.* 1999;3:181–93.
21. Whitaker, L.L., Su, H., Baskaran, R., Knudsen, E.S., and Wang, J.Y. Growth suppression by an E2F-binding-defective retinoblastoma protein (RB): contribution from the RB C pocket. *Mol Cell Biol.* 1998;18:4032–42.

22. Trimarchi, J.M., and Lees, J.A. Sibling rivalry in the E2F family. *Nat Rev Mol Cell Biol.* 2002;3:11–20.

23. Russo, A.A., Jeffrey, P.D., Patten, A.K., Massague, J., and Pavletich, N.P. Crystal structure of p27Kip1 cyclin-dependent-kinase inhibitor bound to the cyclin A-Cdk2 complex. *Nature.* 1996;382(6589):325–31.

24. Morgan, D.O. Cyclin-dependent kinases: engines, clocks, and microprocessors. *Annu Rev Cell Dev Biol.* 1997;13:261–91.

25. Morris, M.C., Gondeau, C., Tainer, J.A., and Divita, G. Kinetic mechanism of activation of the Cdk2/cyclin A complex. Key role of the C-lobe of the Cdk. *J Biol Chem.* 2002;277:23847–53.

26. Sherr, C.J., and Roberts, J.M. CDK inhibitors: positive and negative regulators of $G_1$-phase progression. *Genes Dev.* 1999;13:1501–12.

27. Puri, P.L., Maclachlan, T.K., Levrero, M., and Giordano, A. The intrinsic cell cycle from yeast to mammals. In: Stein, G.S., Baserga, R., Giordano, A., and Denhardt, D.T. (Eds.). *The Molecular Basis of Cell Cycle and Growth Control.* New York: John Wiley and Sons, 1999, pp. 15–79.

28. Koh, J., Enders, G.H., Dynlacht, B.D., and Harlow, E. Tumour-derived p16 alleles encoding proteins defective in cell-cycle inhibition. *Nature.* 1995;375:506–10.

29. Hara, E., Smith, R., Parry, D., Tahara, H., Stone, S., and Peters, G. Regulation of p16CDKN2 expression and its implications for cell immortalization and senescence. *Mol Cell Biol.* 1996;16:859–67.

30. Hall, M., and Peters, G. Genetic alterations of cyclins, cyclin-dependent kinases, and Cdk inhibitors in human cancer. *Adv Cancer Res.* 1996;68:67–108.

31. Russo, A.A., Tong, L., Lee, J.O., Jeffrey, P.D., and Pavletich, N.P. Structural basis for inhibition of the cyclin-dependent kinase Cdk6 by the tumor suppressor p16INK4a. *Nature.* 1998;395:237–43.

32. Brotherton, D.H., Dhanaraj, V., Wick, S., Brizuela, L., Domaille, P.J., Volyanik, E., Xu, X., Parisini, E., Smith, B.O., Archer, S.J., Serrano, M., Brenner, S.L., Blundell, T.L., and Laue, E.D. Crystal structure of the complex of the cyclin D-dependent kinase Cdk6 bound to the cell-cycle inhibitor p19INK4d. *Nature.* 1998;395:244–50.

33. Sherr, C.J. Cancer cell cycles. *Science.* 1996;274:1672–7.

34. Blagosklonny, M.V., Wu, G.S., Omura, S., and el-Deiry, W.S. Proteasome-dependent regulation of $p21^{WAF1/CIP1}$ expression. *Biochem Biophys Res Commun.* 1996; 227:564–9.

35. Pagano, M., Tam, S.W., Theodoras, A.M., Beer-Romero, P., Del Sal, G., Chau, V., Yew, P.R., Draetta, G.F., and Rolfe, M. Role of the ubiquitin-proteasome pathway in regulating abundance of the cyclin-dependent kinase inhibitor p27. *Science.* 1995; 269:682–5.

36. Yew, P.R. Ubiquitin-mediated proteolysis of vertebrate G1- and S-phase regulators. *J Cell Physiol.* 2001;187:1–10.

CHAPTER

# 13 CELL CYCLE CHECKPOINTS

The cell cycle is a well-orchestrated process in which each molecular participant plays its role once, exactly, and in the correct sequence. The players include cyclins, cyclin-dependent kinases, and their inhibitors. The many proteins involved include Rb, E2Fs, universal kinase inhibitors (P21, P27, and P57), and the INK-4 family of inhibitors. The inadequacies of the molecular cooperative effort, caused by defective nutrition or genotoxins, are a threat to cellular integrity. If these molecular defects are not corrected, the cell may die (apoptosis) or proliferate in a disorderly fashion (cancer in the case of many somatic cells), cause autoimmunity (in the case of memory lymphocytes), or initiate genetic disorders (in the case of germ cells).

To prevent defective proliferation or death, several control posts are placed along the cell cycle path that leads from $G_1$ to cell division. Thus, as the cell goes through the cycle, it passes through several checkpoints that determine whether or not the cycle continues. Checkpoints make sure that all required molecular associations and activities are executed accurately at each given step of the cycle, before the next step is engaged. Passage through three cycle checkpoints have been investigated: from $G_1$ to S, from $G_2$ to M, and the spindle checkpoint. An S checkpoint is believed to exist as well.

It is not surprising that much of what is known about checkpoints is derived from studies on cell cycle arrest after DNA damage. In such cases, the operation of any checkpoint depends on the sensing of the damage (to DNA or chromosomes) and transduction of the message through a chain of molecules that links damaged DNA to the checkpoint and switches it off. In yeasts, RAD 17, RAD 9, RAD 26, Hus 1, and other molecules contribute to switching the cycle off and on. Each of these molecules or group of molecules have homologues in mammalian cells. For example, the structure of RAD 1 and RAD 9 is similar to that of the PCNA clamp that brings the polymerase to DNA. RAD 17 is similar to the replication factor C (RFC) and governs the interaction between RAD 1, RAD 9, and Hus 1 with the damaged DNA.

Whatever the primary sensor of the damage may be, ATM and ATR (or other kinases) are activated when the DNA damage or chromosomal segregation is sensed. Available evidence suggests, however, that other pathways for the checkpoint exist. For example, ATR may be activated when damage occurs during the $G_1$/S phase and ATM may be activated when damage is manifested during the $G_2$ phase. The first is mainly associated with UV damage (and methylmethane sulfonate) and the second with ionizing radiation (1–3).[1]

## $G_1$/S Transition Checkpoint

Prior to activation of the $G_1$/S transition, both Rb and cMyc play critical roles. Cyclin D–Cdk4 kinase activity phosphorylates Rb, and E2F is freed to activate the transcription of target genes, among them that of the cyclin E gene, which is achieved by the cooperative activity of E2F and c-Myc. The product of the cyclin E gene complexes with Cdk2 and activates it. Activation of the cyclin E–Cdk2 complex is believed to constitute the rate-limiting step in the cycle's progress. Keeping this pathway in mind, two potential sites of attack for the $G_1$/S transition delay are conceivable: Rb phosphorylation close to the restriction point, at the start of the pathway that leads to S, and activation of Cdk2 close to the point of entrance in the S phase.

The phosphorylation sites of cyclin D1–Cdk4 or D1–Cdk6 are potential targets for cyclin kinase inhibitors, among which P21 is prominent. Consequently, it was suspected and later demonstrated that the cell cycle arrest in $G_1$ involves P53 activation and P21 induction and transcription (**FIGURE 13-1**).

More extensive investigation of the $G_1$/S checkpoint revealed the following:

- The P53-dependent response is too slow to explain the rapid arrest of the $G_1$/S progression after DNA damage.
- The P53 response does not account for the Cdk2 inhibition, since DNA damage blocks the activity of the cyclin E–Cdk2 complex in the absence of P53 and P21.

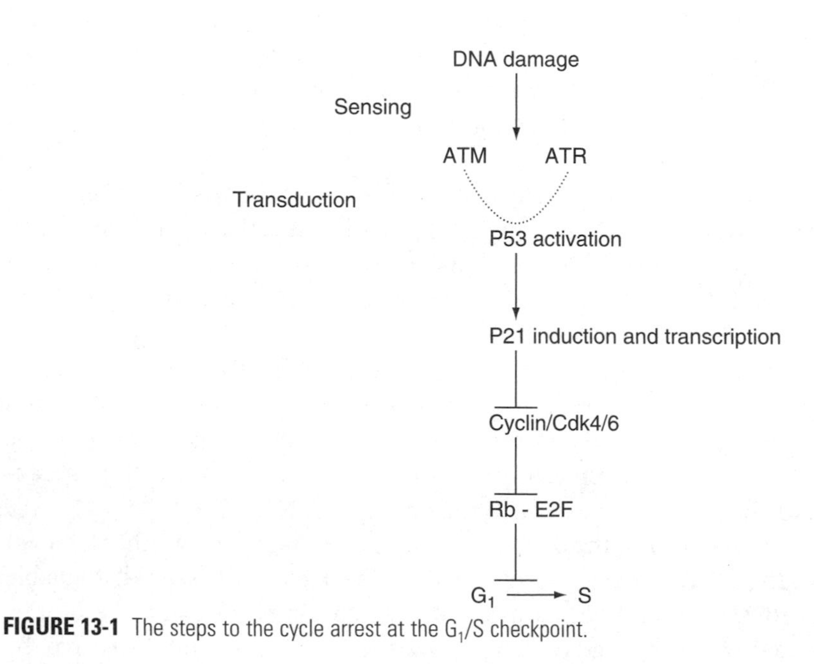


**FIGURE 13-1** The steps to the cycle arrest at the $G_1$/S checkpoint.

[1]After UV exposure, most of the damage affects the pyrimidine bases. Other less frequent damage includes single and double breaks.

These findings led to the notion that the arrest of $G_1$/S progression after DNA damage occurs in two distinct pathways: a rapid P53 independent and a slow P53 dependent pathway. This process is referred to as the two-wave checkpoint.

The rapid response to DNA damage takes minutes and does not involve gene transcription or protein synthesis; instead, it is modulated by phosphorylation–dephosphorylation and proteolytic reactions. This P53 independent pathway involves the ATM/ATR gene products, the Chk1 and Chk2[2] kinases, the cdc25A phosphatase, and Cdk2. Dephosphorylation of the phosphotase activates cyclin E–Cdk2, which is required for entrance into the S phase in the cell cycle.

In unstressed cells, cdc25A[3] dephosphorylates the tyrosine 14 and tyrosine 15 of Cdk2 and thereby activates the cyclin E–Cdk2 complex. Cyclin E activation triggers the binding of cdc45 to DNA. The cdc45 protein starts the formation of the DNA prereplication complex and recruitment of DNA polymerase. Thus, in the presence of DNA damage, ATM (after x-radiation) and ATR (after ultraviolet radiation) phosphorylate Chk1 and Chk2, respectively, and ATR phosphorylates cdc25A. The phosphorylation of cdc25A (at serine 123, at least in the case of Chk2) marks the phosphatase for ubiquitination and its rapid proteosomal degradation. The absence of the phosphatase leaves Cdk2 in the phosphorylated state and the cyclin E–Cdk2 remains inactive; the recruitment of cdc45 to chromatin is impaired, and the cycle is unable to enter the S phase (**FIGURE 13-2**). This pathway is believed to operate in all cell types.[4]

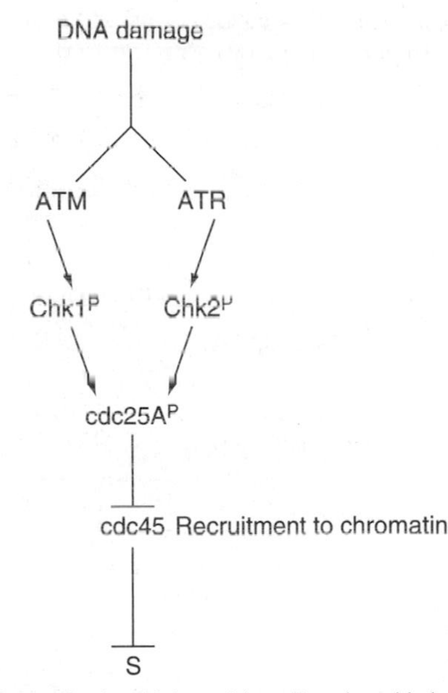


**FIGURE 13-2** First wave in $G_1$/S checkpoint activation. (Adapted from Bartek and Lukas [4]).

---

[2]The Chk2 kinase, first identified in yeast (*S. Pombe*), was shown to be required for cell cycle arrest; Chk2 is the mammalian homolog of Cds1 in *S. Pombe* and Rad-3 in *S. cerevisiae*. This kinase is mutated in the Li-Fraumeni syndrome.

[3]The cMyc-Max heterodimer binds to the Myc binding site of the gene that encodes the cdc25A protein, a phosphatase that plays a critical role in the $G_1$/S transition. At least in fibroblasts, cdc25A induction coincides with c-Myc dependent activation of cyclin D/Cdk4 and cyclin E/Cdk2.

[4]This does not exclude the existence of other contributory checkpoint pathways involving, for example, the anaphase-promoting complex.

This immediate response to DNA damage is followed by stabilization of the cycle arrest. The second wave shares with the first pathway the functions of sensing of DNA damage, transduction by the ATM/ATR, and the Chk1/Chk2 phosphorylation, but differs from the first pathway by the P53 activation, P21 transcription, and direct inhibition of the cyclin E/Cdk2 activity. Thus, ultimately both the first and second waves in the $G_1$/S arrest process target cyclin E/Cdk2 (**FIGURE 13-3**) (4).

However, the pathways may vary depending upon the type of DNA damage. For example, in response to double strand breaks, the tumor suppressor ARF is induced and ARF strengthens P53 stabilization by cytoplasmic sequestration of MDM2.

## $G_2$/M Checkpoint

Compared with the $G_1$/S checkpoint, the $G_2$/M checkpoint is incompletely understood. It involves P53, P21, and GADD45, at least in some cases, but functions in others through a P53-independent pathway. It is likely that several other molecules, some suspected and others still unknown, are involved in the process. This section discusses the presently accepted pathway(s) for the regulation of the $G_2$/M checkpoints, and briefly considers the role of several proteins also suspected to be involved.

The function of the $G_2$M checkpoint is a first step in preventing the segregation of damaged chromosomes. The identified genes that control the $G_2$/M checkpoint include those that encode the ATM, Cdk1, Cdk2, the 14-3-3 protein family, PRb, E2F, P53, and P21 proteins (5, 6). Cyclin B and Cdc2, the traditional maturation promoting factor (MPF) are at the core of the $G_2$/M transition (**FIGURE 13-4**). While Cdc2 levels remain constant during most of the cell cycle, the levels of cyclin B1 vary in a cell cycle-dependent fashion and rise primarily in $G_2$/M (7).

Before considering the steps involved in the $G_2$/M checkpoint after genotoxic stress, let's examine the regulation of the $G_2$/M transition in unstressed cells. This essentially involves three steps: condensation of cyclin B1 and Cdc2, phosphorylation of the

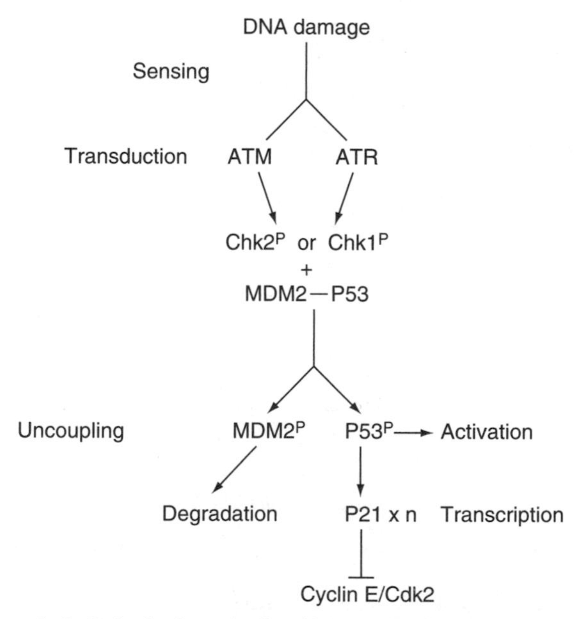


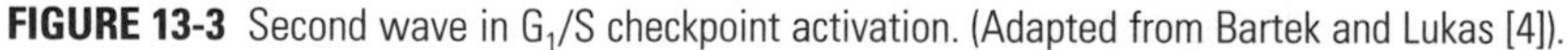

**FIGURE 13-3** Second wave in $G_1$/S checkpoint activation. (Adapted from Bartek and Lukas [4]).

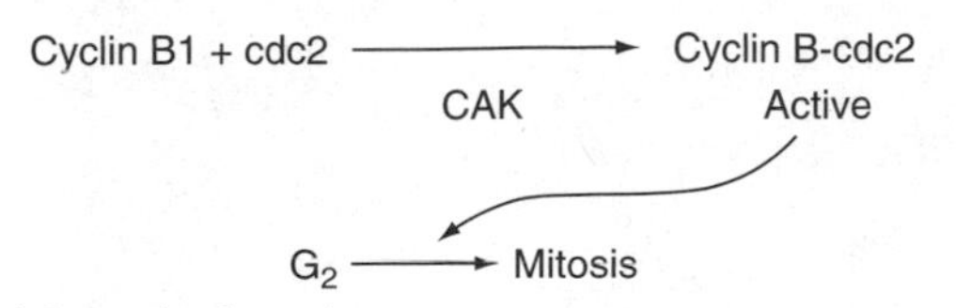


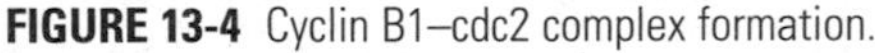
**FIGURE 13-4** Cyclin B1–cdc2 complex formation.

complex, and its dephosphorylation by cdc25c. Cyclin B1 and cdc2 form a cyclin B1-cdc2 complex, and the CAK activating kinase (CA K) is involved in the reaction (**FIGURE 13-4**).

The function of cyclin B1-cdc2 is regulated by kinases (Myt 1 and Wee 1) and a phosphatase cdc25C. The two kinases hyperphosphorylate the cyclin B1–cdc2 complex and thereby inactivate it. The complex is dephosphorylated by the cdc25C phosphatase, and then hypophosphorylated cyclin B1–cdc2 is activated and the $G_2$ phase transits into the M phase (**FIGURE 13-5**).

cdc25C itself is regulated; it is phosphorylated by Chk1 or Chk2 at serine 216. The phosphorylation modulates the binding of cdc25C and the protein 14-3-3. The cdc25C–14-3-3 complex is transferred from nucleus to cytoplasm where it is sequestered, thus leaving the cyclin B1-cdc2 complex in the active hypophosphorylated state (**FIGURE 13-6**).

In summary, the cdc25C phosphatase plays a critical role in regulation of the $G_2$/M transition checkpoint. On the basis of the above molecular interactions, the $G_2$/M transition pathway can be outlined (**FIGURE 13-7**). A dual pathway for the activation of the $G_2$/M checkpoint has been described: one is P53 dependent and the other is P53 independent (Figure 13-7) (8, 9).

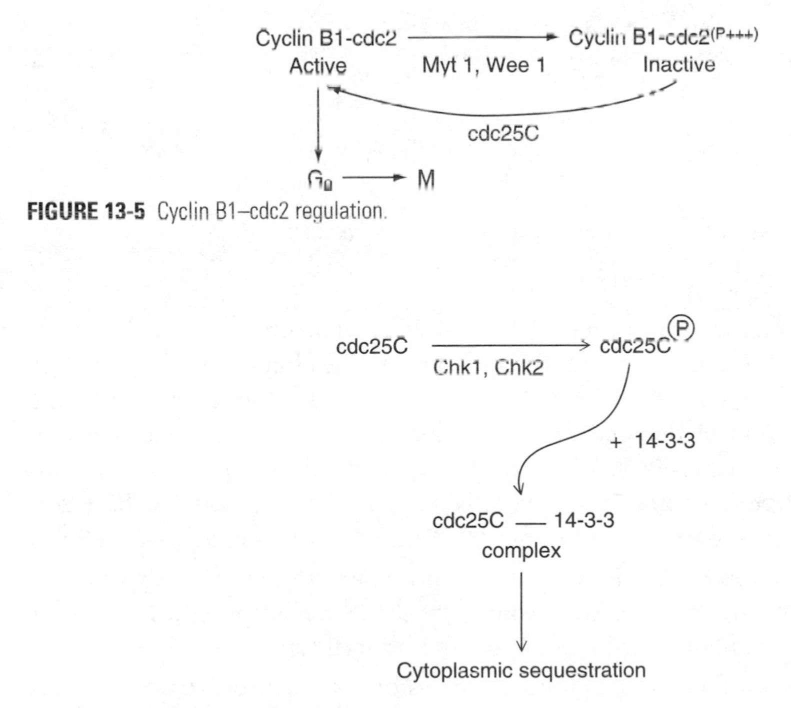


**FIGURE 13-5** Cyclin B1–cdc2 regulation.



**FIGURE 13-6** Modulation of the cdc25C function by 14-3-3.

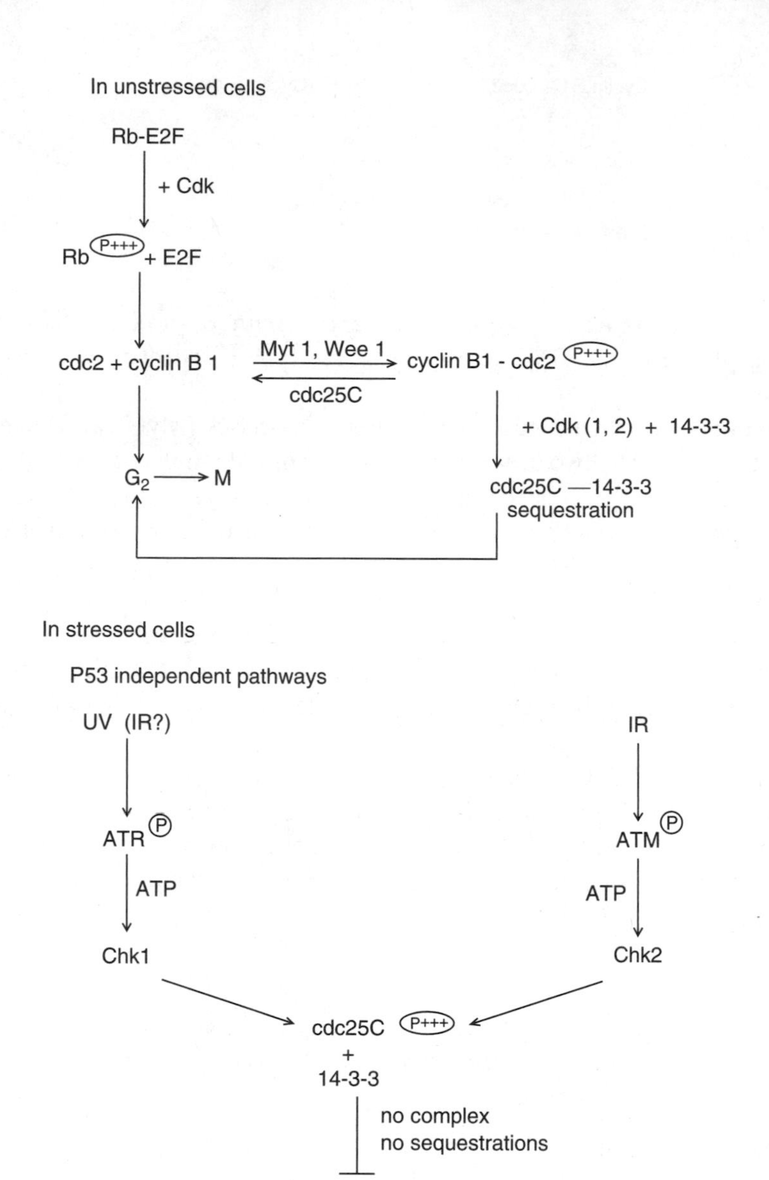


**FIGURE 13-7** Schematic representation of the steps to the $G_2$/M arrest in stressed cells.

This is only a schematic illustration of the potential mechanism of the $G_2$/M checkpoint. There are other proteins involved depending upon the circumstances. Crawford and Piwnica-Worms (6) investigated gene expression during progress from S to G2, which includes the G2 DNA damage checkpoint in cells deficient in functional P53. The same authors used Hela cells in which the *p53* gene is intact, but its product P53 is rapidly degraded (see reference 7). U2OS cells were used as controls. In the U2OS cells, P53 was stabilized and induced P21 after DNA damage. The process is as follows:

- In Hela cells, the levels of P53 are low, and do not rise with time after exposure to radiation. In contrast, P53 rises significantly in U2OS cells exposed to radiation. P21 responds to irradiation in the same way in both cell types.
- A transcriptional profile of the genomic expression in synchronized cells before and after $\gamma$ radiation revealed:

1. An increase in eight genes expressed in a cell cycle-dependent manner in nonirradiated cells, including cyclin B1, several cdcs, and protein kinases.
2. No rise, after irradiation, of genes associated with the induction of DNA damage. However, the expression of a large number of genes decreases significantly for reasons that need clarification. For example: decrease of the expression of some genes (e.g., those that encode histones) may be directly related to the delay in the completion of the S phase, but a decrease in the expression of other genes may be a collateral consequence of the cell cycle arrest. Finally, a decrease in expression of some genes may simply reflect growth failure.

In any event, results show that the response to DNA damage in P53-deficient cells fails to elicit the anticipated expression of genes whose products are associated with damage response, such as DNA repair genes. There is evidence that the RAD50, the RINT, the GADD45, and the CARB proteins might at some point also be involved at the $G_2$/M checkpoint (10).[5]

In RAD50 null mice, the phenotype is associated with embryonic death and radiation hypersensitivity. RAD50 forms a triplex with two proteins, MRE11 and NBS 1. The MRE11-RAD50-NBS 1 complex is involved in multiple cellular functions related to maintenance of the chromosomes; structural and functional integrity during mitosis, and meiosis, including recombination and double strands break repair, telomere stabilization, and checkpoint control. Another protein (RINT-1) binds to RAD50 through its C terminal, but the binding only takes place during the S and $G_2$/M phases. It is significant that the expression of truncated RINT-1 in mouse fibroblast cells impairs the proper function of the $G_2$/M checkpoint after genotoxic stress. Together, these findings suggest that RAD50 and RINT-1 may be involved in the control of the checkpoint.

The $G_2$/M checkpoint is defective in murine GADD45 null cells. In contrast, overexpression of GADD45 in human fibroblasts blocks the cells at the $G_2$/M transition. However, this defect occurs only after exposure to UV radiation or methylmethane sulfonate (MMS), but not after X-radiation.[6] GADD45 is a (165-amino acid) protein[7] induced in a P53-dependent fashion in the presence of various genotoxic agents, such as ionizing and UV radiations, alkylating agents, and growth factors' withdrawal) (11). The molecule is apt to bind cdc2 (among other proteins, e.g., P21, PCNA). GADD45 null mice show centrosome amplification after exposure to genotoxins, most likely because of the failure of the $G_2$/M checkpoint. Indeed, GADD45 arrests the $G_2$/M transition in a manner dependent on the presence of P53 but which does not require P21 (12). The inactivation of the checkpoint results from inhibition of the cyclin-cdc2 kinase, possibly because of a subtle but unidentified molecular interaction between GADD45 and cdc2. This subtle interaction may involve a specialized sequence (the DEDDDR residues) (13).

[5]Rad 50 (153 kDa) is a protein highly conserved from yeast to humans (50% identity within their N and C terminals). It is a member of a family of proteins involved in the structural maintenance of chromosomes (SMCs), and contains two Walker NTP binding domains, one in the N and another in the C terminal (Walker A and B), joined by leucine heptad repeats. Mutations in the Walker A domain protein abolish the protein's function.

[6]Differences in responses after UV and ionizing radiation exposure are not entirely surprising if the effects of both types of radiation are compared. Indeed, the ionizing radiations trigger three special events: (a) more strand breaks than UV radiation, and (b) an ATM-dependent response. Ataxia patients are much more sensitive to ionizing radiations (but not to UV exposure) than normal persons. (c) The activation of the $G_2$/M checkpoint response may be different after exposure to ionizing radiation from that after UV or MMS exposure.

[7]GADD45 is discussed in more detail in Chapter 16.

A protein discovered by McShea and associates is also suspected of being involved in cell cycle regulation during the $G_2$/M phase (14). The protein, CARB, regulates the function of cyclin B1 and binds to P21. It is associated with the centrosome and forms a complex with cyclin B1 (cyclin B1–CARB). Such association is preponderant in the absence of P21. When overexpressed in primary murine cells, CARB inhibits cell growth and death. In the presence of P21, CARB binds to the C terminal of the kinase inhibitor. CARB's overexpression in $P21^{+/+}$ cells have little effect on cell growth probably because the CARB's binding to P21 frees cyclin B1 and makes cyclin B1 available for progress of the cycle through the $G_2$/M phase. If true, this suggests that in addition to arresting the cell cycle in $G_1$/S, P21 also activates the $G_2$/M transition by freeing cyclin B1 from CARB. Therefore, it has been proposed that in $p21^{-/-}$ cells the arrest in $G_1$/S is eliminated, but the passage from $G_2$ to M remains blocked because of the binding of CARB to cyclin B1 (14).

In conclusion, in addition to those proteins traditionally involved in the $G_2$/M transition, others may contribute to the $G_2$/M checkpoint, some by binding to cdc2 (GADD45) and others by not binding to P21 (and thereby permitting cyclin B1 inhibition through CARB). In addition, RAD50 and RINT may contribute to the $G_2$/M checkpoint through still unknown mechanisms (15). It cannot be ruled out that even more proteins, still unknown, may contribute to the $G_2$/M checkpoint.

A somewhat unexpected finding is observed in the liver of young mice defective in ERCC1 (excision repair cross complement in gene 1). ERCC1 functions in NER; it forms a complex with XPE that excises the DNA sequence 5′ to the DNA lesion. The $G_2$/M cell cycle is arrested in hepatocytes of ERCC1-defective mice. The arrest is associated with polyploidy (endoreplication) and a rise in P53 in the absence of external DNA damaging factors. The mechanism of the process is unclear; however, one plausible interpretation was proposed by Nunez and associates (16). The absence of ERCC1 does not block the cells' progression phase, but the cells' progression may be incorrect or incomplete. This triggers the arrest in $G_2$M, which is associated with endoreplication of DNA (**FIGURE 13-8**). Thus, polyploidy could constitute a form of protection against propagation of cells harboring damaged DNA. But then, why is liver DNA damaged in the absence of obvious exogenous DNA damaging agents? Nunez

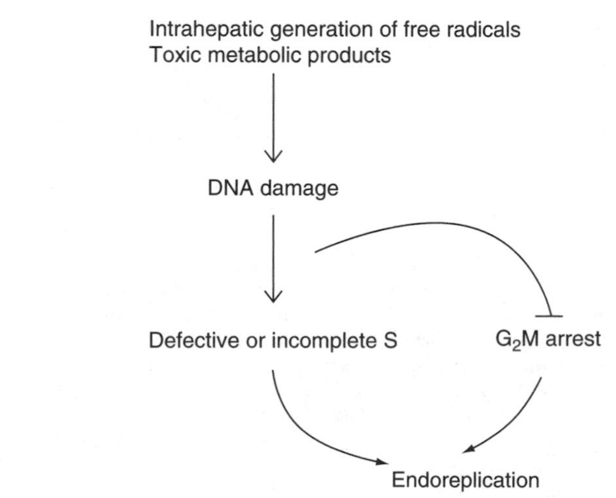

**FIGURE 13-8** Putative origin of polyploidy in livers of ERCC1-deficient young mice.

et al. proposed that intrahepatic genotoxic agents (free radicals and toxic metabolites) are responsible for the potential defective DNA replication. Whether such a mechanism for DNA polyploidy in liver after regeneration, or in hypertrophic hearts, or in cancer cells can be generalized is unknown (16).

## Intra-S Checkpoint

Checkpoints secure close intelligence between early events (e.g., $G_1$) and a later event (e.g., S) in the cell cycle. The purpose of this intelligence is to make sure that all previous tasks are completed satisfactorily and in no way compromise the integrity of the daughter cell, in particular, the integral transfer of the genetic material. There are two critical steps during the cell cycle that mainly concern the distribution of genetic material to the daughter cell: DNA replication and chromosome segregation. The checkpoints surrounding the S phase of the cycle ($G_1$/S and $G_2$M) have already been considered. Two others exist: the intra-S and the spindle checkpoints.

An intra-S checkpoint was suspected and has been partially described. Double strand breaks caused by ionizing radiations trigger ATM activity and the phosphorylation of a multitude of downstream kinases: Chk2, BRCA1, cdc25A, NBS1, MRE11, and RAD 50. Prior phosphorylation and release of E2F is believed to be required for the assemblage and phosphorylation of the NBS1–MRE11–RAD50 complex. Although the interactions and the exact sequence of these operations remain unclear, the net result is a delay but never a complete arrest of the S phase (5). P53 involvement in the intra-S checkpoint remains uncertain. P53 involvement could lead to apoptosis (17).

## Spindle Checkpoint

The maintenance of the integrity of the genome is far from assured even after integral DNA synthesis. In the latter steps of mitosis, the process must also provide each daughter cell with an exact copy of the duplicated chromosome (chromatids).

Separation of chromatids takes place in an extraordinary bouquet that is most elegant in its final morphological manifestations, and very complex and precise in its molecular preparation. The mechanics and the morphology of this process are not discussed in this chapter (18). Suffice it to point out that among the morphologically detectable actors of mitosis, two are critical to the spindle checkpoint: the kinetochore[8] (18) and the mitotic spindle.

### Kinetochore

After completion of the S phase, the cell's chromosomes number has doubled. Thus, the parent cell is tetraploid, but each daughter cell must return to euploidy before the next S phase. Euploidy in daughter cells can only be secured if each pair of sister chromatids are properly aligned in the metaphase. Daughter cells become anaploid if one or two sister chromatids fail to segregate to one of the daughter cells. Failure of segregation leads in the first case to a one-chromosome and in the second to a two-chromosome loss.

[8]The kinetochore is made of a group of specialized proteins (some that are structural and others that are catalytic), which bind to a specialized segment of the chromosome: the centrosome. It recognizes unattached chromosomes and sends a signal to block the cell cycle in metaphase. Proteins of the kinetochore sense the tension that the spindle exerts on it.

The role of the spindle checkpoint is to guaranty equal distribution of chromosomal material to each daughter cell. For that purpose the checkpoint must, during the transition between metaphase and anaphase, secure that each chromatid[9] pair (19):

- is present on the spindle equatorial plate;
- is separated into two identical chromosomes;
- and that each chromosome is transferred, intact, to only one of the daughter cells.

Note that it takes only one lagging chromatid to block the cycle in metaphase.

Our knowledge of the proteins involved in implementing the checkpoint at the metaphase-anaphase transition comes from studies of mutated budding yeast (*Saccharomyces cerevisiae*) (20, 21) and *Drosophila melanogaster.* In yeast, seven genes involved in the checkpoint were identified: *BUB1, BUB2,* and *BUB3; MAD1, MAD2,* and *MAD3; MPS1*).

## Regulation of the Spindle Checkpoint

Replication of chromosomes starts at the DNA level, and is only completed when each chromatid is separated from its sister and each daughter cell has received an equal set of identical chromosomes. A complex sequence of steps leads from DNA synthesis to telophase. These steps are only known in part and will not be discussed in this chapter. However, it should be made clear that cyclin and cyclin-dependent kinases are essential and paramount in the sequence. This section focuses on the segregation of sister chromatids, a central event in the spindle checkpoint only understood in part and mainly in yeast at that. However, there is no doubt that unequal distribution of chromosomes in daughter cells often causes clonal proliferation of cells with acquired survival advantages as is the case in transformation and cancer. Fortunately, the failure of some sister chromatids to assemble on the metaphase plate can be avoided by delaying chromatid segregation and giving them more time to join the plate. Physically ionizing radiations, chemically genotoxins cause such a delay, most likely to allow repair of DNA or chromosomal damage.

In yeast at least 30 and probably more genes have been shown to be involved in chromosome segregation (22). Some of these genes encode proteins with homologues in *Xenopus, Drosophila,* mice, and humans. However, most of our understanding of the process, albeit incomplete, comes from studies in yeast. Yet, several laboratories have identified key proteins or protein complexes involved in chromosome segregation and thereby have made it possible to present a schematic molecular sequence for the process. This molecular sequence includes: the kinetochore, the anaphase promoting complex securin, or MSP1, separin or ISP1, cohesins, and Polo-like kinases (23).

## Some Key Proteins in Chromatid Segregation

### Anaphase Promoting Complex

The anaphase promoting complex (APC), also called cyclosome because of its complexity and its large molecular weight (~1,500 kDa), plays a critical role in regulating the metaphase-anaphase transition. APC is a multiprotein complex made of at least eight subunits, including cyclins and a subunit with ubiquitin ligase properties ($E_3$

---

[9]The chromatid is one of a pair of identical chromosomes (each an exact copy of the other) that emerged at the end of the S phase. Thus, whenever chromatids fail to attach to the spindle, the kinetochore sensor proteins detect the decrease in tension normally generated by the attachment and signal a delay in chromosome segregation. It is suspected that phosphorylation by protein kinases is associated with the amplification of the kinetochore signal.

enzyme). A primary, but not exclusive, function of the APC ubiquitin ligase activity is to degrade mitotic cyclins and anaphase inhibitors, thereby permitting the transition from metaphase to anaphase. APC catalyzes the transfer of ubiquitin to its substrates (cyclins and anaphase inhibitors). The cyclins contain a destruction box near the N terminal (at least in yeast). The proteolysis of the mitotic cyclins starts during anaphase and persists through the $G_1$ phase, but ceases when other cyclins are again activated to trigger a new S phase.

The activity of APC itself is regulated. In 1997, Cohen-Fix and Koshland reviewed the role of APC and proposed a mechanism of activation, referred to as the primed model (23, 24). The regulation of APC varies with the phase of the cell cycle. During metaphase, APC forms a complex with cdc20 (also called Fizzi),[10] is a conserved pro tein and is a member of the WD40 family of proteins. After anaphase, cdc20 is replaced by HCTIP (or Fizzi related protein). As a result, other late mitotic proteins, including cdc20, are digested by the HCTIPI-APC complex (23).

In summary, the anaphase promoting complex harbors a well regulated ubiquitin ligase that promotes the proteolysis of several proteins involved in the progress from the metaphase to the anaphase transition. Thus, APC is essential for sister chromatid separation, but functions only indirectly, by degrading securin (25). It cannot be ruled out that APC may also be regulated by phosphorylation or dephosphorylation by kinase or phosphatase.

## Cohesins

Cohesins are components of a family of proteins involved in structural maintenance of chromosomes or SMCs. In yeast, cohesins include two SMCs, SMC1 and SMC3, and two sister chromatids cohesion proteins (SCCs) SCC1 and SCC2. These four cohesin subunits are required for function (26). Homologues have been found in some eukaryotes (*Xenopus* and humans) (26). Cohesins are found at specific sites on the chromosomes and on centromeres. In yeast, cohesins bind to DNA with a preference for the A-T regions and sequences located between genes (intergenic) (27, 28). In higher eukaryotes the interaction of cohesin and DNA remains unclear, but cohesins certainly associate with chromatin during the cell's interphase and start to dissociate during prophase (29, 30). Losada and Hirano have described the mode of binding of human cohesin to chromosomal DNA (26).

At metaphase, most of the cohesin has dissociated from the chromosomes except for that bound to the kinetochore. The dissociation of the remaining cohesin occurs in anaphase and requires a special protease, separin or Esps. At the start of anaphase, separin (see below) becomes active (31), but the sister chromatids' separation remains complete until the scissile SCC1 subunit[11] of cohesin has been degraded (31–33).

## Securin and Separin

Securin and separin are two other molecules that play an essential role in the metaphase/anaphase transition checkpoint in yeast and mammals. PdS1 (or securin) delays the sister chromatid separation in cells challenged with toxins that attack the spindle (taxol, colchichine). Securin was later found in humans. Moreover, it was shown to be identical to the pituitary tumor-transforming gene (*PTTG*), a potential

[10]MAD2 could also activate $APC^{Cd20}$ by triggering activation of APC by cdc20 or by directly acting on APC.
[11]SCC1 N terminal arginine is targeted for ubiquination.

oncogene (34). The function of securin in the spindle transition checkpoint became clear when it was discovered that securin (Pds1) binds to a caspase-like cysteine protease separin (EsP1) (35, 36). Scc1, the subunit of cohesins is the critical substrate of separin (37).

### Polo-Like Kinase

Polo-like kinases (PLKs) are a conserved family of threonine-serine kinases that play multiple roles in the cell cycle; they rise and fall with the phases of mitosis. They are believed to physically interact with cohesins prior to their degradation (38).

### Cdk1

Cdk1 plays a central role in mitotic progression. The kinase activates $APC^{cdc20}$, which degrades mitotic cyclins and thereby triggers anaphase. However, the cell does not complete mitosis as long as Cdk1 is not itself inactivated. Once mitotic cyclins have been degraded APC activity (which is dependent on cdc20 for activation) drops and cdc20 is replaced by Cd1, which in turn is inactivated by SCDKs, and mitotic cyclins re-accumulate. The sum of these observations and others (not discussed) led to a model for the regulation of sister chromatid's separation during the metaphase/anaphase transition and for activation of the checkpoint.

Once the last chromatid has reached the spindle, its presence is sensed by the kinetochore and a signal, whose nature has not yet been identified, is forwarded, telling the cycle to enter anaphase. The anaphase promoting complex $APC^{cdc20}$ is activated and proteins including securin, among others that prevent chromatid's segregation, are degraded by the ubiquitin proteolytic pathway. The degradation of securin frees separin, which in turn attacks the scission subunit of cohesin. The chromatids separate and one of a pair migrates to the opposite pole. However, if near the end of metaphase one of the chromatids has failed to reach the spindle, the spindle checkpoint is activated:

- Kinetochore proteins sense the lack of tension imposed by the spindle.
- A signal is sent to the spindle checkpoint molecular machinery (Bul1, Bul2, Bul3, and Mad1, Mad2, Mad3, and MpS1-EsP1 are involved in yeast and securin/separin, $APC^{cdc20}$, and others in several eukaryotes).
- In yeast, Mad2 (and possibly other proteins) leaves the checkpoint complex, binds to the $APC^{cdc20}$ complex, and inhibits its protease activity.
- Degradation of the anaphase inhibitor (Pds1 and Esp1) in yeast or the securin-separase complex in several eukaryotes prevents the remaining cohesin degradation, chromatids are not separated, and anaphase is delayed.

Once the situation is corrected by attaching the kinetochore of the wandering chromosome to the spindle microtubules, the checkpoint component, which inhibits the $APC^{cdc20}$ activity, leaves the scene (Mad2 in yeast). The anaphase inhibitor (Esp1[12] in yeasts) is inactivated. In higher eukaryotes separin is released from securin. Ultimately, the cohesin associated with the kinetochore is cleaved and chromatids separate. It is possible that cohesin cleavage may also require phosphorylation by polo-kinase (PLK) (further details are in 22, 24, 39).

[12]Esp1, the anaphase inhibitor protein is sequestered by Pds1 when freed Esp1 cleaves the Scc1-Med1, a component of the cohesin molecular complex.

### P53 and P21 and the Spindle Checkpoint

When exposed to microtubular poisons (colchicine, taxol) $P53^{-/-}$ fibroblasts (MEF cells) become polyploid; their chromosomes replicate without dividing (endoreplication). In contrast, $P53^{+/+}$ fibroblasts arrest in metaphase after exposure to the same poisons and do not show polyploidy (40). These findings linked to that of the association of P53 with the centrosome support the notion that P53 is involved in the spindle checkpoint and is required for mitotic arrest whenever the spindle malfunctions.

However, mitosis is not delayed in $P53^{+/+}$ cells compared with $P53^{-/-}$ cells (41, 42), and there is no increase in the level of P53 during metaphase or anaphase, but P53 rises prior to $G_1$ in tetraploid cells (9). P21 induction by P53 inhibits cyclin E/Cdk2 activity after mitotic poisoning and prevents endoreplication. These observations suggest that the role of P53 in arresting the cycle in $G_1$ extends to cells that have suffered damage to the mitotic spindle, not by intervening at the spindle checkpoint, but by inhibiting DNA replication in the $G_1$ phase (43).

## Conclusion

The normal cycle is mainly triggered by growth factors. These growth factors bind to receptors and send a signal(s) to start or resume the cycle. The signals are transduced to the cell's nuclear and cytoplasmic targets, which engage special molecular machineries that prepare the cell for replication.[13] Duplication of the genomic material stored in the chromosomes is the crowning event of the process. The cycle is divided into four phases: two Gaps ($G_1$ and $G_2$), and two critical periods distinct and detectable by microscopic and biochemical techniques, the DNA synthesis (the S phase), and the magic of mitosis (the M phase). The cell prepares for DNA synthesis through $G_1$, for mitosis through $G_2$. These divisions of the cycle are convenient because they allow investigators to focus on critical events of the cycle, but in reality the cycle is a continuum during which molecular events flow into a well orchestrated sequence. Along the path from $G_1$ to the telophase, there are checkpoints located at the entrance of each new phase of the cycle. Passage of one phase to another is put into gear by special molecules, the cyclin kinases. These proteins are cyclically activated and inactivated. The mechanisms for activation may be slow (e.g., activation of gene expression, transcription, and translation) or rapid (e.g., phosphorylation, dephosphorylation, or acetylation-deacytylation). Inactivation of the cycle involves appearance of inhibitors of the kinases (the P21 and the INK4 families of inhibitors) or proteolytic degradation, for example, the ubiquitin proteosomal pathway.

Between $G_1$ (or $G_0$) and telophase, there are several checkpoints located at the transition from one phase to another; they include the $G_1$/S, the $G_2$/M and the spindle checkpoints. These checkpoints may be activated to delay or arrest the cycle. The most studied of the checkpoints is that occurring at the transition of the $G_1$ to the S phase of the cycle; less known are the $G_2$/M and the spindle checkpoints. The purpose of these control posts is to assess the progress of the cycle up to the point of transition, to: (a) make sure that energy sources and the capacity to provide the building blocks

[13]Regulation of cell growth during the cell cycle is still poorly understood. Cyclin D/Cdk-4/Cdk-6 is believed (among other factors) to contribute to the rise of protein, lipid, and carbohydrate synthesis, which are needed for coupling growth to the cycle.

required in the next phase are available, and (b) make sure that the genetic material newly assembled during DNA synthesis and stored in the chromosome is accurately duplicated and faithfully distributed to the chromatids, and that the chromatids are divided equally among the progeny. Whenever the cell fails to complete its task during a previous phase and risks thereby to transmit a defective genome, the cycle is arrested or delayed at the checkpoint. Arrest may provide sufficient time for damage repair, thus allowing the cycle to resume its course, or it may be followed by the cell's demise (apoptosis). Each of the checkpoints has its own mechanism of operation, one or more biochemical pathways that link the signal to the endpoint: namely, the cell cycle arrest. Although knowledge of the pathway of the three checkpoints is still incomplete, common denominators have emerged. Thus, despite the uniqueness of each pathway, some steps are similar or identical for some or all of them. Each pathway can be divided into three steps: sensing the defect; transduction of the signal; and response of the effector molecules. The sensor is not known, but the transduction involves (among other molecules) special kinases (ATM, ATR, DNAPK, BRCA 1 and BRCA 2, Chk1, Chk2, etc.). The effectors block DNA synthesis, ($G_1$/S) entry into mitosis ($G_2$/M), or chromosomal segregation (the spindle checkpoint).

## References and Recommended Readings


1. Sanchez, Y., Wong, C., Thoma, R.S., Richman, R., Wu, Z., Piwnica-Worms, H., and Elledge, S.J. Conservation of the Chk1 checkpoint pathway in mammals: linkage of DNA damage to Cdk regulation through Cdc25. *Science.* 1997;277:1497–501.
2. Liu, Q., Guntuku, S., Cui, X.S., Matsuoka, S., Cortez, D., Tamai, K., Luo, G., Carattini-Rivera, S., DeMayo, F., Bradley, A., Donehower, L.A., and Elledge, S.J. Chk1 is an essential kinase that is regulated by Atr and required for the G(2)/M DNA damage checkpoint. *Genes Dev.* 2000;14:1448–59.
3. Guo, Z., Kumagai, A., Wang, S.X., and Dunphy, W.G. Requirement for Atr in phosphorylation of Chk1 and cell cycle regulation in response to DNA replication blocks and UV-damaged DNA in *Xenopus* egg extracts. *Genes Dev.* 2000;14:2745–56.
4. Bartek, J., and Lukas, J. Mammalian G1- and S-phase checkpoints in response to DNA damage. *Curr Opin Cell Biol.* 2001;13:738–47.
5. Elledge, S.J. Cell cycle checkpoints: preventing an identity crisis. *Science.* 1996; 274:1664–72.
6. Crawford, D.F., and Piwnica-Worms, H. The G(2) DNA damage checkpoint delays expression of genes encoding mitotic regulators. *J Biol Chem.* 2001;276:37166–77.
7. King, R.W., Jackson, P.K., and Kirschner, M.W. Mitosis in transition. *Cell.* 1994;79:563–71.
8. Stewart, Z.A., and Pietenpol, J.A. p53 Signaling and cell cycle checkpoints. *Chem Res Toxicol.* 2001;14:243–63.
9. Abraham, R.T. Cell cycle checkpoint signaling through the ATM and ATR kinases. *Genes Dev.* 2001;15:2177–96.

10. Alani, E., Subbiah, S., and Kleckner, N. The yeast RAD50 gene encodes a predicted 153-kD protein containing a purine nucleotide-binding domain and two large heptad-repeat regions. *Genetics.* 1989;122:47–57.
11. Zhan, Q., Carrier, F., and Fornace, A.J., Jr. Induction of cellular p53 activity by DNA-damaging agents and growth arrest. *Mol Cell Biol.* 1993;13(7):4242–50. Erratum in: *Mol Cell Biol.* 1993;13(9):5928.
12. Wang, X.W., Zhan, Q., Coursen, J.D., Khan, M.A., Kontny, H.U., Yu, L., Hollander, M.C., O'Connor, P.M., Fornace, A.J. Jr., and Harris, C.C. GADD45 induction of a G2/M cell cycle checkpoint. *Proc Natl Acad Sci USA.* 1999;96:3706–11.
13. Yang, Q., Manicone, A., Coursen, J.D., Linke, S.P., Nagashima, M., Forgues, M., and Wang, X.W. Identification of a functional domain in a GADD45-mediated G2/M checkpoint. *J Biol Chem.* 2000;275:36892–8.
14. McShea, A., Samuel, T., Eppel, J.T., Galloway, D.A., and Funk, J.O. Identification of CIP-1-associated regulator of cyclin B (CARB), a novel p21-binding protein acting in the G2 phase of the cell cycle. *J Biol Chem.* 2000;275:23181–6.
15. Taylor, W.R., and Stark, G.R. Regulation of the G2/M transition by p53. *Oncogene.* 2001;20:1803–15.
16. Nunez, F., Chipchase, M.D., Clarke, A.R., and Melton, D.W. Nucleotide excision repair gene (ERCC1) deficiency causes G(2) arrest in hepatocytes and a reduction in liver binucleation: the role of p53 and p21. *FASEB J.* 2000;14:1073–82.
17. Gottifredi, V., Shieh, S., Taya, Y., and Prives, C. From the cover: p53 accumulates but is functionally impaired when DNA synthesis is blocked. *Proc Natl Acad Sci USA.* 2001;98:1036–41.
18. Shimoda, S.L., and Solomon, F. Integrating functions at the kinetochore. *Cell.* 2002;109:9–12.
19. Shah, J.V., and Cleveland, D.W. Waiting for anaphase: Mad2 and the spindle assembly checkpoint. *Cell.* 2000;103:997–1000.
20. Burke, D.J. Complexity in the spindle checkpoint. *Curr Opin Genet Dev.* 2000;10:26–31.
21. Amon, A. The spindle checkpoint. *Curr Opin Genet Dev.* 1999;9:69–75.
22. Page, A.M., and Hieter, P. The anaphase-promoting complex: new subunits and regulators. *Annu Rev Biochem.* 1999;68:583–609.
23. Lee, J.Y., and Orr-Weaver, T.L. The molecular basis of sister-chromatid cohesion. *Annu Rev Cell Dev Biol.* 2001;17:753–77.
24. Cohen-Fix, O., and Koshland, D. The metaphase-to-anaphase transition: avoiding a mid-life crisis. *Curr Opin Cell Biol.* 1997;9:800–6.
25. Zachariae, W., and Nasmyth, K. Whose end is destruction: cell division and the anaphase-promoting complex. *Genes Dev.* 1999;13:2039–58.
26. Losada, A., Yokochi, T., Kobayashi, R., and Hirano, T. Identification and characterization of SA/Scc3p subunits in the *Xenopus* and human cohesin complexes. *J Cell Biol.* 2000;150:405–16.
27. Blat, Y., and Kleckner, N. Cohesins bind to preferential sites along yeast chromosome III, with differential regulation along arms versus the centric region. *Cell.* 1999;98:249–59.

28. Tanaka, T., Cosma, M.P., Wirth, K., and Nasmyth, K. Identification of cohesin association sites at centromeres and along chromosome arms. *Cell.* 1999;98:847–58.
29. Sumara, I., Vorlaufer, E., Gieffers, C., Peters, B.H., and Peters, J.M. Characterization of vertebrate cohesin complexes and their regulation in prophase. *J Cell Biol.* 2000;151:749–62.
30. Losada, A., and Hirano, T. Intermolecular DNA interactions stimulated by the cohesin complex in vitro: implications for sister chromatid cohesion. *Curr Biol.* 2001;11:268–72.
31. Waizenegger, I.C., Hauf, S., Meinke, A., and Peters, J.M. Two distinct pathways remove mammalian cohesin from chromosome arms in prophase and from centromeres in anaphase. *Cell.* 2000;103:399–410.
32. Uhlmann, F., Lottspeich, F., and Nasmyth, K. Sister-chromatid separation at anaphase onset is promoted by cleavage of the cohesin subunit Scc1. *Nature.* 1999;400:37–42.
33. Uhlmann, F. Chromosome cohesion and segregation in mitosis and meiosis. *Curr Opin Cell Biol.* 2001;13:754–61.
34. Zou, H., McGarry, T.J., Bernal, T., and Kirschner, M.W. Identification of a vertebrate sister-chromatid separation inhibitor involved in transformation and tumorigenesis. *Science.* 1999;285:418–22.
35. Uhlmann, F., Wernic, D., Poupart, M.A., Koonin, E.V., and Nasmyth, K. Cleavage of cohesin by the CD clan protease separin triggers anaphase in yeast. *Cell.* 2000;103:375–86.
36. Nasmyth, K. Separating sister chromatids. *Trends Biochem Sci.* 1999;24:98–104.
37. Alexandru, G., Uhlmann, F., Mechtler, K., Poupart, M.A., and Nasmyth, K. Phosphorylation of the cohesin subunit Scc1 by Polo/Cdc5 kinase regulates sister chromatid separation in yeast. *Cell.* 2001;105:459–72.
38. Yanagida, M. Cell cycle mechanisms of sister chromatid separation; roles of Cut1/separin and Cut2/securin. *Genes Cells.* 2000;5:1–8.
39. Jallepalli, P.V., and Lengauer, C. Chromosome segregation and cancer: cutting through the mystery. *Nat Rev Cancer.* 2001;1:109–17.
40. Di Leonardo, A., Khan, S.H., Linke, S.P., Greco, V., Seidita, G., and Wahl, G.M. DNA rereplication in the presence of mitotic spindle inhibitors in human and mouse fibroblasts lacking either p53 or pRb function. *Cancer Res.* 1997;57:1013–9.
41. Khan, S.H., and Wahl, G.M. p53 and pRb prevent rereplication in response to microtubule inhibitors by mediating a reversible G1 arrest. *Cancer Res.* 1998;58:396–401.
42. Khan, S.H.; Moritsugu, J., and Wahl, G.M. Differential requirement for p19ARF in the p53-dependent arrest induced by DNA damage, microtubule disruption, and ribonucleotide depletion. *Proc Natl Acad Sci USA.* 2000;97:3266–71.
43. Janus, F., Albrechtsen, N., Dornreiter, I., Wiesmuller, L., Grosse, F., and Deppert, W. The dual role model for p53 in maintaining genomic integrity. *Cell Mol Life Sci.* 1999;55:12–27.
44. Tapon, N., Moberg, K.H., and Hariharan, I.K. The coupling of cell growth to the cell cycle. *Curr Opin Cell Biol.* 2001;13:731–7.

# CELL CYCLE ARREST

# 14

The cell cycle is a most important phase in the lifespan of the cell in vitro and in vivo. In vitro, cells can proliferate indefinitely; in vivo they seldom do. However, like the cells in vitro, cells in vivo need to repair damage to their genetic material, but they may also choose to differentiate, senesce (grow old), or die. DNA repair, differentiation, senescence, and cell death are all associated with cell cycle arrest.

Up to this point, we have considered arrest of the cycle triggered by DNA or chromosomal damage. However, one cannot escape wondering whether the signals that emerge at each of these moments of the cell's life—when arrest of the cycle is imperative—are the same. If these signals are different, what is the genomic program that makes them so? This may appear to be a farfetched question, especially because few data are available. Yet one starting point is to examine whether P53 and P21 are always needed for cell cycle arrest.

## Cycle Arrest Dependence on P21

It is appealing to assume that mitotic arrest after exposure of cells to ultraviolet (UV) and γ radiations or to genotoxins allows for DNA repair, and thereby prevents replication of damaged DNA and defective chromosomes (see Chapters 12 and 13) and that binding of P21 or other cdk inhibitors of cyclin kinase are critical to the arrest of the cell cycle. However, this elegant mechanism for arresting the cell cycle only raises more questions. Among them are:

- Can P53 arrest of the cycle in $G_1$ occur without P21 induction?
- Can P21 activation occur independently of P53 activation?
- Is P21 absolutely required for the $G_1$ arrest?

Deng and associates (1) exposed fibroblasts of P21-null mice to nocodazole or colcemid, agents that prevent spindle formation. After such treatment the cells are unable to enter a new S phase because they are arrested in $G_1$. Thus, at least in this case, P53 uses a P21-independent pathway to arrest the cells in $G_1$. Activated monocytes or macrophages secrete cytokinases among them the interleukin-1 (IL-1) polypeptide. IL-1 arrests growth in human melanoma cells (A375C6). The growth arrest is associated with the hypophosphorylation of the Rb protein and takes place in $G_1$. Inhibition of wild-type P53 by a dominant negative mutant or/by inhibition of *p21* expression does not prevent cell cycle arrest. These findings led Nalca and

Rangnekar to conclude that the cell cycle arrest induced by interleukin (IL-1) is independent of both wild-type P53 and P21 (2).

When P53 null cells (HL-60) are exposed to tumor necrosis factor-alpha (TNF-$\alpha$) the *p21* gene is markedly induced in monocytes prior to differentiation and apoptosis. Actinomycin inhibits the rise of P21. Therefore, the activation must be at the level of transcription. Since P53 is absent, the activation of the *p21* gene is independent of that of *p53* (3).

Diploid fibroblasts exposed to $\gamma$ radiation show a relatively long P21 induction that is P53-dependent and associated with the arrest of the cycle in the $G_1$ phase. The arrest persists as long as double strands breaks (even a few) are present in the DNA (4, 5). However, this response of normal diploid cells is very different from that of malignant cells, even if the malignant cells contain wild-type P53. Indeed, in some fibrosarcoma cells the arrest is of short duration and in others (i.e., colon and breast cancer cells) there is no arrest in $G_1$ at all (6). These findings suggest that the ability to sustain the cell cycle arrest after ionizing irradiation varies with cell type and the type of DNA repair. The above findings are in agreement with those of Wani and associates, who demonstrated down-regulation of P21 expression in various human cancer cell types after exposure to UV radiation (7). Moreover, the down-regulation is independent of the P53 status of the cells and the cell cycle arrest in $G_1$ even in absence of P21. Finally, Wani et al. showed that the removal of anti-benzo(d)pyrene-diol-epoxide DNA adducts require P53 for global genomic repair (GGR) but not for transcription coupled repair (TCR) (7).

Haaparjarvi and colleagues (8) investigated the effects of ultraviolet C (UVC) irradiation on *p21* gene expression in both $p53^{+/+}$ and $p53^{-/-}$ fibroblasts. UVC irradiation arrests the cell cycle in the P53 null and the wild-type cells, but while the P53 null die (apoptosis?), the wild-type cells survive and reenter the cycle. Evaluation of the P21 messenger RNA and protein revealed that, even in P53 null cells, both are increased up to sixfold at lower doses of UVC, but P21 mRNA and proteins do not rise at higher doses of UVC. Thus, it appears that P53, if not absolutely required for cell cycle arrest, is needed for survival and probably DNA repair. In addition, it seems that P21 induction can be independent of P53, but P21 persistant induction is not an absolute requirement for cell cycle arrest.

In summary, cell cycle arrest at the $G_1$ checkpoint sometimes can be independent of both P53 and P21. The requirement for P53 for cycle arrest may vary with the mode of repair response engaged (e.g., global excision repair [GER] vs. transcription exicision repair). The requirement for activation of P21 can be independent of P53HLA 60 cells). The response to UV and ionizing radiations at the $G_1$ checkpoint may vary with the cell type (diploid or cancerous) and with the type of radiation. It appears that neither P53 nor P21 are always necessary requirements for cell cycle arrest at the $G_1$ checkpoint.

These findings, although few among many, are examples of experiments in which P53 and P21 have been shown to be not always indispensable for arrest of the cell cycle. These studies imply the existence of redundancy in the control of the cell cycle checkpoints. In the case of P53, other members of the family could be drafted such as P73 and P63. Similarly, in the case of the P21 family either P27 or P57 could come into play, but there is also another family of cyclin inhibitors—the INK4 family—that can be called into action. These considerations are not purely academic when it comes to targeted gene therapy in cancer. They illustrate the need for identifying, in each case, the molecular signal to the target genes involved in the checkpoint control and the triggering of apoptosis.

## P21 and DNA Repair

If there is little doubt that P53 contributes to some, if not most, forms of DNA repair (including NER repair) (8, 9), the steps between the P53 signal and the cell cycle arrest, the repair process, and resumption of DNA synthesis or apoptosis are still not entirely clarified (10, 11).[1]

The role of P21 in DNA repair, and in particular nucleotide excision repair (NER), is complex. P21 may contribute to DNA repair indirectly by arresting the cell cycle and/or directly by contributing to the repair process through binding to PCNA. PCNA is a polymerase δ–associated protein. The polymerase is one of the enzymes involved in the polymerization step of nucleotide excision repair. PCNA binds to P21 and this binding could either facilitate the NER process (12) or inhibit it by neutralizing PCNA (13).

Cooper, Balajee, and Bohr investigated the role of the P21 protein in NER (14) and demonstrated that the C terminal of the P21 protein, but not its N terminus, inhibits NER in vitro and in vivo. Binding of PCNA to P21 is responsible for the inhibition, and the inhibition can be relieved by addition of purified PCNA to the in vitro system (14, 15).

Therrien and colleagues demonstrated that NER is after exposure to UV radiation enhanced in $P53^{-/-}$ $P21^{-/-}$ cancer cells compared to $P53^{-/-}$ ($P21^{+/+}$) cancer cells, thus increasing the survival advantages of P21-deficient cells in cancer associated with P53 defects (16).

Knowledge of the direct contribution of P21 to DNA repair was advanced by elegant studies in Hanawalt's laboratory. Ford and Hanawalt showed that the loss of P53 function is associated with a substantial decrease in nucleotide excision repair, but mainly of the GGR component at the exclusion of the TCR (17, 18). P53 is known to activate the transcription of P21 and GADD45, but what is not known is which of these two proteins (or others) play the most significant role in NER. Cells genetically deficient (–/–) in P53, P21, and GADD45 were exposed to ultraviolet light, and their capacity to repair DNA was determined by measuring unscheduled DNA synthesis and removal of photoproducts from either genomic DNA or from the *dhfr* gene. While $P21^{-/-}$ cells showed little or no decrease in NER compared to wild type cells, $P53^{-/-}$ and $GADD45^{-/-}$ cells were significantly defective in NER (19).

These experiments and others confirm that:

- The P53 transcription factor activates P21 transcription in NER.
- P21 arrests the cell cycle in NER.
- P21 binds to PCNA.
- P53 is required for GGR but not TCR.

However, these data do not prove that P21 is required for GGR nucleotide excision repair, nor do they explain how P53 contributes to GGR. Do P21 and P53 act indirectly by activating the $G_1/S$ checkpoint or directly by transcription control of protein(s) involved in GGR (but not in TCR)? In an attempt to solve the problem, Ford's group used several cell lines, some wild-type for P53 ($P53^{+/+}$) and others harboring a mutant

[1]There is evidence that P53 directly contributes to the expression of molecules involved in DNA repair, such as the P48 product of the *XP* gene proliferating cell nuclear antigen (PCNA) and ribonucleotide reductase.

P53 ($P53^{-/-}$) (20). The cells were divided into three groups: $P53^{+/+}$-$P21^{+/+}$; $P53^{+/+}$-$P21^{+/-}$ (heterozygous for P21); and $P53^{+/+}$-$P21^{-/-}$ (homozygous for P21). These cells were irradiated with UVC and levels of DNA repair were measured by following the rate of removal of DNA adducts (6-4 pyrimidine dimers and cyclobutane pyrimidine dimers [CPDs]). Both levels of TCR and GGR were the same in P21 deficient and P21 competent cells, thus demonstrating that P21 is not required for either GGR or TCR. These data were confirmed by overexpression of P21 in P53 mutants, $P53^{-/-}$, and $P21^{+/+}$ (20). A definitive explanation for the molecular mode of action of P53 in GGR is still lacking; however, the above findings suggest that such a role is not linked to the arrest of the cell cycle by P21, but to modulation of the special function of proteins in GGR. Such selective function may depend on the mode of transcription *trans*activation in GGR compared to that of TCR.

In conclusion, despite P21 induction by P53, its ability to arrest the cell cycle, and its capacity to bind to PCNA, P21 has no role in NER repair whether it be GGR or TCR. This does not exclude the possibility of P21's intervention in other forms of DNA repair (e.g., recombination repair); however, P53 is required for GGR, but in what capacity is still unknown. Moreover, if cell cycle arrest is a requirement for DNA repair, which P53 inducible protein is responsible for the arrest? Obviously, as proposed by Ford et al., GADD45 is a likely candidate. These various (and selected) findings further confirm that several pathways may exist for arresting the cycle in the $G_1$ or $G_2$ phase, some of which are dependent on the activation of P53 and the induction of P21, others in which the induction of P21 is independent of P53, and still others that require neither P53 nor P21. The differences may depend on cell type, nature of the DNA damage, mode of DNA repair, and the level of exposure to the genotoxins.

## MYC/MAX Family and Differentiation

In the course of organogenesis or terminal differentiation, precursor cells stop proliferating because in most cases cell proliferation (and, therefore, immortalization and cancer) are with few exceptions incompatible with differentiation. The nuclear protein C-myc is the counterpart of the retroviral transforming gene *V-myc* isolated from avian leukemia viruses. Thus, the acute chicken myeloma virus MC29 contains a sequence similar to that of the human oncogene *myc*. There are three evolutionary conserved genes in the *myc* family: *c-myc*, *n-myc*, and *l-myc*. The product of the *myc* gene, the Myc protein, contributes to cell proliferation, differentiation, and apoptosis.[2] Myc regulates the expression of several genes, causes cell transformation, and induces various types of cancers in a range of animals. Despite this constellation of functions, c-Myc's mRNA has a short half-life (15 to 20 minutes) and its transcription activity is weak. Consequently, it is not surprising that the molecular mechanism by which c-Myc exerts its multiple functions remain somewhat elusive.

The molecular mechanisms of transcription activation have only begun to be understood. First, there are multiple steps involved, such as: chromatin remodeling, pre-initiation, recruitment of transcription factors and RNA polymerase, exclusion of

[2]This function will be discussed in further detail later in this chapter; suffice it to point out that Myc contributes to switching the cell from life to death.

repressors and corepressors, among others. Each of these steps often requires large protein complexes. Second, the mode of interaction between proteins and DNA needs to be better understood. The c-Myc protein (439 amino acids) contains several domains in the NH and the C terminals:

- The NH terminal (the transactivating domain consisting of ~150 amino acids) includes two conserved Myc boxes: MBI (amino acids 44–53) and MBII (amino acids 128–148). MBI is an activating domain, and MBII is in part an activating and in part a repressing domain.
- The molecule also contains a nuclear localization domain.
- The C terminal sequence harbors a basic helical loop helix and a leucine zipper domain that is involved in protein-protein association (bHLHZ).
- The DNA activating site for c-Myc includes the canonical nucleotide sequence referred to as the E-Box.
- C-Myc, heteromerizes with Max (in the C terminus). The Max protein contains a bHLH-zip in its C terminus, but no transactivation sequence. Myc-Max dimerization occurs through their respective bHLH-zip (21). While Myc has a short half-life, the expression of which is triggered by specific growth signals, Max is expressed constitutively (22).
- Other functional molecules known to bind to Myc include TRAAP (in the N terminus), and molecules believed to be involved in chromatin remodeling (at the BHLHZ site) and also gene repression.

The Myc-Max association generates a molecular complex with: 1) DNA binding capacity to the E-Box; and 2) with transcription activation and facilitation of specific coactivators of transcription and repressors (see below).

Max is a promiscuous molecule, however; it also associates with other BHLHZ molecules: the Mad family (4 of them are Mad1, Mad2, Mad3, and Mad4), and the closely related molecules Mnt1, and ROX (23),[3] Mga (24).[4] Max heteromerizes with these molecules through their respective BHLHZ domains. The complexes contribute to, but are not sufficient for, repression of the Myc targeted genes.

Members of the Mad family and their homologues complex with Max but not Myc. The Mad/Max complexes lack a transcription activation domain and do not bind to the E-Box sites. The mammalian homologues of the SIN3 yeast transcription repressors, mSIN3A and mSIN3B, were found to bind to Mad and function as coexpressors (25, 26); mSin3A is part of a large multiple protein complex that contains stoichiometric amounts of histone deacetylases (HDAC1 and HDCA2). The transcriptional repression of the Mad proteins requires that such deacetylase activity be effective (**FIGURE 14-1**).

Eilers and colleagues (27) investigated the interaction between the paired alipathic helices (PAH 2) present in mSin3A and the "SIN3 interaction domain" (SID) that is present in Mad. For that purpose, they studied the structure of SID and introduced mutations in its sequence. The amino terminal sequence of Mad contains 35 amino acids required for recruitment of the mSin3A–HDAC complex (27–29). Of those, a 13 amino

---

[3]Rox heteromerizes with Max and weakly homodimerizes. Ros-Max binds preferentially to the CACGG site instead of the canonical E-Box sequence. Repression requires at least the Sin3 corepressor. Rox is highly expressed in quiescent fibroblasts and decreases rapidly at the origin of the cell cycle.

[4]Mga interacts with Max; it possesses a bHLHZ motif and has an mSinA reaction domain (SID), yet it shows no obvious similarities with Myc or Max proteins.

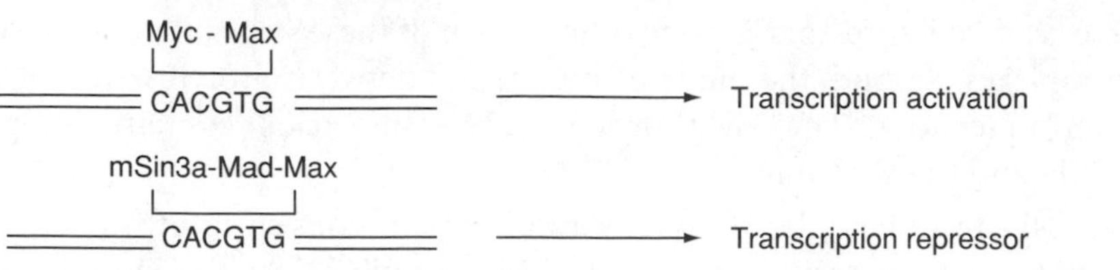

**FIGURE 14-1** Interaction of Myc-Max and mSin3a-Mad-Max with DNA in transcription activation and repression. See text for details.

acid sequence (8–20) of Mad suffices for the SID-PAH2 interaction. The interaction site adopts an amphipathic α helix structure in solution and its hydrophobic face makes contact with mSin3A. The exact mode of the molecular interaction between PAH2 and SID is not yet clear, but it has been suggested that the two helices (A and B) of the pair contribute to form a cleft that serves as a receptacle for the hydrophobic phase of SID (27). Note that, like Max, mSin3a α B are expressed constitutively (in contrast to the Myc and Mad families).

In conclusion, the Myc-Max complex contributes both to transcription activation (and thereby to cell proliferation) and to transcription suppression (and thereby to differentiation).

This oversimplified view cannot explain either the existence of the multiple transcription targets of Myc or the selective gene repression. The latter mechanisms have been clarified in part by the discoveries of the Myc dependent recruitment of TRRAP that links Myc to 1) histone acetyl transferase and 2) the recruitment of deacetylases, by mSin3a.

The N terminal amino acid sequence of c-Myc (150 amino acids is required for transcription activation) contains two highly conserved motifs or Myc boxes: MBI and MBII. Myc biological functions are lost by mutations of MBII, thus establishing the importance of the latter in the Myc's operations. MBII binds TRRAP,[5] a component of a larger complex that includes histone acetyl-transferase activity. Thus, TRRAP recruits the histone acetyltransferase hc-CN5 to c-Myc (31, 32).

## c-Myc, the Cell Cycle, and Differentiation

In 1999, Bouchard et al. demonstrated that c-Myc induces cyclin D2 and thereby contributes to cell cycle progression and sequestration of the cyclin inhibitor p27 (33). Bouchard et al. further established that in vivo, Myc binds to the cyclin D2 promoter and induces the expression of not only cyclin D2, but also of the acetylation of histone H4. Both events take place in a single nucleosome. Moreover, binding and induction require an integral MBII site in the N terminus of c-Myc and the recruitment of TRRAP. Frank et al. further demonstrated that after TRRAP recruitment, Myc induces acetylation of histone in seven target genes upon serum stimulation of Rat1 fibroblasts (34). In 1995, Ayer and colleagues had shown that mSin3a, a highly conserved corepressor, associates with the N terminal of Mad (26). Later it was established that mSin3a is part of a polypeptide complex that contains histone deacetylases HDAC1 and HDAC2 (35).

[5] *Saccharomyces cerevisiae* requires histone acetyl transferase (HAT) for specific gene transcription, and GCN 5 is such a HAT. It is part of a molecular coactivator complex (SAGA) that includes the SPT-ADA-GCN5 acetyl transferases. Two GCN5 homologues have been found in mammalian cells: PCAF, which forms a molecular complex with 20 polypeptides; and GNSL. Each of the two is associated with different polypeptide complexes (31).

Prior to cell differentiation, the Myc/Max complex is replaced on the promoter by the Mad/Max complex; TRRAP is released while bound histone deacetylase rises (33). After reviewing these and other findings, Eisenman proposed a most plausible mechanism for the role of Myc in gene expression through transcription and repression of target genes (36). The model serves to explain the functions of Myc in cell proliferation and cell differentiation. In essence, Myc and Mad contribute to the regulation of histone acetylation and deacetylation after binding to the E-Box.

During transcription activation and cell cycle progression:

- The Myc-Max heterodimer binds the CACGTA sequence of the E-Box.
- The Myc Box II (MBII) recruits the TRRAP coactivator which brings the acetyltransferase (GCN5) to Myc.
- GCN5 acetylates the nucleosomal histones located at the E-Box and in adjacent areas.
- In the presence of additional factors (not discussed) transcription activation of the target gene takes place.
- Further activation of other target genes may occur by displacement of the Myc/Max complex along the DNA sequence.

During transcription repression and differentiation:

- Formation of the complex Mad/Max occurs through their respective bHLHZ motifs at the E-Box of the target gene.
- Recruitment of mSIN3 and the associated histone deacetylase complex occurs.
- Histone deacetylation and gene silencing is completed prior to differentiation (36–38).

It remains to be seen whether the genes (*p21*, *p27*, *p57*, and *ink4*) whose products arrest the cell cycle are up-regulated prior to differentiation by the mechanism described above and are inactivated prior to cycle progression.

In theory, c-Myc can both induce or repress expression of the *p21* gene or other cyclin-cdk inhibitors, and thereby either cause the cycle to arrest and allow senescence and differentiation to take place, or it may turn off the checkpoint and allow cell proliferation. If so, *p21* becomes critical to the cell's choice between cell growth and differentiation. Succinctly, when the *p21* gene is induced, the cell either differentiates or senesces, and when repressed the cell divides. However, by repressing cell cycle regulatory genes (e.g., *p21* [39][6] *Gadd45* and members of the *ink4* family), c-Myc allows cell transformation to take place (40, 41).

## Cyclin, Cyclin Kinases, Cyclin Inhibitors, and Differentiation

Not unlike transcription initiation, there are multiple mechanisms in gene repression:

- They may involve limited or extended segments of DNA.
- They may or may not be reversible.
- In many (if not all) cases, they involve chromatin reorganization through acetylation and phosphorylation.

[6]The mechanism of repression of the *p21* gene needs clarification. It may involve the formation of the complex mSinB, Mad, and Max, and bind to the E-Box sequence, or it may (at least in some cases) form complexes with Sp1/Sp3 by reacting with the zinc finger domain of the latter proteins, thereby sequestering the S1/S3 factor proteins containing a zinc finger domain that recognizes a GC rich region located upstream of the transcription site (39).

During differentiation, genes controlling cell cycle progression are repressed and those that determine specific cell structure and function expressed. This section considers two examples of regulation of differentiation by molecules critical to the operation of the cell cycle: the function of cyclin-Cdks in the cycle's arrest, and that of Rb in chromatin silencing.

The study of differentiation requires that a stable and consistent marker for the triggering of the process be available. Members of the myogenic transcription factor have been used to follow differentiation and regeneration of skeletal muscle, and Myo D is a member of that family. There is vast literature on the subject, so only a brief review of the role of cyclin and cyclin kinases in the emergence and expression of Myo D will be discussed (42). A family of muscle-specific transcription factors (MsFs) controls muscle differentiation. Included in this family are: Myo D, myogenin, Myf 5, MRF 4, and Myf 6. These proteins have in common a classic helix loop helix domain (70 amino acids).

Myo D is a member of the skeletal myogenic transcription factor family expressed in dividing cells in the presence of growth factors prior to the occurrence of differentiation. Upon withdrawal of the growth factor, the cycle arrests and the cells differentiate (43). Cell cycle arrest and the trigger for differentiation are thus clearly coupled. The molecular mechanisms that control the coupling are not entirely known, but it has been shown that:

- Myo D transcription activation is inhibited by activators of the cycle: cyclin D/cdk complexes (44, 45).
- The cell cycle arrest coincides with the induction of the *p21* gene by Myo D (46, 47).
- Overexpression of E2F1 in myoblast cultures after withdrawal of mitogens restores cell replication and inhibits differentiation (48).
- Overexpression of P21 or 16 INK-4A in growth factor stimulated cultures in which these factors were not withdrawn, arrests the cycle and triggers differentiation (44).
- Overexpression of P57 stabilizes Myo D by inhibiting the cyclin E–Cdk2 kinase in proliferating myeloblasts (49). Raynaud and associates propose that the stabilization is not due to P57's capacity to act as a cyclin inhibitor, but rather to a direct interaction between P57 and Myo D involving the conserved helix of P57 (50).

In summary, molecules that activate the cycle (cyclin-cdks and E2F) interfere with differentiation whereas molecules that arrest the cycle (cdk inhibitors) stimulate differentiation, and the two processes (cycle and differentiation) act in strict coordination. Ren and associates (51) developed a method that allows one to identify the binding of specific transcription factors to the DNA promoter sites in the global genome. By combining the binding of the factor to the target gene with the expression of that gene in the phenotype, it becomes possible to characterize the function(s) of the transcription factor. The method was used in fibroblasts to identify the target genes of two E2F factors: E2F1, a gene activator, and E2F4, a repressor. The study revealed that approximately 1200 genes are expressed at the start of the cycle after mitogenic stimulus. E2F4 binds to the promoters of 127 of those genes in primary fibroblasts, but E2F1 can also bind to the promoter of 50 of those genes. The promoter of the gene, therefore, may receive contradictory signals—activation by E2F1 and repression by E2F4. The target genes under study could be divided into three groups depending on the molecules encoded: those involved in DNA replication and chromatin rebuilding, those involved in cell cycle damage checkpoints, and those involved in DNA repair. These functions are interrupted during quiescence and activated by mitotic stimuli (**FIGURE 14-2**) (for details see reference 51). In conclusion, a gene that contributes to cell proliferation when activated by the appropriate transcription factor (e.g., E2F1) may also contribute to cell cycle arrest and differentiation when a repressor (e.g., E2F4) is bound to its promoter.

Experiments on cancer cells demonstrate that many findings made in vitro have a counterpart in vivo. Overexpression of cyclin kinase inhibitors (P21, P27) triggers differentiation of neuroblastoma cells. P21 overexpression combined with vitamin $D_3$ stimulates differentiation of monocytic leukemic cells. The expression of two cytokinin inhibitors, combined with the administration of an inhibitor of Ckd2, stimulates the differentiation of murine erythroleukemic cells (52–55).

In vivo molecular triggers that inactivate and maintain the arrest of the cell cycle in favor of differentiation are unknown. In culture, the addition or withdrawal of growth factors, and the overexpression or inhibition of critical molecules (cyclins or cyclin kinase inhibitors) can switch paths from replication to differentiation. In vivo, the nature of these extracellular stimuli is varied (hormones, vitamins, cytokinases, etc.) and, moreover, they must be strictly regulated to secure the harmonious development and maintenance of homeostasis in an organ. For most cells, the switch from replication to differentiation is irreversible, but the factor(s) that maintain that state are not known. The existence of a "timer" or clock that regulates the process has been postulated, but at this point its molecular construction is difficult to imagine (**FIGURE 14-3**) (55).

Although early investigations targeted P21, it soon became clear that other cyclin kinase inhibitors also participate in cell cycle arrest after DNA damage differentiation and senescence. Attempts to understand cell cycle arrest in differentiation revealed that molecules or molecular complexes involved in the cell cycle contributed to the cell's progress from telophase to differentiation. The notion that the regulators of the cell cycle may also contribute to differentiation and that regulators of differentiation in turn may contribute to the cell cycle not only survived further investigations, but was reinforced. Thus, the two processes are closely correlated and the mechanisms that master these concerted efforts between cell replication and terminal differentiation have only begun to be explored.

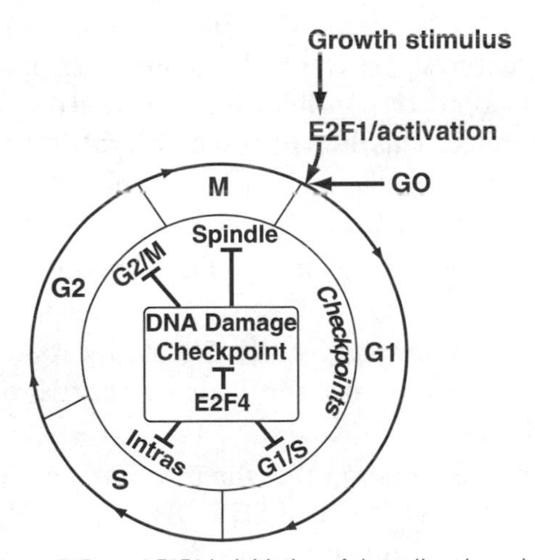


**FIGURE 14-2** Interaction between E2F1 and E2F4 in initiation of the cell cycle and quiescence (modified from Ren et al. [51]).

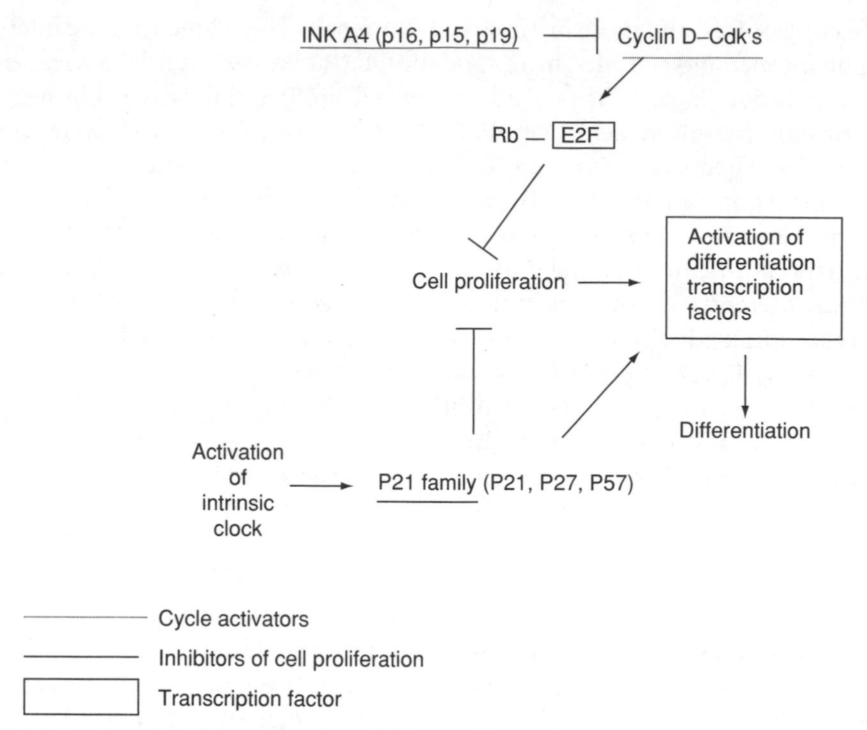


**FIGURE 14-3** An adaptation of the schematic model for coupling of cell proliferation and differentiation proposed by Zhu and Skoultchi (55).

## Rb Protein and Differentiation

One of the first requirements for differentiation after cell cycle arrest is the silencing of genes that encode the proteins that activate the cell cycle. The nucleosome forms the unit structure of chromatin. Its DNA wraps around histones and forms the core, a compact mass, and histone tails (notably histone 3) protrude from that core. The activity of chromatin is modulated in part by relatively simple biochemical events: phosphorylation, dephosphorylation, acetylation, and methylation. Two proteins, SUV39H1 and the heterochromatin protein (HP1), contribute to, among other processes, the formation of heterochromatin. Heterochromatin is an extended modification of chromatin structures that cause chromatin silencing (56, 57). However, in most cells the silencing of genes responsible for cell cycle arrest, prior to differentiation, needs to be subtle and limited. This does not exclude a role for the SUV39H1 and HP1, but it emphasizes the need for regulation. The Rb protein contributes to the latter (58).

Nielsen and associates investigated the process and showed that (58):

- Repression of two Rb regulated genes (cyclin A and cyclin E) requires SUV39H1 for silencing.
- Silencing of the cyclin E promoter requires H3 methylation and HP1 binding, thus suggesting that SUV39H1 is needed for silencing of other or all genes regulated by Rb that are involved in the cell cycle.
- Most important is the observation that the process does not extend to surrounding chromatin even in the presence of SUV39H1 and HP1, two proteins involved in the formation of heterochromatin.
- Excessive cell proliferation is associated with SUV39H1 loss.

In conclusion, Rb can arrest the cycle in two ways: (1) by binding to E2F and thereby repressing the promoter of the *E2F* gene, and (2) by recruiting proteins that convert chromatin to heterochromatin. However, in the presence of Rb, the pathway for extended silencing by heterochromatin is converted into a limited silencer of selected genes.[7] Despite the progress made, the complete molecular mechanism that limits the process of silencing is only partly known.

Studies of bone differentiation provide another example of the role of *Rb* and *p21* in differentiation. When it became clear that *Rb* plays a critical role in cell cycle regulation and that the expression of its gene may be defective in cancers other than retinoblastomas, it was suggested that the increased incidence of osteosarcoma[8] in patients with retinoblastoma also results from an *Rb* gene defect. Investigations by Thomas and associates (59) have shown that wild-type *Rb* contributes to bone differentiation (**FIGURE 14-4**). In 1997, Ducy and associates (60) discovered an important transcription regulator (cBFA1) of osteoblast differentiation. Mice devoid of both genes encoding the factor die at birth from respiratory failure because of a complete lack of ossification including that of the rib cage. Heterozygotes survive but have a condition resembling human cleidocranial dysplasia (61). cBFA1 regulates the expression of at least two genes encoding proteins needed for bone matrix development and calcification: osteopontin and osteocalcin (60).

Only Rb (among the pocket proteins) regulates bone differentiation. After eliminating the role of other regulators of bone differentiation (BMP-2 and alkaline phosphatase) and providing evidence that Rb binds to the cBFA1 factor, Thomas et al. proposed a plausible mechanism for the regulation of bone differentiation (59):

- After Rb inactivation through hyperphosphorylation, the cBFA1 factor recruits Rb and binds it to specific promoters (the osteocalcin gene is believed to be a prime target) and enhances transcription of the gene.

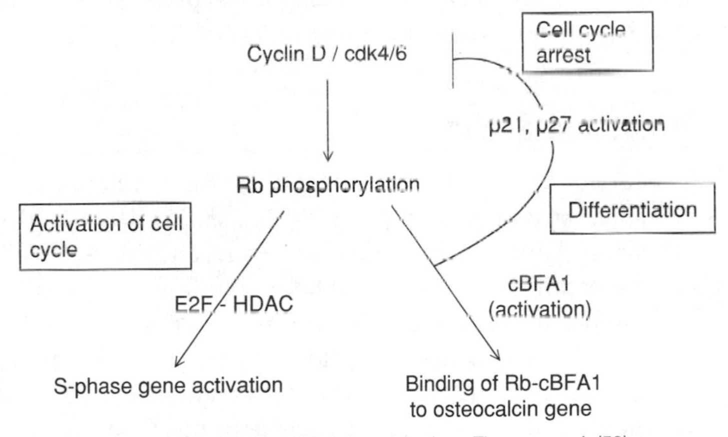


**FIGURE 14-4** Role of Rb in bone differentiation. Based on data from Thomas et al. (59).

---

[7]It cannot be ruled out that deacetylase also contributes to the silencing, for example, by deacetylating lysine 9 of histone H3, thereby allowing its methylation.

[8]Retinoblastoma treatment often involves radiation therapy. Treatment of hereditary retinoblastoma by such (or other) means is sometimes associated with the development of a second cancer, mainly osteosarcoma. For a long time it was believed that the latter was induced by ionizing radiations. However, the incidence of osteosarcoma in retinoblastoma patients is so great (~500 times that in the general population) that a genetic cause was suspected. Moreover, the incidence of retinoblastoma was also shown to be higher in victims of osteosarcoma.

- Because differentiation is associated with cycle arrest, Thomas et al. (59) suggested that in some way Rb or cBFA1 (or both) activate expression of members of the p21 family (p21 or p27) in the differentiating osteoblasts and thereby arrest the cell cycle.

The latter assumption is in keeping with other findings made on muscle differentiation (62, 63).

In conclusion, factors that regulate the cell cycle such as cyclins and the Rb protein also play a key role in the regulation of differentiation.

## Cell Cycle Arrest and Quiescence

Cells in culture most often enter a new cycle ($G_1$) soon after the last mitosis is completed. Under special circumstances they may be rendered quiescent and no longer divide or take a long time before they divide again. They are then said to enter a $G_0$ phase. It was often assumed that $G_0$ is an extension of $G_1$. However, it is a special phase in the cell's life. In vivo, many cells (e.g., hepatocytes, kidney cells, pancreatic cells) remain in $G_0$ until a special stimulus forces them into $G_1$ (e.g., partial hepatectomy for the hepatocyte).

The molecular events that keep the cells in $G_0$ are bound to be complex. The report of Ogawa and associates has opened the field, however. The arrest in $G_0$ is associated with the silencing of chromatin whose DNA is normally transcribed during $G_1$and S. During $G_1$, cyclin-cdks phosphorylate Rb and thereby release E2F ($E2F_1$ to $E2F_5$) to activate target genes whose products trigger the cycle. During $G_0$ the Rb independent $E2F_6$ is activated. It functions as a repressor by recruiting silencing factors to chromatin at sequence sites that respond to E2F and Myc.

Among the proteins recruited by $E2F_6$ are Myc and Max, a histone methyl transferase that uses histone H3 as substrate and modifies its lysine 9, (Eu-HMtase I or euchromatin methyltransferase) and the transcription repressor HP1γ. HP1γ binds to the methyl residue 9 of H3 and silences chromatin (64, 65).

## Cell Cycle Arrest and Senescence

The *Oxford Dictionary of Biochemistry and Molecular Biology* defines senescence as: "The process of aging." It is true that the word is used to describe the process of aging in cells, organs, and entire organisms. However, senescence of cells in culture has assumed a special meaning, which is not to say that it is unrelated to the process of aging. This section briefly introduces the notion of senescence and then focuses on cell cycle arrest as one of its manifestations. Senescent cells in culture reveal a special phenotype that is expressed by their permanent exit from cell replication, their morphology, and the activity of their genome. The process is still only partly understood, but available evidence strongly suggests that events observed in vitro occur in vivo and are related to aging, and possibly to cancer and apoptosis (66).

### Replicative Senescence

The concept of replicative senescence originated in 1961 with an important experiment of Hayflick and Moorhead, who found that human cells in vitro cannot proliferate indefinitely (67). After 40 to 50 cell divisions (the Hayflick number[9]) the cell arrests its cycle and assumes a particular morphology (67, 68). This phenomenon is referred to as replicative senescence. Evidence suggesting a relationship between replicative senescence and aging among other similarities includes:

- A correlation between replicative capacity and the age of the animal from which it is derived. Fibroblasts obtained from human embryos compared with those obtained from a 40-year-old human have a replicative capacity (or number of doublings) of 50 (for embryo cells) to 30 (for the adult).
- Correlation between the number of doublings and the lifespan of the species. Cells of mice with a lifespan of approximately two years have a Hayflick number close to 10. Cells of humans with a maximum lifespan of approximately 110 years have a Hayflick number of close to 50.
- Cells of young patients with the Werner syndrome do not exceed 20 doublings (69).

There are several molecular changes associated with senescence, which include:

- Telomere shortening.
- Defective response to growth factors.
- Activation of pathways that trigger cell cycle arrest.
- Formation of extrachromosomal circles (EAC) in yeast (70).[10]

Senescent cells are also characterized by several phenotypic manifestations:

- Irreversible cell cycle arrest in $G_1$.
- Resistance to apoptosis (71).
- Changes in cell morphology. Senescent cells are larger, flatter, and more irregular than normal but metabolically active cells. Special staining or electron microscopy shows an increased number of lysosomes and decreased activities of the endoplasmic reticulum (ER) and the Golgi apparatus (**FIGURE 14-5**).
- Changes in cell function.

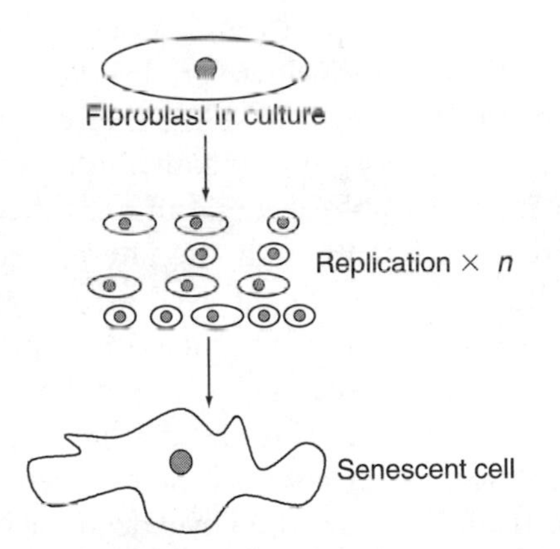


**FIGURE 14-5** Schematic representation of senescence, where *n* equals 40 to 50 in humans.

---

[9]The Hayflick number is the replicative capacity of the cell in vitro, the potential number of cell division. The number of cells doubles after each division, and thus increases exponentially with the total number of replications. After 50 replications the total number of cells equals $2^{50}$.

[10]Extrachromosomal ribosomal DNA circles (ERC) precede the onset of senescence. The event has no counterpart, as yet, in mammalian cells.

Manifestation of the latter reveals a selective alteration in the pattern of protein synthesis. Some increase (e.g., collagenase and other matrix proteases in senescent dermal fibroblasts), others decrease or can no longer be induced by the normal stimulus (e.g., 17-α-reductase). Clearly, the morphologic and functional changes reflect an abnormal pattern of differentiation. Of particular interest for the detection of senescent cells is the expression of "senescent (cells)-associated β galactosidase" (SAβ galactosidase), a marker that serves to identify senescent cells. Senescent cells thereby have been demonstrated to exist in skin, where the SAβ galactosidase clearly separates them from quiescent and terminally differentiated cells (72). Further details are in the excellent review of Campisi and the references therein (73).

## Other Modes of Induction of Senescence and Non-Replicative Senescence

Although this section focuses on replicative senescence, there are other modes of induction of senescence, namely oncogene activation and reactive oxygen species (74).

The *ras* oncogene cannot transform primary cell lines; it makes them senesce. Although it is not clear whether this mode of senescence is identical to that following multiple cell replications, the morphologic phenotype is similar, including the expression of SAβ glycosidase.

Senescence is also delayed in cell cultures grown under low oxygen tension. The appearance of hypoxic cells in the target tissue constitutes a major impediment to the successful eradication of cells by radiation therapy. The cells are not only radioresistant, they also retain their ability to divide (75). In contrast, cells exposed to reactive oxygen species (ROS) show premature replicative senescence and telomerase shortening. Whether these cells manifest true senescence or lose their ability to survive optimally in a hyperoxic or ROS-containing medium is unclear (76). The changes may mimic aging. This observation is in keeping with the accepted free radical theory of aging (77).

## Pathways to Senescence

Senescent cells are euploid. Therefore, the replication that precedes senescence, the last, is integral and complete, but all attempts to enter a new cycle are inhibited somewhere in $G_1$ most likely at the $G_1/S$ transition. However, this blockage of DNA synthesis does not exclude a rearrangement of gene expression either before or soon after the arrest in the $G_1$ phase. Shifts in gene expression may contribute to the phenotypic manifestations of senescence that may resemble ageing, but are different from those of quiescence or terminal differentiation. Although the altered pattern of gene expression varies with the origin of the senescent cell, cell cycle arrest in $G_1/S$ is the most common denominator in all senescent cells. Therefore, the critical question raised by the senescent phenotype is, "What are the molecules involved in the arrest of the cell cycle?" The answer is still incomplete, but two sets of molecules have been invoked: the telomeres and the cyclin kinase inhibitors. Truncation of the telomeres is associated with senescence (and with ageing). Truncation is caused by a decrease in telomerase activity and, indeed, restoration of the telomerase's activity leads to both elongation of telomeres and extension of cell replication (78).

## Telomere Shortening and Senescence

The end of the linear chromosome is made of special nucleoprotein structures, the telomeres. Telomeres secure the stability of the chromosomes and prevent their end-to-end fusion. The function of telomeres can be more easily understood by considering yeast plasmids. All viable yeast plasmids are circular, becasue if they were not, the chromosomes would be attacked by deoxyribonucleases. Moreover, the ends of chromosomes are sticky and therefore apt to react with pieces of broken chromosomes, thereby disturbing their equal distribution in the progeny by causing insertions or deletions. Thus, telomerization of the chromosome prevents their stickiness or accessability to DNases. Telomerases are ribonucleoproteins. Their RNA component is short (150 to 200 bases), but critical; it serves as a template for the repeated deoxyribonucleotide sequences. The enzymic portion of the molecule thus serves as a reverse transcriptase (RNA → DNA), a function that is reflected in the name, telomerase reverse transcriptase (TERT). The human gene is referred as *hTert.* Further details about this process can be found in the review by Collins (79). Several proteins are required for telomere's synthesis, and telomerases are prominent among them.

Telomeres form a long chain of sequences repeated in tandem. The TTAGGG sequence is found in human telomeres. In human cells, senescence takes place when telomeres reach a critical length. The signal for the senescence phenotype involves the P53/Rb pathway for cell cycle arrest. Indeed, inactivation of P53 or Rb by viral oncoproteins forestalls senescence and immortalizes the cells. The telomeric repeat fragment (tcTRF) is made of a sequence located upstream of the telomere and repeated sequences that span over 12 to 15 kb in germ cells or, depending upon the donor's age and cell type, 8 to 10 kb in somatic cells. Human fibroblasts lose 50 telomeric base pairs per doubling and at senescence the length of the telomere (tcTRF) has dropped to 4 to 6 kb. This is because most somatic cells, in contrast to germ cells and immortalized cells, express little or no telomerase. When the tcTRF of cells in culture drops to 2 kb, the cells enter a "crisis" (death) unless the telomere's length increases. Telomerase can resurrect the telomeres and lead them to immortality (cancer). Thus, most immortalized cells harbor active telomerase. In addition, telomerases are considered targets of choice for killing cancer cells as once they are devoid of telomerase, the cells are left to senesce or die (reviewed in 80).

The coincidence of telomeric amputation associated with each cell replication and the appearance of the senescence phenotype was bound to give birth to the telomeric shortening hypothesis to explain not only senescence, but also cell aging in general. As pointed out by Campisi (73), this may well be one of the most appealing hypotheses for the irreversible arrest after repeated cell replication (also referred to as the division counting mechanism) (81). The evidence that supports this notion includes:

- The observation that the replicative lifespan of cells derived from older persons is shorter than that obtained from younger persons.
- Senescence is, in some cells, reversed by ectopic expression of human telomerase (hTERT) (82, 83).

However, telomerase cannot by itself prevent telomere shortening or senescence, nor can the expression of hTERT in certain cell types (e.g., keratinocytes) overcome senescence (82, 83). Therefore, other molecular responses must be placed in gear to induce senescence at least in some cells.

## Cyclin Kinase Inhibition and Senescence

P53 and the members of the CIP family of cyclin kinase inhibitors (P21, P27, and P57) as well as members of the INK4 family have been implicated in cell cycle arrest. It is possible that more than one cyclin kinase inhibitor is involved in senescence and it also cannot be ruled out that the cyclin kinase inhibitor varies with the type of cell that senesces.

Alexander and Hinds, on the basis of their own work and that of others, proposed a model for the persistent arrest in $G_1$ associated with senescence (84). $Rb^{-/-}$ cells proliferate indefinitely, restoration of wild-type Rb allows senescence to take place, indicating that Rb is required for senescence. Alexander and Hinds found that the cyclin kinase inhibitor P27 accumulates posttranscriptionally in the presence of Rb, and that P27 accumulation and subsequent senescence do not require Rb-E2F interaction. Indeed, they can be induced by Rb pocket mutants that impair the binding of Rb to E2F. P27 blocks the cycle in $G_1$ by binding to cyclin E. The model proposes a two step pathway leading to senescence. First, there is an acute cell cycle arrest in the $G_1$/S phase that is E2F-dependent and, at least in human cells, P53-P21–dependent. The second step secures persistent arrest in $G_1$ associated with P27 posttranscriptional accumulation[11]; it is Rb-dependent but E2F-independent. The P27 accumulation inhibits cdk2 and the persistent arrest evolves into the typical flat cell phenotype of senescent cells (**FIGURE 14-6**).[12]

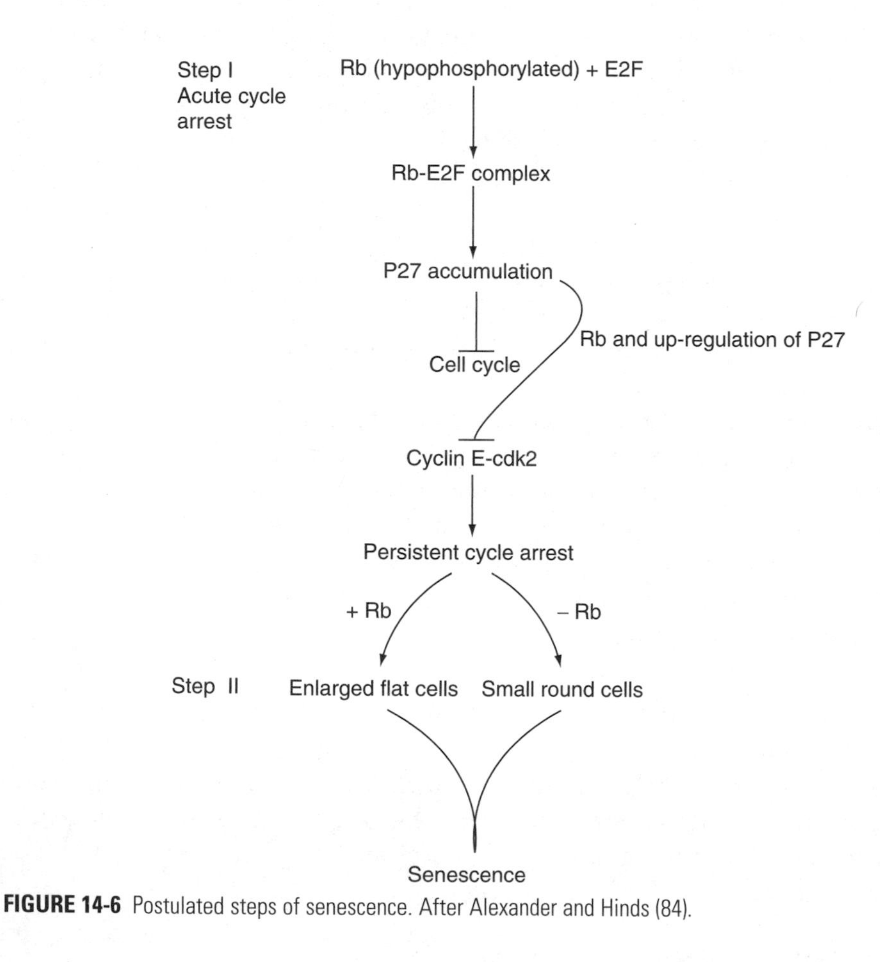


**FIGURE 14-6** Postulated steps of senescence. After Alexander and Hinds (84).

---

[11]The mechanism of the posttranscriptional accumulation of P27 is unknown.
[12]Caused by the alteration of the normal cytoskeletal signals.

Step I Hypophosphorylation Rb complexes with E2F and impairs its transcription activation, and thereby causes acute cell cycle arrest.

Step II Persistent cycle arrest is secured by the Rb up-regulation of P27 and subsequent inhibition of cyclin E–Cdk2 (84).

Thus, expression of the pathway to senescence starting with telomeric shortening requires Rb. However, ectopic expression of P21 or P27 can induce replicative cell cycle arrest without causing the morphological changes associated with it. This suggests that the two pathways—that of the cycle arrest and that of the morphological phenotype—are uncoupled (84). Such uncoupling also has been observed in older P21-deficient fibroblasts, indicating that P21 is required for arrest of the cycle, but not for other manifestations of the senescence phenotype (85).

The elegant pathway that leads to cell cycle arrest associated with senescence outlined above does not exclude the existence of a different sequence of events, particularly in non-replicative senescence.

## Ras and Senescence

Ras, a member of the p21$^{GTPase}$[13] superfamily associated with cell membranes, is activated by an upstream tyrosine kinase operating as a G-protein coupled receptor. This upstream signal is transmitted to various targets through a sequence of kinases, such as the mitogen-activated protein (MAP) kinase kinase kinase (RAF), the MAP kinase-kinase (MEK), and MAP kinase (ERK) (**FIGURE 14-7**).

The MAP kinase cascade exists in yeast, *Drosophila*, *C. elegans*, and humans (among other yeasts and eukaryotes). The targets vary; in humans they trigger cell proliferation and differentiation in vivo and, therefore, it is not surprising that Ras can function as an oncogene.

In cell lines of human fibroblasts, ectopic Ras overexpression causes immortalization. In contrast, Ras overexpressed in primary cell lines causes the senescence phenotype to appear; this phenotype is associated with accumulation of P53 and P16 (86). Since both Raf and Mek also are able to induce senescence, transduction of the signal

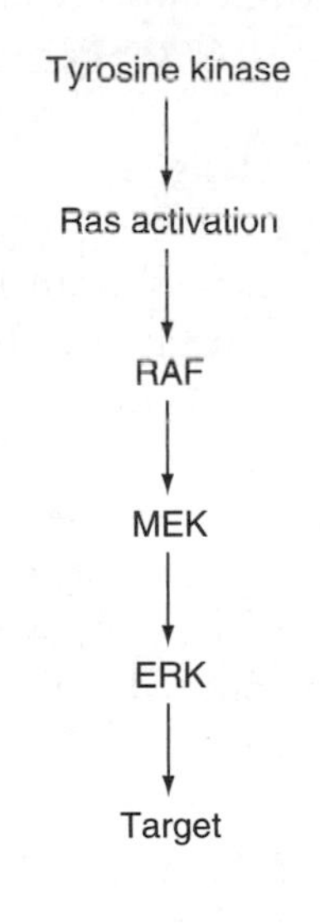


**FIGURE 14-7** Simplified Ras pathway.

[13]Not to be confused with the P21$^{CIP}$ family.

to P53 and P16 must involve the MAP kinase pathway (Figure 14-7) (87, 88). On the basis of these findings and others, Lundberg et al. (74) proposed a general model for the induction of senescence in cells in culture. Telomere shortening is assumed to be involved only in replicative senescence. Furthermore, it is proposed that cell cycle arrest that leads to senescence is caused by P53 stabilization induced by ARF. $P19^{ARF}$ (in mice; $p14^{ARF}$ in humans) binds to, sequesters, and inhibits MDM2. As a result, P53 is not degraded, but is stabilized, activated, and induces cell cycle arrest that is followed by senescence (89).

Several laboratories sought to clarify the molecular pathway that connects cell cycle and senescence (86, 90), and a consensus has emerged. In mice the sequence involves $P19^{ARF}$, P53, and Rb. Inhibitors that interfere with any step of the pathway cause the cells to proliferate. Co-expression of cMyc or overexpression of cyclins induced by the product of the human gene (*hDRILL*) trigger the cells' escape from Ras induced senescence (90), probably by interruption of the sequence to senescence downstream via the $19^{ARF}$/P53 pathway (91).

The role of ARF in Ras-induced compared to Ras-replicative senescence has been investigated also. Ras and E2F up-regulate the small G proteins ARF and causes premature senescence (92, 93).[14]

ARF (P14 ARF in humans) blocks MDM2 degradation and thereby leads to P53 accumulation and signals arrest of the cell cycle. However, up-regulation of the *arf* gene in human fibroblasts induces senescence without significant changes in P53 levels. Yet ARF is unable to cause senescence in absence of P53 and its downstream effector P21 (94, 95). These observations led Wei et al. (96) to investigate the role of ARF in the sequence that leads to senescence, downstream to telomeric shortening. The authors constructed a $P53^{-/-}$ – $P21^{-/-}$ human fibroblast cell line to determine whether ARF plays a role in replicative or Ras-induced senescence. They found that ARF is not up-regulated in replicative senescence. Contrary to ARF, Ras can induce growth arrest in $P53^{-/-}$ and $P21^{-/-}$ cells. In cells in which the *p21* and *p16* genes are present, Ras induces both genes, but the induction of *p21* is ARF-independent. Thus, even in premature senescence ARF is not indispensable. In summary, it appears that ARF is not required in the pathway that links telomeric shortening to replicative senescence of human fibroblasts (**FIGURE 14-8**). Thus, the pathway can be summarized as follows: telomeres truncation activates P53 (96)[15] and P53 induces P21, which in turn inhibits cyclin D–Cdk4/Cdk6; Rb–E2F stays complexed and the cycle arrests. It is, however, not excluded that P16 may be involved in an alternative pathway leading to replicative senescence of human fibroblasts (97), nor can the participation of ARF in replicative senescence of cells other than fibroblasts be ruled out. The expression of P16, ARF, and P21 in senescent hepatic cells derived from a primary culture and immortalized hepatic tumor cells was compared. In hepatic cells, the expression of P16, ARF, and P21 disappeared during cell proliferation but were associated with senescence (93). In contrast, in hepatic tumor cells the expression of P16, ARF, and P21 persisted during proliferation. Since the expression of these genes further increased after exposure to UV radiation, it is likely that the P53 pathway was operating normally in tumor cells (94, 95).

In summary, in addition to Rb, P53, and P21, two other molecules are also involved in the irreversible cell cycle arrest associated with senescence, namely P16 and ARF. Sequence may vary depending upon whether it is caused by repeated replication or induced prematurely by Ras.

[14]In contrast, c-Myc and E1, which also activate ARF, ultimately cause apoptosis.

[15]Note that the mechanism of P53 activation in the replicative senescence pathway does not result from stabilization as is the case after DNA damage. In senescence, it appears that P53 functions as a transcription activator for P2, and it is possible that the mechanism is universal.

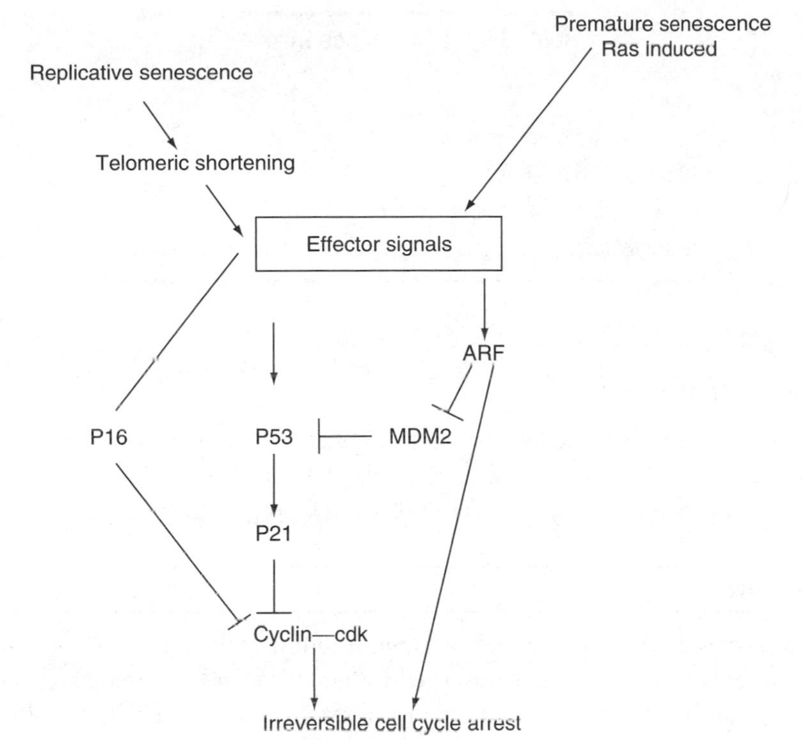


**FIGURE 14-8** Pathway to irreversible cell cycle arrest. Adapted from Lundberg et al. (74) and Wei et al. (96).

## PML and Senescence

The oncogene Ras induces senescence; however, the pathway linking Ras to senescence is not always clear. Parson et al. (98) sought to clarify it in response of the tumor suppressor gene *PML*, which is located on chromosome 15.[16] In normal cells, PML is found in subnuclear bodies or PODs (PML-related oncogene domains). PMLs are believed to play a role in transcription regulation.

In MEF cells, Ras up-regulates PML expression, thereby inducing P53-dependent senescence. P53 PML and an acetyltransferase (CBP transferase) form a complex. P53 in turn is acetylated at serine 382(K382), causing the cycle to arrest and senescence to follow. The cells enlarge, flatten, become β acidic galactosidase-positive and fail to incorporate bromodeoxyuridine. Induction of senescence fails in P53- and P21-deficient cells, demonstrating that the suppressor gene and the cyclin inhibitor are both required for Ras-PML induction of senescence; of course, the presence of intact PODS is also required.

Note that in immortalized cells PML may induce apoptosis. Whether these two pathways (senescence and apoptosis) are separate or overlap is not yet known, but a better understanding of the sequence of events that lead to each may help to find ways to encourage the cell to choose between apoptosis or senescence. Such knowledge could help in the design of more appropriate targets for cancer therapy.

[16]Acute promyelocytic leukemia is associated with a t15:17q21:q11-22 locus, which includes the gene that encodes PLM1 and the gene that encodes the retinoic receptor RARA. The fusion products of PML remain dissociated from the PODS.

**Table 14-1 Senescence in vivo**

| Organs |
| --- |
| Human prostate benign hyperplasia |
| Rabbit carotid arteries |
| Liver chronic hepatitis hepatocarcinoma |
| Neuroblastoma |

### Senescence In Vivo

Availability of the SAβ galactosidase test enabled a search of senescence in vivo in human prostate benign hyperplasia (99), rabbit liver chronic hepatitis hepatocarcinoma (100), and neuroblastoma (101) (**TABLE 14-1**).

## Conclusion

Currently, senescence is known as a phenomenon primarily but not exclusively observed in cells in culture. First observed by Hayflick and his associates (68), it was recognized as an irreversible limit to the proliferative capacity of cells in culture and an impediment to immortalization, an observation that led to the notion that senescence evolved to suppress cancer.

The concept of replicative senescence was extended to similar if not identical phenotypes induced by stress superimposed to the labor of replication. Thus, DNA damage, reactive oxygen species, and oncogenes may induce repeated replication and to ultimately arrest prematurely during the life of the cell line. Senescence is associated with the expression of a specific morphological, functional and molecular phenotype. The causative molecular events are not yet completely known. Cells enter an irreversible inability to replicate because of a permanent block of the cell cycle. At the same time, they undergo a reorganization of gene expression that causes a new morphologic appearance (large flattened cells expressing the SAβ galactosidase), a different functional state and resistance to apoptosis. They cannot be immortalized and, therefore, they will not become cancerous. They are found in primary cultures of plants and animals with long lifespan and they are seen in vivo. The molecules involved in the pathway that leads from replication to senescence are still under investigation. A consensus is emerging: at least in replicative senescence, the process may be triggered by telomere shortening caused in part by a drop in telomerase activity. The cell cycle is arrested most likely at the $G_1/S$ checkpoint. P53, ARF, the Rb protein, and cyclin inhibitors (P21, P16, and P18), and probably others, are involved.

## References


1. Deng, C., Zhang, P., Harper, J.W., Elledge, S.J., and Leder, P. Mice lacking p21CIP1/WAF1 undergo normal development, but are defective in G1 checkpoint control. Cell. 1995;82(4):675–84.

2. Nalca, A., and Rangnekar, V.M. The G1-phase growth-arresting action of interleukin-1 is independent of p53 and p21/WAF1 function. *J Biol Chem.* 1998; 273:30517–23.
3. Yoshida, K., Murohashi, I., Hirashima, K. p53-independent induction of p21 (WAF1/CIP1) during differentiation of HL-60 cells by tumor necrosis factor alpha. *Int J Hematol.* 1996;65:41–8.
4. Di Leonardo, A., Linke, S.P., Clarkin, K., and Wahl, G.M. DNA damage triggers a prolonged p53-dependent G1 arrest and long-term induction of Cip1 in normal human fibroblasts. *Genes Dev.* 1994;8:2540–51.
5. Huang, L.C., Clarkin, K.C., and Wahl, G.M. Sensitivity and selectivity of the DNA damage sensor responsible for activating p53-dependent G1 arrest. *Proc Natl Acad Sci USA.* 1996;93:4827–32.
6. Pellegata, N.S., Antoniono, R.J., Redpath, J.L., and Stanbridge, E.J. DNA damage and p53-mediated cell cycle arrest: a reevaluation. *Proc Natl Acad Sci USA.* 1996;93:15209–14.
7. Wani, M.A., Zhu, Q., El-Mahdy, M., Venkatachalam, S., and Wani, A.A. Enhanced sensitivity to anti-benzo(a)pyrene-diol-epoxide DNA damage correlates with decreased global genomic repair attributable to abrogated p53 function in human cells. *Cancer Res.* 2000;60:2273–80.
8. Haapajarvi, T., Kivinen, L., Heiskanen, A., des Bordes, C., Datto, M.B., Wang, X.F., and Laiho, M. UV radiation is a transcriptional inducer of p21(Cip1/Waf1) cyclin-kinase inhibitor in a p53-independent manner. *Exp Cell Res.* 1999;248:272–9.
9. Smith, M.L., Chen, I.T., Zhan, Q., O'Connor, P.M., and Fornace, A.J. Jr. Involvement of the p53 tumor suppressor in repair of U.V.-type DNA damage. *Oncogene.* 1995;10:1053–9.
10. Shivakumar, C.V., Brown, D.R., Deb, S., Deb, S.P. Wild-type human p53 transactivates the human proliferating cell nuclear antigen promoter. *Mol Cell Biol.* 1995;15:6785–93.
11. Tanaka, H., Arakawa, H., Yamaguchi, T., Shiraishi, K., Fukuda, S., Matsui, K., Takei, Y., and Nakamura, Y. A ribonucleotide reductase gene involved in a p53-dependent cell-cycle checkpoint for DNA damage. *Nature.* 2000;404:42–9.
12. McDonald, E.R. 3rd; Wu, G.S.; Waldman, T., and El-Deiry, W.S. Repair defect in p21 WAF1/CIP1 –/– human cancer cells. *Cancer Res.* 1996;56:2250–5.
13. Waga, S., Hannon, G.J., Beach, D., and Stillman, B. The p21 inhibitor of cyclin-dependent kinases controls DNA replication by interaction with PCNA. *Nature.* 1994;369:574–8.
14. Cooper, M.P., Balajee, A.S., and Bohr, V.A. The C-terminal domain of p21 inhibits nucleotide excision repair in vitro and in vivo. *Mol Biol Cell.* 1999; 10:2119–29.
15. Pan, Z.Q., Reardon, J.T., Li, L., Flores-Rozas, H., Legerski, R., Sancar, A., and Hurwitz, J. Inhibition of nucleotide excision repair by the cyclin-dependent kinase inhibitor p21. *J Biol Chem.* 1995;270:22008–16.
16. Therrien, J.P., Loignon, M., Drouin, R., and Drobetsky, E.A. Ablation of p21waf1cip1 expression enhances the capacity of p53-deficient human tumor cells to repair UVB-induced DNA damage. *Cancer Res.* 2001;61:3781–6.

17. Ford, J.M., and Hanawalt, P.C. Expression of wild-type p53 is required for efficient global genomic nucleotide excision repair in UV-irradiated human fibroblasts. *J Biol Chem.* 1997;272:28073–80.

18. Therrien, J.P., Drouin, R., Baril, C., and Drobetsky, E.A. Human cells compromised for p53 function exhibit defective global and transcription-coupled nucleotide excision repair, whereas cells compromised for pRb function are defective only in global repair. *Proc Natl Acad Sci USA.* 1999;96:15038–43.

19. Smith, M.L., Ford, J.M., Hollander, M.C., Bortnick, R.A., Amundson, S.A., Seo, Y.R., Deng, C.X., Hanawalt, P.C., and Fornace, A.J. Jr. p53-mediated DNA repair responses to UV radiation: studies of mouse cells lacking p53, p21, and/or gadd45 genes. *Mol Cell Biol.* 2000;20:3705–14.

20. Adimoolam, S., Lin, C.X., and Ford, J.M. The p53-regulated cyclin-dependent kinase inhibitor, p21 (cip1, waf1, sdi1), is not required for global genomic and transcription-coupled nucleotide excision repair of UV-induced DNA photoproducts. *J Biol Chem.* 2001;276:25813–22.

21. Blackwood, E.M., and Eisenman, R.N. Max: a helix-loop-helix zipper protein that forms a sequence-specific DNA-binding complex with Myc. *Science.* 1991; 251:1211–7.

22. Ayer, D.E., and Eisenman, R.N. A switch from Myc:Max to Mad:Max heterocomplexes accompanies monocyte/macrophage differentiation. *Genes Dev.* 1993;7:2110–9.

23. Meroni, G., Reymond, A., Alcalay, M., Borsani, G., Tanigami, A., Tonlorenzi, R., Nigro, C.L., Messali, S., Zollo, M., Ledbetter, D.H., Brent, R., Ballabio, A., and Carrozzo, R. Rox, a novel bHLHZip protein expressed in quiescent cells that heterodimerizes with Max, binds a non-canonical E box and acts as a transcriptional repressor. *EMBO J.* 1997;16:2892–906.

24. Hurlin, P.J., Steingrimsson, E., Copeland, N.G., Jenkins, N.A., and Eisenman, R.N. Mga, a dual-specificity transcription factor that interacts with Max and contains a T-domain DNA-binding motif. *EMBO J.* 1999;18:7019–28.

25. Schreiber-Agus, N., Chin, L., Chen, K., Torres, R., Rao, G., Guida, P., Skoultchi, A. I., and DePinho, R.A. An amino-terminal domain of Mxi1 mediates anti-Myc oncogenic activity and interacts with a homolog of the yeast transcriptional repressor SIN3. *Cell.* 1995;80:777–86.

26. Ayer, D.E., Lawrence, Q.A., and Eisenman, R.N. Mad-Max transcriptional repression is mediated by ternary complex formation with mammalian homologs of yeast repressor Sin3. *Cell.* 1995;80:767–76.

27. Eilers, A.L., Billin, A.N., Liu, J., and Ayer, D.E. A 13-amino acid amphipathic alpha-helix is required for the functional interaction between the transcriptional repressor Mad1 and mSin3A. *J Biol Chem.* 1999;274:32750–6.

28. Hassig, C.A., Fleischer, T.C., Billin, A.N., Schreiber, S.L., and Ayer, D.E. Histone deacetylase activity is required for full transcriptional repression by mSin3A. *Cell.* 1997;89:341–7.

29. Laherty, C.D., Yang, W.M., Sun, J.M., Davie, J.R., Seto, E., and Eisenman, R.N. Histone deacetylases associated with the mSin3 corepressor mediate mad transcriptional repression. *Cell.* 1997;89:349–56.

30. Zhang, Y., Iratni, R., Erdjument-Bromage, H., Tempst, P., and Reinberg, D. Histone deacetylases and SAP18, a novel polypeptide, are components of a human Sin3 complex. *Cell.* 1997;89:357–64.

31. Martinez, E., Palhan, V.B., Tjernberg, A., Lymar, E.S., Gamper, A.M., Kundu, T.K., Chait, B.T., and Roeder, R.G. Human STAGA complex is a chromatin-acetylating transcription coactivator that interacts with pre-mRNA splicing and DNA damage—binding factors in vivo. *Mol Cell Biol.* 2001;21:6782–95.

32. McMahon, S.B., Wood, M.A., and Cole, M.D. The essential cofactor TRRAP recruits the histone acetyltransferase hGCN5 to c-Myc. *Mol Cell Biol.* 2000; 20:556–62.

33. Bouchard, C., Thieke, K., Maier, A., Saffrich, R., Hanley-Hyde, J., Ansorge, W., Reed, S., Sicinski, P., Bartek, J., and Eilers, M. Direct induction of cyclin D2 by Myc contributes to cell cycle progression and sequestration of p27. *EMBO J.* 1999;18:5321–33.

34. Frank, S.R., Schroeder, M., Fernandez, P., Taubert, S., and Amati, B. Binding of c-Myc to chromatin mediates mitogen-induced acetylation of histone H4 and gene activation. *Genes Dev.* 2001;15:2069–82.

35. Kornberg, R.D., and Lorch, Y. Twenty-five years of the nucleosome, fundamental particle of the eukaryote chromosome. *Cell.* 1999;98:285–94.

36. Eisenman, R.N. Deconstructing myc. *Genes Dev.* 2001;15:2023–30.

37. Grandori, C., Cowley, S.M., James, L.P., Eisenman, R.N. The Myc/Max/Mad network and the transcriptional control of cell behavior. *Annu Rev Cell Dev Biol.* 2000; 6:653–99.

38. Baker, S.J., and Curran, T. Oncogenic transcription factors: FOS, JUN, MYC, MYB, and ETS. In: Peters, G., and Vousden, K.H., eds. *Oncogenes and Tumour Suppressors.* Oxford, UK: Oxford University Press; 1997:155–85.

39. Gartel, A.L., Ye, X., Goufman, E., Shianov, P., Hay, N., Najmabadi, F., and Tyner, A.L. Myc represses the p21(WAF1/CIP1) promoter and interacts with Sp1/Sp3. *Proc Natl Acad Sci USA.* 2001;98:4510–5.

40. Warner, B.J., Blain, S.W., Seoane, J., and Massague, J. Myc downregulation by transforming growth factor beta required for activation of the p15(Ink4b) G(1) arrest pathway. *Mol Cell Biol.* 1999;19:5913–22.

41. Mitchell, K.O., and El-Deiry, W.S. Overexpression of c-Myc inhibits p21WAF1/CIP1 expression and induces S-phase entry in 12-O-tetradecanoylphorbol-13-acetate (TPA)-sensitive human cancer cells. *Cell Growth Differ.* 1999; 10:223–30.

42. Sabourin, L.A., and Rudnicki, M.A. The molecular regulation of myogenesis. *Clin Genet.* 2000;57:16–25.

43. Tapscott, S.J., Lassar, A.B., and Weintraub, H. A novel myoblast enhancer element mediates MyoD transcription. *Mol Cell Biol.* 1992;12:4994–5003.

44. Skapek, S.X., Rhee, J., Spicer, D.B., and Lassar, A.B. Inhibition of myogenic differentiation in proliferating myoblasts by cyclin D1-dependent kinase. *Science.* 1995;267:1022–4.

45. Guo K., and Walsh, K. Inhibition of myogenesis by multiple cyclin-Cdk complexes. Coordinate regulation of myogenesis and cell cycle activity at the level of E2F. *J Biol Chem.* 1997;272:791–7.

46. Guo, K., Wang, J., Andres, V., Smith, R.C., and Walsh, K. MyoD-induced expression of p21 inhibits cyclin-dependent kinase activity upon myocyte terminal differentiation. *Mol Cell Biol.* 1995;15:3823–9.
47. Halevy, O., Novitch, B.G., Spicer, D.B., Skapek, S.X., Rhee, J., Hannon, G.J., Beach, D., and Lassar, A.B. Correlation of terminal cell cycle arrest of skeletal muscle with induction of p21 by MyoD. *Science.* 1995;267:1018–21.
48. Wang, J., Helin, K., Jin, P., and Nadal-Ginard, B. Inhibition of in vitro myogenic differentiation by cellular transcription factor E2F1. *Cell Growth Differ.* 1995; 6:1299–306.
49. Reynaud, E.G., Pelpel, K., Guillier, M., Leibovitch, M.P., and Leibovitch, S.A. p57(Kip2) stabilizes the MyoD protein by inhibiting cyclin E-Cdk2 kinase activity in growing myoblasts. *Mol Cell Biol.* 1999;19:7621–9.
50. Reynaud, E.G., Leibovitch, M.P., Tintignac, L.A., Pelpel, K., Guillier, M., and Leibovitch, S.A. Stabilization of MyoD by direct binding to p57(Kip2). *J Biol Chem.* 2000;275:18767–76.
51. Ren, B., Cam, H., Takahashi, Y., Volkert, T., Terragni, J., Young, R.A., and Dynlacht, B.D. E2F integrates cell cycle progression with DNA repair, replication, and G(2)/M checkpoints. *Genes Dev.* 2002;16:245–56.
52. Kranenburg, O., Scharnhorst, V., Van der Eb, A.J., and Zantema, A. Inhibition of cyclin-dependent kinase activity triggers neuronal differentiation of mouse neuroblastoma cells. *J Cell Biol.* 1995;131:227–34.
53. Liu, M., Lee, M.H., Cohen, M., Bommakanti, M., and Freedman, L.P. Transcriptional activation of the Cdk inhibitor p21 by vitamin D3 leads to the induced differentiation of the myelomonocytic cell line U937. *Genes Dev.* 1996;10:142–53.
54. Adachi, M., Roussel, M.F., Havenith, K., and Sherr, C.J. Features of macrophage differentiation induced by p19INK4d, a specific inhibitor of cyclin D-dependent kinases. *Blood.* 1997;90:126–37.
55. Zhu, L., and Skoultchi, A.I. Coordinating cell proliferation and differentiation. *Curr Opin Genet Dev.* 2001;11:91–7.
56. Lachner, M., O'Carroll, D., Rea, S., Mechtler, K., and Jenuwein, T. Methylation of histone H3 lysine 9 creates a binding site for HP1 proteins. *Nature.* 2001; 410:116–20.
57. Bannister, A.J., Zegerman, P., Partridge, J.F., Miska, E.A., Thomas, J.O., Allshire, R.C., and Kouzarides, T. Selective recognition of methylated lysine 9 on histone H3 by the HP1 chromo domain. *Nature.* 2001;410:120–4.
58. Nielsen, S.J., Schneider, R., Bauer, U.M., Bannister, A.J., Morrison, A., O'Carroll, D., Firestein, R., Cleary, M., Jenuwein, T., Herrera, R.E., and Kouzarides, T. Rb targets histone H3 methylation and HP1 to promoters. *Nature.* 2001;412:561–5.
59. Thomas, D.M., Carty, S.A., Piscopo, D.M., Lee, J.S., Wang, W.F., Forrester, W.C., and Hinds, P.W. The retinoblastoma protein acts as a transcriptional coactivator required for osteogenic differentiation. *Mol Cell.* 2001;8:303–16.
60. Ducy, P., Schinke, T., and Karsenty, G. The osteoblast: a sophisticated fibroblast under central surveillance. *Science.* 2000;289:1501–4.
61. Mundlos, S., Otto, F., Mundlos, C., Mulliken, J.B., Aylsworth, A.S., Albright, S., Lindhout, D., Cole, W.G., Henn, W., Knoll, J.H., Owen, M.J., Mertelsmann, R.,

Zabel, B.U., and Olsen, B.R. Mutations involving the transcription factor CBFA1 cause cleidocranial dysplasia. *Cell.* 1997;89:773–9.

62. Schneider, J.W., Gu, W., Zhu, L., Mahdavi, V., and Nadal-Ginard, B. Reversal of terminal differentiation mediated by p107 in Rb-/-muscle cells. *Science.* 1994; 264:1467–71.
63. Novitch, B.G., Mulligan, G.J., Jacks, T., and Lassar, A.B. Skeletal muscle cells lacking the retinoblastoma protein display defects in muscle gene expression and accumulate in S and G2 phases of the cell cycle. *J Cell Biol.* 1996;135:441–56.
64. Ogawa, H., Ishiguro, K., Gaubatz, S., Livingston, D.M., and Nakatani, Y. A complex with chromatin modifiers that occupies E2F- and Myc-responsive genes in G0 cells. *Science.* 2002;296:1132–6.
65. La Thangue, N.B. Transcription. Chromatin control—a place for E2F and Myc to meet. *Science.* 2002;296:1034–5.
66. Hayflick, L. SV40 and human cancer. *Science.* 1997;276:337–8.
67. Hayflick, L., Moorhead, P.S. The serial cultivation of human diploid cell strains. *Exp Cell Res.* 1961;25:585–621.
68. Hayflick, L. The limited in vitro lifetime of human diploid cell strains. *Exp Cell Res.* 1965;37:614–36.
69. Kirkwood, T.B. Towards a unified theory of cellular ageing. *Monogr Dev Biol.* 1984;17:9–20.
70. Johnson, F.B., Marciniak, R.A., and Guarente, L. Telomeres, the nucleolus and aging. *Curr Opin Cell Biol.* 1998;10:332–8.
71. Wang, E., Lee, M.J., and Pandey, S. Control of fibroblast senescence and activation of programmed cell death. *J Cell Biochem.* 1994;54:432–9.
72. Dimri, G.P., Lee, X., Basile, G., Acosta, M., Scott, G., Roskelley, C., Medrano, E.E., Linskens, M., Rubelj, I., Pereira-Smith, O., et al. A biomarker that identifies senescent human cells in culture and in aging skin in vivo. *Proc Natl Acad Sci USA.* 1995;92:9363–7.
73. Campisi, J. Replicative senescence and immortalization. In: Stein, G., Baserya, R., Giordono, A., Denhardt, D. *The Molecular Basis of Cell Cycle and Growth Control.* New York, NY: Wiley-Liss Publication; 1999;8:348–373.
74. Lundberg, A.S., Hahn, W.C., Gupta, P., and Weinberg, R.A. Genes involved in senescence and immortalization. *Curr Opin Cell Biol.* 2000;12:705–9.
75. Steel, G.G., and Peacock, J.H. Why are some human tumours more radiosensitive than others? *Radiother Oncol.* 1989;15:63–72.
76. von Zglinicki, T., Saretzki, G., Docke, W., and Lotze, C. Mild hyperoxia shortens telomeres and inhibits proliferation of fibroblasts: a model for senescence? *Exp Cell Res.* 1995;220:186–93.
77. Walford, R.L. The extension of maximum life span. *Clin Geriatr Med.* 1985; 1:29–35.
78. Vaziri, H., and Benchimol, S. Reconstitution of telomerase activity in normal human cells leads to elongation of telomeres and extended replicative life span. *Curr Biol.* 1998;8:279–82.
79. Collins, K. Mammalian telomeres and telomerase. *Curr Opin Cell Biol.* 2000; 12:378–83.

80. de Lange, T., and Jacks, T. For better or worse? Telomerase inhibition and cancer. *Cell.* 1999;98:273–5.

81. Levy, M.Z., Allsopp, R.C., Futcher, A.B., Greider, C.W., and Harley, C.B. Telomere end-replication problem and cell aging. *J Mol Biol.* 1992;225:951–60.

82. Kiyono, T., Foster, S.A., Koop, J.I., McDougall, J.K., Galloway, D.A., and Klingelhutz, A.J. Both Rb/p16INK4a inactivation and telomerase activity are required to immortalize human epithelial cells. *Nature.* 1998;396:84–8.

83. Dickson, M.A., Hahn, W.C., Ino, Y., Ronfard, V., Wu, J.Y., Weinberg, R.A., Louis, D.N., Li, F.P., and Rheinwald, J.G. Human keratinocytes that express hTERT and also bypass a p16(INK4a)-enforced mechanism that limits life span become immortal yet retain normal growth and differentiation characteristics. *Mol Cell Biol.* 2000;20:1436–47.

84. Alexander, K., and Hinds, P.W. Requirement for p27(KIP1) in retinoblastoma protein-mediated senescence. *Mol Cell Biol.* 2001;21:3616–31.

85. Dulic, V., Beney, G.E., Frebourg, G., Drullinger, L.F., and Stein, G.H. Uncoupling between phenotypic senescence and cell cycle arrest in aging p21-deficient fibroblasts. *Mol Cell Biol.* 2000;20:6741–54.

86. Serrano, M., Lin, A.W., McCurrach, M.E., Beach, D., and Lowe, S.W. Oncogenic ras provokes premature cell senescence associated with accumulation of p53 and $p16^{INK4a}$. *Cell.* 1997;88:593–602.

87. Zhu, J., Woods, D., McMahon, M., and Bishop, J.M. Senescence of human fibroblasts induced by oncogenic Raf. *Genes Dev.* 1998;12:2997–3007.

88. Lin, A.W., Barradas, M., Stone, J.C., van Aelst, L., Serrano, M., and Lowe, S.W. Premature senescence involving p53 and p16 is activated in response to constitutive MEK/MAPK mitogenic signaling. *Genes Dev.* 1998;12:3008–19.

89. Sherr, C.J., and Weber, J.D. The ARF/p53 pathway. *Curr Opin Genet Dev.* 2000;10:94–9.

90. Peeper, D.S., Shvarts, A., Brummelkamp, T., Douma, S., Koh, E.Y., Daley, G.Q., and Bernards, R. A functional screen identifies hDRIL1 as an oncogene that rescues RAS-induced senescence. *Nat Cell Biol.* 2002;4:148–53.

91. Lloyd, A.C. Limits to lifespan. *Nat Cell Biol.* 2002;4:E25–7.

92. Bates, S., Phillips, A.C., Clark, P.A., Stott, F., Peters, G., Ludwig, R.L., and Vousden, K.H. p14ARF links the tumour suppressors RB and p53. *Nature.* 1998;395:124–5.

93. Palmero, I., Pantoja, C., and Serrano, M. p19ARF links the tumour suppressor p53 to Ras. *Nature.* 1998;395:125–6.

94. Pomerantz, J., Schreiber-Agus, N., Liegeois, N.J., Silverman, A., Alland, L., Chin, L., Potes, J., Chen, K., Orlow, I., Lee, H.W., Cordon-Cardo, C., and DePinho, R.A. The Ink4a tumor suppressor gene product, p19Arf, interacts with MDM2 and neutralizes MDM2's inhibition of p53. *Cell.* 1998;92:713–23.

95. Zhang, Y., Xiong, Y., and Yarbrough, W.G. ARF promotes MDM2 degradation and stabilizes p53: ARF-INK4a locus deletion impairs both the Rb and p53 tumor suppression pathways. *Cell.* 1998;92:725–34.

96. Wei, S., Wei, S., and Sedivy, J.M. Expression of catalytically active telomerase does not prevent premature senescence caused by overexpression of oncogenic Ha-Ras in normal human fibroblasts. *Cancer Res.* 1999;59:1539–43.

97. Stein, G.H., Drullinger, L.F., Soulard, A., and Dulic, V. Differential roles for cyclin-dependent kinase inhibitors p21 and p16 in the mechanisms of senescence and differentiation in human fibroblasts. *Mol Cell Biol.* 1999;19:2109–17.
98. Parson, M., Carbone, R., Sebastiani, C., Cioce, M., Fagioli, M., Saito, S., Higashimoto, Y., Appella, E., Ninucci, S., Pandolfi, P.P., Pelleci, P.G. PML regulates P 53 acetylation and premature senescence induced by oncogene Ras. *Nature.* 2000;6792:207–10.
99. Choi, J., Shendrik, I., Peacocke, M., Peehl, D., Buttyan, R., Ikeguchi, E.F., Katz, A.E., and Benson, M.C. Expression of senescence-associated beta-galactosidase in enlarged prostates from men with benign prostatic hyperplasia. *Urology.* 2000; 56:160–6.
100. Paradis, V., Youssef, N., Dargere, D., Ba, N., Bonvoust, F., Deschatrette, J., and Bedossa, P. Replicative senescence in normal liver, chronic hepatitis C, and hepatocellular carcinomas. *Hum Pathol.* 2001;32:327–32.
101. Wainwrigh, L.J., Lasorella, A., Iavarone, A. Distinct mechanisms of cell cycle arrest control the decision between differentiation and senescence in human neuroblastoma cells. *Proc Natl Acad Sci USA.* 2001;98:939–940.

CHAPTER

# 15 P53 TRANSACTIVATION OF GADD45 AND MDM2

## GADD Family

The GADD family is made of relatively small acidic nuclear proteins (GADD45, MydD118, and CR6, also referred to as GADD45a,b,c) (1). The *gadd* genes are activated and transcribed when cells are under various kinds of stress (**TABLE 15-1**). Although the function of the GADD proteins is not entirely clear, there is good evidence that they arrest the cycle at the $G_2$/M checkpoint (see above). The GADD45a protein is up-regulated by P53, PCNA, and P21. It binds to cyclin B–Cdk2 and activates the $G_2$/M checkpoint. The transduction of the signal from GADD45 to the checkpoint is likely to involve the MTK1/MEKK4 kinase pathway (1). GADD45 harbors two self-association sites, one in the N terminal (amino acids 33-61) and another in 40 C-terminal amino acids. The protein can oligomerize (2); hydrophobic interaction brings about homo-dimerization. Despite these laborious studies the exact function of GADD45a remains unclear. Knockout GADD45 mice are normal during development but later have an increased incidence of tumors, mainly lymphomas, because of increased susceptibility to physically or chemically induced mutations (3). However, GADD45 mutations are generally rare in human cancers. GADD45 is an inducible gene-encoding protein that couples DNA repair and cell cycle arrest (4) after exposure to agents that cause DNA damage. Whether GADD45 plays a role in apoptosis is still unclear.

The upstream activator of GADD45 varies but in some cases the induction of GADD45 requires P53. Indeed, no $G_2$/M arrest is observed in fibroblasts devoid of a functional P53 because of overexpression of mutated P53 or because the cells were derived from patients affected with Li-Fraumeni syndrome (5). In other situations (as is the case with delta-12 prostaglandin $J_2$), the cell cycle arrest induced by GADD45 is P53 independent (6). Lee and Lathange have shown that P73 activates the *gadd45* genes more effectively than P53. The inverse is true for P21 induction, in which case P53 is the most effective inducer (7). BRCA1 induction of GADD45 was demonstrated by Harkin et al. (8), but the signal transduction pathway that links BRAC1 to the GADD45 induction remains unknown (9).

The role of GADD45 in cell cycle arrest has been discussed (see Chapter 14). In contrast to p21, which is a universal cdk/cyclin inhibitor that contributes to both the $G_1$/S

**Table 15-1 Inducers of GADD45**

| Inducers of GADD45 |
|---|
| Radiation (ultraviolet [UV], X, and $\gamma$ radiation) |
| Alkylating agents |
| Reactive oxygen species (ROS) |
| Serum starvation |
| Hypoxia |
| Hypoosmolarity in kidneys |

and the $G_2$/M arrest, GADD45 acts only on the $G_2$/M checkpoint in cells stressed by genotoxic agents (10). Thus, GADD45 inhibits the cdk2-cyclin B complex but not the cdk2-cyclin E complex (5, 11). Expression of antisense GADD45 impairs the $G_2$M arrest in UV and methyl-methane sulfonate (MMS) stressed cells but not in X-irradiated cells. Lymphocytes obtained from GADD45 knockout mice retain the ability to block the $G_2$/M checkpoint after exposure to ionizing radiation. These findings indicate that the switch that controls the opening or closure of the $G_2$/M checkpoint answers to different signals that vary with cell type and cellular stress (6). All three members of the GADD45 family bind to proliferating cell nuclear antigen (PCNA) (12, 13). The PCNA interacting sequence involves amino acids 1 to 46 and amino acids 100 to 127 of the middle region of the molecule. The reacting sites of the GADD45s are located in the C terminal of the molecules. The capacity of the GADD45s to arrest the cell cycle is impeded by their association with PCNA. Despite extensive literature on the subject, the exact contribution of GADD45 to apoptosis is still not clear (4).

GADD45 activity is modulated by c-Myc. c-Myc overexpression triggers cell proliferation and c-Myc null cells only grow slowly. However, the expression of most target genes of c-Myc is unchanged in null cells except for the *Cad* and the *gadd45* genes. Therefore, it was suggested that these two genes play a special role in growth induced by c-Myc (14). Amudson and colleagues have shown that the Myc induction of GADD45 is impaired after exposure to MMS and UV irradiation and, therefore, it has been proposed that c-Myc may contribute to the development of the cancer phenotype by interfering with growth arrest imposed by GADD45 (15). After genotoxic stress, the interaction between c-Myc on GADD45 is believed to result from the binding of c-Myc to GC rich sequences at the promoter site of GADD45.

In conclusion, GADD45 is a member of a larger family of acidic nuclear proteins. It is induced after exposure to genotoxic agents, among them MMS and UV radiation, but not ionizing radiations (Table 14-1). Induction may involve P53 and P73 but not exclusively. GADD45s arrest the cycle at the $G_2$/M checkpoint. The binding of the C terminal of the GADD45 molecules to PCNA impedes its cell cycle inhibitory functions. Transcription of GADD45 is modulated by c-Myc, but the significance of the interaction between c-Myc and the GADD45s remains to be clarified.

## MDM2 Protein

P53 is a short-lived protein that is barely detectable in normal cells. DNA damage activates P53, causing a rapid rise in the functional activity and the level of the protein that results in $G_1$ arrest or apoptosis. Thus, in crisis P53 helps correct the damage by facilitating DNA repair and, if needed, letting the cells die. High levels of P53, however, would be inconvenient in the absence of DNA damage. Therefore, mechanisms securing low levels of P53 exist all through embryologic development and later in adult proliferating tissues. Moreover, after damage wild-type P53 controls its own levels by activating the *mdm-2* gene. The gene encodes a 90 kDa protein, MDM2 (murine double minute clone 2). The protein binds to P53 and inhibits the expression of other target genes normally induced by wild-type P53 (e.g., P21), and thereby triggers an elegant negative feedback mechanism. Thus, MDM2 allows the cells to exit in $G_1$ when they contain P53 and the Rb proteins.

Amplified MDM2 immortalizes embryonic cells and is associated with the appearance of gliomas and sarcomas. The exact mechanism of immortalization and transformation by MDM2 is not clear. MDM2 binds to both P53 and Rb proteins and thereby inhibits their regulatory growth; such functions offer a plausible explanation for the mode of action of amplified MDM2 (16–18). In contrast, MDM2 is required for development of the embryo. Mice defective in the *mdm-2* gene die early in embryogenesis, but the embryo can be rescued by simultaneous inactivation of P53 possibly by preventing cell cycle arrest by P53 (19, 20).

## MDM2 and the Function of P53

MDM2 is a 90 kDa protein composed of 429 amino acids. The sequence contains a ring finger domain, a nuclear localization signal domain, and an acidic transcriptional activation domain (discussed in Chapter 9). MDM2 is the product of a proto-oncogene that is amplified in 30 to 40% of sarcomas.

Wild-type P53 binds specifically to the *mdm2* gene and activates its transcription, thereby generating a feedback loop. P300 is involved in the process of MDM2 transcription (21, 22). P300 is a transcription coactivator that binds specifically to the P53 transactivation domain, and it also contributes to the activation of p21 and Bax promoter regions (23, 24). The P300 molecule possesses three domains flanked by adjacent sequences: C/H1 in the N terminal, C/H2 in the central, and C/H3 in the C terminal regions. The protein has a specific binding site for MDM2 in the C/H1 domain and the immediately adjacent sequences. Mutants of P53 and MDM2 molecules unable to bind the C/H1 do not degrade P53. However, P300 truncated mutants that bind to MDM2 (but not to P53) stabilize endogenous P53 (25).

The nuclear protein P53 is phosphorylated at serines 15 and 20, and this phosphorylation is believed to prevent P53 nuclear export, thereby facilitating P53's main function in the nucleus: its transcription activation. As mentioned earlier, *mdm2* is one among many genes activated by P53. The MDM2 ring finger protein functions as a negative regulator. It binds to P53 and the ring finger is required for binding (26, 27). The MDM2 protein binds to P53 and, thereby, interferes with various P53 activities:

- MDM2 binds to wild-type P53 at the site of its transcription activation domain (28) and thereby blocks its DNA binding.
- Interferes with P53's ability to activate transcription.
- Abolishes P53 antiproliferative and proapoptotic functions (28)
- Allows an exit from $G_1$ in P53- and Rb-positive cells. Thus, after binding to P53 MDM2 activates P53 degradation and abolishes its transcription transactivity (e.g., P21) and restores the cell proliferative activity.

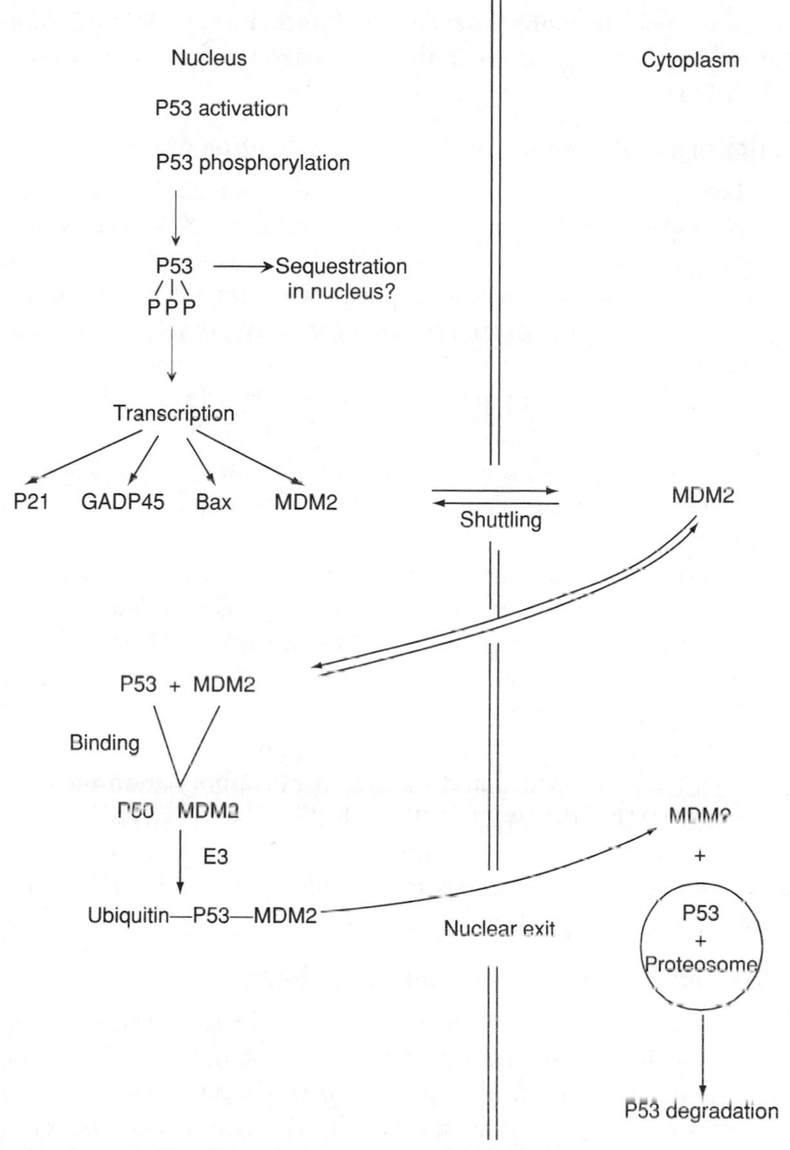


**FIGURE 15-1** Regulation of P53 function through the shuttling of MDM2 from nucleus to cytoplasm.

The MDM2 protein includes an E3 ubiquitin ligase function that ubiquitinates P53; ubiquitinated P53 is exported from nucleus to cytoplasm where it is delivered to the proteosome for degradation (29), and proteosome inhibitors prevent P53 degradation (18, 30; reviewed in 31, 32). P53 nuclear export is required for P53 degradation. The nuclear localization signal of P53 guides the shuttling of P53 from cytoplasm to nucleus (**FIGURE 15-1**).

## Role of the MDM2 Ring Finger

The ring finger of MDM2 is located in its C terminus and contains a C3HC4 motif. Mutations within the ring finger of MDM2 inhibit both P53's ubiquitination and P53's export to the cytoplasm by MDM2. However, ubiquitination of P53 is not sufficient for securing P53 degradation. Blocking the export of ubiquitinated P53 leads to its accumulation and sequestration in the nucleus (33). On the other hand, P53 cannot be exported from the nucleus in the absence of ubiquitination by MDM2. Indeed, mutations in the MDM2 ring finger domain that impair P53 ubiquitination also prevent its nuclear export (34).

## Phosphorylations of P53 and MDM2, and Stabilization of P53

P53 is phosphorylated in vitro at threonine 18 and serine 20, by caseine kinase I and chk1 and chk2, respectively. Phosphorylation at threonine 18 is required for P53 transcription activation activity (P21, GADD45, MDM2, etc.) Phosphorylation of serine 20 coincides with P53 accumulation after exposure to ultraviolet or ionizing radiation (35, 36). Because these two phosphorylations are early events, they are suspected to play a role in the dissociation of P53 from MDM2 and the regulation of P53 stability. However, there are other sites of phosphorylation (serines 15, 33, 37) whose contributions to the function of P53 remain obscure. Aschroft and Vousden suggest that P53 phosphorylation is not an absolute requirement for stabilization. They suspect that modulation of other molecules that contribute to the regulation of P53 might also affect its stabilization (37).

Regulatory phosphorylations are not restricted to P53; they also occur on MDM2. The ataxia telangiectasia-mutated (ATM) protein phosphorylates MDM2 in cells exposed to ionizing radiations or genotoxins. Moreover, MDM2 phosphorylation precedes the P53 accumulation caused by DNA damage, suggesting that in addition to activating P53, ATM may further contribute to P53 stabilization by MDM2 phosphorylation (38). Rapid phosphorylation of MDM2 by ATM after exposure to agents that damage DNA was confirmed and the site of phosphorylation identified, at least in vitro. S395 of MDM2 is the target of the early ATM phosphorylation after exposure of cells to ionizing radiation (39). In summary, DNA damage may contribute to P53 accumulation and activation indirectly by phosphorylation of MDM2 (at S395) and directly by phosphorylating P53 (at T20).

## P53 Shuttles Between the Nucleus and Cytoplasm

In cells free of stress the levels of P53 are low in the nucleus, because much of the P53 is bound to the nucleolus or other nuclear molecular structures (40). The binding occurs in an RNA dependent fashion since RNAse solubilizes the nucleolar P53. In cells stressed by transcription inhibitors (UV radiation or others), there is no induction of P53 and preexisting P53 becomes freely diffusible. In this case, P53 is believed to flood the nucleus and thereby become freely accessible to DNA targets. In contrast, after proteosomal inhibition, P53 remains sequestered and colocalizes with MDM2 in the nucleolus.

Once in the nucleus, P53 is not entirely secure; MDM2, like P53, possesses a nuclear export signal (NES) that allows it to move back and forth between cytoplasm and nucleus. When the time is ripe MDM2 binds to P53 and activates its NES. P53 is then brought to nuclear export receptors and ultimately is expelled from the nucleus, still bound to MDM2 (41). The P53 molecule also harbors two nuclear export signals, one in the C terminus (between amino acids 320 and 355) and another in the N terminus (between amino acids 11 and 27) (42, 43).[1] P53 ubiquitination is required for transport from the nucleus to the cytoplasm (44, 45).

---

[1]Note that the phosphorylation sites threonine 18 and serine 20 are located within the NES of P53. Moreover, MDM2 with defective nuclear localization signal (NSL) or NES fail to degrade P53.

### c-Jun–N (Amino) Terminal Kinase (JNK)

JNK binds to the central domain of P53 (between amino acids 97 and 116); it functions as an adapter molecule that contributes to the assemblage of the E3/ubiquitin ligase complex, and thereby targets P53 for ubiquitination and degradation. JNK, which is constitutively expressed, regulates P53 activity mainly in unstressed cells and in a manner that is independent of MDM2, which needs to be induced (27).

### ARF and MDM2 Regulation

Chromosomes 5 in the rat, 4 in mice, and 9 in humans harbor the CDKN2A locus (chromosome 9p21 in humans). The single gene encodes two proteins, the transcript $p16^{INK4\alpha}$ and the alternative reading frame (ARF). The gene contains four exons: exon 1β, exon 1α, exon 2, and exon 3. The "transcript" RNA derives from exons 1α, 2, and 3, while ARF derives from exons 1β, 2, and 3. Although derived from the same locus, the amino acid sequence of $p16^{INK4\alpha}$ reveals no homology with ARF. $p16^{INK4\alpha}$ functions as a specific inhibitor of cyclin D–Cdk4/Cdk6. $19^{ARF}$ in mice and $p14^{ARF}$ in humans binds to MDM2 and inhibits the E3 ligase, the ubiquitination of P53 and its nuclear export. P53 is stabilized as a result. The ARF-MDM2 complex is retained in the nucleolus (46, 47).

### MDM2 and SUMO-1

Yeast and humans harbor genes that encode a family of small ubiquitin-like proteins. Among them is the yeast SMT3 in *Saccharomyces cerevisiae* and the mammalian SUMO-1 (small ubiquitin-related modifiers) (48, 49). SUMO-1's amino acid sequence is 48% identical to that of SMT3p and 17% identical to ubiquitin. It binds to a number of proteins (mainly nuclear proteins) after they have been translated. Although not completely known, the pathway of conjugation is reminiscent, albeit distinct, from that of ubiquitin conjugation (50, 51).[2] The function(s) of SUMO proteins are still being investigated; however, the common denominator that has emerged suggests that they are small ubiquitin-like proteins that bind other target proteins, mainly nuclear proteins, for the purpose of modifying their function. Among the functions modulated by SUMO-1 are: protein-protein interactions, protein localization, and protein stabilization.

It has been suggested that SUMO-1 inhibits ubiquitination of MDM2 and that "sumoylated" MDM2 is less susceptible to degradation than "unsumolyated" MDM2. Some of what is known includes:

- MDM2 is modified by SUMO.
- The site of binding of SUMO in MDM2 is unknown.
- MDM2 bound to SUMO is more stable than the naked molecule.
- Ultraviolet and γ radiations dissociate MDM2 from SUMO in a dose-dependent manner.
- Polyubiquitination of MDM2 inhibits the MDM2 E3 ligase and thus protects P53 against degradation.
- In contrast, MDM2-SUMO conjugation stimulates P53 ubiquitination and its degradation (52).

[2]The bindings of these small ubiquitin-related modifiers (SUMO), like ubiquitin, require an E1 enzyme, an E2 enzyme, and a family of E3 factors. The E1 enzyme in the presence of adenosine 5′-triphosphate (ATP) activates the SUMO by thioester bond formation that is transferred to the E2 conjugating enzyme and ultimately linked to the target lysine. With the E3 enzyme, the SUMO ligase transfers the SUMO to an ultimate target that needs to be modified.

Too little is known about the role of "sumoylation" to evaluate its role in modifying MDM2, but its existence and the fact that it reacts with MDM2 requires further investigation.

In conclusion, the *mdm2* gene is a target of the P53 transcription factor. The product, the MDM2 protein, binds to P53, ubiquitinates it, and translocates it from nucleus to cytoplasm where it is delivered to proteosomes for degradation. The interaction between P53 and the MDM2 serves as a feedback loop for the negative regulation of P53. This elegant control mechanism is itself modulated by other proteins, some that provide an alternative pathway for P53 degradation (JNK), and others that interfere with the degradation process (ARF by binding to MDM2 and preventing ubiquitination and nuclear export of P53). SUMO-1 binds to MDM2, thereby preventing MDM2 degradation and stimulating P53 ubiquitination and its degradation.

## References


1. Takekawa, M., and Saito, H. A family of stress-inducible GADD45-like proteins mediate activation of the stress-responsive MTK1/MEKK4 MAPKKK. *Cell.* 1998;95:521–30.
2. Kovalsky, O., Lung, F.D., Roller, P.P., and Fornace, A.J. Jr. Oligomerization of human Gadd45a protein. *J Biol Chem.* 2001;276:39330–9.
3. Hollander, M.C., Kovalsky, O., Salvador, J.M., Kim, K.E., Patterson, A.D., Haines, D.C., and Fornace, A.J. Jr. Dimethylbenzanthracene carcinogenesis in Gadd45a-null mice is associated with decreased DNA repair and increased mutation frequency. *Cancer Res.* 2001;61:2487–91.
4. Sheikh, M.S., Hollander, M.C., and Fornace, A.J. Jr. Role of Gadd45 in apoptosis. *Biochem Pharmacol.* 2000;59:43–5.
5. Wang, X.W., Zhan, Q., Coursen, J.D., Khan, M.A., Kontny, H.U., Yu, L., Hollander, M.C., O'Connor, P.M., Fornace, A.J. Jr., and Harris, C.C. GADD45 induction of a G2/M cell cycle checkpoint. *Proc Natl Acad Sci USA.* 1999;96:3706–11.
6. Ohtani-Fujita, N., Minami, S., Mimaki, S., Dao, S., and Sakai, T. p53-Independent activation of the gadd45 promoter by Delta12-prostaglandin J2. *Biochem Biophys Res Commun.* 1998;251:648–52.
7. Lee, C.W., and La Thangue, N.B. Promoter specificity and stability control of the p53-related protein p73. *Oncogene.* 1999;18:4171–81.
8. Harkin, D.P., Bean, J.M., Miklos, D., Song, Y.H., Truong, V.B., Englert, C., Christians, F.C., Ellisen, L.W., Maheswaran, S., Oliner, J.D., and Haber, D.A. Induction of GADD45 and JNK/SAPK-dependent apoptosis following inducible expression of BRCA1. *Cell.* 1999;97:575–86.
9. Mullan, P.B., Quinn, J.E., Gilmore, P.M., McWilliams, S., Andrews, H., Gervin, C., McCabe, N., McKenna, S., White, P., Song, Y.H., Maheswaran, S., Liu, E., Haber, D.A., Johnston, P.G., and Harkin, D.P. BRCA1 and GADD45 mediated G2/M cell cycle arrest in response to antimicrotubule agents. *Oncogene.* 2001;20:6123–31.
10. Zhan, Q., Antinore, M.J., Wang, X.W., Carrier, F., Smith, M.L., Harris, C.C., and Fornace, A.J. Jr. Association with Cdc2 and inhibition of Cdc2/Cyclin B1 kinase activity by the p53-regulated protein Gadd45. *Oncogene.* 1999;18:2892–900.

11. Zhang, Y., and Xiong, Y. Mutations in human ARF exon 2 disrupt its nucleolar localization and impair its ability to block nuclear export of MDM2 and p53. *Mol Cell.* 1999;3:579–91.

12. Vairapandi, M., Azam, N., Balliet, A.G., Hoffman, B., and Liebermann, D.A. Characterization of MyD118, Gadd45, and proliferating cell nuclear antigen (PCNA) interacting domains. PCNA impedes MyD118 and Gadd45-mediated negative growth control. *J Biol Chem.* 2000;275:16810–9.

13. Azam, N., Vairapandi, M., Zhang, W., Hoffman, B., and Liebermann, D.A. Interaction of CR6 (GADD45gamma ) with proliferating cell nuclear antigen impedes negative growth control. *J Biol Chem.* 2001;276:2766–74.

14. Bush, A., Mateyak, M., Dugan, K., Obaya, A., Adachi, S., Sedivy, J., and Cole, M. c-myc null cells misregulate cad and gadd45 but not other proposed c-Myc targets. *Genes Dev.* 1998;12:3797–802.

15. Amundson, S.A., Zhan, Q., Penn, L.Z., and Fornace, A.J. Jr. Myc suppresses induction of the growth arrest genes gadd34, gadd45, and gadd153 by DNA-damaging agents. *Oncogene.* 1998;17:2149–54.

16. Xiao, Z.X., Chen, J., Levine, A.J., Modjtahedi, N., Xing, J., Sellers, W.R., and Livingston, D.M. Interaction between the retinoblastoma protein and the oncoprotein MDM2. *Nature.* 1995;375:694–8.

17. Oliner, J.D., Kinzler, K.W., Meltzer, P.S., George, D.L., and Vogelstein, B. Amplification of a gene encoding a p53-associated protein in human sarcomas. *Nature.* 1992;358:80–3.

18. Haupt, Y., Maya, R., Kazaz, A., and Oren, M. Mdm2 promotes the rapid degradation of p53. *Nature.* 1997;387:296–9.

19. Jones, S.N., Roe, A.E., Donehower, L.A., and Bradley, A. Rescue of embryonic lethality in Mdm2-deficient mice by absence of p53. *Nature.* 1995;378:206–8.

20. Montes de Oca Luna, R., Wagner, D.S., and Lozano, G. Rescue of early embryonic lethality in mdm2-deficient mice by deletion of p53. *Nature.* 1995; 378:203–6.

21. Barak, Y., Juven, T., Haffner, R., and Oren, M. mdm2 expression is induced by wild type p53 activity. *EMBO J.* 1993;12:461–8.

22. Wu, X., Bayle, J.H., Olson, D., and Levine, A.J. The p53-mdm-2 autoregulatory feedback loop. *Genes Dev.* 1993;7:1126–32.

23. Gu, W., and Roeder, R.G. Activation of p53 sequence-specific DNA binding by acetylation of the p53 C-terminal domain. *Cell.* 1997;90:595–606.

24. Lill, N.L., Grossman, S.R., Ginsberg, D., DeCaprio, J., and Livingston, D.M. Binding and modulation of p53 by p300/CBP coactivators. *Nature.* 1997;387:823–7.

25. Grossman, S.R., Perez, M., Kung, A.L., Joseph, M., Mansur, C., Xiao, Z.X., Kumar, S., Howley, P.M., and Livingston, D.M. p300/MDM2 complexes participate in MDM2-mediated p53 degradation. *Mol Cell.* 1998;2:405–15.

26. Honda, R., and Yasuda, H. Association of p19(ARF) with Mdm2 inhibits ubiquitin ligase activity of Mdm2 for tumor suppressor p53. *EMBO J.* 1999;18:22–7.

27. Fuchs, S.Y., Adler, V., Buschmann, T., Yin, Z., Wu, X., Jones, S.N., and Ronai, Z. JNK targets p53 ubiquitination and degradation in nonstressed cells. *Genes Dev.* 1998;12:2658–63.

**28.** Oliner, J.D., Pietenpol, J.A., Thiagalingam, S., Gyuris, J., Kinzler, K.W., and Vogelstein, B. Oncoprotein MDM2 conceals the activation domain of tumour suppressor p53. *Nature.* 1993;362:857–60.

**29.** Ashcroft, M., Taya, Y., and Vousden, K.H. Stress signals utilize multiple pathways to stabilize p53. *Mol Cell Biol.* 2000;20:3224–33.

**30.** Kubbutat, M.H., Jones, S.N., and Vousden, K.H. Regulation of p53 stability by Mdm2. *Nature.* 1997;387:299–303.

**31.** Brown, J.P., and Pagano, M. Mechanism of p53 degradation. *Biochim Biophys Acta.* 1997;1332:O1–6.

**32.** Maki, C.G., Huibregtse, J.M., and Howley, P.M. In vivo ubiquitination and proteasome-mediated degradation of p53(1). *Cancer Res.* 1996;56:2649–54.

**33.** Geyer, R.K., Yu, Z.K., and Maki, C.G. The MDM2 RING-finger domain is required to promote p53 nuclear export. *Nat Cell Biol.* 2000;2:569–73.

**34.** Boyd, S.D., Tsai, K.Y., and Jacks, T. An intact HDM2 RING-finger domain is required for nuclear exclusion of p53. *Nat Cell Biol.* 2000;2:563–8.

**35.** Sakaguchi, K., Saito, S., Higashimoto, Y., Roy, S., Anderson, C.W., and Appella, E. Damage-mediated phosphorylation of human p53 threonine 18 through a cascade mediated by a casein 1-like kinase. Effect on Mdm2 binding. *J Biol Chem.* 2000;275:9278–83.

**36.** Shieh, S.Y., Taya, Y., and Prives, C. DNA damage-inducible phosphorylation of p53 at N-terminal sites including a novel site, Ser20, requires tetramerization. *EMBO J.* 1999;18:1815–23.

**37.** Ashcroft, M., and Vousden, K.H. Regulation of p53 stability. *Oncogene.* 1999; 18:7637–43.

**38.** Khosravi, R., Maya, R., Gottlieb, T., Oren, M., Shiloh, Y., and Shkedy, D. Rapid ATM-dependent phosphorylation of MDM2 precedes p53 accumulation in response to DNA damage. *Proc Natl Acad Sci USA.* 1999;96:14973–7.

**39.** Maya, R., Balass, M., Kim, S.T., Shkedy, D., Leal, J.F., Shifman, O., Moas, M., Buschmann, T., Ronai, Z., Shiloh, Y., Kastan, M.B., Katzir, E., and Oren, M. ATM-dependent phosphorylation of Mdm2 on serine 395: role in p53 activation by DNA damage. *Genes Dev.* 2001;15(9):1067–77.

**40.** Rubbi, C.P., and Milner, J. Non-activated p53 co-localizes with sites of transcription within both the nucleoplasm and the nucleolus. *Oncogene.* 2000;19:85–96.

**41.** Roth, J., Dobbelstein, M., Freedman, D.A., Shenk, T., and Levine, A.J. Nucleo-cytoplasmic shuttling of the hdm2 oncoprotein regulates the levels of the p53 protein via a pathway used by the human immunodeficiency virus rev protein. *EMBO J.* 1998;17:554–64.

**42.** Stommel, J.M., Marchenko, N.D., Jimenez, G.S., Moll, U.M., Hope, T.J., and Wahl, G.M. A leucine-rich nuclear export signal in the p53 tetramerization domain: regulation of subcellular localization and p53 activity by NES masking. *EMBO J.* 1999;18:1660–72.

**43.** Zhang, Y. A p53 amino-terminal nuclear export signal inhibited by DNA damage-induced phosphorylation. *Science.* 2001;292:1910–5.

**44.** Gottifredi, V., and Prives, C. Molecular biology. Getting p53 out of the nucleus. *Science.* 2001;292:1851–2.

45. Stott, F.J., Bates, S., James, M.C., McConnell, B.B., Starborg, M., Brookes, S., Palmero, I., Ryan, K., Hara, E., Vousden, K.H., and Peters, G. The alternative product from the human CDKN2A locus, p14(ARF), participates in a regulatory feedback loop with p53 and MDM2. *EMBO J.* 1998;17:5001–14.
46. Lohrum, M.A., Ashcroft, M., Kubbutat, M.H., and Vousden, K.H. Identification of a cryptic nucleolar-localization signal in MDM2. *Nat Cell Biol.* 2000;2:179–81.
47. Weber, J.D., Taylor, L.J., Roussel, M.F., Sherr, C.J., and Bar-Sagi, D. Nucleolar Arf sequesters Mdm2 and activates p53. *Nat Cell Biol.* 1999;1:20–6.
48. Glickman, M.H., and Ciechanover, A. The ubiquitin-proteasome proteolytic pathway: destruction for the sake of construction. *Physiol Rev.* 2002;82:373–428.
49. Mahajan, R., Delphin, C., Guan, T., Gerace, L., and Melchior, F. A small ubiquitin-related polypeptide involved in targeting RanGAP1 to nuclear pore complex protein RanBP2. *Cell.* 1997;88:97–107.
50. Takahashi, Y., Kahyo, T., Toh-E.A., Yasuda, H., and Kikuchi, Y. Yeast Ull1/Siz1 is a novel SUMO1/Smt3 ligase for septin components and functions as an adaptor between conjugating enzyme and substrates. *J Biol Chem.* 2001;276:48973–7.
51. Johnson, E.S., and Blobel, G. Ubc9p is the conjugating enzyme for the ubiquitin-like protein Smt3p. *J Biol Chem.* 1997;272:26799–802.
52. Buschmann, T., Fuchs, S.Y., Lee, C.G., Pan, Z.Q., and Ronai, Z. SUMO-1 modification of Mdm2 prevents its self-ubiquitination and increases Mdm2 ability to ubiquitinate p53. *Cell.* 2000;101:753–62.

CHAPTER

# 16 THE P53 FAMILY AND APOPTOSIS

Two relatives of P53, P63 and P73, were discovered in 1997. There are similarities but also important differences between these family members. They have the following attributes in common.[1]

- An acidic aminoterminal that includes the transactivation domain.
- A core domain that includes the DNA binding site.
- The binding domain is highly conserved in all three members of the family. The human P63 and P73 share 60% amino acid identity with the DNA binding site of P53. In P53, this is the site of most cancer mutations (1).
- A carboxyterminal required for oligomerization.

P63 and P73 share some fundamental functional properties of P53, such as transactivation of some of the P53 target genes and induction of apoptosis when exogeneously expressed (2).

## Difference Within the Family

P53 is encoded by a locus on chromosome 1(1p 35) and P63 by a locus on 3q:2-7-29. Despite their similarities, these relatives reveal functional individuality in that sometimes they act like P53 and at other times they oppose the P53 activity. The complexity is a consequence of the variable genomic expression (discussed below). The structure of the *p53* gene was discussed in Chapter 10. It is a straightforward gene with a single promoter that encodes a 393 amino acid protein. The major differences between the *p53* and its cousins *p63* and *p73* are that:

- *p63* and *p73* contain two independent promoters: P1, which is located in the 5′ untranslated region (UTR), and P2 in intron 3.
- The *p63* and *p73* genes are highly susceptible to alternative splicing.

Because of these differences in gene expression, a number of proteins are generated, six of which have been identified. This chapter focuses on the proteins encoded by the *p63* gene.[2] They are divided into two categories on the basis of the structures of their aminoterminal:

[1]A homologue of P53 has been described in *C. elegans*: Cep 1 required for DNA damage-induced cell death.

[2]A similar distribution of variant proteins has been found for the P73 group (TA P73 and ΔN P73). It is assumed that a single extended gene is at the origin of both the *p63* and *p73* genes.

- One group (which engages the first promoter) contains an acidic aminoterminus, the truncated aminoterminus (TA)-p63 α, β, and γ.
- The other (which engages the second promoter) has a truncated aminoterminus, ΔN-p63, devoid of the acidic region.

The splice variants of the TA and ΔN groups differ by their COOH terminus. Each group yields an α, β, and γ isoform, the characteristics of which include:

- The α and β isoforms are both longer than the γ isoform. They contain a sterile A motif (SAM)-like sequence. The motif was discovered in yeast and *Drosophila* proteins and is believed to secure protein–protein interactions (3). The α isoform also possesses a post-SAM sequence (~70 amino acids) that is needed for inhibition of transactivation of TA P63α and TA P73α.
- The shortest isoform is the γ isoform, which resembles TP53.
- TP73 can also undergo alternative splicing at exon 2, thereby generating a TA-deficient protein (p73 Ex2 del or TA P73α)

## Function of P63 and P73 Proteins

As may be anticipated, the functions of P63 and P73 vary depending upon which promoter is engaged. There are numerous reports on the functional similarities between TA isotypes of P63 and P73 and the classical P53, some of which are:

- Ultraviolet (UV) light induces TA p63 expression (1).
- Genotoxic agents, such as ionizing radiation and chemotherapeutic agents including taxol, cause P73 phosphorylation (4).
- P63 and P73 transactivate P53 target genes including *p21*, *gadd45*, *Bax*, and *mdm2* (5, 6).[3]
- P63 and P73, when exogenously expressed, cause apoptosis (1).
- P73 and P63 contribute to the development and the progression of cancer (see Chapter 26, Cancer and Apoptosis) (7–9).

## Function of the ΔN Group of P63/P73 Isotypes

These isotypes are characterized by a truncated transactivation domain, the presence of a core containing DNA binding domains, and an oligomerization domain. The combination of these structural events deprives these molecules of the P53 transactivation potential without altering the ability of the isotypes to form heterotetramers with P53 monomers, dimers, or trimers, or to bind to P53 consensus DNA binding sites that may inhibit P53 functions. The importance of these potential inhibitory functions of P53 by the truncated aminoterminal of ΔN isotypes is only of significance if the isotypes are expressed in vivo in substantial amounts. Indeed, the ΔN isotypes of P63 account for almost all of the P63 present in epithelial cells of skin, cervix, urogenital tract, and myoepithelium of the breast. Similarly, in the developing and mature brain the ΔN isotypes of P73 are preferentially expressed over TA isotypes (ΔN expression is 20 times greater than TA expression, ΔN/TA = 20).

[3]The function of MDM2 binding to P73 differs from that of its binding to P53. Binding to P73 does not ubiquitinate P73 but blocks its association with the P300-CBP cotranscription coactivator and, thereby, prevents P73's access to the target gene.

The significance of these findings with respect to the evaluation of functions identified for P53 is puzzling. For example, inhibition of P53 proapoptotic functions during development is likely to cause disturbance of homeostasis of the cell population. Consequently, it can be expected that overexpression of ΔN will favor cell proliferation (hyperplasia or cancer) even in the presence of wild-type P53, unless the ΔN P63-P73 isotypes have decreased and the P53, TA P63, and/or TA P73 isotypes have increased in the course of development, thus favoring the proapoptotic function. Only detailed and quantitative examination of the transcripts of the *p53*, *p63*, and *p73* genes in development and cancer cells will help to resolve the conundrum.

Comparison of genetically engineered knockout mice has begun to provide some clues, for example, P53-dependent apoptosis rises in $P73^{-/-}$ mice neurons (10, 11). There is also evidence of changes (under some circumstances) in the ratio of ΔN isotypes/P53. The ΔN P63 levels decrease in keratinocytes exposed to UV radiation, thus awakening the P3 proapoptotic function (12).

P63 null mice not only die soon after birth, but also show severe developmental anomalies (13, 14) including:

- Limb truncation
- Complete absence of skin
- Craniofacial anomalies
- Defective development of the hair, teeth, lacrymal, sweat and salivary glands, breast, and prostate glands
- Aberrations of the urogenital tract, stomach, and tongue.

Humans carrying TA P63 mutations within a DNA binding domain present with:

- Ectrodactyly
- Ectodermal dysplasia
- Facial clefts (EEC syndrome)

The common denominator for these diverse lesions resides in their fundamental structure. They are stratified epithelia that emerge from a basal layer containing stem cells that differentiate into glandular or epithelial cells. P63, in particular ΔN P63, is expressed in basal cells of human epithelia, but disappears as the cells differentiate, suggesting that P63 plays a role in securing stem cells' renewal (15–17). The article by Reis-Filho and Schmitt was a major source used in this description of the P53 family of proteins (16).

P73 null mice differ substantially from P63 null mice. They survive and may reach adulthood despite rather severe deficits involving neurogenesis, fluid dynamics in the cerebrospinal fluid (CSF), loss of control of pheromonal signals and hippocampal dysgenesis.

P53 null mice show:

- Deficient fluid dynamics in the upper respiratory tract.
- Hydrocephalus (believed to result from hypersecretion of CSF by the choroid complex supplemented by mucous respiratory hypersecretions, which become infected and turns purulent).
- Abnormal reproductive and social behavioral patterns controlled by pheromonal signals. The mice have no interest in mating and avoid other "social" interaction.

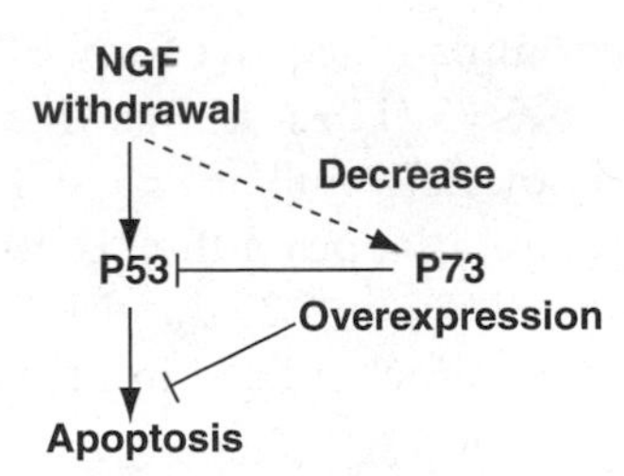


**FIGURE 16-1** The role of P53 and P73 in apoptosis after NGF withdrawal.

- Severe hippocampal dysgenesis caused by selective loss of Cajal-Retzius cells. Those are large bipolar neurons expressing p73 and a glycoprotein reelin, indispensable for neural migration in the cortex. This defect is associated with memory loss.

In conclusion, P73 is a functional protein more involved in regulating cellular activities than protecting against cancer or causing apoptosis.

In contrast to TA P73, ΔN P73 plays an important antiapoptotic role in the developing brain where it contributes to the adjustment of its morphogenesis. Survival of sympathetic neurons requires nerve growth factor (NGF) and withdrawal of NGF signals P53 induction and apoptosis; in contrast, the levels of ΔN P73 drop. Viral delivery of ΔN P73 prevents apoptosis (10).

Posniak and associates showed that the ratio of truncated P73 (ΔN subtype) levels to P53 levels is high in developing neurons (10). However, when sympathetic neurons are induced to undergo apoptosis by withdrawal of NGF the level of truncated P73 drops "dramatically" and P53-dependent apoptosis takes over. Apoptosis is prevented by inhibiting the drop of the truncated P73. The study concluded that P73 is an essential antiapoptotic agent during development (10) (**FIGURE 16-1**).

Yang and McKeon tabulated known information assembled on the functions of P63 and P73 isotypes in the induction of transactivation of P53 target genes, apoptosis, and inhibition of P53 functions (2).

In summary, the P73 and P63 isotypes are involved in gene transactivation, induction of apoptosis, and inhibition of P53 function. Only TA isomers of P63 and P73 transactivate P53 target genes; TA P63β and γ are more effective than TA P73β and γ isotypes. The α isotypes of both PTA63 and TPA72 have comparatively few activities. It has been proposed that this is caused by the inhibition of transactivation of the extended C terminus. Induction of apoptosis seems to correlate with target gene transactivation capacity. Thus, TA P63β and γ are more effective than TA P73 isotypes. Only the ΔN P63/73 isotypes inhibit P53 transactivation of target genes and apoptosis either by blocking the P53 DNA consensus binding sites or by forming ineffective heterodimers. The role of the P63 and P73 isotypes in cancer is discussed with that of the role of P53 in cancer (see Chapter 26).

## P53 Cousins and P53 Independent Apoptosis

T cell receptor (TCR)-induced cell death is a process that involves strong stimulation of the T cell receptor of cycling T cells and causes apoptosis. The process is P53 independent; however, cell death still occurs between late $G_1$ and the beginning of S phases, that is, the $G_1$/S checkpoints. However, if the transcription of antiapoptotic proteins is inhibited, T cells proliferate and the spleen is enlarged. Moreover, the condemned T cells induce p73, a mediator of apoptosis.

Introduction of E2F1 and a dominant negative form of P73 restore apoptosis in cells whose TCR is stimulated during $G_1$/S (E2F2, E2F4). In contrast, effectors that stimulate cell proliferation or P53 do not. E2F1 null or P53 null T cells are unable to undergo TCR-induced cell death. Therefore, it appears that TCR-induced cell death occurs at the late $G_1$ checkpoint and depends upon E2F1 and P73, but not P53. E2F1 initiates apoptosis and P73 is a downstream target of E2F1 in a common pathway to T cell death (18, 19).

Because P63 and P73 (at least some isotypes of the latter) are known to activate P53 target genes, Flores and associates investigated their role in P53-independent apoptosis caused by DNA damage (20). P53-deficient E1A mouse embryo fibroblasts are markedly resistant to apoptosis when treated with DNA damaging agents. For the purpose of clarifying the cause of resistance to apoptosis, Flores et al. engineered ($P63^{-/-}$ and $P73^{-/-}$) mouse embryonic cells expressing E1A and treated them with doxyrubin. Fibroblasts sensitized to undergo apoptosis through the expression of adenovirus E1A oncogene deficient in both *p63* and *p73*, but containing wild-type P53, fail to go into apoptosis after exposure to doxorubicin. This inability to undergo apoptosis is linked to the incapacity of P53 to bind to the promoters of some target genes. Indeed, in absence of P63 and P73, P53 does not bind to the promoters, whereas P63 can bind to the promoters even in the absence of P53. The study concluded by suggesting that P53 targets two types of genes: some activated by P53 alone (e.g., *p21* and *mdm2*); others require the presence of P53, P63 or P73 for activation (e.g., *Bax*, *Noxa*, and *PARP*). Whether overexpression of P53 and P73 might prevent transformation or sensitize cancer cells to chemotherapy remains to be determined (20).

In summary, TCR-induced apoptosis in T cells requires E2F1 and P73 rather than P53, and E1A fibroblasts deficient in P63 and P73 but not P53 are resistant to apoptosis.

## How P53 Triggers Apoptosis

The cell population of an organ (in the absence of immigration of lymphocytes, polymorphonuclear, monocytes, or stem cells) is regulated by cell proliferation, differentiation, and cell death. Differentiation is often characterized by the overproduction of one or more proteins, for example, hemoglobin in red cells and keratin in the skin's epithelial cells. After the preprogrammed lifespan is completed, red cells are phagocytized and the keratinocytes are shed. Although not exclusive[4] to red cells and keratinocytes, this form of cell loss is not the only means by which an organ loses cells. There is also apoptosis, an irreversible programmed cell death that may occur "spontaneously," for example during development, or be caused by various forms of cellular damage. Any interference with this homeostatic process causes excessive cell proliferation (hypertrophy or cancer), survival of damaging cells (mutated cells that cause cancer, cells that attack the self), or degeneration of essential cells (degenerative diseases, i.e., Parkinson and Alzheimer disease).

When cells are damaged, they are at the crossroad between life and death. P53 is the most studied "gatekeeper"; it arrests the cycle and determines whether the cell will proliferate or engage in apoptosis. The decisions to be made at this fatal crossroad raise at least two critical questions: First, after the cell cycle arrest, what molecular events emerge that determine whether the cell will enter apoptosis or resume the cell cycle? Second, after the cell arrest caused by DNA damage, what molecular events lead to apoptosis?

---

[4]It also occurs in polymorphonuclear cells and many epithelial cells, such as in the intestine.

## Choice Between Cycle Resumption and Apoptosis

Before attempting to answer the question of whether the cell enters apoptosis or resumes the cell cycle after cell cycle arrest, it is useful to briefly review how the cycle is arrested. DNA damage is sensed, P53 is activated, and the cell cycle is arrested at one of the checkpoints, for example, the $G_1$/S boundary. The cyclin/cdk or cyclin-dependent kinase complexes are inhibited either by P53 transactivation of the *p21* gene, the P21 protein, or by transactivation of genes that encode other kinase inhibitors, for example P16. Because of kinase inhibition, the Rb protein remains hypophosphorylated and continues to sequester transcriptor regulator(s), among them E2F/DP(s). As a result, the cell is unable to progress from $G_1$ to S and the cycle is arrested. The choice is now between apoptosis or resumption of the cell cycle. Very little is known about what determines this choice.

The work of Xie and associates begins to answer this problem (21). Plk3, a member of the family of poloprotein kinase, phosphorylates cdc25C on serine 216 and thereby prevents the phosphatase from binding to 14-3-3-3. This interferes with sequestration of cdc25C in the cytoplasm. The same kinase also phosphorylates P53 at serine 20, at least in vivo (21). In addition, it is known that the kinase can induce chromatin condensation and apoptosis (22). Bahassi et al. inserted these findings in the traditional pathway of P53 activation after DNA damage caused by UV, x-rays, and reactive oxygen species (ROS), and provided a model for the potential role of the polokinase in the decision to engage in apoptosis or resume the cell cycle after cell cycle arrest (22). In this model the functions of P53 and polokinase are linked. Polo-like kinase 3 (Plk3) enhances apoptotic function by phosphorylating P53 on serine 20 and restores cell cycle progression by phosphorylation of cdc25C phosphatase on serine 216.

Polyak et al. approached the problem by using a technique known as serial analysis of gene expression (SAGE). An immortalized cell line devoid of functional P53 was subjected to forced P53 expression with the aid of an adenovirus vector (23). The study then examined the 7202 transcript in P53-minus cells and P53-plus cells. The focus was on genes that were overexpressed or underexpressed by a factor of 10 in $P53^{+/+}$ cells compared to $P53^{-/-}$ cells. Only 34 genes differed to that extent; 14 were overexpressed and 20 underexpressed. Some of those genes were known to be P53 induced, for example, the *p21* gene, which is actively induced in cells containing P53 but absent on $P53^{-/-}$ cells. Most of the other genes that were overexpressed encoded proteins with oxidoreductase activity or genes that contribute to the rise of ROS. This led to a three step model for P53-induced apoptosis: activation of transcription of redox-related genes, accumulation of ROS, and triggering of mitochondrial degradation leading to caspase activation (23, 24). To date, this kind of experiment suggests a mechanism by which P53 could lead to apoptosis, and although nothing is known as yet about the regulation of the choice between apoptosis and cycle resumption, it suggests that the SAGE approach may provide a good tool for identifying the molecules involved in the process. In conclusion, too little is known about the molecules that determine whether the cell should resume the cycle or die for a complete model of the sequence of events to be presented. However, it is possible to anticipate the extent of the complexities on the basis of the known factors involved.

The nature of the genotoxic agent (e.g., UV vs. ionizing radiation) is most likely to be one of these factors; the extent and the nature of the damage inflicted (e.g., formation of adducts and single or double strand breaks) is another factor that may determine the choice between life and death. The capacity of the cell to secure integral repair is also critical. Repair potential is likely to vary with the nature of the insult and the stage in the cell cycle, for example, before S, between $G_2$ and M, or during the interphase, and the absence or presence of transcription at the site of damage.

Depending on the type and the stage in their life cycle, cells will have different quantitative levels of the multiple components involved in restoration of normality. For example, cells respond differently according to their ability to induce the P53 target genes. Are these genes activated all at once or sequentially? Moreover, the exclusion of one transcript (e.g., p21) may not necessarily exclude the possibility that other molecules assume redundant functions. Finally, the lifespan of P53 may vary considerably from one cell type to another depending in part on the function of MDM2.

## P53 and Apoptosis

Apoptosis is a destructive process in which a predetermined orderly activation of special proteases, the caspases, are put in gear. To date, two different pathways have been identified. Both include initiator and effector caspases. Initiator caspases such as caspase 8 and 9 are activated by upstream signals triggered by cell damaging agents. They are able to activate effector caspases and thereby unleash a proteolytic cascade that ultimately kills the cell. In the first pathway, members of the tumor necrosis factor (TNF) family of cytokines (prominent among them TNF-$\alpha$ or Fas) bind to membrane receptors (TNFR). The receptor activated by the ligand recruits adaptive proteins, and a protein scaffold is constructed by the interaction of special protein domains present in receptors and adaptor molecules (e.g., the death domains). The adaptor molecules then interact with each other or caspase 8 through special death effector domains. Activation of caspase 8 triggers that of caspase 3 and the entire caspase cascade explodes. In the second pathway, the mitochondrion is the target. Stimuli, and there is a great variety of them, attack mitochondria and disturb the rapport between antiapoptotic (Bcl-2, $Bclx_L$, Mcl I) and proapoptotic (Bax, Bid, Bim) proteins. This unbalance leads to disturbance of the membrane permeability potential and releases mitochondrial proteins in the cytosol, in particular, cytochrome c. Cytochrome c, Apaf-1, and caspase 9 form a complex (see Chapter 7). Caspase 9 is activated and in turn stimulates caspase 3 and the caspase cascade is triggered.

### Consequences of DNA Damage

Genotoxic stress, in particular that caused by exposure to UV and ionizing radiation, is likely to be sensed by one or more of a group of enzymes: ATM, ATR, BCRA1, DNA-PK, Ref-1, and PARP-1. After genotoxic stress, the P53 that was latent in unstressed cells is activated through several posttranslational modifications, mainly phosphorylation, at various sites. Moreover, P53 transcription rises in the presence of the transcription coactivator p300 CBP (an acetyl transferase) and the level of P53 in the cell also rises, resulting in an accumulation of activated P53 into the nucleus. However, active P53 does not remain unchallenged, because P53 self-regulates its activity by transactivating the *mdm2* gene and levels of the MDM2 protein in the cell.

P53 itself is further modified by acetylations and sumoylations, and becomes susceptible to ubiquitination and proteosomal degradation. P53 operates in at least at two checkpoints ($G_1$/S and $G_2$/M) by arresting the cycle through cyclin kinase inhibition. This cycle arrest in most cases is believed to be transient and followed either by complete and integral DNA repair and resumption of faithful DNA replication, or faulty DNA repair and proliferation of potentially offending cells that cause cancer or autoimmunity and, finally, apoptosis or resumption of the cycle.

Assuming that a death sentence is linked to the extent of unrepaired DNA damage, what steps lead to apoptosis after the sentence? The downstream steps (caspase[s] activation and cytochrome c release) are known, but the upstream events that take place between recognition of the damage (its nature and severity) and the determination to kill cells remain unknown.

Among the suspects are:

- Bax, whose gene is a P53 target (25, 26).
- E2F-1: DP-1, an inducer of P53 that was shown to override survival factors and trigger apoptosis (27–29).
- Up-regulation of the insulin-like growth factor (IGF) binding protein, whose gene is a target for P53. Such up-regulation could of course impair at least one survival signal, that of IGF-1 (30, 31).
- P53AIP: a P53 target gene that encodes the P53-regulated apoptosis-inducing protein (p53AIP1) located in mitochondria (32–35).

Assuming that the answer to the above question is known, the next question is: at which level of the sequence of events (briefly reviewed above) is programmed cell death initiated? Among the potential sites for initiation of apoptosis are: the balance between antiapoptotic and proapoptotic proteins, direct activation of caspases, and opening of the mitochondrial permeability transition pore.

## P53, Bax Induction, and Apoptosis

In unstressed cells, Bax is in equilibrium with Bcl-2 and Bclx, and with Mcl1 under stress. Bax is an intracellular membrane protein found in the membranes of the nucleus, the mitochondria, and the ER, where Bax can exist in the form of heterodimers with Bcl-2. In stressed cells, Bax is induced often in a P53-dependent fashion, and when in excess it forms homodimers that cause apoptosis (36). Thus, at least in some circumstances, competitive dimerization (Bax-Bax/Bax-Bcl-2/Bax-Bclx$_L$ or Bax-Mcl-1) regulates the passage from survival to apoptosis (37, 38).

In the choroid plexus, Bax is an obligatory effector of P53-dependent apoptosis. The requirements for transactivation of the *bax* gene by P53 have been investigated (39).

The P53 transactivation domain is located in its aminoterminus. The domain contains specific DNA binding sites. The sequence of these sites include consensus sequences (palindromic decamers 5′-Pu-Pu-Pu-C-A/T-A/T-A/T-G-Py-Py-Py-3′ and pentameric sequences 5′-Pu-Pu-Pu-C-A/T (40, 41). The transcription of some genes targeted by P53 (*p21, bax, mdm-2, gadd45,* and others) requires that P53 bind to the promoters of the target gene (30). However, Thornborrow and Manfredi showed that the binding requirement of P53 to the promoter site of *bax* differs from that required for *p21* (39). In addition, the transactivation response element for *p21* (2 consensus *p53* half sites), Bax, requires an additional transcription response element, a six base-sequence (GGGGGT-3′), adjacent to the two P53 half sites. After deletion or mutation of the GT sequence, P53-dependent transcription of *bax* is impossible. The GT box contributes to the recruitment of the Sp1 transcription cofactor.

Although a mechanism of P53-induced apoptosis causing imbalance between the antiapoptotic (Bcl-2-Bclx) and proapoptotic (Bax) proteins is reasonable and has been clearly demonstrated to exist under some experimental conditions (42), it may not operate in all cases. For example, P53 regulation of Bax expression in mice varies with cell type. It is high in spleen, thymus, small intestine, lung, Purkinje cells, and cortical neurons, but low in kidney, heart, liver and most of the brain (43).

The imbalance between proapoptotic and antiapoptotic proteins could be explained, in the absence of Bax expression, by activation of other antiapoptotic proteins. Indeed, other antiapoptotic proteins have been reported to be responsible for such imbalance. Knudson and Korsmeyer reported that in Bax-deficient mice, DNA damage correlates with an expected rise in apoptosis in thymocytes (44). Moreover, when a Bax transgene is introduced in a P53-deficient mouse, it does not overcome the resistance to apoptosis caused by the absence of P53. Such findings suggest that other genes encoding proapoptotic proteins, including BH3-only proteins, might be involved in tilting the balance of antiapoptotic and proapoptotic proteins in favor of the latter (44).

## P53, Bak, Noxa, and Puma

At least two other antiapoptotic proteins have been shown to play a role in P53-induced apoptosis: Bak and Noxa. The antineoplastic agent cisplatin induces conformational changes in Bak that activate its proapoptotic function. But even if either Bax or Bak contribute to TNF-α–induced apoptosis, they are not always needed to secure the inhibition of cell transformation induced by P53 (45), suggesting that Bax and Bak are not the only targets in the P53-independent apoptosis pathway. Noxa, a 103 amino acid protein, contains two 9-amino acid homologous sequences to the BH3 motif. Noxa, a relative of the BH3-only family, was found in humans and mice.

The expression of Noxa mRNA is constitutive in at least some tissues (brain, thymus, spleen, and kidney) and rises markedly after exposure to x radiations, a rise that coincides with that of MDM2. Moreover, Noxa-deficient mice are also x-radiation resistant (46). These findings are in keeping with the notion that the transactivation of the *noxa* gene is P53-dependent and, indeed, the *noxa*'s promotor sequence is a P53 responsive element. E2F1 is believed to contribute to the upstream activation of Noxa; its downstream response involves the mitochondrial pathway for apoptosis (47, 48).

Puma is another P53-induced proapoptotic BH3-only protein that responds to ionizing radiations and their stimuli. It is believed to be a critical participant in P53-dependent and P53-independent apoptosis (49, 50). Puma activates the mitochondrial apoptotic pathway after inducing conformation changes in Bax and as its translocates to mitochondria (49, 50).

## P53AIP

Oda et al., while seeking to clone *p53* binding sequences directly from the human genome discovered an unidentified protein, P53AIP1, whose mRNA rises after genotoxic stress in a P53-dependent way (32). The P53AIP1 (P53 regulated apoptosis inducing protein 1) is only induced when P53 is phosphorylated at serine 46. The UV-induced protein accumulates in mitochondria where it binds to Bcl-2 and activates the release of cytochrome c, unleashing the caspase cascade. Inhibition of P53AIP1 stops apoptosis. The protein is not expressed after triggering apoptosis by the TNF family of ligands. When p53AIP1 is ectopically expressed, it dissipates mitochondrial $\Delta\psi m$ and apoptosis follows. Activation of the *P53AIP1* gene is likely to be regulated by P53 after its phosphorylation on Ser-46. Oda et al. proposed an elegant mechanism for the regulation of apoptosis by P53 that involves a switch at the phosphorylation sites. Early after exposure to cellular stress, P53 is phosphorylated at least at two sites, serine 15 and serine 20, but not at serine 46. At this stage the phosphorylated P53 (minus the serine 46 phosphorylation) induces target genes involved in cell cycle arrest (e.g., *p21*), DNA repair, and MDM2. Once the cell senses that the damage cannot be repaired (possibly through P53 exhaustion by MDM2) a serine 46 kinase is activated, and P53 transcription activation of target genes switches from those involved in survival to that of P53IAP1, which in turn depresses mitochondrial $\Delta\psi m$ and causes apoptosis. Thus, substitution of serine 46 abrogates P53's ability to induce apoptosis (32).

P53A1P1 (induced after exposure to x-rays or UV in a P53-dependent manner), once released, triggers the cytochrome c pathway but not the death receptor pathway (33).[5]

## PERP

Attardi and associates investigated events that occur upstream to P53 and the release of cytochrome c, which is followed by the caspase cascade. They discovered *PERP* (p53 apoptosis effector related to PMP22) (51). The human *PERP* gene is located on chromosome 6 (6q24). PERP is a tetraspan transmembrane protein, a member of the well established family of similar proteins (PMP-22/Gaz).[6] PMP22/gaz3 is a structural component of myelin, but more pertinent to this discussion is the fact that it can induce $G_1$ arrest in Schwann cells (52). No less significant is the observation that fibroblasts in which PMP22/gaz3 is overexpressed die; however, Bcl-2 and caspase inhibitors can prevent apoptosis. PERP is P53-induced in the fibroblasts of mouse embryo and causes apoptosis, but it is unable to induce apoptosis in $P53^{-/-}$ cells. Moreover, it does not contribute to apoptosis triggered by UV exposure. The presence of the P53 binding site suggests that PERP functions downstream to induce apoptosis in the embryo and, indeed, expression of PERP in $P53^{-/-}$ cells causes apoptosis. However, P21 is not consistently required to prevent cancer. P21 null mice develop normally, their incidence of cancer is not increased, but they are defective in $G_1$ checkpoint control (33).

## INK and P53-Independent Apoptosis

Small cell carcinoma of the lung is highly sensitive to ionizing radiation and chemotherapeutic agents even in the presence of mutated P53, suggesting that a P53-independent pathway can be responsible for apoptosis. Indeed, small cell carcinoma cell lines, after exposure to genotoxins, undergo typical apoptosis after arrest at the $G_1$/S and $G_2$/M checkpoints. Blocking of cellular proliferation by Fas/CD95 antibodies and blockage of *c-myc* by antisense polynucleotides prevent apoptosis, suggesting that c-Myc and Fas/CD95 are required for P53-independent apoptosis (53). Presently, the full story of the sequence of steps that leads from DNA damage to apoptosis independently of P53 is unknown. The $p19^{arf}$ gene has been shown to induce apoptosis; the pathway may be P53-dependent in some cases and P53-independent in others (54).

## Arf and the P53 Pathway

The INK-4a/ARF locus found on chromosome 1 encodes $P19^{ARF}$ and $P16^{Ink4a}$ (55). Both $p19^{ARF}$ and $P16^{Ink4a}$ prevent development of certain cancers, but by different mechanisms. In the first, P19 impairs function of MDM2 by inhibiting the ubiquitin

[5]Whether p53A1P1 in an adenoviral vector will be effective in triggering apoptosis in cancer cells remains to be seen (33).

[6]The gene of PMP22/gaz3 is mutated in human demyelinating hereditary neuropathies, including Charcot-Marie-Tooth disease.

ligase and relocating it into the nucleolus, thereby enhancing the function of P53. Thus, in presence of genotoxic stress, P53 is activated and several proteins are induced, among them P21 (which causes cell cycle arrest), Bax, Noxa, Puma, and P53AIP. In the second mechanism, P16 inhibits cyclin D1/cdk4/cdk6 and thereby causes cell cycle arrest at the $G_1$/S checkpoint. However, studies using triple knockout mice illustrate the potential complexity of the pathway. Triple knockout mice ($P53^{-/-}$, $ARF^{-/-}$, and $MDM2^{-/-}$) develop more cancers than double knockout mice ($P53^{-/-}$ and $MDM2^{-/-}$) or single $P53^{-/-}$ knockout mice (56). Thus, ARF is able to impair cancer development independently of P53 and MDM2. Overexpression of $p19^{ARF}$ in embryonic fibroblasts defective in ARF, P53, and MDM2 ($ARF^{-/-}$, $MDM2^{-/-}$, or $P53^{-/-}$) causes cell cycle arrest by itself, findings suggesting that $p19^{ARF-/-}$ induces cell cycle arrest without involving the P53-MDM2 feedback loop.

$ARF^{-/-}$ / $P53^{-/-}$ / $MDM2^{-/-}$ —overexpression of ARF→ cell cycle arrest; rise in number of cancers

Is this rise in cancers seen in the above experiments caused by a reduction in the incidence of apoptosis? Tsuji and colleagues used an adenovirus-mediated $p19^{ARF}$ expression to investigate the role of MDM2 in apoptosis (57). $p19^{ARF}$ expression induced apoptosis in both P53-proficient and P53-deficient cells ($ARF^{-/-}$ and $P53^{-/-}$ cells). Co-expression of MDM2 inhibited apoptosis in P53-proficient cells but not in P53-deficient cells.

Under these experimental conditions, apoptosis induction does not require signals in addition to adenoviral-mediated $p19^{ARF}$ expression, and the observation that MDM2 coexpression inhibited apoptosis induction by $p19^{ARF}$ in P53-intact cells indicates that the use of the adenoviral expression tool is not by itself the cause of apoptosis. Finally, the response of P53-proficient and P53-deficient cells to the adenoviral induction of $p19^{ARF}$ clearly indicates that the P53-independent pathway for apoptosis does not require MDM2 (57). Unfortunately, details of the signal transduction steps for the P53-independent pathway are still unknown.

In conclusion, P53 is an important determinant in signaling apoptosis after infliction of DNA damage to the cell. The molecules or the molecular sequence that transduce the signal are not known with certainty, although components of the pathways are suspected. The possibility that the pathway to the determination and transduction of apoptosis after DNA damage may vary with the type of damage, its severity, and the cell type cannot be ruled out; indeed, P53-independent pathways to apoptosis have been described. If apoptosis is critical to the destruction of irremediably damaged cells (sheltering an unstable genome that may secure persistent survival advantages), it becomes obvious that knowledge of each step of these events is critical not only to the prevention of cancer, but also to the control of its progression.

## References and Suggested Reading


1. Jost, C.A., Marin, M.C., and Kaelen, W.G. Jr. p73 is a simian [correction of human] p53-related protein that can induce apoptosis. *Nature.* 1997;389:191–4.
2. Yang, A., and McKeon, F. P63 and P73: mimics, menaces and more. (Review). *Nat Rev Cell Biol.* 2000;1:199–207.

3. Chi, S.W., Ayed, A., and Arrowsmith, C.H. Solution structure of a conserved C-terminal domain of p73 with structural homology to the SAM domain. *EMBO J.* 1999;18:4438–45.
4. White, E., and Prives, C. DNA damage enables p73. *Nature.* 1999;399:34–5.
5. Scharnhorst, V., Dekker, P., van der Eb, A.J., and Jochemsen, A.G. Physical interaction between Wilms tumor 1 and p73 proteins modulates their functions. *J Biol Chem.* 2000;275:10202–11.
6. Yang, A., Kaghad, M., Wang, Y., Gillet, E., Fleming, M.D., Dotsch, V., Andrews, N.C., Caput, D., and McKeon, F. p63, a p53 homolog at 3q27-29, encodes multiple products with transactivating, death-inducing, and dominant negative activities. *Mol Cell.* 1998;2:305–16.
7. Corn, P.G. Transcriptional silencing of the p73 gene in acute lymphoblastic leukemia and Burkitt's lymphoma is associated with 5′CpG island methylation. *Cancer Res.* 1999;59:3352–6.
8. Han, S., Semba S., and Abe T. Infrequent somatic mutations of the P73 gene in various human cancers. *Eur J Surg Oncol.* 1999;25:194–8.
9. Melino, G., Lu, X., Gasco, M., Crook, T., and Knight, R.A. Functional regulation of p73 and p63: development and cancer. *Trends Biochem Sci.* 2003;28:663–70.
10. Pozniak, C.D., Radinovic, S., Yang, A., McKeon, F., Kaplan, D.R., and Miller, F.D. An anti-apoptotic role for the p53 family member, p73, during developmental neuron death. *Science.* 2000;289:304–6.
11. Yang, A., Walker, N., Bronson, R., Kaghad, M., Oosterwegel, M., Bonnin, J., Vagner, C., Bonnet, H., Dikkes, P., Sharpe, A., McKeon, F., and Caput, D. p73-deficient mice have neurological, pheromonal and inflammatory defects but lack spontaneous tumours. *Nature.* 2000;404:99–103.
12. Liefer, K.M., Koster, M.I., Wang, X.J., Yang, A., McKeon, F., and Roop, D.R. Down-regulation of p63 is required for epidermal UV-B-induced apoptosis. *Cancer Res.* 2000;60:4016–20.
13. Yang, A., Schweitzer, R., Sun, D., Kaghad, M., Walker, N., Bronson, R.T., Tabin, C., Sharpe, A., Caput, D., Crum, C., and McKeon, F. p63 is essential for regenerative proliferation in limb, craniofacial and epithelial development. *Nature.* 1999;398:714–8.
14. Mills, A.A., Zheng, B., Xiao-Jing, W., Vogel, H., Roop, D.R., and Bradley, A. p63 is a p53 homologue required for limb and epidermal morphogenesis. *Nature.* 1999;398:708–13.
15. Tiberio, R., Marconi, A., Fila, C., Fumelli, C., Pignatti, M., Krajewski, S., Giannetti, A., Reed, J.C., and Pincelli, C. Keratinocytes enriched for stem cells are protected from anoikis via an integrin signaling pathway in a Bcl-2 dependent manner. *FEBS Lett.* 2002;524:139–44.
16. Reis-Filho, J.S., and Schmitt, F.C. Taking advantage of basic research: p63 is a reliable myoepithelial and stem cell marker. (Review). *Adv Anat Pathol.* 2002;9:280–9.
17. Ellisen, L.W., Ramsayer, K.D., Johannessen, C.M., Yang, A., Beppu, H., Minda, K., Oliner, J.D., McKeon, F., and Haber, D.A. REDD1, a developmentally regulated transcriptional target of p63 and p53, links p63 to regulation of reactive oxygen species. *Mol Cell.* 2002;10:995–1005.

18. Lissy, N.A., Davis, P.K., Irwin, M., Kaelin, W.G., and Dowdy, S.F. A common E2F-1 and p73 pathway mediates cell death induced by TCR activation. *Nature.* 2000;407:642–5.

19. Levrero, M., De Laurenzi, V., Costanzo, A., Gong, J., Wang, J.Y., and Melino, G. The p53/p63/p73 family of transcription factors: overlapping and distinct functions. *J Cell Sci.* 2000;113:1661–70.

20. Flores, E.R., Tsai, K.Y., Crowley, D., Sengupta, S., Yang, A., McKeon, F., and Jacks, T. p63 and p73 are required for p53-dependent apoptosis in response to DNA damage. *Nature.* 2002;416:560–4.

21. Xie, S., Wu, H., Wang, Q., Cogswell, J.P., Husain, I., Conn, C., Stambrook, P., Jhanwar-Uniyal, M., and Dai, W. Plk3 functionally links DNA damage to cell cycle arrest and apoptosis at least in part via the p53 pathway. *J Biol Chem.* 2001;276:43305–12. (E-pub).

22. Bahassi el M., Conn, C.W., Myer, D.L., Hennigan, R.F., McGowan, C.H., Sanchez, Y., and Stambrook, P.J. Mammalian Polo-like kinase 3 (Plk3) is a multifunctional protein involved in stress response pathways. *Oncogene.* 2002;21:6633–40.

23. Polyak, K., Xia, Y., Zweier, J.L., Kinzien, K.W., and Vogelstein, B. A model for p53-induced apoptosis. *Nature.* 1997;389:300–7.

24. Wyllie, A. Clues in the p53 murder mystery. *Nature.* 1997;389:237–8.

25. Maxwell, S.A., Acosta, S.A., Tombusch, K., and Davis, G.E. Expression of Bax, Bcl-2, Waf-1, and PCNA gene products in an immortalized human endothelial cell line undergoing p53-mediated apoptosis. *Apoptosis.* 1997;2:442–54.

26. McCurrach, M.E., Connor, T.M., Knudson, C.M., Korsmeyer, S.J., and Lowe, S.W. Bax-deficiency promotes drug resistance and oncogenic transformation by attenuating p53-dependent apoptosis. *Proc Natl Acad Sci USA.* 1997;94:2345–9.

27. Wu, X., and Levine A.J. p53 and E2F-1 cooperate to mediate apoptosis. *Proc Natl Acad Sci USA.* 1994;91:3602–6.

28. Hiebert, S.W., Packham, G., Strom, D.K., Haffner, R., Oren, M., Zambetti, G., and Cleveland, J.L. E2F-1:DP-1 induces p53 and overrides survival factors to trigger apoptosis. *Mol Cell Biol.* 1995;15:6864–74.

29. Hsieh, J.K., Fredersdorf, S., Kouzarides, T., Martin, K., and Lu, X. E2F1-induced apoptosis requires DNA binding but not transactivation and is inhibited by the retinoblastoma protein through direct interaction. *Genes Dev.* 1997;11:1840–52.

30. Buckbinder, L., Talbott, R., Velasco-Miguel, S., Takenaka, I., Faha, B., Seizinger, B.R., and Kley, N. Induction of the growth inhibitor IGF-binding protein 3 by p53. *Nature.* 1995;377:646–9.

31. Butt, A.J., and Williams, A.C. IGFBP-3 and apoptosis—a license to kill? *Apoptosis.* 2001;6:199–205.

32. Oda, K., Arakawa, H., Tanaka, T., Matsuda, K., Tanikawa, C., Mori, T., Nishimori, H., Tamai, K., Tokino, T., Nakamura, Y., and Taya, Y. Links p53AIP1, a potential mediator of p53-dependent apoptosis, and its regulation by Ser-46-phosphorylated p53. *Cell.* 2000;102:849–62.

33. Matsuda, K., Yoshida, K., Taya, Y., Nakamura, K., Nakamura, Y., and Arakawa, H. p53AIP1 regulates the mitochondrial apoptotic pathway. *Cancer Res.* 2002; 62:2883–9.

34. Ko, L.J., and Prives, C. p53:puzzle and paradigm. (Review). *Genes Dev.* 1996; 10:1054–72.

35. Levine, A.J. p53, the cellular gatekeeper for growth and division. *Cell.* 1997; 88:323–31.

36. Miyashita, T., and Reed, J.C. Tumor suppressor p53 is a direct transcriptional activator of the human bax gene. *Cell.* 1995;80:293–9.

37. Sedlak, T.W., Oltvai, Z.N., Yang, E., Wang, K., Boise, L.H., Thompson, C.B., and Korsmeyer, S.J. Multiple Bcl-2 family members demonstrate selective dimerizations with Bax. *Proc Natl Acad Sci USA.* 1995;92:7834–8.

38. Oltvai, Z.N., and Korsmeyer, S.J. Checkpoints of dueling dimers foil death wishes. *Cell.* 1994;79:189–92.

39. Thornborrow, E.C., and Manfredi, J.J. The tumor suppressor protein p53 requires a cofactor to activate transcriptionally the human BAX promotor. *J Biol Chem.* 276:15598–608.

40. El-Deiry, W.S., Kern, S.E., Pietenpol, J.A., Kinzler, K.W., and Vogelstein, B. Definition of a consensus binding site for P53. *Nat Genet.* 1992;1:45–9.

41. Funk, W.D., Pak, D.T., Karas, R.H., Wright, W.E., and Shay, J.W. A transcriptionally active DNA-binding site for human p53 protein complexes. *Mol Cell Biol.* 1992; 12:2866–71.

42. Martin, L.J., and Liu, Z. Injury-induced spinal motor neuron apoptosis is preceded by DNA single-strand breaks and is p53- and Bax-dependent. *J Neurobiol.* 2002; 50:181–97.

43. Bouvard, V., Zaitchouk, T., Vacher, M., Duthu, A., Canivet, M., Choisy-Rossi, C., Nieruchalski, M., and May, E. Tissue and cell-specific expression of the p53-target genes: bax, fas, mdm2 and waf1/p21, before and following ionizing irradiation in mice. *Oncogene.* 2000;19:649–60.

44. Knudson, C.M., and Korsmeyer, S.J. Bcl-2 and Bax function independently to regulate cell death. *Nat Genet.* 1997;16:358–63.

45. Degenhartdt, K., Chen, G., Lindsten, T., and White E. BAX and BAK mediate p53-independent suppression of tumorigenesis. *Cancer Cell.* 2002;2:193–203.

46. Shibue, T., Takeda, K., Oda, E., Tanaka, H., Murasawa, H., Takaoka, A., Morishita, Y., Akira, S., Taniguchi, T., and Tanaka, N. Integral role of Noxa in p53-mediated apoptotic response. *Genes Dev.* 2003;17:2233–8. E-pub.

47. Seo, Y.W., Shin, J.N., Ko, K.H., Cha, J.H., Park, J.Y., Lee, B.R., Yun, C.W., Kim, Y.M., Seol, D.W., Kim, D.W., Yin, X.M., and Kim, T.H. The molecular mechanism of Noxa-induced mitochondrial dysfunction in p53-mediated cell death. *J Biol Chem.* 2003;278:48292–9. E-pub.

48. Hershko, T., and Ginsberg, D. Up-regulation of Bcl-2 homology 3 (BH3)-only proteins by E2F1 mediates apoptosis. *J Biol Chem.* 2004;279:8627–34. E-pub.

49. Villunger, A., Michalak, E.M., Coultas, L., Mullauer, F., Bock, G., Ausserlechner, M.J., Adams, J.M., and Strasser, A. p53- and drug-induced apoptotic responses mediated by BH3-only proteins puma and noxa. *Science.* 2003;302:1036–8. E-pub.

50. Liu, F.T., Newland, A.C., and Jia, L. Bax conformational change is a crucial step for PUMA-mediated apoptosis in human leukemia. *Biochem Biophys Res Commun.* 2003;310:956–62.

51. Attardi, L.D., Reczek, E.E., Cosmas, C., Demicco, E.G., McCurrach, M.E., Lowe, S.W., and Jacks, T. PERP, an apoptosis-associated target of p53, is a novel member of the PMP-22/gas3 family. *Genes Dev.* 2000;14:704–18.
52. Zoidl, G., and D'Urso, D., Blass-Kampmann, S., Schmalenbach, C., Kuhn, R., and Muller, H.W. Influence of elevated expression of rat wild-type PMP22 and its mutant PMP22Trembler on call growth of NIH3T3 fibroblasts. *Cell Tissue Res.* 1997;287:459–70.
53. Supino, R., Perego, P., Gatti, L., Caserini, C., Leonetti, C., Colantuono, M., Zuco, V., Carenini, N., Zupi, G., and Zunino, F. A role for c-myc in DNA damage-induced apoptosis in a human TP53-mutant small-cell lung cancer cell line. *Eur J Cancer.* 2001;37:2247–56.
54. Zindy, F., Eischen, C.M., Randle, D.H., Kamijo, T., Cleveland, J.L., Sherr, C.J., and Roussel, M.F. Myc signaling via the ARF tumor suppressor regulates p53-dependent apoptosis and immortalization. *Genes Dev.* 1998;12:2424–33.
55. Bates, S., Phillips, A.C., Clark, P.A., Stott, F., Peters, G., Ludwig, R.L., and Vousden, K.H. p14ARF links the tumour suppressors RB and p53. *Nature.* 1998;395:124–5.
56. Weber, J.D., Jeffers, J.R., Rehg, J.E., Randle, D.H., Lozano, G., Roussel, M.F., Sherr, C.J., and Zambetti, G.P. p53-independent functions of the p19(ARF) tumor suppressor. *Genes Dev.* 2000;14:2358–65.
57. Tsuji, K., Mizumoto, K., Sudo, H., Kouyama, K., Ogata, E., and Matsuoka, M. p53-independent apoptosis is induced by the p19ARF tumor suppressor. *Biochem Biophys Res Commun.* 2002;295:621–9.

SECTION

# Life and Death of the Cancer Cell

## Nature of Cancer

The cancer cell acquires survival advantages over its neighbors, and it proliferates disorderly in the midst of a population of cells that faithfully obey the cycle of life and death. The cancer cell has no respect for natural barriers; it crosses basal membranes and invades surrounding tissues. It penetrates the walls of blood vessels, migrates, and settles in distant tissues (lymph nodes, liver, kidney, bone) where it pursues its reckless disorderly growth. Fortunately, all cancers do not always grow fast and the aggressiveness varies with the type of cancer. But in its worst form, cancer cells evolve into a population of quasi-immortal cells that penetrate and destroy the surrounding tissues (invasion) and find their way to distant organs (metastasis).

How does it all start? A combination of environmental and genetic damage induces a single cell to live the reckless freedom of unicellular cells, and like them grow and grow into a population of "immortals" that ultimately die with the host. All cells of a cancer are derived from a single mutated progenitor. Together they form a clonal population that does not derive from or evolve into a mosaic of cells of different types. Thus, gene mutation and cloning are two major hallmarks of cancer. What makes a progenitor? The earliest morphologic reports and description of cancer cells recognized the significance of the nuclear changes. Chromatin, the molecular complex that make the chromosome, is more abundant in the cancer cell than in normal cells and mitoses are more frequent.

## Cancer and Chromosomes

As early as 1914, Bovary suspected that cancer might be linked to chromosomal alterations (1). Later, DNA was shown to be a major component of the chromosome. The capacity of the cancer cell to divide and the role of chromosomes in cell division suggested the likelihood that the primary injury of cancer resided in chromosomal damage. However, sophisticated studies of chromosomes proved difficult if not impossible until the 1950s, when the 46 human chromosomes were identified.

In 1956, Jerome Lejeune made the first association between a chromosome and disease with the discovery of the chromosome 21 trisomy in Down syndrome. Later it was shown that Down syndrome is associated with an increase in the incidence of leukemia (2).

In the middle of the 1980s, high resolution of Giemsa-stained chromosomes revealed specific chromosomal defects in leukemia and solid tumors through translocation, deletion, or amplification. For example, in Burkitt disease a *myc* sequence shifts from chromosome 8 to chromosome 14.

## DNA and Cancer

Once the major role played by DNA in cells' gene expression and cell replication was suspected, a great deal of research was undertaken, albeit often crude, on the potential changes in DNA associated with cancer initiation. The studies concerned overall changes in DNA amounts in the nucleus (using UV light microscopes, or Feulgen reaction). Yet no differences in cancer cell DNA chromatographic profiles or base compositions were detectable at that time. A significant breakthrough that linked cancer, chromosomes, and DNA appeared in 1960 with the report of the discovery of the Philadelphia chromosome by Nowell and Hungeford (3). Their study described a consistent alteration in chromosome 22 in myelogenous leukemia. The chromosome was shorter than its normal homologue. Approximately half of the long arm was missing, a loss of 0.02 pg of DNA (0.27% of the DNA content per nucleus or $2 \times 10^7$ nucleotides pairs).

These pioneering studies implicated DNA alterations in the origin of the cancer progenitor cell. Investigators focused their initial concern on the cells' ability to divide and to replicate DNA without stopping. Consider the extent of the problem. It is estimated that the human body contains 30 trillion cells; they contain 23 pairs of chromosomes and the size and shape of each pair varies. The DNA strand in the largest chromosome (ch1), when extended, is five inches long. Clearly, chromosomal replication is, in mammalian cells, a complex process that requires strict regulation and the process takes time. Moreover, some normal cells renew themselves until dead, for example, epithelial cells, white blood cells, etc. Other cells, like hepatocytes, differentiate and remain in $G_0$ to move into $G_1$ and divide only when challenged, for example, by partial hepatectomy. Finally, cells like muscle cells and neurons seldom or never divide.

## Oncogenes and Suppressor Genes

The notion that the initiation of a cancer cell results from a loss of the control(s) of cell proliferation was not new as this was obvious from gross and microscopic observations. That the tumor grows beyond the confines of normal tissue and the microscopic presence of mitotic figures was highly suggestive. The discovery that oncogenes were related to growth factors (e.g., platelet-derived growth factor [PDGF] and the v-sis-transforming factor) for the first time provided a molecular mechanism for the gain in the cells' capacity to divide. The notion that uninterrupted cell proliferation might be caused by the loss of genes whose function is to arrest growth (tumor suppressor genes) came to light only later.

## Viruses Cause Cancer

The very concept of an oncogene and the name originated when it was shown that DNA viruses and retroviruses (RNA viruses) could transform cells in vitro. Our understanding of the early events in carcinogenesis emerged from a return to viral oncology. The history of viral oncology was most aptly told by Rauscher and Shimkin (4).

## Historical Overview

The search for the pathogenesis of cancer not surprisingly was stimulated after Pasteur's and Koch's discoveries of the bacterial origin of infectious diseases. Pasteur's studies on rabies demonstrated the existence of diseases caused by ultramicroscopic agents (not detectable under the light microscope) that pass through a porcelain filter, namely viruses. Several attempts to find viruses, microbes, or even parasites causing cancers, however laborious and enthusiastic, failed to be confirmed until Peyton Rous (1910) of the Rockefeller Institute reported the transfer of chicken sarcoma by cell free extracts (5). However, many scientists believed that the sarcoma produced by Rous resulted from the transfer of intact cells that had passed through the filter, and Rous' experiments were not confirmed for decades, perhaps because of inadequate techniques used by others. Moreover, the transfer of cancer was restricted to chicken and therefore believed to be exceptional. Not until Duran-Reynals (6) of Yale University transferred the Rous virus to other fowl did the viral origin of cancer become accepted. Indeed, the notion that cancer could be caused by viruses had died out between 1911 and 1942 despite two important discoveries: Bittner's demonstration that mouse breast cancer could be transmitted by the mother's milk (7), and Shope's transfer of a papilloma from wild-type to domestic rabbits (8).

## Chemicals Cause Cancer

These pioneering studies and many others were later followed by the discovery of several other oncogenic viruses, but pure chemical carcinogens had been discovered and they quickly stole predominance in the research field, at least until the discovery of the polyoma virus (9).

Long before viruses were discovered clinicians were aware that chemicals cause cancer. Even before Percival Pott (1775) reported his seminal observations on the association of tar with cancer of the groin in chimney sweepers, snuff had been incriminated by Lynch in cancer of the nasopharynx (10). In 1918 Yamagiwa and Ichikawa were the first to demonstrate that tar applied to the ear of rabbits cause cancer (11). Yet the field of research in chemical carcinogenesis opened widely only after 1930, when Sir Ernest Kennaway succeeded in inducing skin cancer with a pure chemical, 1,2,5,6 dibenzanthracene (12). Substitutions in the molecules were investigated, other polycyclic hydrocarbons were tried, and a theory of the molecular requirements for effective carcinogens was proposed by Pullman and Pullman: the K region theory (13).

Elisabeth and James Miller demonstrated that chemical carcinogens bind to protein (14). Pullman predicted that the binding of carcinogens to protein most likely involved their K region and Charles Heidelberger, who was among the first to use isotopes in biological research, confirmed the Pullman prediction using labeled carcinogens (15).

The in vivo binding of carcinogens to proteins was shown to require enzymes, microsomal mixed function oxidases, which convert procarcinogenic molecules into active carcinogens, an epoxide. This epoxide could either bind to macromolecules (proteins and DNA) or be detoxified. The binding to DNA attracted the interest of investigators because it was in keeping with the notion of somatic mutations. In the meantime, it had been shown that aniline, azobenzene, and fluorene derivatives were carcinogens. The addition of amino azotoluene to the diet of mice caused liver cancer (16), the first parenchymal cancer produced by ingestion. Foods contaminated with acetylamino-fluorene cause cancer of the esophagus and the liver, and aniline causes cancer of the bladder (14).

Chemical carcinogens introduce permanent changes in the progenitor cell (initiation) that are transmitted from one cell generation to another (cloning) and, therefore, were suspected to alter the genome. A general theory for chemical carcinogenesis was proposed: the procarcinogen is converted to an active carcinogen that ultimately binds to DNA or proteins. Elisabeth and James Miller proved the existence of such a pathway for several carcinogens, among them 2-acetamide fluorene (17). The theory was simple: carcinogens are mutagens that cause somatic mutations directly responsible for cancer. However, it needed to be established that carcinogens were direct mutagens (e.g., in bacteria) and could cause transformation of cells in culture. Early results were not convincing. An ingenious method invented in Bruce Ames' laboratory established that carcinogens are actually mutagens (18, 19). Ames and associates mixed procarcinogens with microsomes and bacteria, and proved that the procarcinogen was converted to an active mutagen, at least for *Salmonella*. Although the notion that somatic mutations cause initiation in the progenitor cell seemed logical and appealing, the mutated genes still needed to be identified. This was accomplished only after the discovery of oncogenes and suppressor genes.

## Cellular Oncogenes

Persistent investigations of retroviruses ultimately led to the discovery of oncogenes. The genome of retroviruses (RNA viruses) consists of two genetically identical RNA molecules (~8–10 kb), and it includes three genes: *gag*, which encodes viral nucleocapsid protein; *pol*, which encodes a reverse transcriptase; and *env*, which encodes the viral envelope protein. In addition, rapidly transforming retroviruses such as the Rous sarcoma virus possess other elements in their genome. For instance, the Rous sarcoma virus, like some other retroviruses or some DNA viruses, carries the three genes needed for replication (*gag*, *pol*, and *env*) plus the v-*sarc* (located in the 3′ end of the genome). The latter single gene or oncogene v-*sarc* is all that is needed for transformation. V-*sarc* was shown to encode a plasma membrane-bound tyrosine kinase.

Over the years at least 75 retroviruses were discovered that rapidly induce tumors, each carrying a different oncogene. The sequences of these oncogenes have homologues in mammalian cells. However, these homologues, called "proto-oncogenes," cannot induce transformation or cause cancer without alterations (mutations, amplification, or rearrangements).

Identification of the function of proto-oncogenes revealed a common denominator: a role in cell growth. They are homologues of: growth factors, tyrosine kinases, membrane-bound G proteins (Ras), and transcription factors (c-Myc).

In summary, proto-oncogenes:

- Have been highly conserved during evolution in mammalian cells.
- Encode normal proteins involved in cell growth and development.
- Cause cancer only when mutated, rearranged, or amplified (reviewed in 20).

In addition to acute retroviruses that transform and rapidly cause tumors, another population of retroviruses exists. These viruses cause transformation by insertion of genetic material within the host chromosomes. Integration of the viral sequence takes place at sites close or adjacent to cellular genes critical to cell proliferation, thereby altering their transcriptional activity. For example, the Philadelphia (Ph) chromosome is the product of a translocation between chromosomes 9 and 22, t(9;22)(q34:q11). The translocation involves the c-*abl* (the normal proto-oncogene homologous to the v-*abl* gene) from chromosome 9 to the site of the *bcr* gene on chromosome 22. Thus, a chimeric protein is generated whose N terminus carries the protein sequence encoded by *bcr* and its C terminus carries the gene encoded by c-*abl*, a tyrosine kinase activated by a B cell receptor. The appearance of the chimera is associated with myelogenous leukemias. More than 35 similar translocations have been detected that cause leukemias, lymphomas, or sarcoma (20).

## Tumor Suppressor Genes

Two categories of gene mutations are at the origin of the clonal replication of a cancer: cell oncogenes and suppressor genes. When mutated, the oncogenes secure a gain in function. These mutations involve growth factors, growth factor receptors, and transducing molecules that carry signals for growth. In contrast, tumor suppressor genes normally down-regulate cell proliferation by cell cycle arrest followed by DNA repair, or activate transcription factors that contribute to quiescence, senescence, or apoptosis. Mutations in these tumor suppressor genes cause them to lose their regulatory function and thereby allow the clone to proliferate constitutively. The list of existing tumor suppressor genes includes:

- Genes involved in cell cycle control (*rb, apc*).
- Genes involved in DNA repair (*msh2, mlh1, pms1, pms2, msh6*).
- Genes involved in cell cycle arrest (*p53, p16, p14*).

In addition to the established list there are also potential candidates, such as pro-apoptotic genes (*bax* or *bad*).

The prototype tumor suppressor gene is the *rb* gene, which is responsible for the retinoblastoma tumor. Retinoblastomas are unique retinal tumors; they occur sporadically (60%) but can also be familial (40%). Knudson noticed that familial tumors are more frequently bilateral and multifocal, and usually occur at an earlier age than the sporadic tumors (21). Cytogenic and biochemical investigations in several laboratories led to the discovery of a 13q14 deletion. The deletion is observed in a subset of patients in association with a decrease of esterase D activity (to approximately 1/2 the normal values) in blood leukocytes and in fibroblasts in culture, indicating that the Rb locus and that of esterase D are genetically linked. These findings and careful statistical analysis of the data led Knudson to propose a "two hit theory" to explain the origin of familial retinoblastoma. The first hit affects a germ cell (this explains the loss of esterase D in the patients' leukocytes and fibroblasts) and the second hit takes place in a somatic cell of the retina in retinoblastoma (or in an osteocyte in osteosarcoma). Genetically speaking, this implies that although susceptibility to retinoblastoma is inherited as an autosomal dominant trait, at the cellular level the allele responsible for susceptibility is recessive (22). Later, a child was observed to have half of the normal levels of esterase D in his lymphocytes and fibroblasts but no esterase D at all in the retinoblastoma cells and no detectable 13q14 chromosomal deletion. This finding is not in conflict with the two hit theory, yet it proves that the deletion of both the esterase D and the Rb allele can be undetectable cytogenetically and are likely to be submicroscopic. Such occurrence was confirmed by refined biochemical analysis of the chromosome (22). The two hit theory led to a broader concept, that of loss of heterozygosity (LOH). LOH implies the loss of the two alleles responsible for cancer initiation by suppressor genes. When a substantial number of retinoblastomas were investigated using chromosome 13 probes and the marker patterns were compared with those obtained from normal samples, it became clear that loss of both RB alleles occurred in more than 60% of the tumors.

Heterozygotes, wherein one chromosome carries the normal allele and the other has lost the allele for the suppressor genes (carriers in other words), do not express the cancer phenotype. Cancer develops only when both alleles of the suppressor genes are lost, and thus heterozygosity is lost (LOH). The easiest way to lose heterozygosity is by losing the chromosome or part of the chromosome carrying the normal allele. The normal allele expression also can be abrogated by duplication of the mutated allele as a result of recombination. In hereditary cancer the mutant allele is derived from the affected parent (23). It did not take long before the two-hit hypothesis was extended to other familial cancers, first to neuroblastoma and Wilms tumor and later to at least 20 genes predisposing to cancer through loss of function.

The gene responsible for the eye cancer retinoblastoma encodes a 928-amino acid phosphoprotein (the Rb protein) that plays a critical role in cell cycle regulation by shuttling between phosphorylation and dephosphorylation. The hypophosphorylated Rb forms a complex with transcription factors (the E2F family of transcription factors) that activates genes that encode, among others, proteins involved in the S phase of the cycle. The *rb* gene is located on chromosome 13, over a 200 kb locus (13q14) that encodes a 110 kDa protein mostly found in the nucleus. The *rb* gene is a suppressor gene expressed in all human and rat tissues.

In humans and some animals, several mutations (no specific point mutations but mainly deletions) and loss of heterozygosity of the Rb protein have been associated with retinoblastoma, in the sequence predicted by Knudson: the two hit hypothesis (21). During the cell cycle the Rb protein inhibits the E2F transcription factors by two different mechanisms: it binds to the E2F transactivation domain and thereby blocks its transcriptional activation. The Rb/E2F complex recruits enzymes (HDACs) that remove acetyl groups from histone octamers, thereby remodeling chromatin (reviewed in 22–23).

## References and Suggested Reading


1. Bovary, G.F.V. *Zur Frage der Enstellung malignerer Tumoren. JENA.* Heidelberg, Germany: Gustave Fischer Verlag; 1914.
2. Fearon. The metabolic and molecular bases of inherited disease. *Tum Suppr Genes* 2000;655–74.
3. Nowell, P.C., and Hungerford, D.A. Chromosome studies on normal and leukemic human leukocytes. *J Natl Cancer Inst.* 1960;25:85–100.
4. Rauscher, F.J. Jr., and Shimkin, M.B. *Viral oncology.* In: Stetten D., Jr., and Carrigan, W.T. *NIH: an Account of Research in Its Laboratories and Clinics.* Orlando, FL: Academic Press; 1984:368–78.
5. Rous, P. A transmissible avian neoplasm (sarcoma of the common fowl). *J Exp Med.* 1910;12:696–706.
6. Duran-Reynals, F. The reciprocal infection of ducks and chickens with tumor-inducing viruses. *Cancer Res.* 1942;2:343–69.
7. Bittner, J.J. Possible relationship of the estrogenic hormones, genetic susceptibility, and milk influence in the production of mammary cancer in mice. *Cancer Res.* 1942;2:710–21.

8. Shope, R.E. Infectious papillomatosis of rabbits. *J Exp Med.* 1933;58:607–24.
9. Gross L. The fortuitous isolation and identification of the polyoma virus. *Cancer Res.* 1976;36:4195–6.
10. Collins, R., Haurer, W., and Clarke, W. *Chirurgical Works of Percival Pott. F.R.S. and Surgeon to St. Bartolemew Hospital.* London, UK: Poternoster Row; 1775.
11. Yamagiwa, K., and Ichikawa, K. Experimental study of the pathogenesis of carcinoma. *J Cancer Res.* 1980;3:1–21.
12. Kenneway, E.L., and Hieger, I. Carcinogenic substances and their fluorescence spectra. *Br Med J*. 1930;1:1044–6.
13. Pullman, A., and Pullman, B. *Cancerisation par les Substances Chimiques et Structure Moleculaire.* Paris, France: Masson; 1955.
14. Miller, E., and Miller, J. Biochemistry of carcinogenesis. *Ann Rev Biochem.* 1959;28:291–320.
15. Heidelberg, C. *The Protein Binding to Hydrocarbon Carcinogenesis. Mechanism of Action.* Wolstenholme, G., and O'Connor, M., eds. London, UK: I. and A. Churchill, Ltd.;1959:179–96.
16. Yoshida, T. Development of experimental hepatoma by the use of Q-aminoazotoluene with particular reference to gradual changes in the liver up to the time of development of carcinoma. *Jap Path Soc Trans.* 1934;24:523–30.
17. Miller, E.C. Studies on the formation of protein-bound derivatives of 3,4-benzpyrene in the epidermal fraction of mouse skin. *Cancer Res.* 1951;11:100–8.
18. Ames, B.N., Gurney, E.G., Miller, J.A., and Bartsch, H. Carcinogens as frameshift mutagens: metabolites and derivatives of 2-acetylaminofluorene and other aromatic amine carcinogens. *Proc Natl Acad Sci USA.* 1972;69:221–5.
19. Ames, B.N., and Whitfield, H.J. Jr., Frameshift mutagenesis in *Salmonella. Cold Spring Harb Symp Quant Biol.* 1966;31:221–5.
20. Park, M. Oncogenes. In: Scriver, C., Beaudet, A., Valle, D., Sly, D., Childs, B., Kinzler, K., and Vogelstein, B., eds. *The Metabolic and Molecular Bases of Inherited Disease,* 8th ed. New York: MacGraw-Hill Medical Publishing Division; 2001:1665–74.
21. Knudson, A.G. Mutation and cancer: statistical study of retinoblastoma. *PNAS.* 1971;68:820–3.
22. Benedict, W.F., Murphee A.L., Banerjee, A., Spina, C.A., Sparkes, M.C., and Sparkes, R.S. Patient with 13 chromosome deletion: evidence that the retinoblastoma gene is a recessive cancer gene. *Science.* 1983;219:973–5.
23. Cavenee, W.K., Nordenskjold, M., Kock, E., Maumenee, I., Squire, J.A., Phillips, R.A., and Gallie, B.L. Genetic origin of mutations predisposing to retinoblastoma. *Science.* 1985;228:501–3.

CHAPTER

# TUMOR TRANSFORMING GROWTH FACTORS AND CANCER

# 17

Several oncogenes related to known growth factors have been identified. The first (v-*sis*) was discovered by comparing the amino acid sequence of the oncogene with the growth factor sequences stored in a computer database. The v-*sis* sequence proved to be related to the B chain of platelet-derived growth factor (PDGF). PDGF is associated with the induction of gliomas and sarcomas. Other growth factors that operate as oncogenes do so through proviral insertion into DNA (INT2 and INT1) or by DNA transfection (K53 and HST, both are FGF homologues; **TABLE 17-1**). The above examples of growth factor mutations illustrate the fact that they can be initiators of cancer.

Although what follows is not directly involved in this discussion on oncogenes, it should be pointed out that the role of growth factors in cancer may be indirect, for example, by stimulating angiogenesis and facilitating cell dissemination.

- Fibroblast growth factor (FGF) and vascular endothelial growth factor (VEGF) are two members of the PDGF family of growth factors that bind to tyrosine kinase receptors. They contribute to vascularization of the tumors in gliomas, hemangioblastomas, and of many, if not most, solid tumors. They also facilitate the establishment of metastases (1, 2).
- The "scatter factor" induces motility of mammary epithelial cells; it is identical to the hepatocyte growth factor (HGF), and is suspected to be involved in the spread of metastases (3). Its role in vivo in cancer remains dubious, however (4). HGF and its receptor (c-Met) are coexpressed in cancer of the pancreas (5).

## Transforming Growth Factor-Alpha

Both transforming growth factor-alpha (TGF-α) and epithelial growth factor (EGF) bind to the EGF receptors and activate the receptor-associated tyrosine kinase. The downstream signaling molecules involve homology 2 (SH2) domains; these SH2 domains recognizes short peptides with a specific sequence containing phosphotyrosine located within the activated receptor. Proteins involved in the recruitment fall into

**Table 17-1 Partial list of oncogenes**

| Type | Name | Location |
|---|---|---|
| Ligands | EGF | Glioblastoma |
| | IGFI | Breast cancer |
| Receptors | NEU | Neuroblastoma |
| | ERB | Bladder, breast cancers |
| | EGFR | Skin squamous cell cancer |
| Protein kinase | RET | Medullary thyroid, Carcinoma and MEN2B |
| Tyrosine kinase | ABL | Activated by translocation 9q34 |
| | CML | Chronic myelogenic leukemia |
| | HRAS | Colon and pancreas cancers |

Partial list of oncogenes causing cancer in humans.

two categories: catalytic and non-catalytic. The catalytic include phospholipase Cγ (PLCγ), phosphatidylinositol-3-kinase (PI3K) and guanosine 5′-triphosphate (GTP)ase activating proteins. The non-catalytic proteins are adaptor proteins that, in cooperation with the above enzymes and sometimes other proteins, build chains that communicate the signal that emerges at the surface to the target in the nucleus or the cytoplasm.

Properties of TGF-α resemble those of EGF, for example, TGF-α promotes the precocious opening of the eyelids in young mice and also plays a physiologic role in development. TGF-α is one of the many ligands of EGFR5. The exact role of TGF-α in tumor biology remains to be clarified. It is believed to be an autocrine factor that sustains the transformed state (6–8).

A comparison of the molecular properties of TGF-α and those of EGF is enlightening. TGF-α is a single chain polypeptide composed of 50 amino acids. Recombinant TGF-α derived from cDNA of renal cell carcinoma reveals that TGF-α results from the cleavage of a precursor 160 amino acids long, with considerable homology between its amino acid sequence and that of the human and rodent peptide, suggesting evolutionary conservation. There is striking homology between TGF-α and EGF. The homology is concentrated around the cystenic residues and is maximal in the third cystenic loop. The presence of the latter is indispensable to binding. The consistency in the position of the cystenic residues in the sequence strongly suggests that the tertiary structure of TGF-α and EGF are the same, and that such a configuration is essential for recognition by, and bonding to, the receptor. Finally, in addition to the homology TGF-α shares with EGF, the TGF-α sequence also presents some homology with serine proteases involved in coagulation and fibrinolysis and with a 140 amino acid polypeptide encoded by the vaccinia virus, also able to generate a mitotic response.

In general, TGF-α can be viewed as a regulator of cell proliferation and differentiation, and as such, it contributes to the development of many cancers in animals and humans. Overexpression of TGF-α in transgenic mice induces epithelial hyperplasia, pancreatic metaplasia and breast cancer (9, 10). TGF-α also has been associated with hepatocarcinoma and is often detected in association with hepatitis B virus (HBV) infection (11, 12).

In transgenic mice, the sustained overexpression of *c-myc* induces hepatocellular cancer (HCC). Coexpression of TGF-α accelerates this process because of unrestrained proliferation of the hepatocytes that results from activation of the RB-E2F pathway and inhibition of the cancer cells' apoptosis. The hyperplasia of the tumor cells is associated with a drop in the mitotic and a rise in the apoptotic activity of the peritumor cells probably as a result of up-regulation of TGF-β1. These findings suggest that after the combined c-Myc–TGF-α stimulation, precancerous cells lose their sensitivity to TGF-β1. This seems to result from decreased expression of TβRII and a drop in levels of P27 in the HCC cells (13). The contribution of TGF-α in mouse models and human cancer has been reviewed by Humphreys and Hennighausen (14).

The molecular mechanisms of the contribution of TGF-α to oncogenesis remain unsettled. If the normal properties of TGF-α (a growth factor that shares the EGF receptor) is kept in mind, it can be anticipated that overexpression of the cell proliferation of normal, precancerous, and possibly cancerous cells will be encouraged. However, the prediction and the actual description of subsequent events are complicated by the multicipliclty of the signal transduction pathways that EGF-α can or does put in gear (protein kinase A, mitogen-activated protein kinase [MAPK], Janus kinase [JAK], Ras/Raf) and their unpredictable interactions in a specific cancer. With time, patience, vigilant research and imaginative technologies, this too will be resolved, and specific targets will be selected for treatment.

## The Tumor Transforming Factor: TGF-β

In 1981, Roberts and Sporn's laboratory discovered that TGF-β induced anchorage-independent growth in fibroblasts (15). This "founding father" of TGF-β over the years was joined by over 40 isoforms in vertebrates and over 10 related molecules in invertebrates (15). The tumor growth factor TGF-β is ubiquitous and is found in embryonal, adult, and tumoral cells where it interacts with specific receptors. Platelets are richest in TGF-β (100–500× more activity than any other tissue) and serve as a source for purification of the intact molecule.

The molecule is critical to embryonic development and homeostasis during the lifespan of vertebrates, implying that TGF-β has a multitude of cellular functions that when expressed secure homeostasis, a balance between cell life and cell death and between the cell proliferation and apoptosis. On the whole, TGF-β is primarily an antiproliferative factor, at least in vivo, and it also stimulates cellular hypertrophy, and differentiation (16). In the adult, these functions are manifested in reproduction, blood pressure control, angiogenesis, inflammation, and wound healing. When TGF-β's function is defective, for example through mutations, it may cause a panoply of diseases including cancers.

The molecule was purified and characterized from platelets (17). TGF-β is a homodimer ($M_r$ 25.000 Da). Available evidence suggests that the amino acid sequence is the same in each monomer and that the monomers are held together by S-S bonds. The molecule is encoded in a cDNA composed of 2527 bp. Upstream to the initiation codon there is a 842 polynucleotide chain. This sequence contains an unusually high G-C content (up to 80% in the 450 nucleotide segment), but the significance of this high G-C content is not yet known. The nucleotide chain spanning between position 192 and 252 is almost entirely made of purine nucleotides, which are similar to sequences observed in myc and insulin-like growth factor-2 (IGF-2) mRNAs. There is, however, no substantial homology between these mRNAs and that of TGF-β.

## TGF-β and Its Binding Proteins

Mature TGF-β is derived from a larger precursor that is approximately 400 amino acids long; of those, only 112 amino acids in the COOH terminal domain constitute the mature TGF-β. The site accessible to the peptidase, Arg-Arg, is located in the precursor immediately upstream of the N terminus portion of the mature TGF-β.

At least three genes are known to encode TGF-β, and their products are referred to as TGF-β1, -β2, and -β3. There is considerable identity between the three genes and the genes are 99% to 100% conserved from chick to humans. These findings strongly suggest that the three TGF-β genes are derived from a common ancestor (18).

After translation on the ribosomes, the precursor is translocated to the lumen of the endoplasmic reticulum. The precursor dimerizes and the dimer is cleaved to yield a new dimer of mature TGF-β chains. The mature TGF-β is excreted in the extracellular matrix where it is stored in a latent form. The mature TGF-β is constrained by inhibitors of the ligand, the latency-associated peptide (LAP) and the latent TGF-β binding peptide, where the first generates a small (LTGF-β) and the second a large (LTBP) complex. In the small complex, the mature 24 κDa homodimer is associated with the 80 κDa LAP to form the LTGF-β complex. The latter either remains stored or can be activated and bind to its receptor. When stored, the small complex (LTGF-β) binds to LTBP-1, one of four isoforms of the extracellular glycoprotein (LTBP-1, -2, -3, -4). The binding results from the formation of S-S bonds between cysteines (19). LTBP-1 not only sequesters LTGF-β in the extracellular matrix, but also modulates its activation (20).

## Type I and Type II Receptors

The receptors, TGFβRI (55 kDa) and TGFβRII (70 kDa), typically have three components: an extracellular domain, a transmembrane domain, and an intracellular domain. The extracellular domain is made of a 150 amino acid sequence. The sequence includes N glycosylation sites and at least 10 cysteine residues that are likely to contribute to the appropriate folding of the molecules. Three of these cysteines form a traditional cysteine cluster and the other cysteines are dispersed in the sequence. The dispersion is more conserved in type I than type II receptors.

There are no structurally remarkable amino acids or amino acid sequences in the intramembrane domain, except that the sequence includes Ser 165 close to the membrane region. Ser 165 is phosphorylated in type I by the constitutive type II receptor. In contrast, the sequence of the intracellular component reveals some special features: the kinetic site in the distal segment of the receptor and a GS sequence upstream of the kinetic site separated from the *trans*membrane domain by a short sequence. The kinase domain in type I and type II TGF-β receptors ensconces a sequence typical of that of traditional serine/threonine protein kinase domains and, indeed, the type I receptor substrate SMAD is phosphorylated by the receptor and autophosphorylation of type II TGF-β receptor can occur in vivo but not in vitro (reviewed in 21).

## TGF-β: The Ligand and the Receptor

TGF-β has a multitude of functions. The ligand TGF-β transfers its signal to a receptor. The latter is linked to the TGF-β target through a signal transducing pathway in which SMAD molecules play a key role. This apparently simple pathway activates a large and varied group of targets in sometimes different organs.

The three isoforms of TGF-β (TGF-β1, -2 and -3) have similar biological activities in vitro. The fate of TGF-β null mice varies with the isoform:

- When born alive, approximately half of TGF-β1$^{-/-}$ mice die soon after birth with marked infiltration of lymphocytes and macrophages in several organs (inflammatory response) (22, 23). The other half of the progeny dies because of developmental defect in hematopoiesis and vasculogenesis.
- TGF-β1–deficient mothers give birth to TGF-β1 null mice with congenital heart defects. However, the progeny can be rescued by maternal TGF-β1 (24). These findings indicate that TGF-β1 has an important function in embryonal development and that TGF-β1 is supplied to the fetus by the mother.
- TGF-β2 null mice die soon after birth with multiple developmental defects that are the characteristic for TGF-β2 defects only (25).
- TGF-β3 null mice die soon after birth mainly because of defective epithelial-mesenchymal interactions responsible for incomplete lung development and cleft palate (26).

Each TGF-β isoform is encoded by a different gene with a different promoter, but each of these genes ultimately leads to the translation of similar functional proteins. These observations further help to clarify the multiplicity of responses to TGF-βs. Similarities in the amino sequence during evolution of the three proteins suggest that the three TGF-β genes are derived from a common ancestor. In the active form, TGF-β and the ligand are held together by hydrophobic bonds and disulfide bonds strung between the subunits. Present evidence indicates that bioactive receptors are heterotetramers (2 type I and 2 type II receptors).

### Binding of Ligand and Receptor

After release from the binding proteins by thrombospondin-1, the TGF-β heterodimer binds to the TGF-β receptors (TGFβR). The family of TGF-β receptors contains four members: TGFβRI, TGFβRII, TGFβRIII, and endogen; most of these family members are of type III. Two modes of binding have been described: in the first the ligand binds to a type III receptor, and the second mode recruits type II and type I receptors. The constitutively-activated type II receptor activates the type I receptor by phosphorylation. Although TGF-β is functionally activated downstream from the type I receptor, type I activation is central to the process of signal transduction because it determines targets specificity (27, 29). After binding of the ligand, TGFβRII in turn is phosphorylated at serines (Ser) 213, 409, and 416. The Ser 213 is located outside the catalytic domain, but its phosphorylation is needed because it allows for autophosphorylation at serines 409 and 416. In the second mode the ligand directly binds to TGF-β receptor type 2 (TGFβRII), which recruits TFG-β receptor type 1 (TGFβRI) (16).

## Mode of Signaling TGF-β: The SMAD Pathway

Activation of the kinase domains of type I and type II receptors engage a transduction signal pathway: the SMAD pathway. The SMAD pathway was first discovered in *Drosophila* and *C. elegans*, and later was extended to humans. The SMADs family is grouped in three classes: receptor regulated, cooperative SMADs, and inhibitor

SMADs. In 1995, MAD (mothers against decapentaplegic homolog 2) was first discovered in *Drosophila.* MAD was the first transducer molecule of TGF-β to be identified. Genes encoding similar proteins were soon identified in worms and mammals and are referred to as SMADs. SMADs 1, 2, 3, 5, and 8 are regulatory (RSMADs) and are directly phosphorylated by the phosphokinase of the type I receptor. Phosphorylation triggers the translocation of RSAMDs to the nucleus. In the nucleus, the RSMADs are associated with the only CoSMAD or SMAD4. The nuclear SMADs form complexes with appropriate binding proteins, corepressors, and coactivators to modulate transcription of the mRNAs of various types of proteins depending on circumstances. Inhibitory SMADs (ISMADs) block the action of the RSMADs and thereby interrupt TGF-β signaling and ultimately its function(reviewed in 29). At first glance, this is a single pathway: two steps between ligand binding to the receptor and transcription activation of protein phosphokinase and activation of SMADs. However, simple pathways, especially if they are ambitious and seek many different critical biological targets, need modulators (inhibitors and activators); the SMAD pathway is no exception. A closer look at the SMAD pathway reveals that RSMADs and CoSMADs have a relatively long sequence (500 amino acids) that ensconce two conserved structural domains (MAD–homology domains), that is, an N terminal (MH1) and a C terminal (MH2) domain, and a linker sequence between MH1 and MH2 (**FIGURE 17-1**). In R-SMADs, the carboxyterminal sequence of MH2 contains a conserved SSxS conserved sequence that is not found in CoSMADs.

Although the structure of the linker remains unresolved, it is known to possess several phosphorylation sites for MAPK's tyrosine kinases. Both MH1 and MH2 assume a globular structure from which protrudes a β hairpin (Bhx) in MH1 and a loop (L3) as well as two-alpha helical structures (αH-1 and αH-2) that emerge in MH2. The hairpin binds to a CAGAC sequence in DNA, L3 connects with the type I receptors, and αH-1 interacts with DNA binding cofactors. MH1 also contains a motif (PY) that interacts with E3 ubiquitin ligase-containing proteins (snurfs; see below). The carboxyterminal 10 residues, SSxS motifs, is the site of the phosphorylation catalyzed by the receptor kinase (30).

SMADs shuttle between the type I TGF-β receptor and DNA. The activated receptor recognizes RSMADs. A basic patch in the RSMAD structure binds to the GS region of the type I receptor (31). Mutations of a histidine residue found at that site (His 331) reduce the TGFβRI affinity for SMAD and the phosphorylation of the SxS motif (32). This elegant model may well be modified when a detailed structure of the TGFβRI and RSMAD complex becomes available. The relationship of RSMAD to the receptors is modulated by at least two other proteins: SARA and SNURFS. SARA guides SMADs 2 and 3 to the endocytized receptor and SNURF regulates the SMAD's transcription transactivation (or repression) most likely through selective ubiquitination.

- SARA (SMAD anchor for receptor activation) contains a membrane localizing phospholipid domain, FYVE (Fablp/YOTP/Vac 1p/EEA1). SARA is located in early endosomes where it binds SMAD2 and SMAD3. Many surface receptors are known to be endocytized upon ligand binding, which is the case for TGFβRI (33).

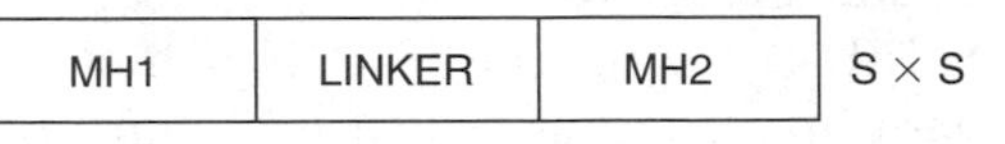


**FIGURE 17-1** Regulatory (RSMADs) and cooperative (CoSMADs) ensconce two conserved structural domains and a linker sequence between MH1 and MH2.

This process is believed to facilitate signaling by adjusting the conformation of the signal complex (in the case at point SMAD2/SMAD3) to that of the receptor complex. SARA is believed to target the SMAD2/SMAD3 to the endocytosed receptor complex. The traditional pathway of endocytic caveolae evolves from early endosomes to late endosomes, and from there either to lysosome for degradation of their content or to the Golgi for posttranslational modification and secretion. But the internalization in early endosomes may have an additional function, namely, to facilitate translocation of the signal transducer and its "anchor" SARA to the target. Evidence suggests that is what happens (34). Moreover, the FYVE domain of SARA is necessary and sufficient for the localization of SARA in the endosome and TG-βRI/SMAD signaling. PtdIns binding to SARA seems to be required for the process (35).

The absolute requirements for endosomal internalization of TGFβRI and the SMAD-SARA complex for TGFβRI signaling were investigated. DiGuglielmo et al. (36) reported that:

- TGFβRI is internalized.
- Inhibition of internalization has no effect on SMAD 2 phosphorylation, on its nuclear translocation, or TGFβRI's modulation of transcription.
- Interference with SARA phosphorylation has no effect on TGF-β signaling.

These findings suggest that neither endosomal internalization nor SARA's phosphorylation are absolutely needed for TGFβRI signaling and modulation of transcription, but they do not exclude that endosomal internalization and SARA's phosphorylation maximally activate signaling. Thus, the simultaneous presence of the receptor and SMAD2/SMAD3 in the endosome amplifies the signals triggered by TGFβRI, namely, the phosphorylation of SMAD2/SMAD3 and ultimately transcription modulation (36).

## The RING Finger Protein, SNURF

The *RING* gene (i.e., really interesting new gene) encodes the RING finger protein. This protein is involved in protein-protein and protein-DNA interactions. Such properties are compatible with roles for RING proteins in oncogenesis development, signal transduction, and apoptosis. The RING finger protein, SNURF (small nuclear ring finger) protein, was first isolated as an androgen receptor binding protein in yeast (37). SNURF and its mammalian counterpart, RNF4 (ring finger protein 4) (38), have been shown to regulate transcription either by operating as a coactivator of various transcription factors (39) or by acting as a repressor. The molecular mechanisms of these antagonistic functions are not completely known. It could involve DNA binding, protein binding, or both. Häkli et al. (40) showed that SNURF possesses double-strand DNA binding properties despite a lack of a specific nucleotide sequence at the binding site. An intact RING structure is not required, but mutations in the SNURF positively-charged amino acid terminus (8-11) impair DNA binding. In addition, SNURF binds to nucleosomes. Such bifunctional activity could explain, at least in part, the role of SNURF in transcription control, because it is compatible with the recruitment of regulatory proteins. Members of the same laboratory established that SNURF and the TATA binding protein (TBP) cooperate to activate an estrogen receptor in human breast cancer cells. The process involves ligand binding to the ERα, a configuration modification of the ERα structure (reposition of helix 12), recruitment of TBP and SNURF, and interaction of the SNURF with both TBP and ERα (41).

Some SNURFs contain an E3 ubiquitin ligase activity that interacts with a specific proline-tyrosine motif present in certain SMADs and contributes to control the intracellular levels of SMAD by degradation through the ubiquitin proteosomal pathway (42).

### Nuclear Translocation of RSMAD

All SMADS are not distributed alike in the cell. Prior to receptor activation, RSMAD proteins are in the cytoplasm while ISMADs tend to be located in the nucleus. After phosphorylation by the receptor, RSMADs need to be detached from the receptor and transferred to the nucleus. The molecular mechanism of the release of RSMAD is partly dictated by the phosphorylation of the SxS motif at the C terminal region of the MH2 domain. The basic pocket and the L3 loop of the RSMAD binds to the GS domain and the L45 loop of the receptor (Figure 17-1). The assumption implies that interference with SxS phosphorylation will cause persistent association of RSMAD and receptor, and this proved to be the case (43). The nuclear localization signal (NSL) is believed to involve either a lysine-rich region found in all SMADs (namely, a solvent exposed sequence, KKLKK) located on helix 2 near the DNA binding hairpin (44) or direct binding of the MH2 domain to the nucleoporins. The phosphorylated SxS motif (SSxS) competes with the GS segment of the receptor for binding of the basic surface pocket of RSMAD and thereby facilitates the release of RSMAD.

SMAD4 travels from cytoplasm to the nucleus either together with RSMAD or on its own. This means that the molecule must include a nuclear import and a nuclear export signal, and the two should not compete. The first involves lysine residues in the $H_2$ helix of MH1 (45). The nuclear export signal was found in the linker region of SMAD4, and the signal is masked by the heteromeric assembly of RSMAD and SMAD4 (46).

All SMADs bind to DNA except SMAD2. The RSMADs, SMAD4 and SMAD1 bind to DNA before they regulate transcription. The contact between SMAD3, SMAD4, and SMAD1 involves the minimal SMAD binding element (SBE) comprised of only four bases (5′AGAC 3′) and the β hairpin of SMADs (44). In addition, SMAD4, SMAD3, and SMAD1 also have been reported to bind to GC-rich sequences. The significance of this second SMAD binding site remains unclear.

## Pleiotropic Functions of TGF-β

TGF-β is ubiquitous and targets many different genes depending on the type of cell and the receptor that is activated. Consider some of the most salient among a multitude of biological responses to TGF-β. TGF-β contributes to mesenchymal cell proliferation (angiogenesis, hematopoiesis, fibrogenesis, and inflammation), epithelial cell proliferation, and lymphocytic apoptosis. At least in vitro, TGF-β causes hematopoietic stem cells to proliferate; however, this is apparently not the case in vivo, although the discrepancy remains unexplained (45).

Most of TGF-β null mice die in utero. A few survive until birth, but they soon develop severe inflammation that is mainly expressed in the cardiovascular system, which causes death. Not much is known about the mechanisms that trigger the process (29). The inflammatory process in the lung causes fibrogenesis in the extracellular matrix associated with a rise in collagen synthesis, decreased collagenase as well as other proteinases and fibroblast proliferation. There is at the same time apoptosis of the alveolar epithelium (47).

The TGF-β signaling pathway involves several steps that are summarized below:

- The ligand binds to the heterotetrameric complex.
- The constitutive type II serine threonine phosphokinase transphorylates the type I receptors on their GS box and activates the receptor complex.
- The ligand receptor complex is engulfed in the early endosome and is anchored by the intermediate of the FYVE domain of SARA to the bilipid membrane.
- Two RSMADs are recruited, the construct "ligand-receptor-SARA-SMAD3-SMAD4" brings about a conformational change in SMAD3 that guides it to its proper position on the type I receptor, thereby allowing its C terminal to be phosphorylated.
- The SMAD3-SMAD4 heterodimers or SMAD3 homodimers are "released" and find their way to the nucleus, where they are believed to form an "adaptor platform" that recruits chromatin modulators, among them histone deacylases (HDAC), transcriptor factors, coactivators (CBP/p300), corepressors (c-Ski), and E3 ubiquitin ligases SNURF (**FIGURE 17-2**).

TGF-β is abundant in reproductive organs and operates in all phases of reproduction: spermatogenesis, ovarian function, embryonal implantation, immunoregulation of pregnancy, placental development, morphogenesis during embryonic life and puberty, and secondary sex organ development. Therefore, it is likely that TGF-β modulates cell growth differentiation and apoptosis (48). The most characteristic cellular functions of TGF-β at the cellular level include:

- Cell cycle arrest in epithelial cells and hematopoietic cells.
- Cell proliferation and differentiation of mesenchymal cells (e.g., in wound healing)
- Extracellular matrix production.
- Immunosuppression (21).

## Cytostatic Effects of TGF-β

This section focuses on two major and most extensively investigated effects of TGF-β: cell cycle arrest by *p15^Ink4b^* and c-*myc* repression. The function of TGF-β in normal epithelia and their cancer (skin, lung, breast, etc.) has been a major target of investigation. Again, on the surface, the role of TGF-β is paradoxical. It arrests cell proliferation in normal epithelial cells, and growth is stimulated in cancer cells.

The arrest of cell proliferation results from transcription transactivation of genes that arrest the cell cycle and transcription repression of genes that encode growth factors. Consider the event at the $G_1$/S checkpoint. The TGF-β signal is transmitted to the SMAD complex associated with the appropriate coactivators, most of which are still unknown. The promoters of the genes that encode P15 (an inhibitor of Cdk2/Cdk6 and of P21, an unspecific inhibitor of cdks) are activated, Rb remains hypophosphorylated and the cycle is arrested.

TGF-β also represses c-*myc*. A complex of at least four proteins is formed in the cytoplasm. SMAD3, E2F4/E2F5, and the corepressor P107 form a complex in the cytoplasm that translocates to the nucleus where it is joined by SMAD4. The construct binds to a specific site on c-*myc* and down-regulates it.

The function of the $p15^{Ink4b}$ gene is modulated in both $G_1$ arrest and c-*myc* repression. In the first instance it is activated; in the second it is repressed because the repression of c-*myc* causes the repression of the $p15^{Ink4b}$ gene. It is believed that

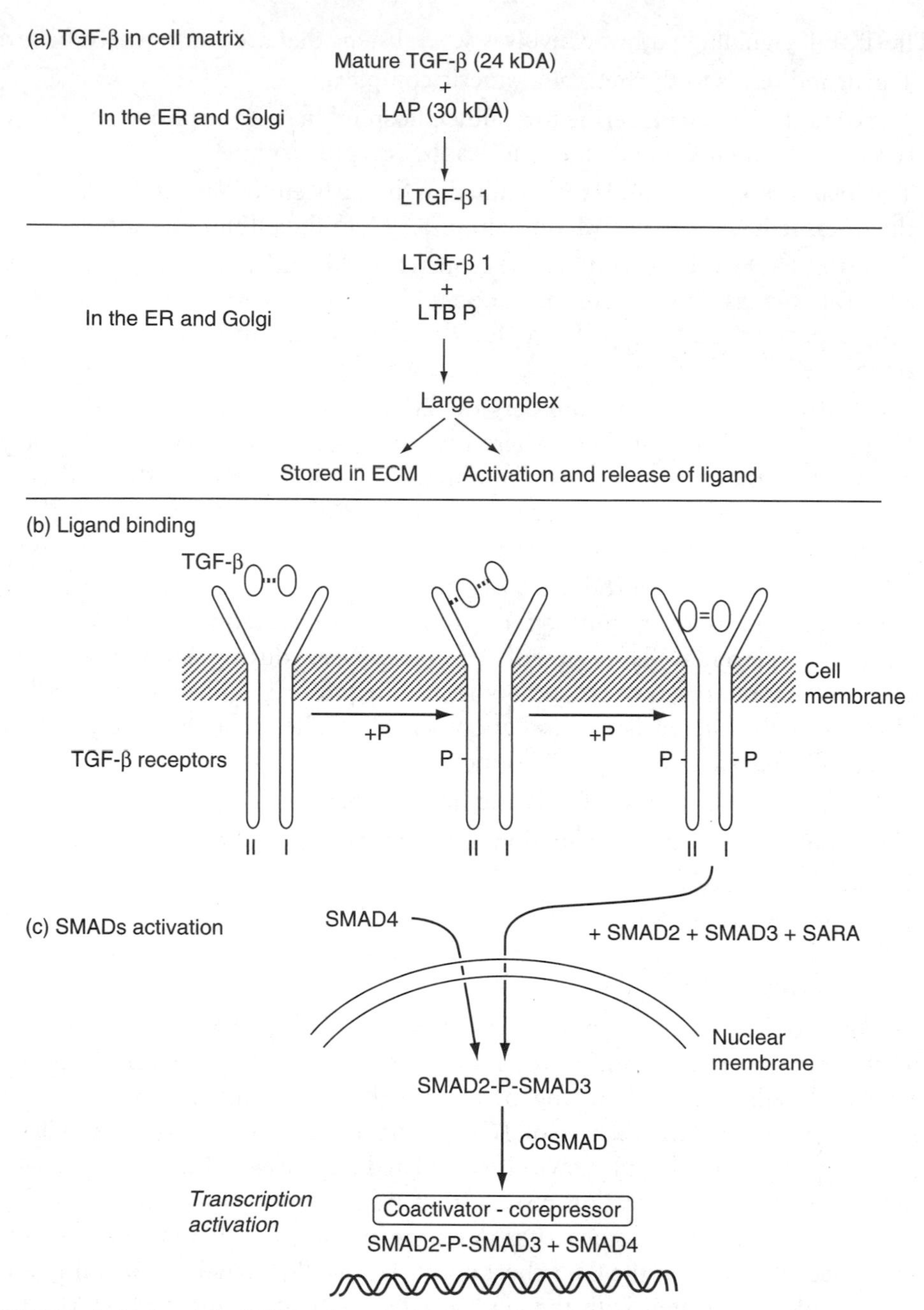


**FIGURE 17-2** Activation of the SMAD pathway by transforming growth factor-β (TGF-β). (a) Activation of the ligand (see p. 268). The mature TGF-β binds to LAP to yield the LTGF-β complex, which is activated and bound to the receptor or stored into a larger complex LTBP and released when needed. (b) Activation of the receptor (second mode, see p. 269). Upon ligand binding to the receptor, the constitutively activated type-II receptor activates the type-I receptor through phosphorylation. (c) Activation of the SMAD pathway (see p. 270). The phosphokinase of the type-I receptor activates the RSMADS, which recruit the CoSMD or SMAD4 and the nuclear SMADs 2 and 3, and the transcription of the mRNA of functional proteins is activated. This last step is modulated by several proteins discussed in the text.

the cofactors involved determine whether the gene is activated or inhibited. However, the molecular mechanisms that lead TGF-β to choose a specific target among a constellation of choices remains, in most cases, mostly elusive. Two mechanisms have been uncovered:

- A repertoire of associated DNA binding proteins (coactivators and corepressors) that combine with SMADs to form complexes able to target specific genes with high affinity.
- Regulatory assistance of other transduction pathways through cross-talk (49, 50).

A family of Ski proto-oncoproteins was discovered (among the members are c-Ski and c-SnoN [Ski-related novel protein N]). When these proteins are overexpressed they induce transformation, and they also have been associated with some types of cancer, for example, melanoma. After activation of TGF-β receptors, ligand binding, and the activation of SMADs that follows, Ski and SnoN form complexes with SMAD4 and the RSMADs SMAD2 and SMAD3, and ultimately interrupt the TGF-β signal. Shi and associates sought to understand the mechanism of inhibition of SMADs by Ski by examining the Ski-MH2 complex in crystal form (51). They discovered that the interaction involves two loops: the L3 of MH2, and a loop in the Ski molecule (the I loop). Moreover, the binding surface of SMAD4 on Ski overlaps with the binding site of RSMAD, thus excluding the RSMAD binding site, and thereby interfering with the progress of the TGF-β signal (51).

The cytostatic function of TGF-β represents only a small component of its total potential of gene responses as was shown by microarrays of a single cell type, namely, epithelial cells. The gene expression repertoire of these cells includes that of members of the Id family of genes. The function of the *Id* family is reminiscent of that of *myc;* they encode transcription regulators that interfere with cell differentiation and stimulate cell proliferation including angiogenesis. Inhibition of *Id* expression arrests cell proliferation. The *Id* gene family is composed of four genes: *Id1, Id2, Id3,* and *Id4* (*Id4* is found only in the brain). These genes are TGF-β and bone morphogenic protein (BMP) responsive. While TGF-β down-regulates, BMP up-regulates the gene's expression (52). These opposite functions are complementary during development. The down-regulation of *Id1* by TGF-β is independent of that of *myc,* which is also down-regulated by TGF-β but much sooner than the *Id* gene.

TGF-β also activates the *ATF3* gene, which encodes a member of the ATF/CREG bZip family of transcription factors (53). ATF3 can function as a transcriptional repressor and is the only member of the family activated by TGF-β. It is consistently found in epithelial cell lines. Could ATF3 be the agent modulating the different responses of *Id* to TGF-β and BMP through signal specificity?

The evidence suggests that:

1. TGF-β increases the levels of transcription of ATF3, but BMP does not.
2. ATF3 is rapidly induced by SMAD3.
3. ATF3 associates with SMAD3, but not with SMAD.
4. The ATF3-SMAD3 complex targets the *Id* gene for repression.

Thus, the transduction pathway of TGF-β differs from that of BPM by its induction of the SMAD3 and by the induction of ATF3, which represses the very gene that was induced by TGF-β. This phenomenon is referred to as a "self-enabling device" in which the primary event generates its own repressor machinery.

Stress (e.g., radiation such as UV and ionizing, heat shock, osmotic shock, TNF-α) triggers the activation of stress responsive kinases (MAP, JNK, and P38), and these in turn induce ATF3 (54). The induction of ATF3 is transient in the absence of TGF-β, but is sustained by the SMAD3–ATF3 association that leads to *Id* repression. Thus, both stress and TGF-β contribute to the generation of ATF3 crosstalk between the

TGF-β–inducing pathway and other inducing pathways, such as MAP, Erk, JNK, P38. The Erk/Map kinase enhances SMAD signals in human mesenchymal cells (55). TGF-β participates in epithelial cell survival through Akt dependent regulation of the Forkhead factor (FKHRL1) (56).

In conclusion, although the cellular responses to TGF-β are multiple and sometimes conflicting, the persistent research of the last two decades has provided an embryonic vision of what is in store for the future. The target cells requirement for the TGF-β ligand is signaled possibly by titration of the isoforms (TGF-β1, TGF-β2, and TGF-β3) messenger RNAs. A new isoform is secreted in the extracellular space where the TGF-β does not immediately enter into action, but is maintained in a latent form before activation. Once activated, the ligand engages the receptor, activates it, and ultimately puts in gear the SMAD family of transducing molecules, which activate transcription factors. The targeting and activation of the appropriate promoters is carefully regulated by various batteries of coactivators and corepressors. The combination of these multiple interacting molecules determines the ultimate nature of the cellular response (48).

## TGF-β and Cancer

TGF-β is a master conductor. It carefully chooses the place and time for each player's moves, but when TGF-β fails, the organ or even the entire organism suffers. Therefore, it is not surprising that TGF-β is also associated with several diseases, including hereditary diseases and cancer (57). This section focuses on cancer.

TGF-β regulates cell proliferation and apoptosis in many cell types, in particular in epithelial, endothelial, and hematopoietic cells. A major function of TGF-β is cell cycle arrest in the $G_1$ phase of the cell cycle. The arrest may involve P15 in a P53-independent way or P21 in a P53-dependent way. The inhibitors (P15 or P21) inhibit the cyclin dependent kinases, and the cyclins A and E, thereby maintaining E2Fs sequestered and unable to activate gene expression, for example, that of c-*myc*.

Cancer may be associated with mutation of critical molecules operating at different levels of the TGF-β pathway: the ligand, the receptor, the SMAD activation, and the modulators of SMAD. The role of TGF-β in cancer has been investigated in mice homozygous and heterozygous for the TGF-β allele. TGF-$\beta^{-/-}$ mice die in utero. In contrast, TGF-$\beta^{+/-}$ mice present a normal phenotype at birth, but have an increased cell turnover in liver and lung, and develop cancers in these organs at a relatively early age. Target deletions of SMAD2 cause early death in utero, while SMAD3 knockout mice survive embryonic life but develop severe immunological defects mainly in the oral mucosa, expressed in the form of multiple abscesses filled with non-pathogenic bacteria in the uncompromised host (58).

In mice and humans, TGF-β can also affect cancer progression in a somewhat indirect fashion, in a biphasic mode. In the early stages, TGF-β acts as a tumor suppressor and may cooperate with P53 to arrest the cell cycle and perhaps cause apoptosis. In the late stage, the cancer cell refuses to respond adequately to TGF-β and cells respond by elaborating more and more TGF-β, which in turn facilitates progression of the cancer by stimulating matrix component production, angiogenesis, and cancer cell proliferation. These findings led to targeting TGF-β for therapy.

Human cancers are associated with several tumor suppressor gene mutations (*TGFβRII*, *SMADS*, *Ski*, *Sno*). One hundred percent of pancreatic cancers and 83% of colon cancer have mutations in one or more components of the TGF-β transduction pathway. Earlier sections of this chapter discussed how deficiencies in the TGF-β function can contribute directly or indirectly to cancer. Mutations that cause loss of the TGFβRII receptor function or expression are associated with epithelial cancers in the colorectum (30%), the stomach 15%, and in endometrial cervix, prostate, breast, liver, pancreas, brain (glioma), and organs of the head and neck. In some cases, the mutations target a 10 consecutive adenine nucleotide repeat on the sequence in the coding region of the receptor gene that is translated into a truncated ineffective receptor. Deletion of the Type II receptor is associated with aggressive cancers of the stomach and T cell lymphoma. Defective type I receptors are associated with 16% of cancer of the breast and several other epithelial cancers: pancreas, bile duct or gallbladder, cervix, and chronic lymphocytic leukemia. The *smad-2* gene is located on chromosome 18p21.1 (~3 MB centromeric to *smad-4*). The gene is composed of 11 exons that encode 467 amino acids. The gene is ubiquitous and mutations are rare. Mutations of SMAD2 are linked to 11% of colorectal cancers and 7% of lung cancers and hepatocarcinoma.

*smad-3* is located on chromosome 15q21-22; it is made of 9 exons and encodes 424 amino acids. No mutations of Smad3 have been reported to date. However, SMAD3 binding to DNA is inhibited by a zinc finger oncoprotein (Evi-1) that interacts with SMAD3 (59).

Mutations of the *smad-4* gene mapped to the chromosome 18 (18q21.1) locus of DPC4 (deleted in pancreatic cancer) emphasized the significance of *smad-4* mutations in gastrointestinal cancer, in particular juvenile polyposis, colorectal cancers, and pancreatic cancers. The *smad-4* gene sequence encompasses 11 exons and encodes a 552 a.a. protein.

Familial juvenile polyposis is an autososmal dominant syndrome. In 1998, Howe and associates detected a germ line mutation on chromosome 18q21-1(FJP), a deletion of codon 414–416 between nucleotides 1372 and 1375. The deletion causes a frameshift mutation that leads to the appearance of a new stop codon at the end of exon 9 of the *smad-4* gene (60). After this seminal observation more mutations were associated with FJP. To date they mainly involve two genes, *smad-4* in and *bmpra-1A* (20% of the cases for each of the two genes). BMPR 1A (the product of the *bmpra-1A* gene) is a member of the TGF-β receptor family. It functions as the bone morphogenetic receptor type 1. The patients who harbor mutations of either of these two genes are at greater risk for gastrointestional polyposis (61).

These observations indicate that *smad-4* functions as a tumor suppressor gene for juvenile polyposis. The molecular mechanism by which the deletions in *smad-4* causes juvenile polyposis is unknown, but it is possible that it allows cells to shunt the antiproliferative function of TGF-β (62).

Somatic mutations of *smad-4* were also found in colorectal cancer, as part of a cascade mutations that involves serveral other genes (see Chapter 24).

The *smad-4* gene is inactivated in approximately 55% of pancreatic cancers. *LOH* at chromosomes 1p, 9p, 17p, 18q (70, 80, 90, and 90%, respectively) were reported in 1996 by Hahn and associates (63). The homozygous deletion of *smad-4* at 18q21 (DPC4) mutations that lead to inactivation of the other allele were discovered later. The latter include nonsense, missense, and frameshift mutations, Mostly located in the *smad-4* gene (90% in the MH2 domain). Inactivation of *smad-4* gene correlates (94%) with immunochemical absence of the protein, but it is only absent in high grade lesions. The progression of the genetic alterations in pancreatic cancer is reminiscent of the pattern seen in colon cancer, in this order: *Kras*, *p16*, *p53*, and *18q21*.

Homozygous deletions have seldom if ever been found in other cancers except cancer of the breast; however, the frequency of intragenic mutation varies in biliary CA 16%, hepatocellular 5%, gastric 3%, head and neck 3%, lung 7%, and acute myelogenous leukemia 17% (11, 64).

In summary, immunohistochemical staining mirrors the inactivation of the *smad-4* gene expression. *smad-4* occurs only late in a few tumors and cancers (juvenile polyposis [JP], cancers of colon and pancreas) and rarely in other cancers, with the exception of breast cancer and myelogenous leukemia.

Little is known of *smad-6,* which is located at chromosome 15q21-22 (the same region as *smad-3*). SMAD7 inhibits TGF-β binding to TGF-βRI and abrogates phosphorylation of SMAD2. Truncated SMAD7 cancels the inhibition of ligand binding.

## Conclusion

TGF-β has been with us for many years and it does many things. It is present in almost every cell from birth to death. During development it emerges as a growth stimulator, and later it controls homeostasis and functions primarily as a growth suppressor. Despite this multitude of functions, it uses a relatively simple transducing pathway. The ligand binds to at least two surface receptors, TGF-βRII and TGF-βRI; TGF-βRII phosphorylates a phosphokinase in TGF-βRI, which in turn calls a small class of molecules, the SMAD family, into play. Some are regulators, some act in cooperation with the regulators, and at least one serves as an inhibitor. In the cytoplasm, SARA guides SMAD2-SMAD3 from cytoplasm to nucleus. In the nucleus, the two SMAD (SMAD1 and SMAD2) join the coactivator SMAD4 and form complexes with cell-specific DNA binding molecules, some of which function as gene activators and others as gene repressors. Harmonious panoply of genes are expressed or repressed, thereby generating the appropriate biological effect. The inhibitor SMAD, once it is engaged, can modulate the process by either blocking transduction or by being removed. This is obviously an oversimplified view of a complex process, and the more that is learned about it, the more characters appear, including a long list of traditional transducing agents (MAKK, RHO, DAXX, etc.), and some still poorly known newcomers like SNURF.

## References and Recommended Reading


1. Fidler, I.J., and Ellis, L.M. The implications of angiogenesis for the biology and therapy of cancer metastasis. *EMBO J.* 1994;16:5775–8.
2. Kim, K.J., Li, B., Winer, J., Armanini, M., Gillett, N., Phillips, H.S., and Ferrara, N. Inhibition of vascular endothelial growth factor-induced angiogenesis suppresses tumour growth in vivo. *Nature.* 1993;362:841–4.
3. Bellusci, S., Moens, G., Gaudino, G., Comoglio, P., Nakamura, T., Thiery, J.P., and Jouanneau, J. Creation of an hepatocyte growth factor/scatter factor autocrine loop in carcinoma cells induces invasive properties associated with increased tumorigenicity. *Oncogene.* 1994;9:1091–9.
4. Birchmeier, W., Brinkman, V., Neimann, C., Meiners, S., DiCesare, S., Naundorf, H., and Sachs, M. Role of HGF/SF and c-Met in morphogenesis and metastasis of epithelial cells. *Ciba Found Symp.* 1997;212:230–40; discussion 240–6.

5. Ebert, M., Yokoyama, M., Friess, H., Buchler, M.W., and Korc, M. Coexpression of the c-met proto-oncogene and hepatocyte growth factor in human pancreatic cancer. *Cancer Res.* 1994;54:5775–8.
6. Reiss, M. Transforming growth factor-β and cancer. In: Gressner, A.M., Heinrich, P.C., and Matern, S., eds. *Cytokines in Liver Injury and Repair.* New York, NY: Kluwer Academic Publishers; 2002:73–104.
7. Twardzik, D.R., Brown, J.P., Ranchalis, J.E., Todaro, G.J., and Moss, B. Vaccinia virus-infected cells release a novel polypeptide functionally related to transforming and epidermal growth factors. *Proc Natl Acad Sci USA.* 1985;82:5300–4.
8. Sporn, M.B., and Roberts, A.B. Autocrine growth factors and cancer. *Nature.* 1985;313:745–7.
9. Sandgren, E.P., Luetteke, N.C. Palmiter, R.D., Brinster, R.L., and Lee, D.C. Overexpression of TGF alpha in transgenic mice: induction of epithelial hyperplasia, pancreatic metaphasia, and carcinoma of the breast. *Cell.* 1990;61:1121–35.
10. Jhappan, C., Stahle, C., Harkins, R.N., Fausto, N. Smith, G.H., and Merlino, G.T. TGF alpha overexpression in transgenic mice induces liver neoplasia and abnormal development of the mammary gland and pancreas. *Cell.* 1990;61:1137–46.
11. Harris, T.M., Rogler, L.E., and Rogler, C.E. Reactivation of the maternally imprinted IGF2 allele in TGFalpha induced hepatocellular carcinoma in mice. *Oncogene.* 1998;16:203–9.
12. Nalesnik, M.A., Lee, R.G., and Carr, B.I. Transforming growth factor alpha (TGFalpha) in hepatocellular carcinomas and adjacent hepatic parenchyma. *Hum Pathol.* 1998;29:228–34.
13. Santoni-Rugiu, E., Jensen, M.R., Factor, V.M., and Thorgeirsson, S.S. Acceleration of c-myc-induced hepatocarcinogenesis by Co-expression of transforming growth factor (TGF)-alpha in transgenic mice is associated with TGF-beta 1 signaling disruption. *Am J Pathol.* 1999;154:1693–1700.
14. Humphreys, R.C., and Hennighausen, L. Transforming growth factor alpha and mouse models of human breast cancer. *Oncogene.* 2000;19:1085–91.
15. Roberts, A.B., Anzano, M.A., Lamb, L.C., Smith, J.M., and Sporn, M.B. New class of transforming growth factors potentiated by epidermal growth factor: isolation from non-neoplastic tissues. *Proc Natl Acad Sci USA.* 1981;78:5339–43.
16. Blobe, G.C., Schiemann, W.P., and Lodish, H.F. Role of transforming growth factor beta in human disease. *N Engl J Med.* 2000;342:1350–8.
17. Assoian, R.K., Grotendorst, G.R., Miller, D.M., and Sporn, M.B. Cellular transformation by coordinated action of three peptide growth factors from human platelets. *Nature.* 1984;309:804–8.
18. Keski-Oja, J., Leof, E.B., Lyons, R.M., Coffey, R.J., and Moses, H.L. Transforming growth factors and control of neoplastic cell growth. *J Cell Biochem.* 1987; 33:95–107.
19. Gleizes, P.E., Beavis, R.C., Mazzieri, R., Shen, B., and Rifkin, D.B. Identification and characterization of an eight-cysteine repeat of the latent transforming growth factor-beta binding protein-1 that mediates bonding to the latent transforming growth factor-beta 1. *J Biol Chem.* 1996;271:29891–6.
20. Govinden, R., and Bhoola, K.D. Genealogy, expression, and cellular function of transforming growth factor-beta. *Pharm Ther.* 2003;98:257–65.

21. Massague, J. TGF-beta signal transduction. *Ann Rev Biochem.* 1998;67:753–91.
22. Kulkarni, A.B., Huh, C.G., Becker, D., Geiser, A., Lyght, M., Flanders, K.C., Roberts, A.B., Sporn, M.B., Ward, J.M., and Karlsson, S. Transforming growth factor beta 1 null mutation in mice causes excessive inflammatory response and early death. *Proc Natl Acad Sci USA.* 1993;90:770–4.
23. Dickson, M.C., Martin J.S., Cousins, F.M., Kulkarni, A.B., Karlsson, S., and Akhurst, R.J. Defective haematopoiesis and vasculogenesis in transforming growth factor-beta 1 knock out mice. *Development.* 1995;6:1845–54.
24. Letterio, J.J., Geiser, A.G., Kulkarni, A.B., Roche, N.S., Sporn, M.B., and Roberts, A.B. Maternal rescue of transforming growth factor-beta 1 null mice. *Science.* 1994;264:1936.
25. Sanford, L.P., Ormsby, I., Gittenberger-de Groot, A.C., Sariola, H., Friedman, R., Boivin, G.P., Cardell, E.L., and Doetschman, T. TGFbeta2 knockout mice have multiple developmental defects that are non-overlapping with other TGFbeta knockout phenotypes. *Development.* 1997;124:2659–70.
26. Kaartinen, V., Voncken, J.W., Shuler, C., Warburton, D., Bu, D., Heisterkamp, N., and Groffen, J. Abnormal lung development and cleft palate in mice lacking TBF-beta 3 indicates defects of epithelial-mesenchymal interaction. *Nat Genet.* 1995; 11:415–21.
27. Feng, X.H., and Derynck, R. A kinase subdomain of transforming growth factor-beta (TGF-bcta) type I receptor determines the TGF-beta intracellular signaling specificity. *EMBO J.* 1997;16:3912–23.
28. Chen, Y.G., Hata, A., Lo, R.S., Wotton, D., Shi, Y., Pavletich, N., and Massague, J. Determinants of specificity in TGF-beta signal transduction. *Genes Dev.* 1998; 12:2144–52.
29. Attisano, L.W.J. Signal transduction by the TGF-beta superfamily. *Science.* 2002;296:1646.
30. Massague, J., and Chen, Y.G. Controlling TGF-beta signaling. *Genes Dev.* 2000;14:627–44.
31. Wu, G.C., Chen, Y.G., Ozdamar, B., Gyuricza, C.A., Chong, P.A., Wrana, J.L., Massague, J., and Shi, Y. Structural basis of Smad2 recognition by the Smad anchor for receptor activation. *Science.* 2000;287:92–7.
32. Huse, M., Muir, T.W., Xu, L., Chen, Y.G., Kuriyan, J., and Massague, J. The TGF beta receptor activation process: an inhibitor- to substrate-binding switch. *Mol Cell.* 2001;8:671–82.
33. Tsukazaki, T., Chiang, T.A., Davison, A.F., Attisano, L., and Wrana, J.L. SARA, a FYVE domain protein that recruits Smad2 to the TGFbeta receptor. *Cell.* 1998;95:779–91.
34. Hayes, S., Chawla, A., and Corvera, S. TGF beta receptor internalization into EEA1-enriched early endosomes: role in signaling to Smad2. *J Cell Biol.* 2002;158: 1239–49.
35. Itoh, F., Divecha, N., Brocks, L., Oomen, L., Janssen, H., Calafat, J., Itoh, S., and Dijke, P.P. The FYVE domain in Smad anchor for receptor activation (SARA) is sufficient for localization of SARA in early endosomes and regulates TGF-beta/Smad signaling. *Genes Cells.* 2002;7:321–31.

**36.** DiGuglielmo, G.M., LeRoy, C., Goodfellow, A.F., and Wrana, J.L. Distinct endocytic pathways regulate TGF-beta receptor signaling and turnover. *Nat Cell Biol.* 2003;5:410–21.

**37.** Moilanen, A.M., Poukka, H., Karvonen, U., Hakli, M., Janne, O.A., and Palvimo, J.J. Identification of a novel RING finger protein as a coregulator in steroid receptor-mediated gene transcription. *Mol Cell Biol.* 1998;18:5128–39.

**38.** Pero, R., Lembo, F., Chieffi, P., Del Pozzo, G., Fedele, M., Fusco, A., Bruni, C.B., and Chiariotti, L. Translational regulation of a novel testis-specific RNF4 transcript. *Mol Reprod Dev.* 2003;66:1–7.

**39.** Lyngso, C., Bouteiller G., Damgaard, C.K., Ryom, D., Sanchez-Munoz, S., Norby, P.L., Bonven, B.J., and Jorgensen, P. Interaction between the transcription factor SPBP and the positive cofactor RNF4. An interplay between protein binding zinc fingers. *J Biol Chem.* 2000;275:26144–9.

**40.** Hakli, M., Karvonen, U., Janne, O.A., and Palvimo, J.J. The RING finger protein SNURF is a bifunctional protein possessing DNA binding activity. *J Biol Chem.* 2001;276:23653–61.

**41.** Saville, B., Poukka, H., Wormke, M., Janne, O.A., Palvimo, J.J., Stoner, M., Samudio, I., and Safe, S. Cooperative coactivation of estrogen receptor alpha in ZR-75 human breast cancer cells by SNURF and TATA-binding protein. *J Biol Chem.* 2002;277:2485–97

**42.** Hakli, M., Lorick, K.L., Weissman, A.M., Janne, Q.A., and Palvimo, J.J. Transcriptional coregulator SNURF (RNF4) possesses ubiquitin E3 ligase activity. *FEBS Lett.* 2004;560:56–62.

**43.** Lo, R.S., Chen, Y.G., Shi, Y., Pavletich, N.P., and Massague, J. The L3 loop: a structural motif determining specific interactions between SMAD proteins and TGF-beta receptors. *EMBO J.* 1998;17:996–1005.

**44.** Chai, J., Wu, W.J., Yan, N., Massague, J., Pavletich, N.P., and Shi, Y. Features of a Smad3 MH1-DNA complex. Roles of water and zinc in DNA binding. *J Biol Chem.* 2003;278:20327–31.

**45.** Larson, J., Blank, U., Helgadottir, H., Björnsson, J.M., Ehinger, M., Goumans, M-J., Fan, X., Levéen, P., and Karlsson, S. TGF-β signaling-deficient hematopoietic stem cells have normal self-renewal and regenerative ability in vivo despite increased proliferative capacity in vitro. *Blood.* 2003;102:3129–35.

**46.** Xiao, Z., Watson, N., Rodriguez, C., and Lodish, H.F. Nucleocytoplasmic shuttling of Smad1 conferred by its nuclear localization and nuclear export signals. *J Biol Chem.* 2001;276:39404–10. (E-pub ahead of print).

**47.** Hodge, S.J., Hodge, G.L., Reynolds, P.N., Scicchitano, R., and Holmes, M. Increased production of TGF-beta and apoptosis of T lymphocytes isolated from peripheral blood in COPD. *Am J Phys Lung Cell Mol Phys.* 2003;285:L492–9.

**48.** Ingman, W.V., and Robertson, S.A. (2002) Defining the actions of transforming growth factor beta in reproduction. (Review). *Bioessays.* 2002;24:904–14.

**49.** Ten Dijke, P., Miyazono, K., and Heldin, C.H. Signaling inputs converge on nuclear effectors in TGF-beta signaling. *Trends Biochem Sci.* 2000;25:64–70.

**50.** Shi, Y., and Massague, J. Mechanism of TGFβ signaling from cell membrane to nucleus. *Cell.* 2003;113:685–700.

51. Wu, J.W., Krawitz, A.R., Chai, J., Li, W., Zhang, F., Luo, K., and Shi, Y. Structural mechanism of Smad4 recognition by the nuclear oncoprotein Ski: insights on Ski-mediated repression of TGF-beta signaling. *Cell.* 2002;111:357–67.
52. Korchynskyi, O., ten Dijke, P. Identification and functional characterization of distinct critically important bone morphogenetic protein-specific response elements in the Id1 promoter. *J Biol Chem.* 2002;277:4883–91.
53. Maekawa, T., Sakura, H., Kanei-Ishii, C., Sudo, T., Yoshimura, T., Fujisawa, J., Yoshida, M., and Ishii, S. Leucine zipper structure of the protein CRE-BP1 binding to the cyclic AMP response element in brain. *EMBO J.* 1989;8:2023–8.
54. Kang, Y., Chen, J., B.A., and Massague, J. A self-enabling TGFbeta response coupled to stress signaling: Smad engages stress response factor ATF3 for Id1 repression in epithelial cells. *Mol Cell.* 2003;11:915–26.
55. Haÿashida, T., deCaestecker, M., and Schnaper, H.W. Cross-talk between ERK MAP kinase and Smad-signaling pathways enhances TGF-β dependent responses in human mesangial cells. *FASEB J.* 2003;17:1576–8. (Epub ahead of print).
56. Shin, I., Bakin, A.V., Rodeck, U., Brunet, A., and Arteaga, C.L. Transforming growth factor beta enhances epithelial cell survival via Akt-dependent regulation of FKHRL1. *Mol Biol Cell.* 2001;12:3328–39.
57. Ingman, W.V., and Robertson S.A. Defining the actions of transforming growth factor beta in reproduction. *Bioessays.* 2002;24:904–14.
58. Cohen, M. TGF beta/Smad signaling system and its pathologic correlates. *Am J Med Genet.* 2003;116A:1–10.
59. Kurokawa, M., Mitani, K., Irie, K., Matsuyama, T., Takahashi, T., Chiba, S., Yazaki, Y., Matsumoto, K., and Hirai, H. The oncoprotein Evi-1 represses TGF-beta signaling by inhibiting Smad3. *Nature.* 1998;394:92–6.
60. Howe, J.R., Roth, S., Ringold, J.C., Summers, R.W., Jarvinen, H.I., Sistonen, P., Tomlinson, I.P., Houlston, R.S., Bevan, S., Mitros, F.A., Stone, E.M., and Aaltonen, L.A. Mutations in the SMAD4/DPC4 gene in juvenile polyposis. *Science.* 1998; 280:1086–8.
61. Howe, J.R., Sayed, M.G., Ahmed, A.F., Ringold, J., Larsen-Haidle, J., Merg, A., Mitros, F.A., Vaccaro, C.A., Peterson, G.M., Giardiello, F.M., Tinley, S.T., Aaltonen, L.A., and Lynch, H.T. The prevalence of MADH4 and BMPR1A mutations in juvenile polyposis and absence of BMPR2, BMPR1B, and ACVR1 mutations. *J Med Genet.* 2004;41:484-91.
62. Merg, A., and Howe, J.R., Genetic conditions associated with intestinal juvenile polyps. (Review). *Am J Med Genet C Semin Med Genet.* 2004;129:44–55.
63. Hahn, S.A., Schutte, M., Hoque, A.T.M.S., Moskaluk, C.A., da Costa L.T., Rozenblum, E., Weinstein, C.L., Fischer, A., Yeo, C.J., Hruban, R.H., and Kern, S.E. *DPC4*, a candidate tumor-suppressor gene at human chromosome 18q21.1. *Science* 1996;271:350–3.
64. Siegel, P.M., and Massague, J. Cytostatic and apoptotic actions of TGF-β in homeostasis and cancer. *Nat Rev Cancer.* 2003;3:801–20.

CHAPTER

# PROTEIN TYROSINE KINASE RECEPTORS AND CANCER

# 18

Growth factors constitute a family of polypeptides that either stimulate or inhibit growth by interacting with receptors. Over 50 such receptors have been identified. Two major classes of receptors exist: protein tyrosine kinase receptors (PTKRs) and cytokine receptors. Each class is divided into subfamilies. This chapter focuses on the tyrosine kinase receptors.

The three common denominators for all the members of this class are:

- An aminoterminal extracellular component
- An intracellular catalytic component, the tyrosine kinase function
- An intramembrane link.

Although intracellular tyrosine kinase is the most constant component of the PTKRs, the extracellular component varies. The main function of the extracellular segment of the receptor is to access the ligand and modulate its conformation in a fashion that allows the faithful transfer of the extracellular signal(s) to the transducing pathway, the first link of which is the tyrosine kinase. Therefore, it can be anticipated that the external component of the receptor will vary with the type of cognate ligand. The connection between ligand and receptor must be without fault to avoid an ineffective or exaggerated response. In the case of growth factor receptors, defaults imply arrest or uninterrupted cell growth (1).

Dimerization of receptors after ligand binding is a common feature of the PTKR. However, if the ligand is a monomer (e.g., EGF), it brings the two receptors together and triggers transphosphorylation. Autophosphorylation, in turn, stabilizes the dimers and provides binding sites for downstream transducing agents. If the ligand is a dimer (e.g., PDGF), it carries two binding sites, one for each receptor, and brings the receptors together with high affinity (i.e., PDGF contains 3 autophosphorylation sites). Several special domains have been associated with the extracellular composition of the PTKRs. They include:

- The cysteine-rich domain (CRD)
- The fibronectin type III domain (FNIII)

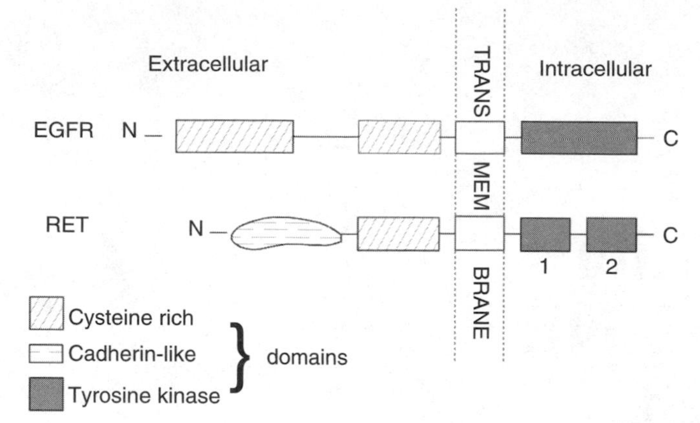

**FIGURE 18-1** Extracellular and intracellular composition of protein tyrosine kinase receptors (PTKRs).

- The IgD immunoglobulin-like domain (IgD)
- The epidermal growth factor domain (EGFD)
- The cadherin-like domain (CadhD)
- The Kringle-like domain (KrinD)
- The discoidin-like domain

In addition, some PTKRs may harbor the following in the extracellular component (**FIGURE 18-1**):

- An acidic Box (AB)
- A leucine-rich domain (LDR)

## PTKR as Oncogenes

When engaged by ligands, protein kinase activated receptors transmit their message through protein tyrosine kinases. Mutations of the receptor often lead to oncogenesis. Several genes linked to cancer are derived from growth factor/protein kinase receptors through gene point mutation, rearrangement, or amplification (**TABLE 18-1**). Receptor dimerization often leads to the acquisition of oncogenic properties. For example, a single point mutation in the transmembrane domain of Neu/Her 2 triggers ligand-independent dimerization of the receptor and constitutive activation of the tyrosine kinase. Such disturbances of the tyrosine kinase function by single mutations are not unique

**Table 18-1 Exon, codon, and protein domains affected by mutation in familial medullary thyroid carcinoma (FTMC) and in MEN2A and MEN2B**

| | Exon | Codon | Domain |
|---|---|---|---|
| FTMC | 13 | 768 | PTK 1 |
| | 14 | 804 | 2 PTK 2 |
| MEN2A | 10 | 609, 611 | Cysteine rich |
| | | 618, 620 | |
| | 11 | 634 | Transmembrane |
| MEN2B | 16 | 918 | PTK 2 |

to NEU/Her 2. The loss of a single cysteine in RET enhances receptor dimerization and is associated with MEN2A (multiple endocrine neoplasia syndrome type 2A) (2).

Rearrangement relocates the PTKR sequence from one locus on the gene to another, thereby placing them in contact with new sequences. Most of such rearrangements involve proteins that trigger protein-protein interactions. As a result, they facilitate receptor dimerization and constitutive kinase activation. This is the case in the rearrangement 10q/21 of the *ret* gene and its fusion with TKR.

In addition, amplification of a PTKR or other proto-oncogene is frequently found in cancer. It obviously provides greater opportunities for dimerization of the receptors (and maybe multimerization) and constitutive activation of the kinase. Amplification of EGFR is associated with cancer of the skin (squamous cell carcinomas and gliomas).

The downstream signaling molecules involve homology 2 (SH2) domains; SH2 domains recognize short peptides with a specific sequence containing phosphotyrosine located within the activated receptor. Proteins involved in the recruitment fall into two categories: catalytic and non-catalytic. The catalytic include phospholipase Cγ (PLCγ), phosphatidylinositol-3-kinase (P13K), and guanosine 5′-triphosphatase (GTPase) activating proteins. The non-catalytic proteins are adaptor proteins. Adaptor proteins cooperate by binding with other proteins, for instance, enzymes and receptors to build a construct able to transmit a signal emerging at the cell surface to target molecules in the cytoplasm or in the nucleus.

## EGFR

The epidermal growth factor receptor (EGFR) is the prototype of protein tyrosine kinase receptors. It is the cellular counterpart, the transforming protein of the erythroblastosis virus (v-ERBB) (3), thus confirming the connection between protein tyrosine kinases and oncogenes. In humans the chromosomal locus of EGFR is 7p13-p12. Mutations, deletions, and amplification of the gene are often associated with human cancers, and the highest incidence of EGFR amplification occurs in glioblastoma (20–40%) (4, 5). EGFR amplification occurs at the rate of approximately 10 to 20% in a variety of other cancers of the lung, head and neck, esophagus, stomach, and breast.

More than 200 studies suggest that a dose-dependent relationship exists between the levels of EGFR expression in several cancers (head and neck, breasts, cervix and non small cell lung cancer), suggesting an association between poor prognosis and the detection of EGFR in the tumor. This is not surprising as, indeed, EGF and other ligands (e.g., TGF-α) that bind to the receptor, after triggering its activation and that of the downstream transduction pathway, stimulate the synthesis of vascular endothelial growth factor (VEGF) and fibroblast growth factor (FGF), two factors indispensable to vasculogenesis.

EGFR or its downstream pathway became therapeutic targets in phase 1, phase 2, and phase 3 trials (6). These studies are not all convincing, in part because of the small number of participants and the range of different assays.

## *ret* Proto-Oncogene

The *ret* proto-oncogene, located on the centromeric region of chromosome 10;10q 11.2 (7, 8), extends over 80 kilobases, possesses 21 exons, and encodes a receptor that spans the plasma membrane of cells derived from the embryonal crest. From the start, the receptor was assumed to function in the manner of a transmembrane protein with tyrosine kinase activity. The extracellular domain of the protein contains five cysteine residues. Sequences upstream from the transcription starting site (–168 to +98 base pairs) contain a promoter sequence with high GC content and four tandem GC boxes, but no TATA boxes.

The encoded protein is made of 114 amino acids and forms a transmembrane receptor composed of: a signal peptide, an extracellular ligand site, a cadherin-like binding domain, a cysteine rich region, a transmembrane region, and a conserved intracellular catalytic site (tyrosine kinases 1 and 2).

Ret is a member of the receptor tyrosine kinase family with which it shares the transmembrane components and the intracellular tyrosine kinase. Its extracellular component shares a cysteine-rich domain with EGFR and *v-erb*. However, the extracellular portion of RET does not contain immunoglobulin-like domains as in FGFR (1, 2, and 3), each of which has three such domains. PDGFR harbors five of them. A cadherin-like structure constitutes the most unique feature of the extracellular domain of RET, but the structure is not found in *ret*'s closest cousin, the *v-erb* (Figure 18-1). Although the exact function of the cadherin domain in *ret* is not yet clear, it is likely to be associated with the well known ability of that molecule to mediate homophilic cell to cell adhesion (9).

The gene is expressed during development of organs and tissues derived from the neural crest located at the origin of the central and peripheral neural system, including ganglion cells (sensory, autonomic, and enteric) and the excretory system. There is structural homology between the *ret* gene and several other proto-oncogenes (MET, C-KIT, and *c-erb*). Moreover, there is a similarity in the location of the mutations in those genes. *Met* mutations are associated with hereditary papillary renal carcinoma (10, 11).[1]

## Ret Mutations in Thyroid Carcinoma: MEN2A and MEN2B

Ret is a tyrosine kinase receptor that plays a role in development of endocrine, neurological and enteric cells. It was discovered by transfecting lymphoma and gastric cancer DNA. The result was a chimeric oncogene generated by rearrangement, hence the acronym ret for "receptor rearranged during transfection" (12). Three such rearrangements have been discovered and they are referred to as ret-PTC 1, 2, and 3. These rearrangements are not found in multiple endocrine neoplasia but are associated with papillary carcinoma of the thyroid. The rearrangement is a somatic event in which the kinase of the *ret* gene is juxtaposed with unrelated sequences that facilitate the dimerization of the receptor and thereby activate *ret* constitutively (13, 14).

Mutations of the *ret* gene that do not involve rearrangement have been detected in multiple endocrine neoplasia, MEN2A and MEN2B (15), and in medullary thyroid carcinoma (MTC). In MEN2A (familial MTC, pheochromocytoma, and hyperparathyroidism), the mutations cause amino acid substitutions mainly in the cysteine-rich domain of the extracellular component of the receptor. Six codons are affected: 609, 611, 618, 620, 630, and 634. The two codons most susceptible to substitution are codons 634 and 618. In patients with MEN2A who present with pheochromocytoma and hyperparathyroidism, it is by far the cysteine of codon 634 that is the most frequently substituted. The latter cysteine can be replaced by any amino acid that will accommo-

---

[1]C-KIT is a receptor tyrosine kinase expressed in bone marrow stem cells. c-kit mutations have been identified in mastocytosis and malignant hematological disease (10). C-erb is a proto-oncogene derived from *v-erb* that was first found in avian erythroblastosis virus. C-erb A encodes a thyroid receptor and *c-erb* encodes an epidermal growth factor (11).

date the codon, and these include tyrosine, tryptophan, serine, glycine, arginine, and phenylalanine. The most frequent substitutions result from the replacement of cysteine to arginine or cysteine to tyrosine (C634R, TGC → CGC; and C634Y, TGC → TAC). A close association between C634R and the incidence of parathyroid hyperplasia has been reported (16) suggesting that the pattern of the syndrome (2A versus 2B) may be influenced by both the amino acids that substitute for the cysteine and the codon involved (17). The correlation between C634R mutation and hyperparathyroidism has, however, been challenged (details in the excellent review by Eng [18]).

MEN2B (FMTC and pheochromocytoma) is less frequent than MEN2A. A methionine-to-threonine substitution (M918T) located in exon 16 (in the lower tyrosine kinase domain) is present in 95% of the cases of MEN2A. A rare alanine-to-phenylalanine substitution (A883F) located in exon 15 has also been detected in MEN2B (18).

In addition to the germ mutations detected in cases of MTC associated with MEN2 A and B, two other germ mutations have been reported in FMTC: substitution of a glutamate-aspartate and a valine-leucine (E768D and V804L) (20, 21). Several of the germ mutations associated with FMTC have also been detected in the somatic cells of sporadic MTC (19–22). A *ret* gene mutated at codon 918 (M918T) was detected in 23% to 86% of sporadic MTCs (23). Marsh and associates demonstrate that the presence of the mutation in both the germ cell and the somatic cell are associated with poor prognosis and chromosomal imbalances. The authors identify several deletions and at least one amplification of chromosomal loci that harbor genes encoding ligands or molecules participating in the signal transduction pathway(s) of the *ret* gene. These genes potentially are able to affect the expression of *ret* (22).

## Ret's Function

It could not be expected that the discovery of the *ret* oncogene in 1987 would open a Pandora's box. Takahashi and Cooper identified the gene as part of a fusion protein caused by chromosomal translocation (24). The gene was cloned and its extracellular domain was found to be unique (25) in that it contained a cadherin-like sequence. With time RET proved to be not only a member of an important family of proteins, that of a receptor tyrosine kinase, but *ret* is also a member of a family of unique receptors. Moreover, the gene is associated with at least four diseases, two multiple endocrine diseases (MEN2A, MEN2B), medullary carcinoma of the thyroid (Table 10-1), and familial Hirschsprung disease. Ret functions as a proto-oncogene (gain of function) in the first three conditions, whereas mutations that induce loss of function cause Hirschprung disease. However, a common denominator links the four diseases: each involves cells derived from the neural crest that either proliferate into cancers (MEN2A, MEN2B medullary carcinoma) or are lost (in familial Hirschsprung disease). With time it became obvious that RET does not operate alone; it is part of a complex of functional molecules including glial cell-line derived neurotrophic factor (GDNF) and glycophosphatidyl inositol receptor (GFR).

GDNF, a survival factor for various neuronal cell populations mapped on chromosome 5p13, was discovered in 1993 (26). GDNF is a member of the transforming growth factor-β (TGF-β) superfamily of proteins and is a potent neuronal survival factor (27). GDNF I, a member of a subfamily of growth factors, was the first found to interact with the RET tyrosine kinase. The GDNFs form a group of neurotrophins that include GDNF, neurturin, persephin, and artenin. These proteins bind to Ret and thus constitute ligands for a previously orphan receptor.

However, in addition to Ret and GDNF, a GFR is also required for transmission of the signal triggered by the neurotrophin. To date there are four glycophosphophatidyl-inositol receptors, namely, GFRα-1, -2, -3, and -4. There is evidence of preferential or exclusive affinity between the GDNF (GDNF, NTN, artenin, and PSP) and the GPI proteins (GFRα-1, -2, -3, and -4, respectively). Thus, Ret functions as a coreceptor with a GPI protein, the other coreceptor. The couple is joined by the ligand, a member of the GDNF family. A mechanism for the association of the complex has been reported. GFRα-1 is attached to a lipid raft inserted in the membrane. The ligand, the GDNF, joins the coreceptor GFRα-1, with high affinity, and GFRα-1 is attached to a lipid raft inserted into the membrane. The ligand-coreceptor complex then recruits and dimerizes the Ret receptor molecules, and the kinase is activated and puts the signal transduction pathway into gear, which promotes cell survival and differentiation (28).

In conclusion, the completed assembly is a dimer of a triple molecular complex. The complex engages signal transduction pathways, for example, RAF or PIK3. The formation of the complex makes it possible for a TGF-β ligand (GDNF) to activate GFR (which is devoid of a tyrosine kinase and therefore cannot autophosphorylate) because the recruitment of Ret provides the kinase.

The roles of the various components of the pathway in cancer are not clear. However, they certainly operate in the pathogenesis of Hirschsprung disease.

The pathogenesis of Hirschsprung disease is fascinating, but it will not be discussed except to point out that a complex of genes, mainly because of mutations of the *ret* gene, causes the cellular defect in Hirschprung disease. Dominant mutations of *ret* have been observed in 50% of members of families with a penetrance of 50% in men and 70% in females. Thus, *ret* is a clear example of a gene that, through the intricacies of genetic mechanisms and of signal transduction pathways, can cause a gain (cancer) or loss (Hirschsprung disease) in function (29).

In conclusion, *ret* is a proto-oncogene, a receptor tyrosine kinase that functions in cooperation with a companion glycosylphospholipid receptor forming a complex that is activated by neurotrophins. The Ret tyrosine kinase forwards a message through transduction pathways (RAF or PIK3) to transcription factors that promote cell proliferation and differentiation. Mutations of the *ret* gene lead to gain of function and cancers of cells derived from the neural crest in the thyroid, adrenals, and parathyroids. Translocation of the *ret* gene is associated with papillary cancer of the thyroid.

Distinction between a genuine familial condition and a sporadic condition is often important for early diagnosis of the disease in other family members. Therefore, genetic consultations are critical. This is the case in medullary carcinoma of the thyroid, a condition for which early diagnosis often leads to a cure and delayed diagnosis might be fatal.

## References and Recommended Reading


1. Fantl, W.J., Johnson, D.E., and Williams, L.T. Signalling by receptor tyrosine kinases. (Review). *Annu Rev Biochem.* 1993;62:453–81.
2. Asai, N., Iwashita, T., Matsuyama, M., and Takahashi, M. Mechanism of activation of the ret proto-oncogene by multiple endocrine neoplasia 2A mutations. *Mol Cell Biol.* 1995;15:1613–9.

3. Downward, J., Yarden, Y., Mayes, E., Scrace, G., Totty, N., Stockwell, P., Ullrich, A., Schlessinger, J., and Waterfield, M.D. Close similarity of epidermal growth factor receptor and v-erb-B oncogene protein sequences. *Nature.* 1984;307:521–7.
4. Ekstrand, A.J., James, C.D., Cavenee, W.K., Seliger, B., Pettersson, R.F., and Collins, V.P. Genes for epidermal growth factor receptor, transforming growth factor alpha, and epidermal growth factor and their expression in human gliomas in vivo. *Cancer Res.* 1991;51:2164–72.
5. Ekstrand, A.J., Sugawa, N., James, C.D., and Collins, V.P. Amplified and rearranged epidermal growth factor receptor genes in human glioblastomas reveal deletions of sequences encoding portions of the N-and/or C-terminal tails. *Proc Natl Acad Sci USA.* 1992;89:4309–13.
6. Levitsky, A. EGF receptor as a therapeutic target. *Lung Cancer.* 2003;41(Suppl): S9–S14.
7. Mulligan, L.M., Kwok, J.B., Healey, C.S., Elsdon, M.J., Eng, C., Gardner, E., Love, D.R., Mole, S.E., Moore, J.K., Papi, L., et al. Germ-line mutations of the RET proto-oncogene in multiple endocrine neoplasia type 2A. *Nature.* 1993;363:458–60.
8. Ishida, O., Zeki, K., Morimoto, I., Yamamoto, S., Fujihira, T., and Eto, S. Germ line mutation in the RET proto-oncogene associated with familial multiple endocrine neoplasia type 2B: a case report. *Jpn J Clin Oncol.* 1995;25:104–8.
9. Schneider, R. The human protooncogene ret: a communicative cadherin? *Trends Biochem Sci.* 1992;17:468–9.
10. Buttner, C., Henz, B.M., Welker, P., Sepp, N.T., and Grabbe, J. Identification of activating c-kit mutations in adult-, but not in childhood-onset indolent mastocytosis: a possible explanation for divergent clinical behavior. *J Invest Dermatol.* 1998;111:1227–31.
11. Schmidt, L., Junker, K., Nakaigawa, N., Kinjerski, T., Weirich, G., Miller, M., Lubensky, I., Neumann, H.P., Brauch, H., Decker, J., Vocke, C., Brown, J.A., Jenkins, R., Richard, S., Bergerheim, U., Gerrard, B., Dean, M., Linehan, W.M., and Zbar, B. Novel mutations of the MET proto-oncogene in papillary renal carcinomas. *Oncogene.* 1999;18:2343–50.
12. Hanks, S.K., Quinn, A., and Hunter, T. The protein kinase family: conserved features and deduced phylogeny of the catalytic domains. *Science.* 1990;241:42–52.
13. Rodriguez, G.A., and Park, M. Dimerization mediated through a leucine zipper activities the oncogenic potential of the met receptor tyrosine kinase. *Mol Cell Biol.* 1993;13:6711–22.
14. Pierotti, M.A., Santoro, M., Jenkins, R.B., Sozzi, G., Bongarzone, I., Grieco, M., Monzini, N., Miozzo, M., Herrmann, M.A., Fusco, A., et al. Characterization of an inversion on the long arm of chromosome 10 juxtaposing D10S170 and RET and creating the oncogenic sequence RET/PTC. *Proc Natl Acad Sci USA.* 1992;89:1616–20.
15. Santoro, M., Carlomagno, F., Romano, A., Bottaro, D.P., Dathan, N.A., Grieco, M., Fusco, A., Vecchio, G., Matoskova, B., Kraus, M.H., et al. Activation of RET as a dominant transforming gene by germline mutations of MEN2A and MEN2B. *Science.* 1995;267:381–3.
16. Mulligan, L.M., Eng, C., Healey, C.S., Clayton, D., Kwok, J.B., Gardner, E., Ponder, M.A., Frilling, A., Jackson, C.E., Lehnert, H., et al. Specific mutations of the RET

proto-oncogene are related to disease phenotype in MEN 2A and FMTC. *Nat Genet.* 1994;6:70–4.

17. Eng, C. Seminars in medicine of the Beth Israel Hospital, Boston. The RET proto-oncogene in multiple endocrine neoplasia type 2 and Hirschsprung's disease. *N Engl J Med.* 1996;335:943–51.
18. Eng, C. RET proto-oncogene in the development of human cancer. (Review). *J Clin Oncol.* 1999; 17:380–93.
19. Bolino, A., Schuffenecker, I., Luo, Y., Seri, M., Silengo, M., Tocco, T., Chabrier, G., Houdent, C., Murat, A., Schlumberger, M., et al. RET mutations in exons 13 and 14 of FMTC patients. *Oncogene.* 1995;10:2415–9.
20. Eng, C., Smith, D.P., Mulligan, L.M., Healey, C.S., Zvelebil, M.J., Stonehouse, T.J., Ponder, M.A., Jackson, C.E., Waterfield, M.D., and Ponder, B.A. A novel point mutation in the tyrosine kinase domain of the RET proto-oncogene in sporadic medullary thyroid carcinoma and in a family with FMTC. *Oncogene.* 1995, 10:509–13.
21. Fink, M., Weinhusel, A., Niederle, B., and Haas, O.A. Distinction between sporadic and hereditary medullary thyroid carcinoma (MTC) by mutation analysis of the RET proto-oncogene. "Study Group Multiple Endocrine Neoplasia Austria (SMENA)." *Int J Cancer.* 1996;69:312–6.
22. Marsh, D.J., Theodosopoulos, G., Martin-Schulte, K., Richardson, A.L., Philips, J., Roher H.D., Delbridge, L., and Robinson, B.G. Genome-wide copy number imbalances identified in familial and sporadic medullary thyroid carcinoma. *J Clin Endocrinol Metab.* 2003; 88:1866–72.
23. Marsh, D.J., Mulligan, L.M., and Eng, C. RET proto-oncogene mutations in multiple endocrine neoplasia type 2 and medullary thyroid carcinoma. (Review). *Horm Res.* 1997;47:168–78.
24. Takahashi, M., and Cooper, G.M. Ret transforming gene encodes a fusion protein homologous to tyrosine kinases. *Mol Cell Biol.* 1987;7:1378–85.
25. Iwamoto, T., Taniguchi, M., Asai, N., Ohkusu, K., Nakashima, I., and Takahashi, M. cDNA cloning of mouse ret proto-oncogene and its sequence similarity to the cadherin superfamily. *Oncogene.* 1993;8:1087–91.
26. Lin, L.F., Doherty, D., Lile, J.D., Bektesh, S., and Collins, F. GDNF: a glial cell line-derived neurotrophic factor for midbrain dopaminergic neurons. *Science.* 1993; 260:1130–2.
27. Henderson, C.E., Phillips, H.S., Pollock, R.A., Davies, A.M., Lemeulle, C., Armanini, M., Simmons, L., Moffet, B., Vandlen, R.A., and Simpson, L.C. [corrected to Simmons, L., et al.]. GDNF: a potent survival factor for motoneurons present in peripheral nerve and muscle. *Science.* 1994;266:1062–4. Erratum 1995; 267:777.
28. Manie, S., Santoro, M., Fusco, A., and Billaud, M. The RET receptor: function in development and dysfunction in congenital malformation. *Trends Genet.* 2001; 17:580–9.
29. Passarge, E. Dissecting Hirschsprung disease. *Nat Genet.* 2002;31:11–2.

CHAPTER

# THE ONCOGENIC GTPases 19

The Ras proto-oncogenes were the first discovered members of a superfamily of guanosine 5′-triphosphatases (GTPases). In mammalian cells alone the family includes at least 70 members, among them Ras, Rab, Rho, and Arf. The Ras family itself counts for closely related members H-Ras, N-Ras, and K-Ras4 A/4B, and over 12 closely related relatives (50% amino acid identity). Genes that encode this family of highly conserved proteins are derived, through transfection, from the Kirsten and the Harvey sarcoma viruses genomes.

In 1982, a human oncogene was discovered that proved to be a mutant of the H-Ras allele. The Ras proteins are homologous to the subunit of the heterotrimeric G proteins and contain guanine nucleotide binding sites (GDP, GTP) (1).

The three-dimensional structure of the guanine nucleotide binding switch has been published (2).

The protein functions as an off and on switch located downstream of the cell surface tyrosine kinase receptor and upstream of the mitogen activated protein (MAP) kinase transduction pathway. In the "on" position Ras functions as a primary participant in a signal transduction pathway. Ras operation requires that it be present at the inner phase of the cell membrane. After translation the Ras protein is translocated to the cell membrane with the help of highly conserved sequences in the N and C terminals of the molecule. Ras exhibits two different catalytic functions, albeit at a very low level. GTP hydrolase and GDP/GTP exchange activity. The GDP/GTP exchange is at the center of Ras operation. It is not surprising, therefore, that it is strictly controlled by at least two other proteins, guanine nucleotide exchange factor (GEF) and the GTPase activating protein (GAP). GEF stimulates the exchange of GDP for GTP and activates Ras, and GAP raises the activity of the GTPase, thereby inactivating the proto-oncogene. Thus, the molecule is in the "on" mode when GTP is bound and in the "off" mode when GDP is bound. GEFs constitute a family of proteins that include GEF, SOS1, and SOS2. They turn on the switch that is turned off by GTPase activating proteins (**FIGURE 19-1**).

## Upstream Signaling to Ras

Many cell surface stimuli activate Ras and among them are growth factors, integrins, hormones, neurotransmitters, and cytokines. The ligand binds to and then activates its cognate receptors; these receptors in turn activate Ras. The activation of Ras by EGFR is the best known. After its activation by the ligand, EGFR dimerizes and self-

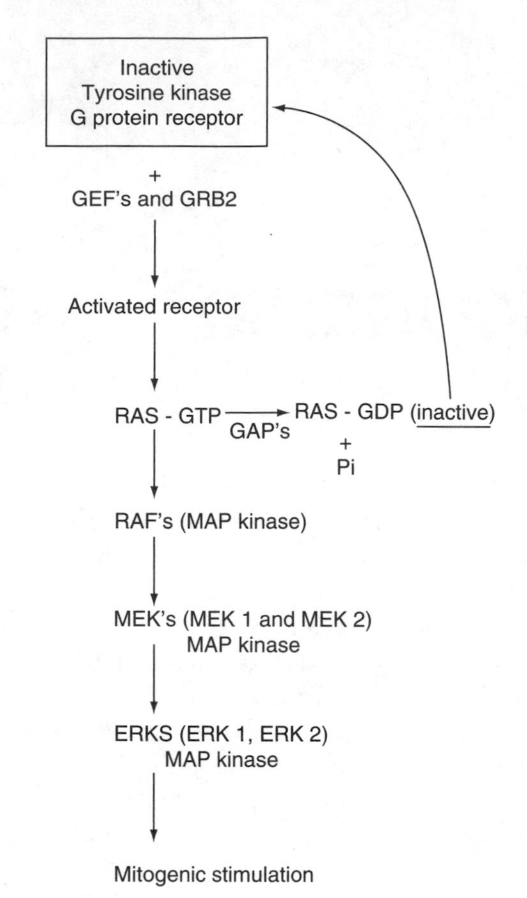


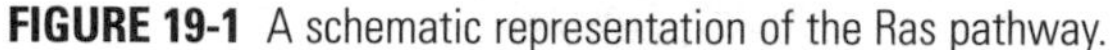
**FIGURE 19-1** A schematic representation of the Ras pathway.

phosphorylates specific tyrosine residues, thereby generating high affinity sites in the receptor for intracellular proteins harboring Sarc homology 2 domains (SH2).

The Gbr2 adaptor molecule contains a central SH2 and two SH3 domains, one on each side of the SH2. It complexes with SoS (a guanine exchange factor) by interfacing its SH3 domain with the proline-rich tail of SoS. After forming a complex with SoS, Gbr2 binds to EGFR through its SH2 domain, at a specific phosphorylated tyrosine site. SoS, now close to Ras, stimulates the exchange of GDP to GTP in the rate-limiting step for Ras activation. In summary, in the normal cell the state of Ras alternates from inactive to active. The binding of the Gbr2-SoS complex to the EGFR stimulates the GDP-GTP conversion and activates Ras. When Ras is to return to inactivity, GAP activates the GTP hydrolase and inactivates Ras.

## Downstream Activation by Ras

The action of GEF is coupled to that of Gbr2, an effector protein containing a SH2 domain that binds to an activated tyrosine kinase growth receptor. The receptor, when stimulated by the ligand (e.g., a growth factor), moves the GEF (SOS) to the plasma membrane. The conversion from GDP to GTP activates Ras and triggers the activation of downstream transduction pathways.

The first step in Ras's activation consists of the recruitment and activation of the serine threonine kinase, Raf, or the closely related proteins Braf or Araf. Activation of the

Ras proteins causes them to be relocated in the plasma membrane. For that purpose Ras binds to Raf at two sites. The first, RBS1, is located in the N terminal site; it includes a cysteine-rich domain found in Raf (Raf-CRD). This binding site is not indispensable for translocation, but it facilitates the complete activation of Raf, stabilizes that activation, and helps dissociate Raf from its entrapment by the 14-3-3 protein (3). Ultimately, the activated Raf triggers the mitogen-activating cascade. A major, but not the only, manifestation of the activation of the MAP pathway is mitogenic stimulation (Figure 19-1). The pathway described above is greatly oversimplified; indeed, progress was made in the understanding of the activation of the Ras in the recent years. For example, although Ras activation is mainly Raf-dependent, this requirement is not absolute as other Ras effectors have been discovered. Moreover, Ras activates several pathways other than the Map kinase cascade, including the RAS-PIK3 and RALGDS pathways (4, 5). RAS-PIK3 inhibits BAD through Akt, and RALGDS (RAL guanine nucleotide dissociation stimulator) contributes to the inhibition of the forkhead inhibitory factor and promotes cell cycle arrest (3, 6, 7).

## Ras and Cancer

Many cancers of various types are associated with mutation, overexpression, or deletion in either Ras, its regulatory proteins (e.g., GAP), or in molecules operating in Ras upstream and downstream pathways. The majority of the mutations (20%) occur in Ras (Kras 85%, Hras 15%, and Nras 1% of the total). They have been detected in cancers including the thyroid, pancreas, colorectum, liver, bladder, kidney, seminoma, lung melanoma, and myelogenous leukemia.

Oncogenic mutation of Kras is often caused by point mutations that result in amino acid substitution at positions 12, 13, and 61. These amino acids are located in the GTPase binding domain. As a result, Ras remains bound to GTP locking the switch in the "on" position, transduction flows uninterrupted, and the cell remains constantly stimulated to proliferate. As mentioned earlier, there are several GAP proteins and among them is neurofibromin. Neurofibromin is encoded in humans on chromosome 17q11, loss of its activity is associated with neurofibromatosis (8).

Mutations in EGFR have been discussed above. These mutations are present in more than 50% of carcinomas, in particular breasts, ovaries, and stomach. Mutations in Braf have been reported in a fraction of melanomas and colorectal cancers.

Events associated with transformation and cancer progression are not limited to cellular proliferation; they also include changes in cell shape and cell motility. The Rho family of small GTPase (Rho, Ras, and Cdc 42) regulates the actin cytoskeleton. Activated Rho contributes to the formation of focal adhesions (9, 10). As pointed out by Park, it cannot be excluded that deregulation of the normal function of the Rho GTPases may play a role in cancer progression (11).

In summary, the Ras pathways are numerous, often incompletely understood, and sometimes in conflict with each other (12–14). However, there is no doubt that RAS is an effective transforming agent and that site-specific mutations are associated with cancer in humans. Which of the various pathways are activated? Logically, one can anticipate that the Raf-Map and that of the Rho family are essential for constitutive cell cycle progression and cancer a cell dispersion (invasion and metastasis). Moreover, it is not excluded that failure of a cell to enter apoptosis, because of mutations in other Ras effector pathways (e.g., PI3K or RALDG), also may contribute to cancer progression.

However, whether the two pathways are activated simultaneously or in sequence, or whether additional mutations other than those of Ras or Rho are required for cancer progression is not clear. Neither is it known what the potential role of other Ras-dependent pathways in carcinogenesis may be. Answers to these questions would undoubtedly help to plan therapy, for example, by targeting components of the signal transduction pathway.

This chapter does not discuss existing or potential therapeutic agents; however, an excellent review of the Ras transducing pathways and a discussion of signal transduction targeted therapy has been published by Downward (15), and the example of interference of transducers with membrane transport is briefly considered.

It is established that once they are synthesized in the ER, the Ras proteins (Hras, Kras, and N Ras) become functional only if they are translocated from the cytosol to the cell membrane (**FIGURE 19-2**). This involves posttranslational modifications mediated by several enzymes, among which is farnesyl transferase (Ftase). Other involved enzymes include: geramyl geranyl transferase (GGTase), methyltransferase, and palmitoyltransferase. The four steps in this process are:

1. Ftase transfers the l5 isoprenoid chain from farnesyl pyrophosphate (FPP) to cysteine 186 (close to the C terminus) of HRAS. This reaction induces Ras to bind to the intracellular membrane through the farnesyl residue.
2. The endopeptidase removes a 3 amino acid sequence (AAX) from the C terminus.
3. The methyl transferase donates a methyl to the new carboxylterminus and the Ras molecule translocates to the plasma membrane.
4. In the last steps, the palmitoyl transferase adds two palmitoyl acid chains to a cysteinic residue close to that which was farnesylated. However, only H RAS and N RAS are palmitoylated; K RAS is not.

Identification of this pathway is significant because inhibition or mutation of the farnesyl transferase blocks the translocation of RAS and thereby keeps it from activating the MAP kinase sequence. A search for inhibitors of the membrane enzymes has been undertaken in several laboratories and clinical trials are now underway (7, 15, 16).

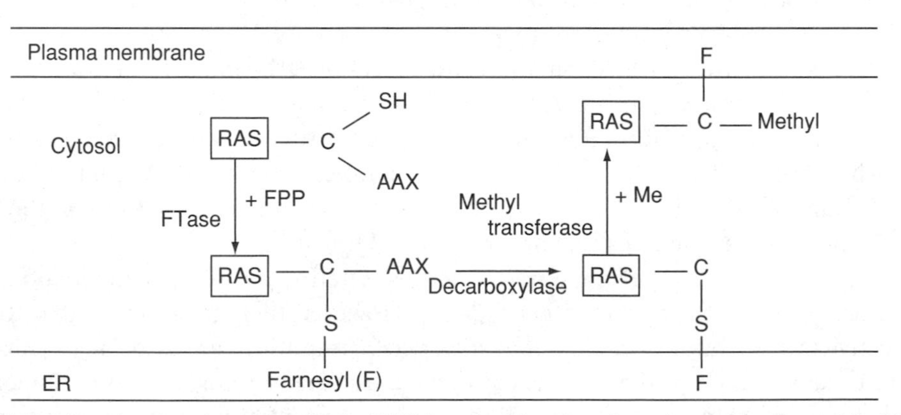


**FIGURE 19-2** Translocation of newly synthesized RAS from the ER to the cell membrane. Mutations of the farnesyl translocase block RAS translocation. Adapted from Julian Downward (15).

## References and Recommended Reading


1. Clapham, D.E. The G-protein nanomachine. *Nature.* 1996;379:297–9.
2. Vetter, I.R., Wittinghofer, A. The guanine nucleotide-binding switch in three dimensions. *Science.* 2001;294:1299–1304.
3. Clark, G.J., O'Brien, J.P., and Der, C.J. Ras signaling and transformation. In: Gutkind, J.S. *Signaling Networks and Cell Cycle Control.* Totowa, NJ: Humana Press.
4. Lambert, J.M., Lambert, Q.T., Reuther, G.W., Malliri, A., Siderovski, D.P., Sondek, J., Collard, J.G., and Der, C.J. Tiam1 mediates Ras activation of Rac by a PI(3) K-independent mechanism. *Nat Cell Biol.* 2002;4:621–5.
5. Malliri, A., van der Kammen, R.A., Clark, K., van der Valk, M., Michiels, F., and Collard, J.G. Mice deficient in the Rac activator Tiam 1 are resistant to Ras-induced skin tumors. *Nature.* 2002;417:867–71.
6. DeRuiter, N.D., Burgering, B., and Bos, J.L. Regulation of the forkhead transcription factor AFX by Ral-dependent phosphorylation of threonines 447 and 451. *Mol Cell Biol.* 2001;21:8225–35.
7. Hancock, J.F. Ras proteins: different signals from different locations. *Nat Rev Mol Cell Biol.* 2003;4:373–84.
8. Boguski, M.S., and McCormick, F. Proteins regulating Ras and its relatives. Review. *Nature.* 1993;366:643–54.
9. Khosravi-Far, R.S.P., Clak, G.J., Kinch, M.S., and Der, C.J. Activation of Rac1, RhoA, and mitogen-activated protein kinases is required for Ras transformation. *Mol Cell Biol.* 1995;15:6443–53.
10. Prendergast, G.C., Khosravi-Far, R., Solski, P.A., Kurzawa, H., Lebowitz, P.F., and Der, C.J. Critical role of Rho in cell transformation by oncogenic Ras. *Oncogene.* 1995;10:2289–96.
11. Park, M. In: Sriver, C.R., Baudet, A.L., Valle, D., Sly, W.S., Childs, B., Kinzler, R.W., and Vogelstein, B., eds. *The Metabolic and Molecular Bases of Inherited Diseases,* 8th ed. New York, NY: McGraw Hill Professional; 2001:645–664.
12. Sternberg, P.W., and Alberola-Ila, J. Conspiracy theory: RAS and RAF do not act alone. *Cell.* 1998;95:447–50.
13. Hunter, T. Oncoprotein networks. *Cell.* 1997;88:333–46.
14. Crews, C.M., and Erikson, R.L. Extracellular signals and reversible protein phosphorylation: what to Mek of it all. *Cell.* 1993;74:215–17.
15. Downward, J. Targetting RAS signaling pathways in cancer therapy. *Nat Rev Cancer.* 2003;3:11–22.
16. Qiu, R.B., Chen J., McCormick, F., and Symons, M. A role for Rho in Ras transformation. *Proc Natl Acad Sci USA.* 1995;92:11781–5.

CHAPTER

# 20 NON-RECEPTOR PROTEIN KINASES AND CANCER

Viral oncogenes such as v-scr encode a non-receptor tyrosine kinase protein that is not located in the membrane (transmembrane) but is linked to it. This association is critical to the oncogenic function. Its disturbance abolishes the transformation capacity of the virus and, therefore, it is likely that the viral protein harbors several structural components critical to its function. Among them are:

- Merystilation of the N-terminal glycine residue is absolutely required for association with the membrane. Mutations of the merystilated site block transduction of a putative membrane signal to the cytoplasmic protein.
- The C-terminus sequence ensconces the catalytic tyrosine kinase activity. It is highly conserved in all viral oncoproteins. SH1 refers to the kinase domain.
- The protein sequence also includes two protein binding domains: SH2 and SH3.

Most viral oncogenes like src cause cancer in animals but are not known to be oncogenic in humans, except for the lck and the ABL tyrosine kinase. In animals and humans, the oncogenic activation results from dysregulation of the kinase through point mutations or deletions in the C-terminal sequence. The prototypes of these oncoproteins belong to members of the src family whose oncogenic functions are restricted to animals. However, the BCR-ABL tyrosine kinase contributes to the oncogenesis of chronic myelogenous leukemia (CML) and acute lymphoblastic leukemia (ALL). The cellular gene *c-abl* is activated in an hematopoietic stem cell by reciprocal chromosomal translocation of *c-abl* from chromosome 9 to chromosome 22 in 90% of CMLs (t9:22, the Philadelphia chromosome). The clonal growth of the mutated cell evolves into leukemia (1, 2).

The lck, another exception among the members of the src family, causes human cancers. It contributes to human T-cell receptor signaling, can transform human T cell lines, and causes T cell leukemia in humans (3).

## The *abl* Proto-Oncogene

The product of *c-abl* proto-oncogene is a ubiquitous tyrosine protein kinase found in both nucleus and cytoplasm. Activated by several external stimuli, mainly growth factors (including PDF through PDGFR), it engenders multiple responses involving the cell cycle, cell death, cytoskeletal reorganization, and cell migration. c-ABL protein's relevance to cancer relates to the consequences of the translocation of its gene, *c-abl*. *c-abl* is the homologue of the viral *v-abl*, a transforming gene found in the sequence of the Abelson murine leukemia virus (a form of cancer that occurs in other animal species).

In 1960, Nowell and Hungerford made a seminal discovery (4) of the fusion of a fragment of chromosome 9 to the major breaking point cluster region on (Mber) on chromosome 22. The translocation associated with 90% of CML. The breakpoint on chromosome 9 occurs upstream of the 5′ first exon of the *c-abl* gene. Chromosome 22 carrying the translocation is referred to as the Philadelphia chromosome. The chimeric gene is transcribed in an 8.5 kb BCR/ABL mRNA, which is further translated into a 210 kDa fusion protein. Another *c-abl* translocation was later discovered to be associated with leukemia, mainly acute myeloid leukemia (AML). The translocation involved the minor breakpoint cluster region (m-bcr). The *c-abl* gene not only fuses with the major breakpoint cluster region (M-bcr) and the m-bcr, but it also fuses with the TEL transcription factor to yield the TEL-ABL chimeric protein (5) in acute and chronic myeloid leukemia. The TEL-ABL is reminiscent of the TEL-PGF beta R fusion product.

## The *c-abl* Gene

Because of its importance in human cancer when translocated, the function of the gene and its products have been investigated aggressively, yet its exact function is partly elusive. The chimeric 210 kDa fusion protein transforms cells in culture and can cause leukemia in mice (6, 7). This observation suggests that the disturbed gene accelerates cell proliferation or interferes with programmed cell death.

A great deal has been learned about the structure of the c-ABL and its components are indicative of its potential role. The N terminal contains the following:

- The Sarc homology domain (SH2 and SH3)
- The tyrosine kinase activity domain
- A proline rich sequence

The proline rich sequence binds special adaptor proteins that also contain SH3 domains (Crk, Grb2, and Nick). The adaptor proteins recruit signaling proteins (among them ABL) through their SH2 and proline rich sequences, and guide them to specific signals. In the case of ABL-Crk, however, the signal appears to inhibit the kinase activity (8).

The C-terminal region contains:

- Three nuclear localization signals (NLS 1, 2, and 3)
- A nuclear export signal (NES)
- Three "high mobility-like regions" (HLBs 1, 2, and 3)
- A binding site for the C-terminal repeated domain of RNA polymerase II
- A, G, and F actin domains related to the cytoplasmic functions of ABL

In conclusion, the N-terminal includes substrates and regulatory domains for the tyrosine protein kinase. The C-terminal contains cell localization factors that facilitate the protein's recruitment at the appropriated functional sites in the nucleus and cytoplasm (9).

Genetic manipulation (i.e., mutations, deletions, etc.) of the SH3 domain often causes ABL inactivation. Two models have been proposed for such inhibition: one that is an extramolecular and the other an intramolecular model. The extramolecular model consists of the association of an inhibitor that binds to both the SH3 domain and the kinetic active site. The intramolecular model involves a direct interaction between the SH3 domain and the C-terminal region. However, it should be kept in mind that the SH3 domain is not always involved in the inhibition of the tyrosine kinase activity.

During the cell cycle, the Rb protein inhibits the tyrosine kinase activity without binding to the SH3 domain; it binds to the ATP lobe of the kinase instead. Among the many SH3 proteins that bind to ABL, only a few inhibit its activity. These findings support the notion of an intramolecular inhibitory mechanism, a mechanism that is known to obtain in other src tyrosine kinases (10).

## c-ABL Activation

Physiologic (TNF) and therapeutic agents that cause apoptosis can also activate the nuclear c-ABL tyrosine kinase. Of course, this does not exclude the possibility that the activated kinase is translocated into the cytoplasm where it exerts at least some of its effect. The c-ABL activation is ATM-dependent and is associated with the phosphorylation of serine 465. Although ATM constitutively binds to the SH3 of c-ABL, the binding is not likely to be directly responsible for the activation, since the latter only occurs after irradiation. Therefore, it appears that the serine 465 phosphorylation rather than the association between the SH3 of c-ABL and ATM is the critical event in the activation process (11). DNAPK also constitutively interacts with c-ABL and activates the kinase after exposure to ionizing radiation. However, it is not yet clear whether the phosphorylation by either ATM or DNA-PK is sufficient to activate c-ABL after exposure to ionizing radiation.

The translocation of *c-abl* to chromosome 22 next to M-bcr induces a chimeric protein that is believed to be directly responsible for the chronic phase of CML. In the chronic phase CML, cells become growth factor-independent, are more resistant to apoptosis, reveal adhesion defects, and trigger various signal transduction pathways. This panoply of events are believed to be caused by activation of the c-ABL. CML is associated with an increased activity of c-ABL, which puts in gear, among others: signal transducer pathways (the phosphoinositide-3-kinase [PI3K]), the protein kinase B or AKT, the Janus kinase (JAK), and activators of transcription (12–14).

These findings are puzzling because of their number and their variability. They reflect the need for either selectivity in the choice of target genes for transcription (or repression) or for the existence of cross talk between several transducing pathways. In any event, these investigations of the downstream molecular effects have shed light on the mechanism by which ABL activates cell proliferation independently of growth factors' stimulation. For example, it could be the result of cross talk of the various transduction pathways that have been reported to be activated (i.e., PI-3K, RAS, PCK, IL-3, and JAK/STAT). It has also been suggested that the activation of the focal adhesion kinase (FAK), which could disturb integrin signals, might contribute to the acceleration of cell proliferation.

The upstream mechanism of c-ABL activation has long remained unknown, but recent studies in the Pendergrast laboratory are beginning to clarify it (15). Interaction between the PDGFs ($\alpha$ and $\beta$) and their receptors (PDGFR A, B, C, or D) leads to the recruitment of an isoform of phospholipase C; in fibroblast the major isoform is PLC-$\gamma$.

PLC-γ1 is activated by phosphorylation at three sites: Tyr771, Tyr773, and Tyr1254. Once activated, PLC-γ1 hydrolyses PtIns (3,4), the product of phosphatidylinositol-3 kinase (pPI-3K; refer to Chapter 21), to yield two second messengers, inositol (3,4,5) phosphate (InsP3) and diacylglycerol (DAG).

In the presence of inactivated PDGFR (ligand-free), c-ABL is kept silent by association with PtdIns(4,5)P2 possibly through direct or indirect binding. After activation of PDGFR by its ligand, PCLγ1 is activated by phosphorylation and PtdIns(4,5)P2 is hydrolyzed. As a result, c-ABL is freed from PtdIns(4,5)P2 and is activated (possibly by releasing a soluble c-ABL inhibitor).

In turn, c-ABL activation causes PLC-γ1 to be inactivated by phosphorylation (at least in vitro) at the site of tyrosine 1003, thereby creating a feedback loop between c-ABL and PLC-γ1 (**FIGURE 20-1**). These complex molecular interactions (simplified here) contribute to the downstream activation of c-ABL's functional targets.

## c-ABL and Apoptosis

c-ABL plays a role in the induction of apoptosis and activation of the kinase is required for this function. Observations that support c-ABL's role in apoptosis include:

- Progenitor B cells from mice deficient in c-ABL are oversensitive to apoptosis induced by either glucocorticoid or growth factor deprivation (16).
- C-ABL–/– cells are resistant to apoptosis (17).
- Cell expressing a c-ABL that carries an inactive protein kinase are resistant to apoptosis induced by ionizing radiation (17).

The induction of apoptosis after DNA damage is often associated with P53 activation. However, after activation c-ABL induces apoptosis in both $P53^{-/-}$ cells and in

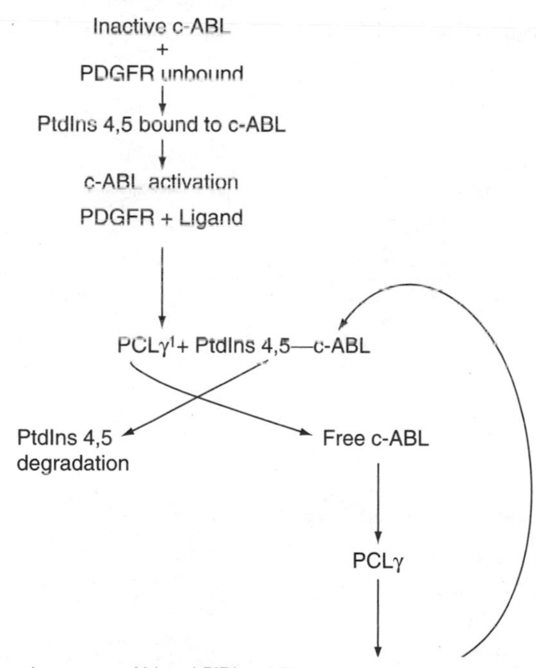


**FIGURE 20-1** Feedback loop between c-Abl and PIDIns 4,5.
[1]Phosphorylation on tyrosine 771, 773, and 1254.
[2]Phosphorylation on tyrosine 1003.

$P53^{+/+}$ in which P53 is degraded by the viral protein (E6). Consequently, it appears that P53 is not absolutely required for apoptosis and may not be a preferred downstream effector for c-ABL. Therefore, it is significant that c-ABL may contribute to the activation of another member of the P53 family.

P73 is structurally very similar to P53, but it differs from P53 functionally in that it is a phosphoprotein, and that the level of its phosphorylation rises after irradiation (18). Gong et al. showed that cisplatinum induces P73 in both P53-deficient and P53+/+ cells. However, P73 is not induced in c-ABL deficient cells. These findings suggest an interaction between c-ABL and P73 (19). P73 was also shown to interact with c-ABL through its proline rich and its SH3 motifs, and to phosphorylate c-ABL on tyrosine 99 in vivo and in vitro. Mutations at the site of phosphorylation abrogate apoptosis after exposure to ionizing radiations. Thus, P73 and C-ABL cooperate in the induction of apoptosis. The molecular mechanism of the process is not entirely clear, but it has been proposed that c-ABL stabilizes P73 and thereby extends P73's half-life and its capacity to induce apoptosis by targeting proapoptotic genes (19).

After DNA damage, the path to DNA repair involves activation of P53 by ATM and synthesis of P21 (the inhibitor of cyclin kinases, among them those that phosphorylate Rb); hypophosphorylated Rb keeps E2F sequestered and the cell cycle is arrested. The arrest may be followed either by DNA repair and resumption of the cycle or by apoptosis after failure of DNA repair. If in most cases the mechanism of cell cycle arrest is understood, the mechanism of apoptosis after P53 activation remains elusive. Among the unanswered questions are the roles of c-ABL and of Rb in the process.

Gong et al. suggest that after DNA damage ATM activates P53, which induces P21, which prevents phosphorylation of Rb, thereby interfering with c-ABL activation and E2F transcription (19). The cell cycle is arrested but apoptosis is not induced because c-ABL is inactive. The road to apoptosis requires ATM, which activates c-ABL, which then activates the P73-P21 pathway. Wang et al. have also proposed a mechanism for sensitization to apoptosis by c-ABL that involves the degradation of Rb by caspases

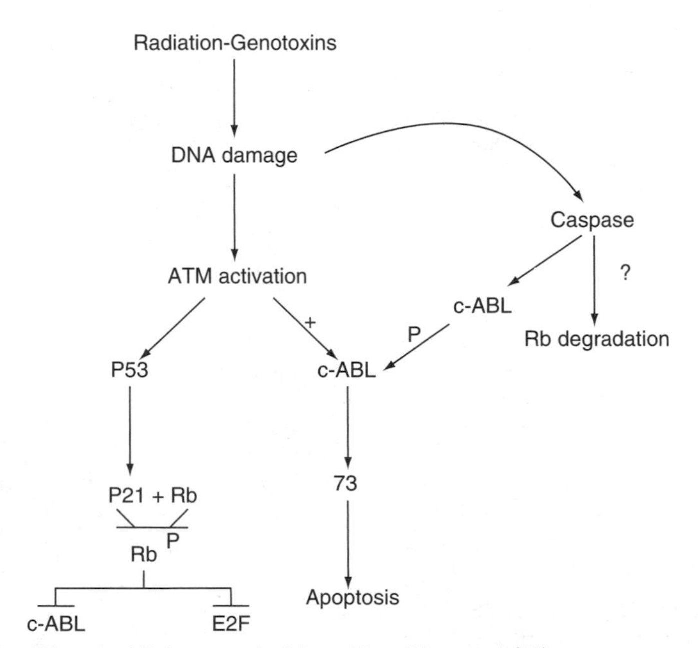


**FIGURE 20-2** Role of Rb and c-ABL in apoptosis. Adapted from Wang et al. (21).

(20, 21). Both physiologic activators and genotoxins cause the caspase degradation of RB, thus releasing c-ABL from inhibition by RB and leaving the activated c-ABL to play its role in apoptosis. This occurs only after cells are committed to enter the S phase (at the end of $G_1$ or the early intra-S phase). The mechanism of caspase activation leading to RB degradation remains unknown (**FIGURE 20-2**).

Although the functional importance of c-Abl in cell physiology should not be underestimated, the major interest in the molecule stems from the consequences of the translocations that are associated with leukemias. Despite the progress of the description of the structure of the molecule and a beginning of understanding of its upstream activation, there is still much to be learned about its downstream signal transduction pathways and their interactions. When translocated, the gene causes an imbalance in the cell population and a resistance to cell death, which hinders therapy (22, 23). Hopefully, a better understanding of the mechanisms of c-Abl signal transduction and its role in the apoptosis of leukemic cells will help in the design of more efficient targeted medication.

## References and Recommended Reading


1. Daley, G.Q., Ben-Neriah, Y. Implicating the bcr/abl gene in the pathogenesis of Philadelphia chromosome-positive human leukemia. *Adv Cancer Res.* 1991;57:151–84.
2. Kurzrock, R., Gutterman, J.U., and Talpaz, M. The molecular genetics of Philadelphia chromosome-positive leukemias. *N Engl J Med.* 1988;319:990–8.
3. Wright, D.D., Sefton, B.M., and Kamps, M.P. Oncogenic activation of the Lck protein accompanies translocation of the LCK gene in the human HSB2 T-cell leukemia. *Mol Cell Biol.* 1994;14:2429–37.
4. Nowell, P., and Hungerford, D.A. Chromosome studies on normal and leukemic human leukocytes. *J Natl Cancer Inst.* 1960;25:85–109.
5. Okuda, K., Golub, T.R., Gilliland, D.G., and Griffin, J.D. p210BCR/ABL, p190BCR/ABL, and TEL/ABL activate similar signal transduction pathways in hematopoietic cell lines. *Oncogene.* 1996;13:1147–52.
6. Daley, G.Q., Van Etten, R.A., and Baltimore, D. Induction of chronic myelogenous leukemia in mice by the P210bcr/abl gene of the Philadelphia chromosome. *Science.* 1990;247:824–30.
7. Heisterkamp, N., Jenster, G., ten Hoeve, J., Zovich, D., Pattengale, P.K., and Groffen, J. Acute leukemia in bcr/abl transgenic mice. *Nature.* 1990;344:251–3.
8. Rosen, M.K., Yamazaki, T., Gish, G.D., Kay, C.M., Pawson, T., and Kay, L.E. Direct demonstration of an intramolecular SH2-phosphotyrosine interaction in the Crk protein. *Nature.* 1995;374:477–9.
9. Wang, J.Y. Integrative signaling through c-Abl: a tyrosine kinase with nuclear and cytoplasmic functions. In: Gutkind, J.S., ed. *Signaling Networks and Cell Cycle Control.* Totowa, NJ: Humana Press; 2000.
10. Barila, D., and Superti-Furga, G. An intramolecular SH3-domain interaction regulates c-Abl activity. *Nat Genet.* 1998;18:280–2.
11. Shafman, T., Khanna, K.K., Kedar, P., Spring, K., Kozlov, S., Yen, T., Hobson, K., Gatei, M., Zhang, M., Walters, D., Egerton, M., Shiloh, Y., Kharbanda, S., Kufe, D.,

and Lavin, M.F. Interaction between ATM protein and c-Abl in response to DNA damage. *Nature.* 1997;387:520–3.

12. Di Bacco, A., Keeshan, K., McKenna, S.L., and Cotter, T.G. Molecular abnormalities in chronic myeloid leukemia: deregulation of cell growth and apoptosis. NCI all Ireland conference. *The Oncologist.* 2000;5:405–15.
13. Daniel, N.N., and Rothman, P. JAK-STAT signaling activated by Abl oncogenes. *Oncogene.* 2000;19:2523–31.
14. Laurent, E., Talpaz, M., Kantarjian, H., and Kurzrock, R. The BCR gene and Philadelphia chromosome-positive leukemogenesis. *Cancer Res.* 2001;61:2343–55.
15. Plattner, R., Irvin, B.J., Guo, S., Blackburn, K., Kazlauskas, A., Abraham, R.T., York, J.D., and Pendergast, A.M. A new link between the c-Abl tyrosine kinase and phosphoinositide signaling through PLC-gamma 1. *Nat Cell Biol.* 2003;5:309–19.
16. Dorsch, M., and Goff, S. Increased sensitivity to apoptotic stimuli in c-abl–deficient progenitor B-cell lines. *Proc Natl Acad Sci USA.* 1996;93:13131–6.
17. Yuan, Z.M., Huang, Y., Ishiko, T., Kharbanda, S., Weichselbaum, R., and Kufe, D. Regulation of DNA damage-induced apoptosis by the c-Abl tyrosine kinase. *Proc Natl Acad Sci USA.* 1997;94:1437–40.
18. Yuan, Z.M., Shioya, H., Ishiko, T., Sun, X., Gu, J., Huang, Y.Y., Lu, H., Kharbanda, S., Weichselbaum, R., and Kufe, D. p73 is regulated by tyrosine kinase c-Abl in the apoptotic response to DNA damage. *Nature.* 1999;399:814–7.
19. Gong, J.G., Costanzo, A., Yang, H.Q., Melino, G., Kaelin, W.G. Jr., Levrero, M., and Wang, J.Y. The tyrosine kinase c-Abl regulates p73 in apoptotic response to cisplatin-induced DNA damage. *Nature.* 1999;399:806–9.
20. Tan, X., Martin, S.J., Green, D.R., and Wang, J.Y. Degradation of retinoblastoma protein in tumor necrosis factor- and CD95-induced cell death. *J Biol Chem.* 1997;272:9613–6.
21. Tan, X., and Wang, J.Y. The caspase-RB connection in cell death. *Trends Cell Biol.* 1998;8:116–20.
22. Kurzrock, R., Kantarjian, H.M., Druker, B.J., and Talpaz, M. Philadelphia chromosome-positive leukemias: from basic mechanisms to molecular therapeutics. Review. *Ann Intern Med.* 2003;138:819–30.
23. Nimmanapalli, R., and Bhalla, K. Mechanisms of resistance to imatinib mesylate in Bcr-Abl-positive leukemias. Review. *Curr Opin Oncol.* 2002;14:616–20.

CHAPTER

# 21

# THE PIK3/Akt PATHWAY IN CANCER AND APOPTOSIS

Eukaryotic cell membranes contain three main groups of lipids: glycerophospholipids, sphingolipids, and cholesterol. Members of the first two classes of lipids are referred to as phospholipids. Phospholipids are unique lipids in that they are polar and ionic. Glycerophospholipids are large molecules containing a long hydrophobic tail and a polar head. They are made of three moieties; two long chain fatty acids (stearate and arachidonic acids) are attached to a head of glycerol phosphate that forms a phosphodiester bond with one of five alcohols: choline, etanolamine, serine, glycerol, and inositol. The glycerol phosphate is linked to the long chain fatty acids through 1,2, diacyl bonds (**FIGURES 21-1** and **21-2**).

Phosphatidylinositols are membrane associated acidic phospholipids. In phosphoinositols, the hexahydroxy-alcohol, inosistol, reacts with phosphorus donated by adenosine 5′-triphosphate (ATP) to form phosphodiesters with one, two, or three hydroxyl groups. For example, 4,5-biphosphate inositol is the source of the 1,4,5-inositol triphosphate, which functions as a second messenger and stimulates the release of calcium from the endoplasmic reticulum (**FIGURE 21-3**).

However, the phosphatidylinositides more directly related to this discussion are those phosphorylated on the 3 OH of inositol by the PI3K kinase, the enzyme that catalyzes the reaction. Phosphoinositide-3-kinases constitute a superfamily of lipid kinases that specifically phosphorylate the 3 OH group of the inositol ring of phosphoinositides (IP).

All phosphoinositides are derived from phosphatidylinositols (PtdIns) (reviewed in references 1 and 2). The inositol of the membrane bound phosphatidylinositols possesses five free hydroxyl groups and phosphorylation of some of these groups generates phosphoinositides. As mentioned, phosphatidylinositol-3-kinase (PI3K) catalyzes the transfers of ATP to yield phosphoinositide-4, phosphate 4,5-biphosphate, 3,4-biphosphate, and 3,4,5-triphosphates (**FIGURE 21-4**).

At first, PI3K was suspected to cause apoptosis directly (3), but with time it became clear that Akt is a major transducer of cell survival and that activation of Akt is mediated by PI3K. PI3K generates higher levels of PtdIns (PtdInks3,4 P2 and PtdIns 3,4,5 P3). D3 phosphoinositides bind in vitro to a large number of targets. These include (detailed in references 4 and 5):

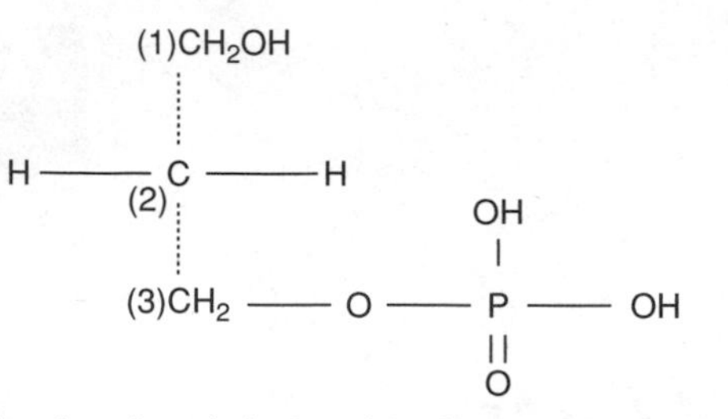

**FIGURE 21-1** The L-glycerol-3-phosphate formula (1, 2, and 3 refer to carbon numbering).

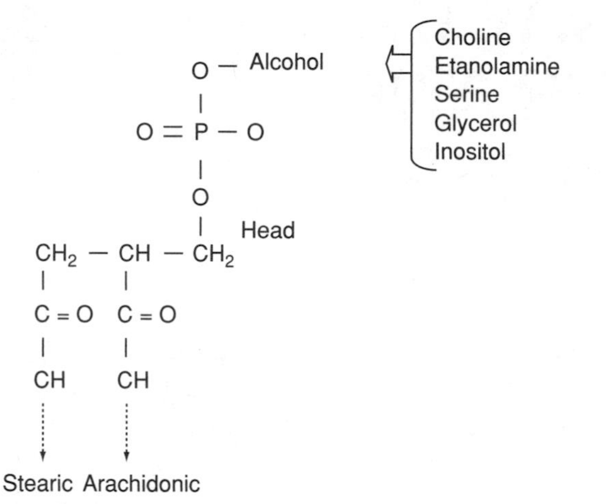


**FIGURE 21-2** Formula of the glycophospholipids.

- The serine/threonine kinase Akt (protein kinase B or Rac-PK)
- Several kinases that possess an SH2 domain, including the Src kinase.
- Protein kinase C enzymes.
- Clathrin adaptor proteins.
- Actin regulatory protein.

As noted by Fruman and Cantley (4), there are some common denominators linking the targets that seem to have emerged from observations of the phosphoinositide pathways in the living cell. Most target proteins are found in the cytoplasm. PI3K's activation generates phospholipids that bind specifically to small domains of the targets. The target proteins are enzymes and they act on their substrates within the context of the cell membrane. Despite advances made in the last two decades, in this very complex chapter of biochemistry much remains to be learned about the intricacies of the function of the PI3 kinases and their substrates in the life cycle of the cells in vivo, during development, growth differentiation, and cell death.

The structure of the PI3K family of kinases is described in references 1, 2, and 4. The PI3K kinases are heterodimers composed of a catalytic unit (110–120 kDa) that connects an N terminal sequence with a regulatory sequence of variable molecular weights (e.g., 85 kDa in PI3K $\alpha$). The three cloned PI3K have a 42 to 58% amino acid identity. They have been found in yeast, worms, and mammals, including humans.

As mentioned earlier, one downstream target of the D3 phosphoinositides is the serine threonine kinase Akt 1 (or Akt/PKB or RacPK). Human Akt 1, Akt 2, and Akt 3 are members of a large family of threonine, serine protein kinases conserved during

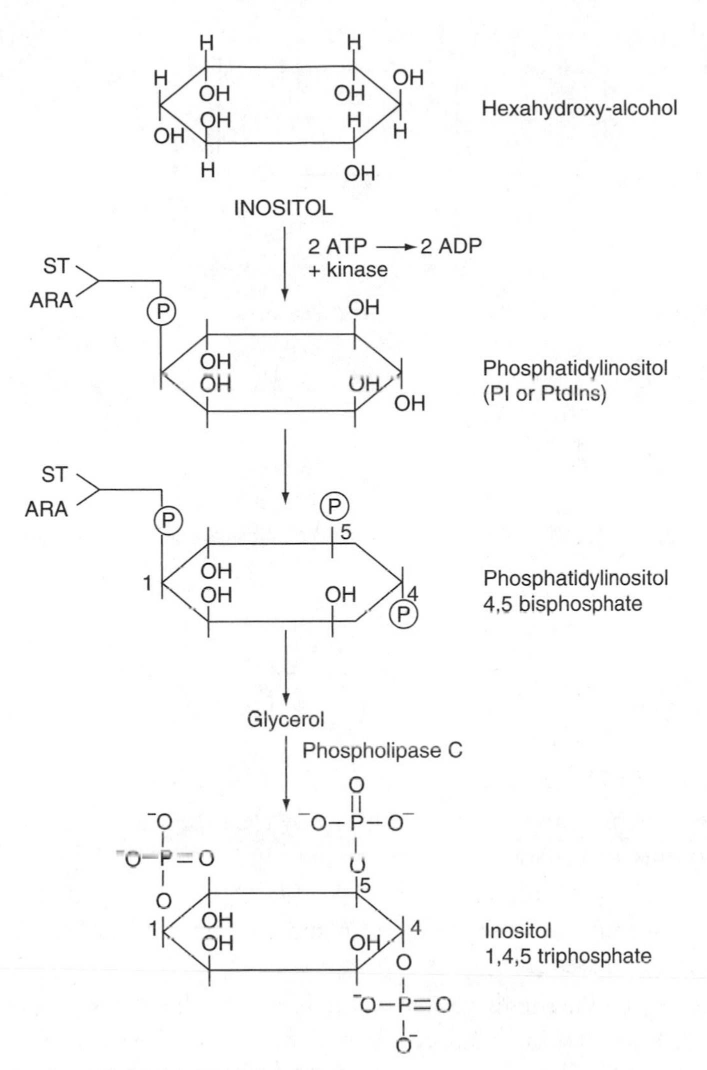


**FIGURE 21-3** Conversion of inositol 1,4,5 triphosphate.

evolution. They are found in *Drosophyila*, *C. elegans*, chicken, mice, and humans (6). The molecule is composed of three segments: an N-terminal pleckstrin domain, a central kinase domain, and a C terminal domain that contains a hydrophobic proline-rich sequence.

The upstream factors that activate Akt are numerous (close to 50 are listed by Datta and colleagues [10]) and variable. They include: growth factors (epithelial growth factor [EGF], fibroblast growth factor [FGF], and platelet-derived growth factor [PDGF]); interleukins (IL)-2, -3, -4, -5, and -8; anti-interferon and anti-FAS antibodies; and numerous toxic agents, among which is $H_2O_2$. Some of these agonists increase Akt activity up to 40-fold. This promiscuous response to stimuli of Akt does not facilitate our understanding of its mode of action. The activation of Akt involves two phosphorylations and at least two enzymes (PI3K and PDK1) are involved in the process.

D3 phosphoinositides bind with high affinity to the pleckstrin domain located in the N terminal of Akt, but full activation of Akt requires its further phosphorylation of threonine 308 and serine 473. The full process of activation is believed to involve the following steps:

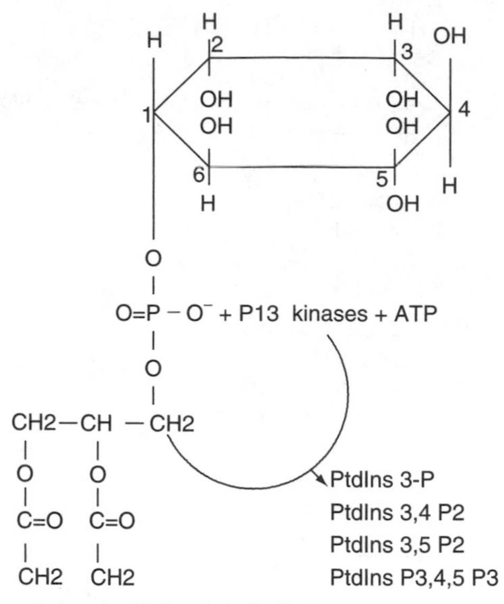


**FIGURE 21-4** Phosphorylation of phosphatidylinositols (PtdIns).

- Constitutive phosphorylation of Akt at serine 124 and threonine 450 in the absence of stimulation.
- Stimulation of P13K by agonists and a rise in PtdIns 3,4,5 P3.
- Akt is then recruited to the cell membrane by the binding of PtdIns 3,4,5 P3 to the pleckstrin domain of Akt.
- Abrogation of an inhibitory effect of the pleckstrin (PH) domain (on the activation Akt) permits the phosphorylation and activation of Akt by membrane-associated protein kinases (7).

The model begged for the discovery of the kinase(s) that activate Akt and it wasn't long before the phosphoinositide-dependent kinase 1 (PDK1) was discovered. PDK1 (67–69 kDa) phosphorylates Akt at T308; it was cloned and found to contain a C-terminal PH domain that mediates binding to Akt. In addition, PtdIns 3,4,5 P3 is needed for the activation of Akt not only to secure recruitment to the membrane, but also for bringing about changes in the molecular conformation of Akt that unmask T308 (8, 9).

PDK1 is located in both cytoplasm and in the surface membrane. Whether it is translocated to the membrane after stimuli is not clear, but electron microscopic studies have not revealed any changes after stimulation of the pathway. The protein kinase that phosphorylates Akt at serine 473 is unknown, but it could be a kinase related to PDK1 (PDK2?).

The downstream pathway of Akt secures cell survival in at least two ways: it activates IKK-$\alpha$ and thereby stimulates the nuclear factor-kappaB (NF-$\kappa$B) survival pathway and it phosphorylates Bad, thereby preventing apoptosis (10).

Most regulatory mechanisms must be controlled and, thus, activation by phosphorylation is regulated by dephosphorylation. Reversible phosphorylation of regulatory protein requires well-regulated dephosphorylating mechanisms. A specific phosphatase could serve such a purpose. Not so long ago, a phosphatase Cdc25 was found to play a central role in the $G_2$/M cell cycle checkpoint. Additional critically functional phosphatases have also been discovered. Therefore, it is not entirely surprising that a phosphatase was discovered that dephosphorylates phosphatidylinositide 3,4,5 triphosphate. Named PTEN, it is an important molecule that functions as a tumor suppressor gene encoded on chromosome 10 (10q22-24) (11).

## PTEN

PTEN (Phosphatase and tensin homolog deleted on chromosome 10), MMAC1 (mutated in multiple advanced cancer), and TEP (transforming growth factor-β [TGF-β] regulated and epithelial cell enriched phosphatase) are the names of the newly discovered suppressor genes often associated with cancer when both alleles are mutated. PTEN is a lipid phosphatase. Human recombinant PTEN dephosphorylates the D3 position of the inositol ring of phosphatidylinositol-3,4,5 triphosphate: PtdIns 3,4,5 P3 or PIP3 (**FIGURE 21-5**) (12, 13). PTEN is mutated in endometrial cancer and glioblastoma; breast, head and neck cancers; melanomas; and prostate and kidney cancers. In endometrial cancer PTEN mutations have been observed at both well differentiated and poorly differentiated stages (14–17). Gliomas are associated with a loss of heterozygosity (LOH) at chromosome 10q23 (18). LOH has also been reported in endometrial breast, thyroid, and prostate cancers. PTEN mutations have been observed in association (~80%) with some rare autosomal dominant hereditary diseases including the Cowden syndrome, Bannayan Zonana syndrome, Lhermitte-Duclos disease, and Bannayan juvenile polyposis coli (19), some of which predispose to cancer. Currently, there is no good correlation between the type of mutations observed in PTEN and the pathologic manifestations.

Mammalian PTENs[1] are proteins with a molecular weight of 40 to 50 kDa. The catalytic function is found in the N-terminal domain. The N terminal contains a PI 4,5 P2 binding motif (Lys/Arg-X4 Lys/Arg-X-Ly/Arg-Lys/Arg), a sequence often found in proteins that regulate actin (13). The C-terminal sequence contains a C2 domain that is frequently found in proteins that participate in phospholipid transduction pathways. The last segment of the C2-terminal sequence contains a PDZ binding domain (Thrp serine-x-valine COOH) that serves to recruit proteins involved in interaction between PTEN and other proteins (Akt?) (20). PTEN also contains sequences whose functions are still unknown but they share identity with tensin and auxillin. These sequences could contribute to PTEN's function in cell adhesion and migration (21).

The structure of the catalytic domain of the phosphatase is reminiscent of that of other tyrosine phosphatases that have been investigated, with the exception that the latter t domain is larger for PTEN, because of the need to accommodate the PtdIns 3,4,5 P3. The role of the C2 domain remains to be clarified, but it is possible that it contributes to facilitate interaction between enzyme and its substrate (e.g., PtdIns 3,4,5 P3). The C terminal domain contains a cluster of phosphorylation sites (S380, T382,

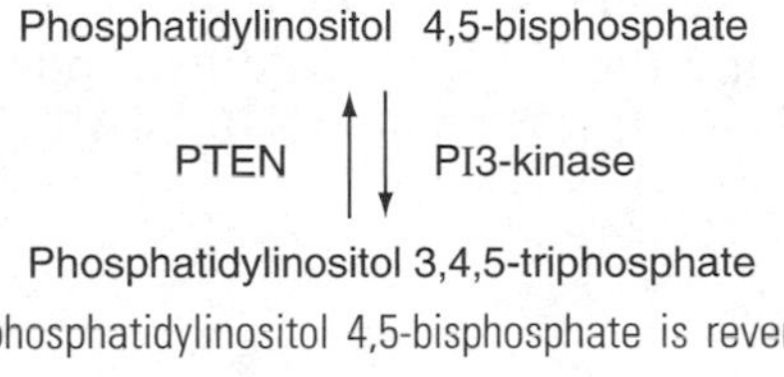

Phosphatidylinositol 4,5-bisphosphate

PTEN ⇅ PI3-kinase

Phosphatidylinositol 3,4,5-triphosphate

**FIGURE 21-5** The conversion of phosphatidylinositol 4,5-bisphosphate is reversed in the presence of the phosphatase PTEN.

[1]A homologue of PTEN has been found in *Drosophila* (Gao et al., *Dev Biol*. 2000;221;401–18).

T383) that contribute to the regulation of the molecule's function and stability. Protein kinase CK2 catalyzes thephosphorylation of the cluster. Dephosphorylation increases the phosphatase activity and accelerates PTEN's degradation by proteosomes (22).

Although no clear common denominator linking the multiple functions of PTEN has emerged, it is established that PTEN plays a role in embryonic development, apoptosis, cell adhesion, and movement:

- During the embryonic development of mice, PTEN levels are low up to day 11, and later they accumulate in every tissue (23).
- PTEN knockout mice do, however, develop early devastating phenotypes that vary with the strain and very likely with the environment, but range from early embryonic death to severe malformations, indicating that the embryo cannot afford losing even the small levels of PTEN present in the first 10 days of embryonic life (24).
- PTEN overexpression interferes with cells spreading, migration on the extracellular matrix proteins, and invasion. These events are associated with fibronectin induction of stress fibers and FAK phosphorylation. The potential relationship between these physiologic events and invasion and metastasis are far from clear, but it cannot be excluded that in absence of PTEN these processes could be relieved from some, if not all, constraints and, therefore, progress rapidly. Increased cell mobility could be the result of dephosphorylation of phosphatidylinositol phosphates, for example, PtdIns 3,4,5 P3 and PtdIns 3,4 P2, or of dephosphorylation of the focal adhesion kinase (FAK) (25).

## PTEN Substrates

An important, if not the principal, substrate of PTEN is the product of the PI3K kinase: PtdIns 3,4,5 P3. The PI3K is most likely activated by cell membrane growth factor receptor and integrins. The phosphatase cleaves the 3′OH phosphate of the inositol of the PtIns 3,4,5 P3 to yield Ptdns 4,5 P2, which has functions different from those of PtIns 3,4,5 P3.

In the absence of PTEN, activity PtdIns 3,4,5 P3 activates a downstream signal transduction pathway: Akt, a serine threonine kinase whose protein substrates regulate the progression of the cell cycle and the cell's proliferation, and in addition suppresses apoptosis. In the presence of PTEN this pathway is regulated by the reversible dephosphorylation of PtdIns 3,4,5 P3. Thus, PTEN and Akt function at the crossroad of life and death. The loss of PTEN function brings about constitutive activation of PtIns 3,4,5 and continuous stimulation of Akt activity, resulting in unfettered cell proliferation and suppression of apoptosis.

PTEN, primarily a lipid phosphatase, can also function as a protein phosphatase, although the latter function needs to be clarified. Keep in mind that several of these phosphoproteins, among them the FAK and the Shc adaptor protein, play a role in the modulation of apoptosis. The molecular mechanisms of interaction between substrates (lipid or proteins) and the phosphatase differ. Indeed, the G129E mutation of PTEN renders it inactive for PtdIns 3,4,5 but does not affect the dephosphorylation of phosphoproteins (26).

## PTEN, Akt, and Apoptosis

Akt is an oncogene isolated from the AKR thymoma (27). The gene products c-Akt/PK B are members of a family of serine protein kinases (in humans, Akt-1, Akt-2, and Akt-3). c-Akt-1 is located on chromosome 14 (14q32), c-akt-2 on chromosome 19

(19q131-132), and the chromosomal location of c-Akt 3 is unknown. All three isoforms were conserved through evolution and are widely distributed in the organism. The kinase has been found in *C. elegans* and *Drosophila.* The activity of Akt-1 has been studied most. Akt kinases are targets of the PI3 kinase.

Akt harbors a 100 to 120 amino acid stretch referred to as the PH domain (for pleckstrin[2] homology domain). The PH domain is found in many cytoplasmic proteins. The uniqueness of the domains does not reside in their sequence homology (the latter is limited), but in their ability to generate similar three-dimensional folds (28, 29). The presence of an intact PH domain in Akt is essential for its binding to the D3 phosphorylated phosphoinositides (**FIGURE 21-6**). Mutations that alter the amino acids of the PH domain that are part of the binding site prevent the association of Akt to the D3 PPIs.

Remember that the activation of Akt occurs in three separate but sequential and interdependent steps. These three steps are briefly summarized:

- Step I: Preactivation of Akt. Akt is constitutively phosphorylated at T450, but the T450's phosphorylation is not required for Akt's activation. It is believed to favor optimal folding of the protein in presence of the appropriate chaperones.
- Step II: Translocation of Akt to the plasma membrane. This step is PI3K-dependent and an integral PH domain is required. The step can be inhibited by wortamin.
- Step III: Phosphorylation of T308–Ser 473, by the phosphoinositide-dependent kinase (**FIGURE 21-7**).

In addition to the pathway described, there is also a PDK1-independent pathway for Akt activation. Coexpression of the integrin linked kinase and Akt brings about the S 473 phosphorylation of Akt (30).

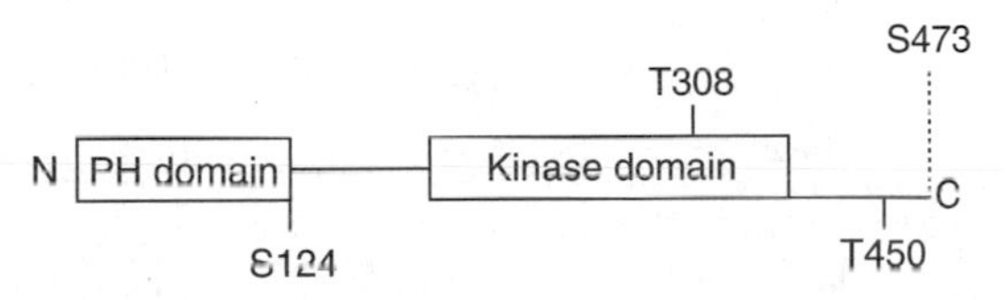


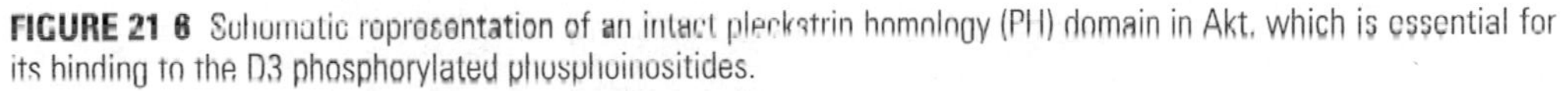
**FIGURE 21-6** Schematic representation of an intact pleckstrin homology (PH) domain in Akt, which is essential for its binding to the D3 phosphorylated phosphoinositides.

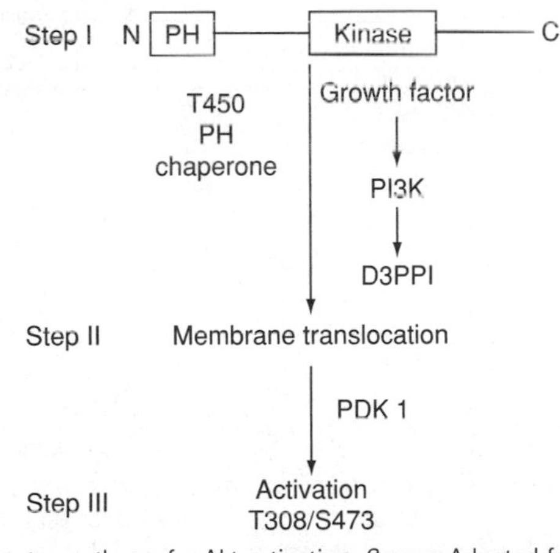

**FIGURE 21-7** Simplified three-step pathway for Akt activation. *Source:* Adapted from Chan et al. (5).

[2]Plextrin is a platelet protein that is phosphorylated by PCK.

What cellular responses are transduced after Akt activation? A multitude if considered separately. They all fall into categories that include the expression of the genome into the phenotype. Among them are transcription, translation, and posttranslation regulation as well as regulation of the cell cycle, metabolic signals (insulin), and apoptosis. This chapter focuses on apoptosis. Documentation of the inhibition of apoptosis by Akt is abundant, but there is no adequate explanation for the mechanism of such inhibition. Clearly, the target must either be a protein whose function in apoptosis is inhibited by phosphorylation (such as Bad, caspase 9, caspase 3, and forkhead proteins) or be the inhibition of a protein whose phosphorylation activates apoptosis (Bcl-2 could be a candidate).

Two plausible but simplified pathways for the inhibition of apoptosis by Akt and apoptosis activation in the presence of PTEN are summarized in **FIGURE 21-8**.

## BAD Phosphorylation and Apoptosis

Unphosphorylated, the proapoptotic Bad binds to Bclx and inhibits its antiapoptotic functions. Phosphorylated Bad binds to the protein 14-3-3 that sequesters Bad into the cytosol. There the phosphorylated proapoptotic agent in some way inhibits the antiapoptotic functions of other molecules (Figure 21-8). Thus, Bad's phosphokinase is one of several proteins that determines whether the cell will live or die (31).

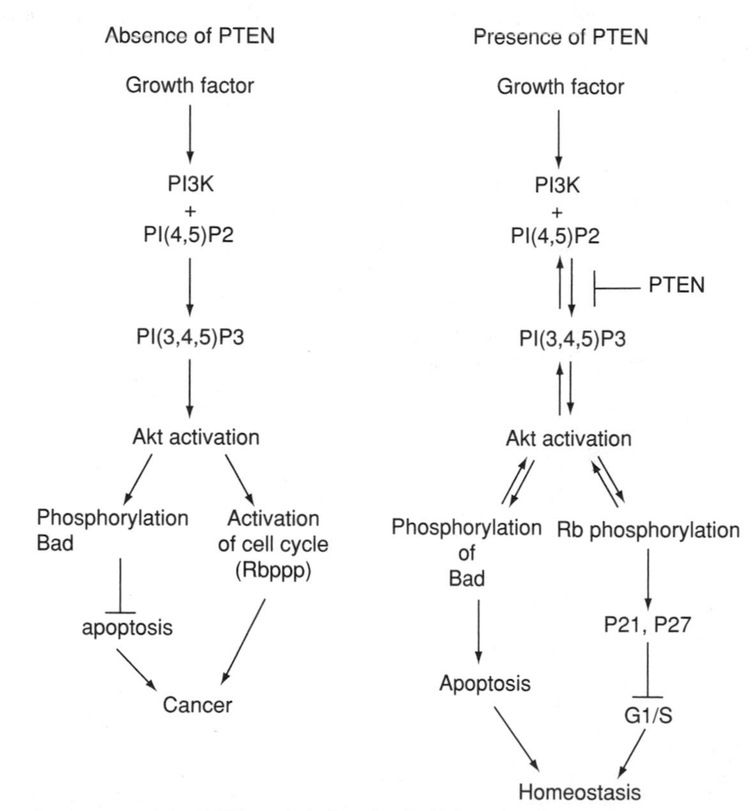


**FIGURE 21-8** In the absence of the PTEN catalytic function (left) the conversion of phosphatidylinositol 4,5-bisphosphate by PI3K to the triphosphate proceeds without interruption, allowing for persistent activation of the gene that encodes Akt and stimulation of cell proliferation. In the presence of the PTEN phosphate (right) the phosphorylation of phosphatidylinositol 3,4,5-triphosphate is reversible allowing for the regulation of the cell cycle and homeostasis. *Source:* Adapted from Muhaena, Taylor, and Dixon (13).

Akt was found to be the kinase or one of the kinases that phosphorylates Bad. Of course, if phosphorylation of Bad is limited to the action of Akt, the finding provides not only an elegant model for selection between survival and apoptosis, but it also offers a rather unique target for modulating the process. Unfortunately, it is not that simple, as other kinases have been shown to phosphorylate Bad, among them the calcium-calmodulin dependent protein kinase, protein kinase A, protein kinase C, and p21 activated kinase.

Does this imply that Bad phosphorylation should not be used as a target to prevent (in degenerative diseases) or to activate apoptosis in cancers? Not necessarily. First, it is unlikely that all these different kinases function at the same time or operate in all cancers. Second, even if several kinases phosphorylate Bad in a given tumor at the same time, the level of phosphorylation by each kinase might be different and vary with time. Therefore, it is conceivable that the kinases could be targeted one by one.

Finally, does the kinase have to be the target in all cases? The answer is also not necessarily, because by targeting the Bad binding sites on 14-3-3 and thereby inhibiting Bad's phosphorylation, it is possible that cell death could be prevented. Of course, such intervention might be helpful in degenerative diseases, but not in cancer (24).

## References and Recommended Reading


1. Rameh, L.E., and Cantley, L. The role of phosphoinositide 3-kinase lipid products in cell function. *J Biol Chem.* 1999;274:8347–50.
2. Fry, M.J. Phosphoinositide 3-kinase signalling in breast cancer: how big a role might it play? (Review). *Breast Cancer Res.* 2001;3:304–12.
3. Yao, R., and Cooper, G.M. Requirement for phosphatidylinositol-3 kinase in the prevention of apoptosis by nerve growth factor. *Science.* 1995;267:2003–6.
4. Fruman, D.A., and Cantley, L.C. P13-kinases: role in signal transduction. In: Gutkind, J.S., and Silvio, J., eds. *Signaling Networks and Cell Cycle Control.* Totowa, NJ: Humana Press; 2000:247–66.
5. Chan, T.O., Rittenhouse, S.E., and Tsichlis, P.N. AKT/PKB and other D3 phosphoinositide-regulated kinases: kinase activation by phosphoinositide-dependent phosphorylation. *Ann Rev Biochem.* 1999;68:965–1014.
6. Coffer, P.J., and Woolgett, J. Molecular cloning and characterization of a novel putative protein-serine kinase related to the cAMP-dependent and protein kinase C families. *Eur J Biochem.* 1991;201:475–81.
7. Franke, T.F., Kaplan, D.R., and Cantley, L.C. PI3K: downstream AKT ion blocks apoptosis. *Cell.* 1997;88:435–7.
8. Alessi, D.R., James, S.R., Downes, C.P., Holmes, A.B., Gaffney, P.R., Reese, C.B., and Cohen, P. Characterization of a 3-phosphoinositide-dependent protein kinase which phosphorylates and activates protein kinase B alpha. *Curr Biol.* 1997; 7:261–9.
9. Stephens, L., Anderson, K., Stokoe, D., Erdjument-Bromage, H., Painter, G.F., Holmes, A.B., Gaffney, P.R., Reese, C.B., McCormick, F., Tempst, P., Coadwell, J., and Hawkins, P.T. Protein kinase B kinases that mediate phosphatidylinositol 3,4,5-trisphosphate-dependent activation of protein kinase B. *Science.* 1998; 279:710–4.
10. Datta, S.R., Brunet, A., and Greenberg, M.E. Cellular survival: a play in three Akts. *Gene Dev.* 1999;13:2905–27.

11. Li, D.M., and Sun, H. TEP1, encoded by a candidate tumor suppressor locus, is a novel protein tyrosine phosphatase regulated by transforming growth factor beta. *Cancer Res.* 1997;57:2124–9.

12. Maehama, T., and Dixon, J.E. The tumor suppressor, PTEN/MMAC1, dephosphorylates the lipid second messenger, phosphatidylinositol 3,4,5-trisphosphate. *J Biol Chem.* 1998;273:13375–8.

13. Maehama, T., Taylor, G.S., and Dixon, J.E. PTEN and myotubularia: novel phosphoinositide phosphatase. *Annu Rev Biochem.* 2001;70:247–79.

14. Tashiro, H., Blazes, M.S., Wu, R., Cho, K.R., Bose, S., Wang, S.I., Li, J., Parsons, R., Ellenson, L.H. Mutations in PTEN are frequent in endometrial carcinoma but rare in other common gynecological malignancies. *Cancer Res.* 1997;57:3935–40.

15. Risinger, J.I., Hayes, K., Maxwell,G.L., Carney, M.E., Dodge, R.K., Barrett, J.C., and Berchuck, A. PTEN mutation in endometrial cancers is associated with favorable clinical and pathologic characteristics. *Clin Cancer Res.* 1998;4:3005–10.

16. Maxwell, G.L., Risinger, J.I., Gumbs, C., Shaw, H., Bentley, R.C., Barrett, J.C., Berchuck, A., and Futreal, P.A. Mutation of the PTEN tumor suppressor gene in endometrial hyperplasias. *Cancer Res.* 1998;58:2500–3.

17. Simpson, L., and Parsons, R. PTEN: life as a tumor suppressor. *Exp Cell Res.* 2001;264:29–41.

18. Davies, M.P., Gibbs, F.E., Halliwell, N., Joyce, K.A., Roebuck, M.M., Rossi, M.L., Salisbury, J., Sibson, D.R., Tacconi, L., and Walker, C. Mutation in the PTEN/MMAC1 gene in archival low grade and high grade gliomas. *Br J Cancer.* 1999;79:1542–8.

19. DiCristofano, A., and Pandolfi, P.P. The multiple roles of PTEN in tumor suppression. *Cell.* 2000;100:387–90.

20. Leslie, N.R., Gray, A., Pass, I., Orchiston, E.A., and Downes, C.P. Analysis of the cellular functions of PTEN using catalytic domain and C-terminal mutations: differential effects of C-terminal deletion on signaling pathways downstream of phosphoinositide 3-kinase. *Biochem J.* 2000;346:827–33.

21. Lee, J.O., Yang, H., Georgescu, M.M., DiCristofano, A., Maehama, T., Shi, Y., Dixon, J.E., Pandolfi, P., and Pavletich, N.P. Crystal structure of the PTEN tumor suppressor: implications for its phosphoinositide phosphatase activity and membrane association. *Cell.* 1999;99:323–34.

22. Vazquez, F., Ramaswamy, S., Nakamura, N., and Sellers, W.R. Phosphorylation of the PTEN tail regulates protein stability and function. *Mol Cell Biol.* 2000;14:5010–18.

23. Podsypanina, K., Ellenson, L.H., Nemes, A., Gu, J., Tamura, M., Yamada, K.M., Cordon-Cardo, C., Catoretti, G., Fisher, P.E., and Parsons, R. Mutation of Pten/Mmac1 in mice causes neoplasia in multiple organ systems. *Proc Natl Acad Sci USA.* 1999;96:1563–8.

24. Yamada, K.M., and Araki, M. Tumor suppressor PTEN: modulator of cell signaling, growth, migration and apoptosis. *J Cell Sci.* 2001;114:2375–82.

25. Tamura, M., Gu, J., Danen, E.H., Takino, T., Miyamoto, S., and Yamada, K.M. PTEN interactions with focal adhesion kinase and suppression of the extracellular matrix-dependent phosphatidylinositol 3-kinase/Akt cell survival pathway. *J Biol Chem.* 1999;274:20693–703.

26. Myers, M.P., Pass, I., Batty, I.H., Van der Kaay, J., Stolarov, J.P., Hemmings, B.A., Wigler, M.H., Downes, C.P., and Tonks, N.K. The lipid phosphatase activity of PTEN is critical for its tumor suppressor function. *Proc Natl Acad Sci USA.* 1998;95:13513–8.
27. Staal, S.P., Hartley, J.W., and Rowe, W.P. Isolation of transforming murine leukemia viruses from mice with a high incidence of spontaneous lymphoma. *Proc Natl Acad Sci USA.* 1977;74:3065–7.
28. Lemmon, M.A., Ferguson, K.M., and Schlessinger, J. PH domains: diverse sequences with a common fold recruit signaling molecules to the cell surface. *Cell.* 1996;85:621–4.
29. Touhara, K., Inglese, J., Pitcher, J.A., Shaw, G., and Lefkowitz, R.J. Binding of G protein beta gamma-subunits to pleckstrin homology domains. *J Biol Chem.* 1994;269:217–20.
30. Yoganathan, T.N., Costello, P., Chen, X., Jabali, M., Yan, J., Leung, D., Zhang, Z., Yee, A., Dedhar, S., and Sanghera, J. Integrin-linked disease (ILK): a "hot" therapeutic target. Biochem Pharmacol. 2000;69:1115–9.
31. Gajewski, T.F., Thompson, C.B. Apoptosis meets signal transduction: elimination of a BAD influence. *Cell.* 1996;87:589–92.

CHAPTER

# 22 TRANSCRIPTION FACTORS AND CANCER

Several transcription factors can transform cells in vitro and some play a definite role in human cancer. Their common denominator is regulation of growth development and survival. This chapter focuses on forkheads, c-Myc, Rb, and the APC proteins.

## Forkheads and Cancer

The proteins of the forkhead family all possess a 100 conserved amino acid binding domain referred to as a forkhead, FOX, or winged domain (1). The winged helix is composed of three branches, each forming compact alpha helix domains. The third (H3) branch binds to consensus recognition sequences in the major DNA groove, and the two other branches also make contact with the DNA. More than 80 such proteins have been found in yeast (*S. cerevisiae*) and in worms (*C. elegans*).

The forkhead proteins are transcription factors that play critical, but also varied roles in fetal, infant, adolescent, and adult life. Not surprisingly, they are closely involved with regulation of cell proliferation and differentiation, and in some cases cancer. The founder gene was discovered by Wiegel and associates in 1989 in *Drosophila*, where its mutation generates anomalies in embryonic development (2). Several of the genes have been cloned, and Larsson et al. established the chromosomal location of six of the human genes by fluorescence in situ hybridization (FISH) and somatic cell hybridization (**TABLE 22-1**) (3). The structure of the forkhead proteins from various sources have been studied in several laboratories (4–7).

The intricacies of the structure of the forkhead are not discussed here. The structure of the DNA binding domain of FREAC-11 is shown in reference (7).

### Forkhead Proteins and Apoptosis

Forkhead transcription factors are found in many if not all tissues. A few of the upstream activators have been identified, such as the hepatocyte nuclear factor-4, the insulin-like growth factor (IGF), erythropoietin, and transforming growth factor-β (TGF-β). The transducing pathways, with few exceptions, include phosphatidylinositols-3-kinase (PI3K), Akt, and phosphoinositides-dependent kinase 1 (PDK1). Consider IGF1; the trophic factor interacts with the IGF receptor, activates the intrinsic tyrosine kinase, the activated tyrosine kinase triggers the Mak/Erk cascade, and

**Table 22-1 Chromosomal localization of the freac-1, -3, -4, -5, -6, and -8**

| Gene | Localization |
|---|---|
| *freac-1* | 16q24 |
| *freac-3* | 6p25 |
| *freac-4* | 5q12-q13 |
| *freac-5* | 9 pericen |
| *freac-6* | 5q34 |
| *freac-8* | 1p32 |

ultimately the *PI3K/Akt* pathways are also activated. The phosphorylation of Akt at T408 and S473 activates the kinase. Akt in turn activates a host of substrates: glycogen synthetase kinase 3, BAD, caspase 9, nuclear factor-kappaB (NF-κB), and the winged finger transcription factors (8–10).

The transcription function of forkhead proteins requires that it be translocated to the nucleus where it binds to the consensus site of the promoter for the gene that encodes the FAS ligand, which ultimately triggers apoptosis.

Akt is able to phosphorylate several forkheads (FKHRL1, FKHR, and AFEX) at three different sites (e.g., Th32, Ser253, and Ser315 in FKHRL1 in Chinese hamster cells) (8). The phosphorylation sequesters the forkhead proteins in the cytoplasm by binding them to the 14-3-3 protein, thereby abrogating their transcription function, for instance, the transcription of proapoptotic proteins FAS, Bad, and Bim. Consequently, the growth factor exerts its survival effect through Akt and stimulates metabolism, cell growth, differentiation, and inhibition of apoptosis (11) (**FIGURE 22-1**). TGF-β, erythropoietin, and nerve growth factor also activate the PI3K/Akt pathway and inhibit transcription activation by the forkhead proteins through their phosphorylation (12–14).

In contrast, nuclear transfer of the forkhead protein activates apoptosis. In the absence of a growth stimulus, Akt remains inactivated and the forkhead protein fails to be phosphorylated; therefore, it can be translocated to the nucleus where it operates as a transcription factor capable of activating proapoptotic genes such as *fas*, *bad*, and *bim* (**FIGURE 22-2**). IGF1 also secures the survival of cerebellar granule neurons by involving the forkhead protein FKHRL1, but by a different mechanism than the one described above.

- FKHRLI is believed to be a transcription activator of Bim.
- Bim is a mediator of Bax-dependent cytochrome C release (see Chapter 7)
- IGF1 phosphorylates FKHRLI through the P3K1/Akt pathway, thereby sequestering it to the cytoplasm and preventing the induction of Bim, hence preventing apoptosis (Figure 22-1).

In conclusion, the P3K1/Akt pathway is essentially a survival pathway mainly because it stimulates the NF-κB pathway and inhibits the effect of the transcriptor factor (the proapoptotic forkhead) by phosphorylating the molecule, thereby blocking the induction of BAD, Bim, FASl, and possibly TRADD (15).

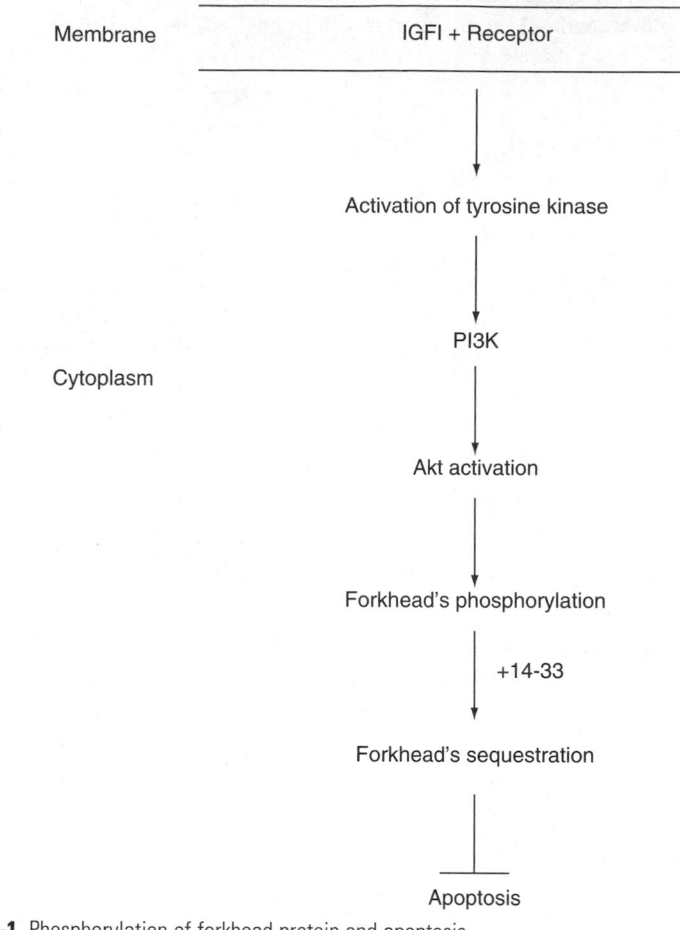


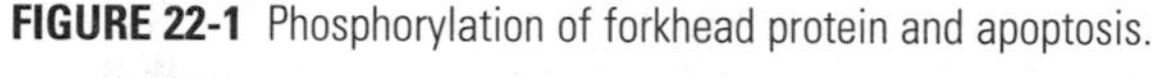

**FIGURE 22-1** Phosphorylation of forkhead protein and apoptosis.

### Forkhead Proteins and Cancer

The translocation of forkhead chimeric proteins have been associated with cancer. The muscle cancer, referred to as alveolar rhabdomyosarcoma, which occurs in children, is associated with a t(2;13)q(35;p14) translocation that results from the fusion of PAX3 and the transactivation domain of FKHR. PAX3 is a transcriptor factor needed for the development of muscle in limbs. The chimera proved to be a stronger transactivator than PAX3 (16,17). The activation of forkhead proteins might prove a useful tool for activation of apoptosis in cancer (Figure 22-2).

## Cancer and c-myc

The function of the *c-myc* gene is discussed in part in Chapter 14. In brief, the Myc protein family regulates gene expression. Myc dimerizes with Max. The two bHLHZ proteins form heterodimers that dimerize to yield a heterotetramer. This heterotetramer binds to E boxes (CACGG) in the target genes through the bHLHZ. Myc recruits various enhancers and modifiers to the target gene and, in part through deacetylation of histones, it up- or down-regulates transcription of target genes and ultimately modulates cell proliferation, differentiation, and programmed cell death.

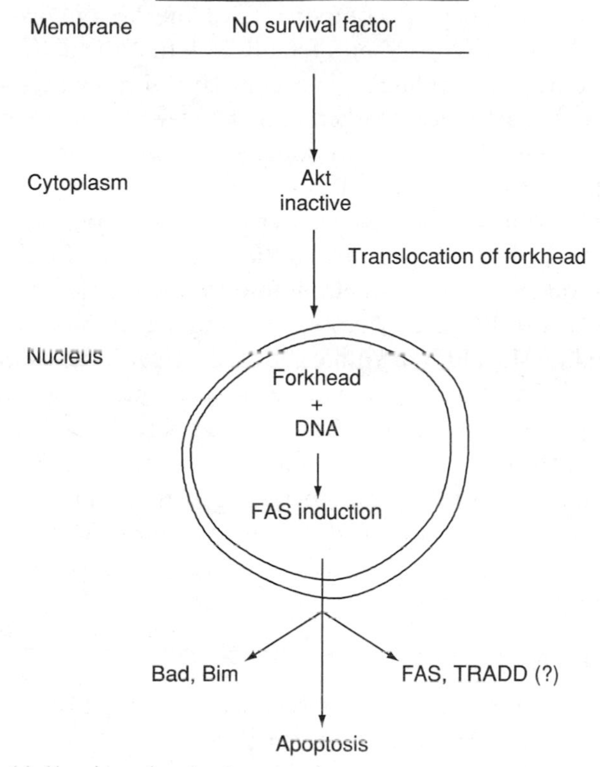


**FIGURE 22-2** The role of forkhead translocation in apoptosis.

The Myc family of proto-oncogenes includes at least seven members: *b-myc, c-myc, n-myc, l-myc, p-myc, r-myc*, and *s-myc*. The products of the *myc* gene are all (except for *b-myc*) basic helix loop helix leucine zipper proteins. The protein includes an N-terminal transactivation domain that binds to DNA nonselectively, with an affinity for the CACGTG sequence. The members of the family are direct regulators of gene expression. Myc does not form homodimers in vivo but exists in the form of heterodimer with Max. The hetrodimer binds to E boxes (CACGTG). Max proteins can be released from Myc to form heterodimers with Mad or Mnt. The Max/Mad or Mnt heterodimer also binds to E boxes.

The functions of the *myc* gene are multiple; the gene controls much of cell metabolism, cell division, cell differentiation, and apoptosis. Levens (18) and others (19, 20) have reviewed the difficulties encountered in identifying the targets activated or repressed by the *myc* gene. Despite much imaginative and laborious research by many groups, our understanding of the mode of activation and repression of the targets is incomplete. Moreover, the targets themselves have not all been identified, and some that are suspected as such actually may not be targets under physiologic or pathologic circumstances.

Briefly, the Myc, Mad (and the Mad-related proteins Mnt and Mya), and the Max family of bHLHZ proteins constitute a network that regulates various cell functions in which the molecules form different heterodimers that bind to the E box sequence.

The Myc, Max, Mad genomic binding was investigated in live *Drosophila* cells (21) by hooking an *E. coli* DNA adenine methyltransferase to each protein. This method allows for methylation of the sequences proximal to the in vivo binding site. The experiments demonstrate that the components of the Myc network bind to a large number of loci (~15% of coding region), establishing that the network interacts extensively with the genome. The binding regions correlate with the presence of an E box C, G repeats, and there is a great deal of interaction between the members of the network, for example, increased levels of d-Max modulate the content of c-Myc, but not that of α-Mnt, with the level of Max expressed.

As expected, induction by the Myc transcription factor is mitogen dependent, at least in normal cells. The binding of Myc to the consensus DNA element in the E box was determined by quantitative chromatin immunoprecipitation. As in an earlier study in which *Drosophila* life cells were used (21), Fernandez and associates identified a wide screen of Myc binding sites in human life cells (22). Again, the finding established an extensive binding of Myc to a very large number of genomic loci within the consensus element of E-box promoters. The Myc binding sites can be divided into two separate clusters, one of high affinity and another of low affinity binding. The high affinity sites were detected in all cell lines (257 genes, including 53 that had been shown to be Myc-regulated in previous studies). However, there may be more Myc binding sites because only one consensus element was targeted (CACGTG) and Myc/Max can bind to at least two other consensus elements (22).

## Oncogenic Functions of *c-myc*

The role of *c-myc* in cancer can only be discussed in the light of what is known about the effects of Myc expression. The oncogenic significance of Myc amplification is linked to its mode of regulation. For example, when Myc heterodimerizes with Max, the heterodimer causes transcription activation of target genes that trigger cell proliferation and the cell never enters $G_0$. In contrast, Max homodimers cause transcription repression and the cell may enter $G_0$ and differentiate or enter quiescence. Amplification causes myc to accumulate in excess of Max and as a result to stimulate cell proliferation.

Heterodimer Myc-Max → Transcription activation → Cell proliferation

Homodimer Max → Rest in $G_0$ → Quiescence

Hence, activation of *c-myc* triggers two antagonistic functions: cell proliferation and apoptosis. The molecular mechanism that determines the choice between the propagation of life and the death sentence is intriguing, but not fully understood. Experiments in vitro and in vivo provide a clue. In vitro, the singular overexpression of c-Myc in fibroblasts causes apoptosis, but the combined overexpression of c-Myc and Bcl2 triggers cell proliferation. The c-Myc induced apoptosis can also be inhibited by cytokinines that stimulate cell growth (e.g., epithelial growth factor [EGF] or insulin-like growth factor 1 [ILGFI]). In vivo overexpression of both c-Myc and Bcl2 in mouse lymphocytes causes the cells to proliferate and lymphomas to appear (23). In contrast, an excess of Max keeps the cells quiescent. Several molecular mechanisms could drive lymphocytic proliferation. Among the most obvious are overexpression of cyclins (cyclin D and E1) and down-regulation of cyclin inhibitors. However, *c-myc* could also induce Bcl2 and thereby directly inhibit apoptosis; alternatively both processes (cell replication and apoptosis) could be activated simultaneously (**FIGURE 22-3**).

Wild-type activation of *c-myc* regulates both cell proliferation and apoptosis. Illegitimate activation of *c-myc* in tumors leads to excessive cell proliferation, either by

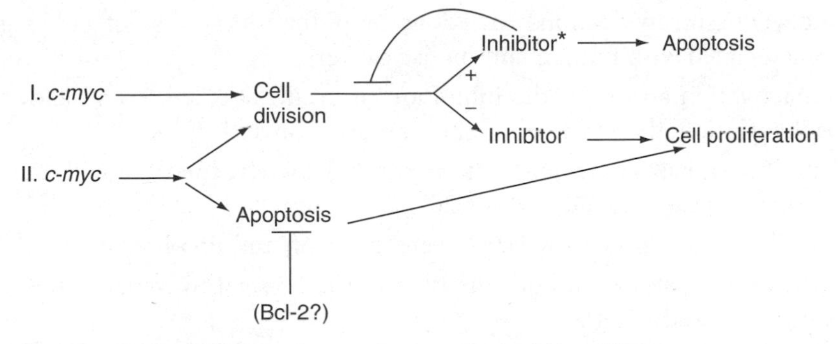


**FIGURE 22-3** The choice between cell proliferation or apoptosis in the presence of cyclin/cdk inhibitors (P21, P27, GADD45, P16) and Bcl-2. Depending upon the stimulus, c-Myc cell proliferation or cell death: (I) the presence or absence of cyclin/cdk inhibitors determines whether the cell proliferates or engages in apoptosis; (II) c-Myc simultaneously activates stimuli for cell proliferation and for Bcl-2 expression. Modified from Ruin, C., and Thompson, C. (46).

accelerating cell progression, by inactivating the inhibitors of cyclin cdk (INK and CIP families of inhibitors) or by inhibiting apoptosis (by inactivating ARF and thereby raising MDM2 expression and P53 destruction). Investigations pursued in Burkitt lymphoma as a prototype helped to clarify this matter (24).

Burkitt lymphoma (BL)[1] is frequently found in children and young adults in Africa, where it is associated with the Epstein-Barr virus. It is manifested by massive lymphocytic proliferation often in the jaw. Its hallmark is a reciprocal translocation of the entire *c-myc* gene next to the transcription enhancer of highly (Ig) heavy chain locus. At least three translocations have been associated with BL—t(8:14), t(2:8), and t(8:22)—with frequencies of 80, 10, and 10%, respectively.

Many of the investigations of the molecular mechanisms by which illegitimate *c-myc* activates cell proliferation have focused on two targets: the cyclin cdk inhibition by the INK4a and the P53 inhibition by ARF (25). In summary:

- A locus on chromosome 9 (9p21) encodes both gene $p16^{INK4a}$ and $p14^{ARF}$.

  The product of $p16^{INK4a}$, INK4a, is a member of a large family of Cdk inhibitors (CKIs); INK4a binds to monomers of Cdkr and Cdk6, and thereby blocks the phosphorylation of RB and arrests the cycle at the $G_1$/S checkpoint.

  Rb + cyclin D/Cdk4/6 → Rb → activation of E2F → INK4a

  The product of $p14^{ARF}$, ARF, binds to MDM2 and thereby keeps P53 from being degraded and allows cell cycle arrest.

  ARF + MDM2 → P53 degradation

  Thus, both the products of $p16^{INK4a}$ and $p14^{ARF}$ may be able to contribute to cell cycle arrest when the two genes are intact, but when mutated they may allow for the cell cycle to bypass the checkpoints.

[1]Burkitt lymphoma is also seen in American and European adults where it is not necessarily associated with Epstein-Barr virus.

- Homozygous incapacitations (deletions or point mutations) of either gene are often associated with human and mouse cancers.
- P16 inactivation abrogates the inhibition of cyclin D/Cdk4/6, and allows for Rb phosphorylation, thereby freeing E2F transcription activators.
- P27 inhibition has been reported to occur and thereby interfere with the cyclin E and Cdk2 cycle arrest (**FIGURE 22-4**) (26).
- INK4 inhibition allows for MDM2 overexpression and P53 inactivation.

In conclusion, an attack on both the Rb and the P53 pathways facilitates cell cycle progression (reviewed in 24).

In reality, INK4a/p16 is frequently inactivated in many cases of Burkitt lymphoma by methylation of its promoter, thereby silencing the gene with abrogation of mRNA transcription and Myc protein translation. The P16 methylation is independent of EBV infection.

On the other hand, homozygous deletion of $p16^{ARF}$ is rare in Burkitt lymphoma (less than 6%). Almost all of the cases in which ARF was inhibited harbored a wild-type P53 (wt P53), and in the majority of cases with a wt P53 either ARF was lost or MDM2 was overexpressed. Since the majority of Burkitt lymphoma cell lines and 30% of BL biopsies harbor mutated P53, it is obvious that the ARF-P53 pathway is often inactivated in Burkitt lymphoma (27). An additional mechanism has been suggested for the inactivation of ARF, its repression by one of two transcription factors BRG-1 (a polycomb-like protein) and Twist. BRG-1 is a member of the transcription modulator family SW/SNF; it has been reported to prevent cell cycle arrest by P53 and apoptosis (28). Twist interferes with the transcription of P53 and that of P21 and MDM2 (two proteins induced by P53). Twist is believed to prevent the P53 induced cell cycle arrest and apoptosis indirectly by modulating the ARF/MDM2/P53 axis (29).

It would be difficult to discuss the role of *c-myc* in each cancer investigated, which include other lymphomas and several solid tumors. Myc is amplified in the following cancers: breast (20%), cervix (25–50%), esophagus (36%), head and neck (7–10%), ovaries (20–30%), hepatocellular (13%), small cell lung (15–20%) (30), and sarcoma(s) and neuroblastoma. N-myc is amplified in neuroblastomas (20–25%).

It is critical to keep in mind that the response to the activation of *c-myc* (wt or illegitimate) may vary from cell type to cell type.

The outcome of *c-myc* activation was investigated in different cell populations in Pelengaris' laboratory using a conditional transgenic expression system (19). The

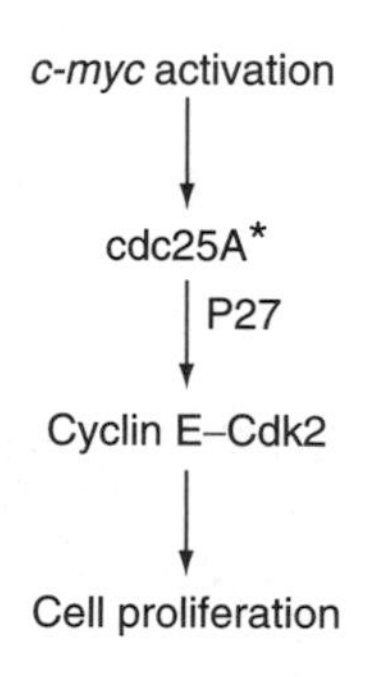


**FIGURE 22-4** Cyclin E-Cdk2 activation by the cdc25A phosphatase.

results vary remarkedly as the cell evolves from the basal to the keratinized. The cells above the basal layer in the epidermis respond by blocking apoptosis, arresting differentiation, and evolving into premalignancy in vivo. In vitro, the same cells derived from the same transgenic mice respond by entering apoptosis and, moreover, cell death is the major outcome in pancreatic β cells in adult transgenic mice (19).

Growth arrest genes (*p21, p27, p15,* and *gadd45*) are up-regulated in *c-myc* null cells and, therefore, it is assumed that in vivo, *c-myc* targets these genes and represses them. The mechanism of repression is not yet known.

### Summary

In cancer the restraints normally imposed upon the size of a given cell population (i.e., keratinocytes, enterocytes, hepatocytes, etc.) are abolished. Cancer cells succeed because they have gained independence from both antiproliferative and mitogenic signals from the host. The cancer cell relies on its defective genes for survival. *c-myc* stimulates both cell proliferation and apoptosis, and when it is defective the balance shifts in favor of cell proliferation. When translocated, *c-myc* causes lymphomas.

Three forms of lymphoma are associated with Burkitt[2] and B cell lymphomas. Gene relocations are responsible for the oncogenic stimulus. All translocations have the same effect: activation of *c-myc*. This activation is believed to be associated with: (1) induction of cyclin D and cyclin E2; (2) inactivation of cyclin inhibitors (P21, P27, P15, P16); (3) inactivation of ARF, overexpression of MDM2, or *p53* mutations; and (4) stimulation of immortalization through activation of the catalytic unit of telomerase (24). C-myc can also be modulated by upstream host factors like ErbB2, the tyrosine kinase overexpressed in breast cancer (31).

The role of *c-myc* in solid cancers is no more clear than in lymphoma. The same questions arise: Does a primary mutation of *c-myc* initiate the cancer or is the mutation a consequence of an already existing hypermutability in a precancerous or a cancer cell? What are the genes targetted by *c-myc*? Are they likely to vary from cancer cell to cancer cell as they do in their wild-type counterpart? The answers to these questions are critical if the transductions pathway of c myc are themselves to be targetted for therapy (reviewed in 32–34).

## Retinoblastoma Pathway and Cancer

Elements of the structure and function of Rb are discussed in Chapter 12; In essence, the RB pathway operates at the center of the regulatory mechanism of the cell cycle.

Mitogenic agents → Cyclin Cdk4/6 → Rb → Rbp → E2F activation → S

At first glance, the role of Rb in cancer is straightforward: Mutations that impair Rb's role in arresting the cell cycle cause uncontrolled cell proliferation. However, there are four problems with this oversimplification. First, it does not readily explain cell proliferation, and it also does not tell us how the proliferating cell degenerates into cancer. Secondly, it ignores the importance of Rb in other phases of cell division. Rb is not only the gatekeeper of the $G_1$/S transition, but it also contributes to the $G_2$/M transition and

[2]In Burkitt lymphoma, the *c-myc* coding element is translocated next to transcription enhancer element of IgH on chromosome 14 q(t8;14(q24;q32). In B cell lymphoma, the IgL locus from chromosome 2 or chromosome 22 is translocated next to the *myc* locus on chromosome 8 t(2;8) (p12;q24) or t(8;22)(q24;q11). *c-myc* is activated in all three cases.

chromosomal segregation during metaphase. Thirdly, c-Rb does not always function alone, as c-Myc also contributes to the regulation of the E2F transcription factors. Finally, the continuity of the pathway from Rb to E2F and to $G_1$/S transition is not without surveillance by other proteins, among them P53 and the CIP and INK4 inhibitors (P21, P27, P57 and P15, P16, P18, P19).

## Rb and the $G_1$/S and $G_2$/M Checkpoints

Rb controls the $G_1$/S checkpoint. When the control fails, the cycle will not arrest after DNA damage, and the presence of damaged DNA in a cell still capable of dividing may yield a mutated progeny. The arrest in $G_1$ phase is associated with the silencing of the E2F target genes in part by: histone deacetylation by HDAC1, HDAC2, HDAC3, and other proteins involved in modulating chromosomal structure (SW12/SNF2) in yeast or their human homologues BRG1 and Brm (35). Moreover, since Rb also contributes to the $G_2$/M checkpoint, regulation of chromosomal segregation may be disturbed. Overexpression of the Rb in cells arrested during the S phase causes $G_2$ arrest (36) in cells in which Rb is impaired and treated with colchicine, vinblastine, or other agents that interfere with spindle building. These findings and others indicate that the integrity of chromosomal stability is compromised when Rb is impaired (37). Such chromosomal instability may cause the accumulation of mutations, some of which may facilitate invasion, angiogenesis, metastasis, etc. This matter deserves further consideration.

## Rb and Chromosomal Segregation

At least in yeast, the expression of exogenous Rb stimulates the fidelity of chromosomal segregation (37, 38). Rb indirectly influences kinetocore proteins through its interraction with mitosin, which is believed to contribute to the assembly of the kinetochore, in part by interacting with Hec1, which is a conserved nuclear protein (350 kDa) with many leucine heptad repeats that regulate several mitotic events. The use of Hec-1–specific antibodies or the introduction of sequence specific mutations indicates that the protein is needed during chromatid separation. Hec-1 also interacts with SMC (structural maintenance of chromosomes) proteins that participate in chromatid cohesion and chromosomal condensation (39). Rb reacts with h-nuc, a protein component of APC (the anaphase protein complex) (40). Rb binding to Hec-1 modulates the chromosomal component of APC and the degradation of APC proteins through the ubiquitin proteosomal pathway. Whether these interactions of Rb affect the maintenance or destruction of proteins of the APC essential to cell proliferation is not entirely clear. In conclusion, Rb, which is the founder of tumor suppressor genes, does not only hold the key to the entrance into the cycle, it also supervises many aspects of the cycle from the stern synthesis of DNA to the anaphase's ballet and the graceful chromosomal exit in metaphase. When Rb sleeps, entrance of $G_1$/S is a free-for-all and the chromatids, sometimes mutilated, cannot find their right place in the equatorial plate, so sick chromosomes are delivered to the cell's progeny and thus cause cancer or cancer susceptibility.

## Other Members of the E2F Axis and Cancer

Rb not only causes cancer directly (via Rb mutations), it also causes cancer indirectly because of mutations or overexpression of molecules that modulate the Rb-E2F pathway. Among these molecules are: members of the E2F family, inhibitors of cyclins and, cyclins themselves. Their role in cancer is briefly summarized:

(A) E2Fs are apparently not associated with cancer in humans probably because of functional redundancy in the family.

(B) The INK4A, Cdk6, and Cdk4 inhibitors. Rb can be hyperphosphorylated in the absence of P16[3]; however, P16 defects are associated with hereditary cancer (familial melanoma and familial pancreatic cancer) and some cancer types associated with somatic mutations (20–30% of breast, lung, pancreas, bladder, etc.).

(C) P19$^{ARF}$, a regulator of P53 activity by the way of MDM2, is the product of the alternative reading of the *p16$^{INKA}$* gene. It is associated (possibly) with familial melanoma and with 15% of several sporadic cancer types.

(D) Cyclins' posttranscriptional modifications can change the expression of P27 and cyclin E. This may explain why normal expression of P27 and cyclin E are associated with breast cancer (41). In contrast, a lower expression of P27 has been reported with cancer of the prostate. P27 could be anticipated to act as a tumor suppressor gene and it does to an extent in mice. Mice$^{-/-}$ for the *P27* gene grow big (30% increased weight), have a gigantism phenotype, and are tumor-prone. However, in humans there is little evidence that *P27* is associated with cancer except for cancer of the breast and prostate, which are two endocrine cancers (42).

(E) The expression of cyclins A, B1, C, D1, D2, D3 and E is increased in 35% of cell lines derived from human breast cancer. A rise in cyclin D1 is an early event in breast tumors, and it is overexpressed in 45% of primary breast cancers. Its overexpression is associated with 76% low grade ductal carcinoma in situ (DCIS), 87% of high grade DCIS, and 83% of invasive ductal carcinoma. D1 is also overexpressed in 80% of invasive lobular cancers of the breast (43, 44).

Rb, first an orphan, soon became the founder and most prominent of a small family of pocket proteins including Rb, P107, and P130. All three are transcriptional regulators that associate with E2F genes. Consequently, their functions markedly overlap with that of Rb, and viruses that attack one also attack the other two. Moreover, the function of all three proteins is deregulated when mutations impair cyclins or cyclin-dependent kinases.

However, there are differences among the pocket proteins:

(1) During the cell cycle, there is a paradoxal behavior between P107 and P130. The first is low in $G_0$ but rises in the S phase, the reverse is true for P130. In contrast, the level of Rb is constant between $G_0$ and $G_1$.

(2) While Rb mainly transactivates E2F4 and E2F5, P107 and P130 seem to prefer EF2 and E2F3. The significance of these different biological functions is not clear, except that it may provide an apparent redundant molecular mechanism.

(3) The pattern of distribution of various pocket proteins varies during embryonic development. While the embryos of Rb null mice die on approximately the

[3]The history of the association of P16 and melanoma is of interest. Since the mid 1980s, it was suspected that melanomas and a marker on chromosome 9 (9p21) were linked, suggesting that a melanoma-linked gene might exist. One hundred melanoma cell lines were used to identify the 9p21 locus. Homozygous deletions clustered around a single site of 9p21 were detected in 60% of cell lines derived from melanomas and two genes present at that site were identified: *p16* and *p15*. By analysis of DNA sequences and introduction of deletions in those two genes, it was established that the mutations (missense, frameshift, unscheduled stops, and splicing) occur in the *p16* sequence in the melanoma cells. These mutations occur in vivo but are rare in sporadic cancers. Intermarriage between a man and a woman from two villages in Holland led to offspring who developed cancer (a melanoma) at the age of 15 and another who died of adenocarcinoma at 55; both offspring were homozygous for a 19 base pair deletion in chromosome 16, and both were also healthy in all other aspects. The finding established that: (1) *p16* activity is not indispensable for normal development; (2) the phenotypic expression of the mutation is variable; (3) *MLM* and *p16* genes are one and the same (45, 46).

thirteenth day of gestation with developmental defects in erythropoiesis, in the nervous system, and in the lens, P107- and P130-deficient mice survive gestation and reveal no detectable phenotypic changes in the adult.

(4) While Rb modulates muscle differentiation, P107 and P130 seem to regulate adipogenesis.

In summary, the three pocket proteins have a great deal in common. Each can contribute to cell cycle arrest, but their functions vary with the cell status of cell growth, cell differentiation, or cell development. In contrast to Rb, which is frequently mutated in various sporadic cancers, P107 and P130 mutations have seldom, if ever, been associated with cancer, unless mutations of related proteins (e.g., P27, P15, P16, and P19) are also present (47).

## *apc* Gene and Its Function

### *apc* Gene and Protein

The *apc* gene is located on chromosome (5q21), and it possesses an open reading frame of 8535 bp, 2644 codons, and 21 exons. The last exon's coding sequence extends over a 657 bp open reading frame. Several alternative spliced forms of *apc* are known, some of which include the coding region. The *apc* gene encodes a very large protein (312 kDa) mostly found in postmitotic epithelial tissue. The most common isoform is composed of 2843 amino acids. When the molecule is dissected, starting from its NH2 toward its COOH terminal, several familiar domains can be identified. Such domains include a number of protein-protein binding sites (some of which may constitute an oligomerization domain), seven heptad repeats, an armadillo repeat, a microtubular binding site, an EB1/RP1 binding site, and a DLG/PTB-BL binding site. In addition, there are three nuclear export signals and two nuclear import signals. Finally, in the COOH terminal are located: a β catenin binding site (a 15 amino acid repeat) a β catenin down-regulator (20 amino acid repeats, each repeat containing an SXXX5 motif, serine threonine phosphorylating sites), and an axon-conducting binding site.

The armadillo (residues 453 to 767) repeat binds to the B56 regulatory subunit of protein phosphatase 2 and the APC stimulated nucleotide exchange factor, two proteins involved in signaling. The size and the numerous interprotein binding sites, the nuclear translocation mechanism, and the catenin binding sites vouch for an important nuclear function that must involve catenins-axin and conduction (48).

### Function of APC Protein

The APC protein operates in the context of the Wnt (pronounced win) pathway. Wnt is an acronym for two genes: the wingless (found in *Drosophila*) and its mouse homologue Int. It has been referred to as a classic morphogen, a diffusible molecule that coordinates the various processes of morphogens of parts of or an entire organ (49). The product of the *apc* gene is a morphogen. Morphogens are molecules that contribute to morphogenesis, which is the generation of the shapes of the whole body or its parts (limbs, wings, organs, etc.). Morphogens are diffusible molecules, most often proteins, that modulate the gene expression of different cell types in a fashion that their growth, differentiation, and (when needed) their elimination evolves in time and sites, in a manner compatible with the programmed development of the whole organism of or any of its parts.

The proteins that contribute to development are numerous in humans and include the TGF-β family, EGF, and PDGF, among many others. Wnt is a prototype of a protein that contributes to development and was present from fly to human in the course of evolution.

In brief, the cysteine-rich glycoprotein Wnt (45–50 kDa) is the ligand that triggers the Wnt pathway. In the presence of a coreceptor (LRP5), Wnt interacts with the serpentine transmembrane receptor (Frizzle F3, 80 kDa). Two inhibitors function at this level: Frizzled related proteins (FRPs) and Dikkopf. The first inhibits the Frizzled receptor and the second blocks the association between the ligand and the receptor. The activated Frizzled receptor in turn interacts with Dishevelled (DVL) that down-regulates the glycogen synthetase kinase (GSK3 kinase), a serine/threonine protein kinase. As a result of the down-regulation of GSK3, some of its substrate, catenin, remains unphosphorylated and able to cooperate with transcription factors (Lef/Tcf) to activate the promoters of target genes.

Under normal circumstances, the pathway is regulated at the level of GSK3 by a trimolecular protein complex: axin, conductin, and APC. APC is a scaffold on which both GSK3β and the axin-conductin complex join. The complex sensitizes the catalytic function of GSK3β; consequently, most catenin is phosphorylated and taken to the merciless proteosome. A functional APC thus contributes, in vertebrates and *C. elegans*, in preventing catenin and the Tcl/Lef proteins from binding to DNA.

Mutations that inactivate APC abrogate this inhibition of the catenin phosphorylation, a situation that allows the unphosphorylated catenin and the Tcf/Lef proteins to bind DNA, activate target genes including *c-myc*, and allow for cell proliferation to take place (**FIGURE 22-5**).

## Mutations in the *APC* Gene

Over 300 germline mutations of the *APC* gene have been identified in kindreds with familial adenomatous polyposis (95%). Nonsense or frameshift mutations encode a truncated protein with an abnormal function and this truncation extends over 142 to 2644 codons. However, most of them are found between codons 1055 and 1309, and mutations upstream codon 200 or downstream codon 1600 are rare (with the exception of codon 2644). In accordance with the two hit theory, the germ line mutation is followed by a somatic hit manifested by loss of heterozygosity (LOH). At least in the case of *APC*, the germline mutation affects the somatic mutation. For germline mutations located upstream of codon 1194 or downstream 1392, the truncated allele appears in a cluster between codons 1286 and 1513. Of note, low risk polyposis is associated with mutations appearing close to or at the 3′ or 5′ end of the *APC* gene.

In familial adenomatous polyposis (FAP), oncogenesis is triggered by APC mutations, which are responsible for overactivation of β catenin/Tcf, at least in the great majority of cases. An association between alteration of the locus 5q and sporadic cancer of the colon was suspected. The discovery of the *APC* gene supported the notion and research was undertaken to find *APC* mutations in sporadic colon cancer. *APC* mutations have been detected in 60% of sporadic colorectal cancers (CRC), and half of those are biallelic. They result from insertions, deletions, point mutations, and frameshift. Most of them occur between codons 1286 and 1513, referred to as the mutation cluster region (MCR), which overlaps with the mutations observed in FAP (50).

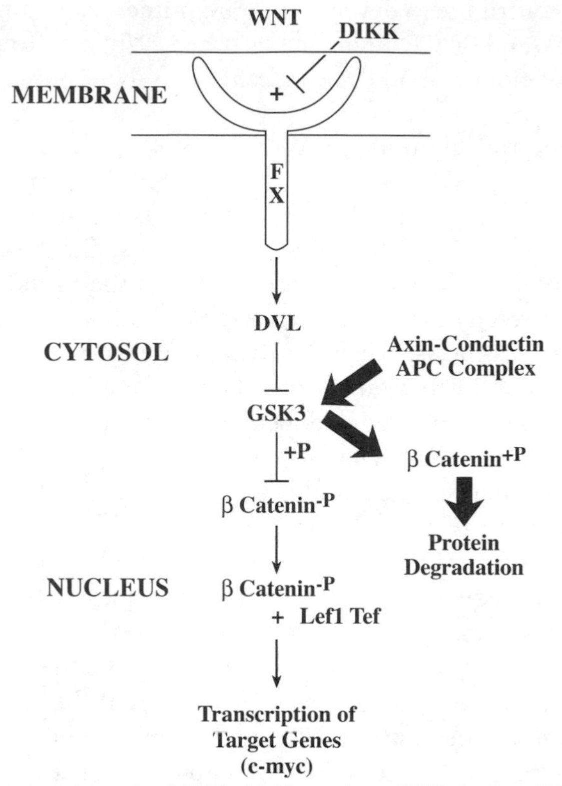


**FIGURE 22-5** Representation of the role of APC in the Wnt pathway. WNT, the ligand binds to Frizzled, the receptor (FX). Two inhibitors may intervene at this point. Dikkopf (DIKK) interferes with the interaction between ligand and receptor, and the Frizzled related protein (FRP) interferes with the function of the receptor. The activated receptor in turn activates dishevelled (DVL), which down regulates the glycogen synthetase kinase 3 (GSK3). As a result catenin remains unphosphorylated and is able, in the presence of Lef1 and TEF, to activate the transcription of target genes, e.g., c-*myc*. The presence of the trimolecular complex APC-axin-conductin upregulates the activity of GSK3, and catenin is phosphorylated and degraded. (Modified after Barish and Williams [48].)

## References and Recommended Reading


1. Lai, E., Clark, K.L., Burley, S.K., and Darnell, J.E. Jr. Hepatocyte nuclear factor 3/fork head or "winged helix" proteins: a family of transcription factors of diverse biologic function. *Proc Natl Acad Sci USA*. 1993;90:10421–3.
2. Weigel, D., Jurgens, G., Kuttner, F., Seifert, E., and Jackie, H. The homeotic gene forkhead encodes a nuclear protein and is expressed in the terminal regions of the *Drosophila* embryo. *Cell*. 1989;57:645–58.
3. Larsson, C., Hellqvist, M., Pierrou, S., White, I., Enerback, S., and Carlsson, P. Chromosomal localization of six human forkhead genes, freac-1 (FKHL5), -3 (FKHL7), -4 (FKHL8), -5 (FKHL9), -6 (FKHL 10), and –8 (FKHL 12). *Genomics*. 1995;30:464–9.
4. Akke, M., Forsen, S., and Chazin, W.J. Solution structure of (Cd2+)1-calbindin D9k reveals details of the stepwise structural changes along the Apo → $(Ca^{2+})$II1 → $(Ca^{2+})$I, II2 binding pathway. *J Mol Biol*. 1995;252:102–21.
5. Starich, M.R., Wikstrom, M., Arst, H.N. Jr., Clore, G.M., and Gronenborn, A.M. The solution structure of a fungal AREA protein-DNA complex: an alternative

binding mode for the basic carboxyl tail of GATA factors. *J Mol Biol.* 1998;277:605–20.

6. Kraulis, P.J., Domaille, P.J., Campbell-Burk, S.L., Van Aken, T., and Laue, E.D. Solution structure and dynamics of ras p21. GDP determined by heteronuclear three- and four-dimensional NMR spectroscopy. *Biochemistry.* 1994;33:3515–31.
7. Van Dongen, M.J.P., Cederberg, A., Carlsson, P., Enerbäck, S., and Wikström, M. Solution structure and dynamics of the DNA-binding domain of the adipocyte-transcription factor, FREAC-1. *J Mol Biol.* 2000;296:351–9.
8. Brunet, A., Bonni, A., Zigmond, M.J., Lin, M.Z., Juo, P., Hu, L.S., Anderson, M.J., Arden, K.C., Blenis, J., and Greenberg, M.E. Akt promotes cell survival by phosphorylating and inhibiting a Forkhead transcription factor. *Cell.* 1999;96: 857–68.
9. Tang, E.D., Nunez, G., Barr, F.G., and Guan, K.L. Negative regulation of the fork-head transcription factor FKHR by Akt. *J Biol Chem.* 1999;274:16741–6.
10. Kops, G.J., de Ruiter, N.D., De Vries-Smits, A.M., Powell, D.R., Bos, J.L., and Burgering, B.M. Direct control of the Forkhead transcription factor AFX by protein kinase B. *Nature.* 1999;398:630–4.
11. Zheng, W.H., Kar, S., and Quirion, R. FKHRL1 and its homologs are new targets of nerve growth factor Trk receptor signaling. *J Neurochem.* 2002;80:1049–61.
12. Shin, I., Bakin, A.V., Rodeck, U., Brunet, A., and Arteaga, C.L. Transforming growth factor beta enhances epithelial cell survival via Akt-dependent regulation of FKHRL1. *Mol Biol Cell.* 2001;12:3328–39.
13. Kashii, Y., Uchida, M., Kirito, K., Tanaka, M., Nishijima, K., Toshima, M., Ando, T., Koizumi, K., Endoh, T., Sawada, K., Momoi, M., Miura, Y., Ozawa, K., and Komatsu, N. A member of Forkhead family transcription factor, FKHRL1, is one of the downstream molecules of phosphatidylinositol 3-kinase-Akt activation pathway in erythropoietin signal transduction. *Blood.* 2000;96:941–9.
14. Zheng, W.H., Kar, S., and Quirion, R. FKHRL1 and its homologs are new targets of nerve growth factor Trk receptor signaling. *J Neurochem.* 2002;80:1049–61.
15. Rokudai, S., Fujita, N., Kitahara, O., Nakamura, Y., and Tsuruo, T. Involvement of FKHR-dependent TRADD expression in chemotherapeutic drug-induced apoptosis. *Mol Cell Biol.* 2002;22:8695–708.
16. Fredericks, W.J., Galili, N., Mukhopadhyay, S., Rovera, G., Bennicelli, J., Barr, F.G., and Rauscher, F.J. 3rd. The PAX3-FKHR fusion protein created by the t(2;13) translocation in alveolar rhabdomyosarcomas is a more potent transcriptional activator than PAX3. *Mol Cell Biol.* 1995;15:1522–35.
17. Bennicelli, J.L., Fredericks, W.J., Wilson, R.B., Rauscher, F.J. 3rd, and Barr, F.G. Wild type PAX3 protein and the PAX3-FKHR fusion protein of alveolar rhabdomyosarcoma contain potent, structurally distinct transcriptional activation domains. *Oncogene.* 1995;11:119–30.
18. Levens, D.L. Reconstructing MYC. *Genes Dev.* 2003;17:1071–7.
19. Pelengaris, S., and Khan, M. The many faces of c-MYC. *Arch Biochem Biophys.* 2003;416:129–36.
20. Gartel, A.L., and Shchors, K. Mechanisms of c-myc-mediated transcriptional repression of growth arrest genes. *Exp Cell Res.* 2003;283:17–21.

21. Orian, A., van Steensel, B., Delrow, J., Bussemaker, H.J., Li, L., Sawado, T., Williams, E., Loo, L.W., Cowley, S.M., Yost, C., Pierce, S., Edgar, B.A., Parkhurst, S.M., and Eisenman, R.N. Genomic binding by the Drosophila Myc, Max, Mad/Mnt transcription factor network. *Genes Dev.* 2003;17:1101–15.
22. Fernandez, P.C., Frank, S.R., Wang, L., Schroeder, M., Liu, S., Greene, J., Cocito, A., and Amati, B. Genomic targets of the human c-Myc protein. *Genes Dev.* 2003;17:1115–29. Epub.
23. Evans, G.I., and Littlewood, T.D. The role of c-myc in cell growth. (Review). *Curr Opin Genet Dev.* 1993;3:44–9.
24. Lindstrom, M.S., and Wiman, K.G. Role of genetic and epigenetic changes in Burkitt lymphoma. *Semin Cancer Biol.* 2002;12:381–7.
25. Sherr, C.J. The INK4a/ARF network in tumour suppression. *Nat Rev Mol Cell Biol.* 2001;2:731–7.
26. Giaccia, A., and Denko, N. Multiple roles for deregulated MYC expression in oncogenesis. In: Ehrlich, M., ed. *DNA Alterations in Cancer.* Natick, MA: Eaton Publishing; 2000:101–19.
27. Lindstrom, M.S., Klangby, U., and Wiman, K.G. p14ARF homozygous deletion or MDM2 overexpression in Burkitt lymphoma lines carrying wild type p53. *Oncogene.* 2001;20:2171–7.
28. Hill, D.A., De La Serna, I.L., Veal, T.M., and Imbalzano, A.N. BRCA1 interacts with dominant negative SW1/SNF enzymes without affecting homologous recombination or radiation-induced gene activation of p21 or Mdm2. *J Cell Biochem.* 2004;91:987–98.
29. Maestro, R., Dei Tos, A.P., Hamamori, Y., Krasnokutsky, S., Sartorelli, V., Kedes, L., Doglioni, C., Beach, D.H., and Hannon, G.J. Twist is a potential oncogene that inhibits apoptosis. *Genes Dev.* 1999;13:2207–17.
30. Hogarty, M.D., and Brodeur, G.M. Gene amplification in human cancers: Biological and clinical significance. In: Scriver, C.R., Beaudet, A.L., Sly, W.S., Valle, D., eds. *The Metabolic and Molecular Bases of Inherited Disease.* New York, NY; McGraw Hill; 2000:597–610.
31. Hortobagyi, G.N. Overview of treatment results with trastuzumab (Herceptin) in metastatic breast cancer. *Semin Oncol.* 2001;28:43–7.
32. Boxer, L.M., and Dang, C.V. Translocations involving c-myc and c-myc function. *Oncogene.* 2001;20:5595–610.
33. Liao, D.J., and Dickson, R.B. c-Myc in breast cancer. *Endocr Relat Cancer.* 2000;7:143–64.
34. Zajac-Kaye, M. Myc oncogene: a key component in cell cycle regulation and its implication for lung cancer. *Lung Cancer* 2001;34(suppl);S43–6.
35. Strober, B.E., Dunaief, J.L., Guha, S, and Goff, S.P. Functional interactions between the hBRM/hBRG1 transcriptional activators and the pRB family of proteins. *Mol Cell Biol.* 1996;16:1576–83.
36. Karantza, V., Maroo, A., Fay, D., and Sedivy, J.M. Overproduction of Rb protein after the G1/S boundary causes G2 arrest. *Mol Cell Biol.* 1993;13:6640–52.
37. Zheng, L., and Lee, W.H. Retinoblastoma tumor suppressor and genome stability. *Adv Cancer Res.* 2002;85:13–50.

38. Nasmyth, K., Peters, J.M., and Uhlmann, F. Splitting the chromosome: cutting the ties that bind sister chromatids. *Science.* 2000;288:1379–85.

39. Koshland, D.E., Guacci, V. Sister chromatid cohesion: the beginning of a long and beautiful relationship. *Curr Opin Cell Biol.* 2000;12:297–301.

40. Chen, P.L., Ueng, Y.C., Durfee, T., Chen, K.C., Yang-Feng, T., and Lee, W.H. Identification of a human homologue of yeast nuc2 which interacts with the retinoblastoma protein in a specific manner. *Cell Growth Diff.* 1995;6:199–210.

41. Porter, P.L., Malone, K.E., Heagerty, P.J., Alexander, G.M., Gatti, L.A., Firpo, E.J., Daling, J.R., and Roberts, J.M. Expression of cell-cycle regulators p27Kip1 and cyclin E, alone and in combination, correlate with survival in young breast cancer patients. *Nat Med.* 1997;3:222–5.

42. Park, M.S., Rosai, J., Nguyen, H.T., Capodieci, P., Cordon-Cardo, C., and Koff, A. p27 and Rb are on overlapping pathways suppressing tumorigenesis in mice. *Proc Natl Acad Sci USA.* 1999;96:6382–7.

43. Oyama, T., Kashiwabara, K., Yoshimoto, K., Arnold, A., and Koemer, F. Frequent overexpression of the cyclin D1 oncogene in invasive lobular carcinoma of the breast. *Cancer Res.* 1998;58:2876–80.

44. Buckley, M.F., Sweeney, K.J., Hamilton, J.A., Sini, R.L., Manning, D.L., Nicholson, R.I., de Fazio, A., Watts, C.K., Musgrove, E.A., and Sutherland, R.L. Expression and amplification of cyclin genes in human breast cancer. *Oncogene.* 1993;8:2127–33.

45. Bergman, W., Gruis, N.A., Frants, R.R. The Dutch FAMMM family material: clinical and genetic data. *Cytogenet Cell Genet.* 1992;59:161–4.

46. Fountain, J.W., Karayiorgou, M., Ernstoff, M.S., Kirkwood, J.M., Vlock, D.R., Titus-Ernstoff, L., Bouchard, B., Vijayasaradhi, S., Houghton, A.N., Lahti, J., et al. Homozygous deletions within human chromosome band 9p21 in melanoma. *Proc Natl Acad Sci USA.* 1992;89:10557–61.

47. Classon, M., and Dyson, N. p107 and p130: versatile proteins with interesting pockets. *Exp Cell Res.* 2001;264:135–47.

48. Barish, G.D., and Williams, B.O. The Wnt signal transduction pathway. In: Gutkind, J.S., ed. *Signaling Networks and Cell Cycle Control.* Totowa, NJ: Humana Press, 2000.53–02.

49. Martinez Arias, A. Wnts as morphogens? The view from the wing of *Drosophila.* (Review). *Nat Rev Mol Cell Biol.* 2003;4:321–5.

50. Kinzler, K., and Vogestein, B. Colorectal tumors. In: Scriver, C., Beaudet, A., Valle, D., Sly, W., Childs, B., Kinzler, K., and Vogelstein, B. *The Metabolic and Molecular Bases of Inherited Disease,* 8th ed. New York, NY: McGraw-Hill Publishers; 2000:1033–62.

CHAPTER

# 23 CANCER AND GENOMIC INSTABILITY

## Roles of the *ATM* and *NBS* Genes

The origin of life remains mysterious (1, 2) and whatever molecules first started the process in the mists of time, somehow RNA or DNA appeared. These molecules harbor moieties that are eager to absorb the ultraviolet light born from the sun. They are also sensitive to ionizing radiations that are also part of our terrestrial environment. Thus, DNA and RNA, children of the sun destined to populate the earth, soon became victims of the sky above and the land below. Photons and physical particles are not the only agents that cause damage to the DNA genome. The very oxygen we breathe can, whenever metabolic pathways fail to consume it effectively, generate reactive oxygen species that damage DNA.

Animals and humans (especially humans) blindly in search of food, drink, and ever greater comfort or—worse—mutual destruction, have contributed abundantly to the pollution of the soil, water, and air with DNA-damaging agents. We eat and drink the bitter price of negligence. Even cell proliferation essential to the development, growth, and maintenance of the living being is not foolproof. DNA synthesis and chromatid or chromosomal separation may fail and cause damage. These insults to the genome would have been fatal to life long ago if DNA repair mechanisms had not appeared in the course of evolution. The repair processes are multiple, each adjusting to different types of DNA insults (e.g., base alterations, strand breaks, crosslinks, insertion, and deletions). Before repair, the lesion must be recognized and the cell cycle must arrest long enough to allow restoration of DNA integrity. If the lesion is serious and the process fails, the consequences are tragic and include genetic errors, cancer, and cell death (**FIGURE 23-1**). Other possible lesions not considered here are atherosclerosis, autoimmune disease, and degenerative diseases.

Epidemiologic, molecular, and genetic investigations of inherited syndromes associated with genome instability and cancer predisposition have provided important new clues on the mechanisms of repair of the genome and the consequence of the loss of genomic integrity. There is a long list of such inherited conditions. The following sections consider oncogenesis associated with the failure of DNA repair of double strand breaks, crosslinks, and missmatches in the Nÿmegen break syndrome, Fanconi anemia, and Bloom syndrome. We also discuss the role of genomic damage in carcinogenesis in patients with hereditary non-polyposis colorectal cancer (HNPCC) and patients

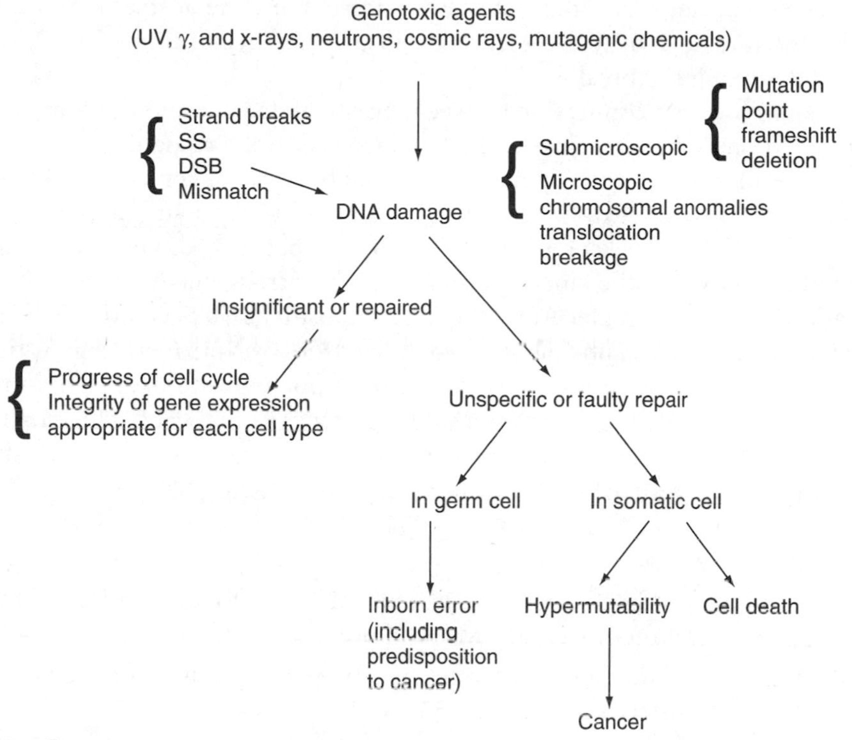


**FIGURE 23-1** Potential consequences of DNA damage.

carrying mutated or deleted *BRCA I* and *II* genes. Xeroderma and single strand breaks were reviewed in Part II of this book.

## Double Strand Breaks, ATM, Nÿmegen Syndrome, and Cancer

Double strand breaks (DSB) are frequent and the most deleterious lesions. One DSB can be fatal to the cell. Double strand breaks may be the consequence of germline mutations (AT and NBS), the direct result of exposure to ionizing radiation, or the indirect result of attempts to repair base damage when encountered on a branch of the replication fork. Among the many requirements for repair are recognition of damage and arrest or delay of the cycle.

### *ATM*, DNA Damage Recognition, and Cell Cycle Arrest

The leader among the sensors of DNA damage is the *ATM* (ataxia telangiectasia mutated) gene. Ataxia telangiectasia (AT) is primarily a disease of double strand breaks. But as this discussion unfolds, it will become obvious that ATM is not the only sensor. Among the relevant symptoms of ataxia telangiectasia are radiosensitivity, immunosuppression (defective recombination), genomic instability, increased death in some cells, and cell proliferation in others (namely, non-Hodgkin's lymphoma, Hodgkin lymphoma, leukemia, and perhaps breast cancer).

Because the mutant is associated with cancer and because AT patients are oversensitive to ionizing radiations, the wild-type *AT* gene is believed to be involved in DSB repair. DSB activates *ATM* by phosphorylation at Ser 1981, and activated *ATM* phosphorylates P53 (Ser 15), MDM2 (Ser 395), and CHK2 on threonine 68. ATM is at the head of a transducing chain that ultimately arrests the cycle at the $G_1$/S checkpoint. Consider the functions of the ATM protein in the presence of limited DNA damage, including double strand breaks.

The repair requires a connection between the site of damage and the complex repair process. ATM has long been suspected to be involved in the recognition of the signal, but the molecular signal at the damage site is not obvious. Bakkenkist and Kastan clarified part of the mystery (3). In resting cells, *ATM* exists as an antiparallel dimer or a multimer and is thereby kept inactive. The bonds between dimer involve the FAT domain of the ATM and the kinetic domain, and this arrangement prevents autophosphorylation of the molecule. After the cells exposure to 0.5 Gy, the FAT of each monomer of the dimer is phosphorylated at Ser 1981 by the kinetic site of the companion monomer. Because the ATM kinase does not join the damaged strand and because ATM is activated by hypotonic swelling or agents that cause chromatin structural change, Bakkenist and Kastan propose that the changes in chromatin structure induced by DSB are responsible for ATM's activation (**FIGURE 23-2**) (3).

The activated ATM attacks its substrates P53, MDM2, and CHK2 (discussed in Chapter 11). This process involves:

1. P53 is phosphorylated on serine 15 and this leads to stimulation of the transcription activation of P21, MDM2, and other proteins.
2. MDM2 is phosphorylated on serine 335, abrogating its ubiquitin ligase function and thereby indirectly stabilizing P53.
3. CHK2 is phosphorylated on threonine 68, and thereby enabled to phosphorylate P53 on serine 20.

On the basis of these findings, it is easy to conceive how ATM activation leads to an arrest of the cell cycle at the $G_1$/S checkpoint (**FIGURE 23-3**).

Does ATM directly contribute to the $G_1$/S checkpoint? P53 is among the earliest proteins phosphorylated by ATM (2).The transcription factor is phosphorylated at Ser 15 and then activated. Molecules that interact with P53, CHK2, and MMD2 are phosphorylated at Ser 395, and at serines 9 and 46 the nuclear export of P53-MDM2 is blocked. CHK2 phosphorylates P53 at Ser 20 and prevents P53 binding to MDM2, thereby interfering with its degradation. Thus, chromatin changes activate ATM, which activates P53, a major target for activation and stabilization first because of its direct phosphorylation by ATM or indirectly[1] by the ATM-dependent phosphorylation of CHK2 and MDM2 (refer to 11).

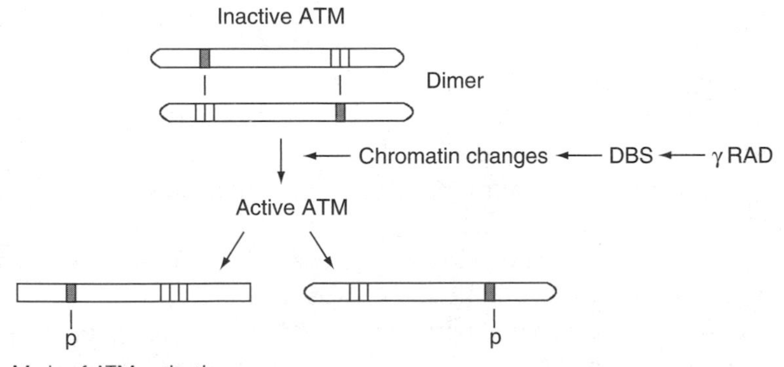

**FIGURE 23-2** Mode of ATM activation.

[1]Other PIKK family members also have been implicated in damage recognition, among them DNAPK.

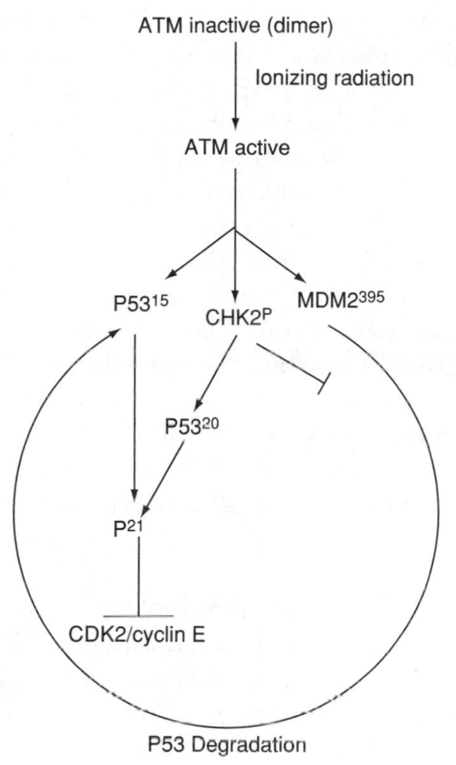


**FIGURE 23-3** ATM-induced cell cycle arrest.

The role of ATM is not restricted to regulation of checkpoints; ATM also regulates the downstream pathway of double strand repair. It phosphorylates BRCA1 complexes and other proteins after DSBs have been inflicted. BRCA1 is phosphorylated at Ser 1387 and Ser 1423. The first phosphorylation is believed to be involved with the $G_1$/S and the second with the $G_2$/M checkpoints. NBSL (which is part of the MER-RAD50–NSB1 complex or MNR) is phosphorylated at Ser 343 and Ser 278. Two serines of the protein SMC-1 (structural maintenance of chromosomes) involved in sister chromatids cohesion are also phosphorylated by ATM (refer to Chapters 11 and 13).

## Function of the *AT* Gene in Normal Cell Physiology

It is often forgotten that ATM's role is not restricted to DNA repair. ATM continually contributes to V(D)J recombination and maintenance of telomeric length. The notion of ATM's maintenance role is based on observation of telomeric shortening and clustering as well as abnormal chromosomal association and abnormal connection between telomeres and the nuclear matrix in ATM-deficient cells (4, 5). Moreover, the removal of endogenous TRF1 causes telomere elongation (6). TRF1 is a protein that interferes with telomere elongation and can also cause apoptosis (7). TRF is ATM-regulated and inhibition of TFR1 in AT cells elongates telomere and prevents apoptosis. Thus, ATM may also be involved in bringing the regulator of telomere length to bind to the telomere.

## ATM and Cancer

The findings discussed above clearly suggest that ATM is important to the maintenance of the integrity of the genome. Therefore, it is not surprising that loss of heterozygosity

(LOH) at 11q-22-23 (including the ATM locus) is associated with some types of leukemia (e.g., T cell prolymphocytic and B cell chronic leukemia) (8). An important but unanswered question is whether heterozygotes for the *ATM* gene, which are found in 1 to 2% of the general population, are over-sensitized to cancer. It is established that the ATM protein is truncated in 70 to 80% of AT affected families. Examination of the *ATM* gene in patients with T cell prolymphocytic leukemia such as B cell lymphoma detected a high incidence of AT missense mutations, mainly in the kinase domain. How can the findings of truncated proteins in AT patients and missense mutations in leukemic patients be reconciled? Gatti and Meyn in the same year proposed what appears to be a logical explanation. There are two classes of heterozygotes: ATtrunc/wtAT and ATmiss/ATmiss. Thus, it appears that the chromosome that harbors the missense mutations in reality functions as a contributor and, therefore, the AT trunc/AT miss are in reality homozygotes for cancer risk (8, 9).

*ATM* and *AT* mutations are associated with lymphoid cancer after LOH, but whether heterozygotes are sensitized to cancer is still in question (10). Moreover, as is the case for many hereditary cancers, the organ specificity of the cancer in AT patients is unsolved.

In conclusion, the *ATM* gene plays a critical role in DNA repair and when mutated compromises genome integrity. The ATM protein senses the chromatin distortion caused by DSB, phosphorylates proteins involved in the $G_1$/S checkpoint (P53) and possibly in the $G_2$/M checkpoint, and contributes to the phosphorylation and the assemblage of several proteins involved in DNA repair (**TABLE 23-1**).

## Nijmegen Breakage Syndrome and Cancer

Not unlike ataxia telangiectasia, the Nijmegen breakage syndrome (NBS) is an autosomal recessive hereditary disease associated with chromosomal instability, cancer predisposition (lymphomas), and radiosensitivity. However, there are differences between the clinical manifestations of NBS and AT, and the most salient difference is the NBS gene function in the repair process. While the product of the *ATM* gene senses the damage, that of the NBS1 gene is believed to contribute to the repair of strand breaks. Because of the differences between the *ATM* and the *nbs1* genes, it is not expected that *nbs1* complements for radiation-induced damage in NBS-AT fusion cells. The NBS1 gene (locus 18q21) encodes a polypeptide with sequence homology (29%) with the yeast

**Table 23-1 ATM substrates**

| Substrates | Sites of Phosphorylation |
|---|---|
| P53 | Serine 15 |
| CHK2 | Threonine 68 |
| MDM2 | Serine 395 |
| BRCA1 | Serines 1367, 1423, 1457, and 1524 |
| NBS1 | Serines 278 and 343 |
| H2AX | COOH terminal |
| SMC1 | Serines 957 and 966 |
| CtP | Serines 664 and 745 |
| 4E:DPI | Serine 112 |

Xrs2 repair protein that complexes with MRE11/RAD50. MRE11/RAD50 is known to be involved in the repair of double strand breaks. Within 30 minutes after exposure to γ radiations, the complex forms foci in the vicinity of double strand breaks. Similarly, NBS1 participates in the NBS1/MRE11/RAD50 complex (NMR). The gene of NBS1 has been cloned; it is located on chromosome 8 (8q21), contains 16 exons, extends over 48,979 bps, and encodes an 85 kDa protein homologous to the Xrs2 of yeast (11). The gene is transcribed into two messenger RNAs (2.6 and 4.4 kb) that differ by the length of their untranslated 3′ region. Most of the identified mutations cluster.

The protein harbors several characteristic domains: a Forkhead associated (FHA) domain (amino acids 24 to 108), a sequence believed to mediate interaction with phosphoproteins, and a BRCT domain (amino acids 109 to 196) similar to that found in the COOH terminal of BRCA1 that operates in DNA checkpoints. FHA/BRCT directly binds to histone γ-H2AX. Both the FHA and BRCT of NBS1 are required for association with H2AX. The FHA/BRCT domains are often conserved in proteins that modulate cell cycle checkpoints. NBS1 also forms a complex with MRE11/hRAD50, which possesses functions as a nuclease (12).

The encoded protein (753 amino acids) has been divided in three functional segments: the N terminus (1–196 amino acids), the central region (278–343 amino acids), and the C terminal (665–693 amino acids). The N terminal includes a forkhead-associated domain (24–108 amino acids) and an adjacent BRCT (108-196 amino acid) domain. FHA and BRCT domains are often found in proteins that regulate cell cycle checkpoints, DNA repair, or both. Functionally, the FHA/BRCT sequence is characterized by its ability to recognize the presence of phosphorylation sites on target proteins. The NBS1 FHA/BRCT binds to histone H2AX after its phosphorylation by ATM. The central region is phosphorylated at two sites, serines 278 and 343 by ATM or ATR, and the histone is believed to operate in signal transduction by modulating the intra-S checkpoint. The C terminal binds to the conserved (from yeast to human) DNA repair duplex MRE11/RAD50 to yield the NBS1/MRE11/RAD50 triplex or NMR. The NMR has three possibly related functions: it participates in homologous recombination; modulates the intra-S checkpoint; and contributes to the maintenance of telomeres (13).

## NMR and Homologous Recombination

NMR is indispensable for homologous recombination after DSB, but probably not for non-homologous end joining. However, it is not excluded that NMR may contribute to mismatch repair (14). Each component of the triplex has a critical role. hRAD 50 functions at least in part as structural components that anchors the triplex at DNA sites located in the vicinity of the DSB. MRE11 is a multiple catalytic unit that harbors: phosphoesterase activity in its N terminal; 3′–5′DSB/DNA exonuclease activity; single stranded DNA endonuclease activity; and DNA unwinding activity. All of these catalytic functions are needed for successful double strand repair. NBS1 guides the triplex to the DNA sites where the triplex contributes to the formation of foci. Two findings support this view. NBS contains two nuclear localization sites and MRE11 is primarily found in the cytoplasm, but is translocated to the nucleus in the presence of NBS. Translocation does not occur in absence of the gene encoding NBS1.

For DSB repair, the MRE11/RAD50 complex needs to be brought into the nucleus. It has been proposed that two hRAD50 molecules are assembled end-to-end to form an antiparallel homodimer that is bent at the central hinge to form an inverted "V." A MRE11 dimer rests on top of the V (**FIGURE 23-4**). When bound over a sequence

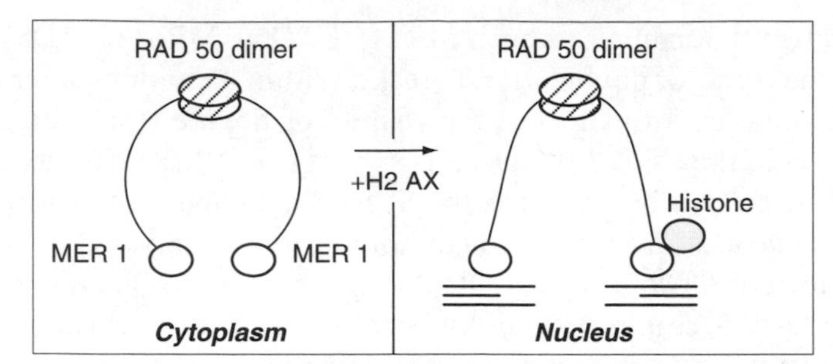


**FIGURE 23-4** The interaction of the MER11-RAD50 with the 3′ and the 5′ ends of the double strand breaks.

containing the DSB, such a structure would restrict the catalytic activities of hRAD50 to the vicinity of the DSB. Once inside the nucleus, the NBS/hMRM/hRAD complex binds a phosphorylated histone γ H2AX and the quaternary complex is ultimately found at the sites of the DSB where it triggers homologous recombination repair.

Within less than 30 minutes after exposure to ionizing radiation, γ H2AX[2] is phosphorylated (mainly by ATM, but to a lesser degree by ATR and DNAPK) in its C terminal to yield γ H2AX. NBS interacts with the histone through its FAH/BRCT motif.

## NBS1 and Cell Cycle Checkpoint

NBS1 contributes to the control of the intra-S checkpoint. Two pathways control the intra-S checkpoint: the SMC1, and the Chk2 and cdc25A pathways. The first is controlled by NBS and the second by ATM. The activation of both is needed to arrest the S phase. NBS1 is likely to be involved only in the first. After DNA damage, ATM phosphorylates SMC1 at serines 957 and 96, but only after NBS1 has been phosphorylated also by ATM at serines 278 and 343. In the second pathway, ATM phosphorylates Chk2, which in turn phosphorylates cdc25A, which once phosphorylated arrests the progress of the intra-S phase (**FIGURE 23-5**) (14). The ability of NBS1 to arrest the intra-S checkpoint explains the appearance of resistant DNA synthesis (RDS) in the absence of NBS1.

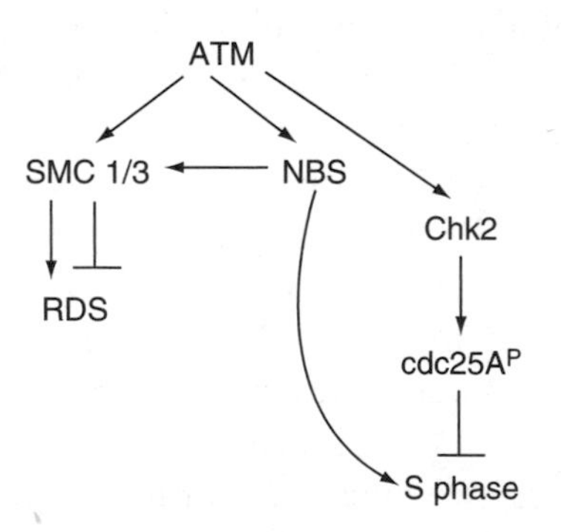


**FIGURE 23-5** The control of the S-phase checkpoint by NBS and the resulting resistant DNA synthesis.

[2]In the nucleosome, the DNA sequence is associated with four histones (H2A, H2B, H3, H4), forming a tetramer. H2A is a member of a larger family of 10 histones encoded by 11 different genes. H2AX is a minor member of the family with a C terminal markedly longer than that of H2A.

### Function of Nuclear Foci

Foci appear after exposure to double strand breaks. They consist of discrete masses of protein complexes detectable by, for example, immunofluorescent staining. At first they were believed to contain only MER11 and hRAD50 (15), but later it was demonstrated that the hMER11/hRAD50 complex is joined by the NBS protein and ultimately associates with γ-H2A foci (16).

## NBS and Telomeres

The telomeres of MBS patients, like those of AT patients, are shortened. This may or may not be related to premature aging. Indeed, if AT patients may age faster, there is no indication that NBS patients do so as well. Yet, the reduction of telomeric length in the cells of NBS patients resembles that observed in AT patients (~200–260/bp per replication). These findings indirectly suggest that the NBS1 must contribute to telomeric lengthening.

NBS1 shares sequence homology (29%) with Xrs2, a yeast protein that is also involved in telomeric elongation. Moreover, NBS1 physically interacts with two mammalian telomeric binding proteins (TFR1 and TFR2), but only during elongation in the S phase, in contrast to MER11, which is found at the interphase. Whatever the exact molecular role of MBS1 in telomeric integrity may be, it is not ruled out that the loss of the *NBS1* gene could facilitate telomeric degradation, an event that contributes to genomic instability.

### NBS and Cancer

Because of their cellular phenotype, radiation sensitivity, chromosomal instability, and defective S checkpoint, defective homologous recombination NBS and ATM are often compared. The clinical phenotypes are, however, significantly different. Patients with AT do not present with microcephaly, bird like face, or growth retardation, and in NBS patients neuronal degeneration and telangiectasia are absent. The two diseases share immunodeficiency and in particular cancer predisposition, however. B cell lymphomas and leukemias are by far the predominant cancers that appear in both NBS and AT patients, and these appear in NBS patients at a relatively earlier age than in AT patients. Although the findings are not entirely conclusive, a predisposition to sporadic cancer seems to exist, especially for breast cancer; otherwise, at least at present, there is no evidence that NBS patients are cancer-prone (3).

## References and Recommended Reading


1. Orgel, L.E. The origin of life—a review of facts and speculation. *Trends Biochem Sci.* 1998;23:491–5.
2. Bada, J.L., and Lascano, A. Perceptions of science. Prebiotic soup—revisiting the Miller experiment. *Science.* 2003;300:745–6.
3. Bakkenist, C.J., and Kastan MB. DNA damage activates ATM through intermolecular autophosphorylation and dimer dissociation. *Nature.* 2003;421:499–506.
4. Pandita, T.K., and Dhar, S. Influence of ATM function on interactions between telomeres and nuclear matrix. *Radiat Res.* 2000;154:133–9.
5. Hande, M.P., Balagee, A.S., Tchirlov, A., Wynshaw-Boris, A., and Lansdorp, P.M. Extra-chromosomal telomeric DNA in cells from Atm(–/–) mice and patients with ataxia-telangiectasia. *Hum Mol Gen.* 2001;10:519 28.

6. van Steensel, B., and de Lange, T. Control of telomere length by the human telomeric protein TRRF1. *Nature.* 1997;385:740–3.
7. Li, B., Oestreich, S., and de Lange, T. Identification of human Rap1: implications for telomere evolution. *Cell.* 2000;101:471–83.
8. Gatti, R.A., Tward, A., and Concannon, P. Cancer risk in ATM heterozygotes: a model of phenotypic and mechanistic differences between missense and truncating mutations. Review. *Mol Genet Metabol.* 1999;68:419–23.
9. Meyn, M.S. Related articles, links ataxia-telangiectasia, cancer and the pathobiology of the ATM gene. *Clin Genet.* 1999;55:289–304.
10. Shilo, Y. ATM and related protein kinases: safeguarding genome integrity. *Nat Rev Cancer.* 2003;3:155–68.
11. Tauchi, H. Positional cloning and functional analysis of the gene responsible for Nijmegen breakage syndrome, NBS1. *J Rad Res.* 2000;41:9–17.
12. D'Amours, D., and Jackson, S. The Mre11 complex: at the crossroads of DNA repair and checkpoint signaling. *Nat Rev Mol Cell Biol.* 2002;3:317–27.
13. Tauchi, H., Kobayashi, J., Sakamoto, S., and Komatsu, K. Nijmegen breakage syndrome gene, NBS1, and molecular links to factors for genome stability. *Oncogene.* 2002;21:8967–80.
14. Gianni, G., Ristori, E., Cerignoli, F., Rinaldi, C., Zani, M., Viel, A., Ottini, L., Crescenzi, M., Martinotti, S., Bignami, M., Frati, L., Screpanti, I., and Gulino, A. Human MRE11 is inactivated in mismatch repair-deficient cancers. *EMBO Rep.* 2002;3:248–54.
15. Maser, R.S., Monsen, K.J., Nelms, B.E., and Petrini, J.H. hMre11 and rRad50 nuclear foci are induced during the normal cellular response to DNA double-strand breaks. *Mol Cell Biol.* 1997;17:6087–96.
16. Kobayashi, J., Tauchi, H., Sakamoto, S., Nakamura, A., Morishima, K., Matsuura, S., Kobayashi, T., Tamai, K., Tanimoto, K., and Komatsu, K. NBS1 localizes to gamma-H2AX foci through interaction with the FHA/BRCT domain. *Curr Biol.* 2002;12:1846–51.

CHAPTER

# CANCER AND GENOMIC INSTABILITY: MISMATCH REPAIR AND CANCER

# 24

Satellite DNA (~10% of total DNA) is made of highly repetitive sequences. The overall sequence of satellite DNA differs enough from most of the remaining cellular DNA that it sediments as a separate band when centrifuged in a buoyant density gradient. Minisatellites and microsatellites are segments of DNA undetectable by centrifugation, but they are also characterized by repetitive sequences. Microsatellites are short base sequences, approximately 50 to 60 bp, which include several repeats less than six bases long. The dinucleotide ($CA_n$) is the most common repeat and the genome contains 50 $\times 10^3$ to $10^5$ such repeats. These are sites at which the DNA polymerase often slips during replication and the larger the repeat, the greater the chance of slippage. When the damage is not repaired, the daughter strand ends up containing too many or too few of the sequences and frameshift mutations develop. Repair of microsatellites takes place through the mismatch repair system, a system first discovered in bacteria and yeast.

During DNA replication, mispairing occurs in almost every $10^3$ to $10^4$ base pair. In most cases, however, proofreading corrects the mispair with the help of 3′ exonuclease activity associated with the polymerase. This reduces the occurrence of mispairing to 1 in $10^{12}$ bp. The mismatch repair system corrects 99.9% of the remaining mispairs. The repair pathway (1, 2) is complex and incompletely understood. Some of the enzymes involved in the pathway (3) are listed in **TABLES 24-1** and **24-2**.

In the presence of a single base mispair, during the S or $G_2$ phase of the cell cycle, hMSH6 and hMSH2 form the complex, hMUTSα. This complex recognizes the mismatch and binds to it. A second heteroduplex is formed between hMSH2 and hMSH3: hMUTSβ. While the hMUTSα has affinity for a single base mispair, hMutβ recognizes loop-out structures that emerge as a consequence of the mispair in a microsatellite.

In the case of a single base mispair, hMUTSα couples to hMUTLα (a heterodimer composed of hPMS2 and hMLH1) to form a tetramer comprised of two heterodimers: hMSH6-hMSH2 and hPMS2-hMLH1. The tetramer corrects the mispair. However,

**Table 24-1 Human homologues of yeast mismatch enzyme**

| Enzyme | Chromosome Locus | kB | Exon | Cancer Associated with Mutations |
|---|---|---|---|---|
| hMLH1 | 3p21.3 | 59 | 19 | |
| hMLH3 | 14q3 | | | |
| hMSH2 | 2p22-p22 | 73 | 16 | HNPCC, colorectal, ovarian |
| hMSH3 | 5q11.2-12 | 154.1 | 16 | Endometrial |
| hMSH4 | 1p31 | | 19 | |
| hMSH5 | 6p21.3 | 23.6 | 25 | |
| hMSH6 | 2p16 | | 10 | General cancer sensibility |
| PMS1 | 2q37-33 | 58.7 | 12 | Endometrium (HNPCC) |
| PMS2 | 7p22 | 16 | 15 | Turest syndrome, colorectal cancer Hereditary non-polyposis type 4 |

**Table 24-2 Complexes functioning in human mismatch repair**

| Complex | | Subunits |
|---|---|---|
| hMUTSα | = | hMSH6/hMSH2 |
| hMUTSβ | = | hMSH2/hMSH3 |
| hMUTLα | = | hPMS2/hMLH1 |

different tetramers operate in the correction of an insertion or a deletion loop, for example, hMH3-hMSH2 and hPMS2-hMLH1 formed by the association of hMUTSβ and HMULTα. Other enzymes involved include helicase II, which separates the strand prior to repair, DNA polymerase and ligase enzymes to secure strand replication and ligation (refer to Chapter 10, Figure 10-4).

## Hereditary Non-Polyposis Colorectal Cancer

Hereditary non-polyposis colorectal cancer (HNPCC) is inherited as an autosomal dominant gene. Victims of the disease develop cancer between 40 and 50 years of age. These cancers are not restricted to the colorectum, as they also occur in the stomach, small intestine, endometrium, ovary, kidney, and (rarely) in the brain. Polyps that arise in HNPCC are different from those seen in polyposis. They are poorly differentiated, often infiltrated with lymphocytes, and actively secrete mucus. Appropriate testing reveals microsatellite instability.

In 90% of HNPCC cases, a germline mutation in a mismatch repair gene is detectable. hMSH2 and hMLM1 are frequent targets for mutations. The heterozygote is asymptomatic and cancer only develops when a somatic mutation appears among the mismatch repair genes (two hit theory, loss of heterozygosity [LOH]). The reasons for the appearance of the somatic mutations in colonic cells, although important, are not yet known, but are suspected to be environmental. Kindred of HNPCC patients can be identified and kept under surveillance by using the Amsterdam criteria (see section on colon cancer below). A definite diagnosis rests with the demonstration of the germline mutation (4).

Slippage at repeats located in a coding region cause frameshift mutations. Despite the numerous repeats found in the genome, some genes are affected more frequently. For example, frameshift mutations affecting the TGFβRII have been observed in germ cells of patients with HNPCC and in sporadic cancers of the colon. IGFIIR and Bax mutations have been found in association with mutations of MSH3 and MSH6. These mutations contribute to cell proliferation by constitutive activation of the receptor (IGFIIR) or interference with apoptosis due to inhibition of the function of Bax. In contrast, mutations of other genes with functions unrelated to cancer progression may have little or no effect on the evolution of the cancer, at least at an early stage. Thus, the significance of the loss of genomic integrity varies with the gene involved.

Some genes are more important to cancer initiation and progress than others, and among these are genes involved in cell proliferation, differentiation, and death. Moreover, mutations of genes involved in DNA repair are more likely to contribute to hypermutability. The existence of somatic frameshift mutations in the coding region of hMSH3 and hMSH6 (two genes involved in mismatch repair) in HNPCC patients are notable, because they do indeed contribute to a hypermutable state (5).

In summary, defects in mismatch repair can trigger a hypermutable phenotype. Loeb proposed that the progression of cancer required that a hypermutable phenotype be established, but no concrete molecular mechanism for such an event was available (16). The discovery of defects of mismatch repair in HNPCC provided the first such mechanism and others have been discovered since. These mechanisms are relevant to genomic instability and the pathogenesis of cancer of the colorectum.

## Genes and Progression of Cancer in the Colorectum

### Sporadic Colorectal Cancer: A Model for Epithelial Carcinogenesis

Cancer of the colorectum is the second leading cause of cancer death in the U.S. with approximately 150,000 cases presenting and 50,000 deaths each year. A higher incidence has been reported in Europe (~210,000 cases with a 50% death rate each year). The cancer becomes evident primarily when patients are in their middle and older age ranges, and 90% are over age 50. Except for its anatomic location, gross appearance, histology, and mode of spreading, until the last few decades little was known about the pathogenesis or the potential of early diagnosis of colorectal cancer (CRC). Germline mutations are responsible for two types of hereditary cancers of the colorectum inherited according to Mendel's laws: familial adenomatous polyposis and hereditary nonpolyposis colorectum cancer. The first (FAP) is caused by mutations of the *APC* gene and the second (HNPCC) by mutations of genes of the mismatch repair pathway.

However, FAP and HNPCCs account only for 5% to 10% of all colorectal cancers; therefore, most colorectal cancers diagnosed routinely are sporadic and not associated with germline mutations. Epidemiologic studies have firmly incriminated environmental factors in the pathogenesis of many, if not all, sporadic CRC. Diet, smoking, and air or water pollution are a few of these factors. The environmental contributions to CRC cannot be ignored, but will be considered only briefly later in this chapter. Whatever the role of the environment, it is established that the evolution of colorectal cancer is linked to sequential somatic mutations, and an understanding of the waves of the somatic mutations provided guidelines for early diagnosis and interventions in CRC.

Inherited susceptibility is estimated to contribute to another 20% of CRCs. When is such susceptibility suspected? It is found in patients with more than one CRC, or in patients with a strong family history of CRC (affecting 2 or more direct family members) who have adenomatous polyps occurring simultaneously or in short succession. Sigmoidoscopies or colonoscopies performed at appropriate times have saved their lives because of an early diagnosis and subsequent surgery or other treatment.

The contributions of the identification of the mutations of the *apc* and the *hnpcc* genes to the pathogenesis of CRC have been discussed earlier in Chapter 22, but because the identification of the HNPCC kindred is subtle and the diagnosis sometimes overlaps with other types of CRCs, the criteria for identifying the kindred (the Amsterdam criteria) (7) are summarized:

1. CRC should be present in at least three relatives, one of whom is a first relative of the others.
2. CRC should occur in at least two generations of the family.
3. One CRC should be diagnosed before the age of 50.
4. FAP patients should be excluded from the kindred.

These criteria are useful for attracting attention to CRC risk, but they do not always strictly correlate with defective mismatch repair. Only 50 to 60% of the families in which the observations are in accordance with the Amsterdam criteria have demonstrated defects of the repair genes. Moreover, such mutations have been detected among patients who do not satisfy these criteria. This is especially true when the patient presents with extracolololonic cancers (8, 9).

The criteria were refined in 1998 (the Bethesda criteria) and again in 2004 (10, 11). Of course, a definitive diagnosis of HNPCC can only be obtained by identifying the genetic mutation in each case. The reader is referred to the excellent reviews of Boland (8) and Fearnhead and associates (9) for further details.

## Progression of Sporadic CRC

Histopathologists have long known that colorectal cancer (like many other cancers such as skin, esophagus, cervix, breast, etc.) evolves in stages, each of which can be associated with a rise in aggressiveness. Although the process is not unique to colorectal cancer, the documentation of anatomic and histopathologic observations coupled to clinical manifestations allow for a clearer definition of each stage of colorectal cancer. These stages are reviewed by Kinzler and Vogelstein and include: dysplastic aberrant crypts foci (12), early adenoma, intermediate adenoma, late adenoma, and carcinoma (13, 14). The carcinomas often progress from well to poorly differentiated and ultimately metastasize.

Thanks to cytogenetics and the identification of specific genes contributing to cancer, in the last decades of the 20th century, somatic mutations associated with critical steps in the chain of events leading to colorectal cancer were defined. This was possible in part because of the long time it takes (20 to 40 years) for CRC to evolve from the earliest precursor (the aberrant crypt focus and or adenoma) to full blown cancer. Some mutations occur at critical phases of the evolution of the CRC (e.g., *apc*, *K-ras*, *smad-4*, *-2*, and *p53*). Others (not all known) have been shown to contribute to cancer progression, though the precise time of their intervention is not clear. In general, it appears that in most cases mutations of the *apc* gene initiate the oncogenic process.

At least in mice, the dysplastic aberrant crypt foci (ACF) are the earliest histologic manifestation of the path to cancer. ACFs were first recognized by Bird et al. (12), who described these crypts as, "Altered lumenal openings with thickened epithelia enlarged compared to that in adjacent crypts." The number of ACFs rises with exposure to carcinogens. ACFs are believed to be the first manifestation of a preneoplastic condition (12 and references therein).

The genes involved in the oncogenic progression of CRC, and their chromosomal locations and functions are already familiar: *apc* (5q21-q22), K*ras* (i2p12.1), *smad-4* (18q21.1), *p53* (17p13.1).(8,9,13). (Refer to Chapters 10, 17–19, and 22.)

## *apc* Gene

The role of the *apc* gene is not restricted to FAP; it also contributes to sporadic colorectal cancers (up to 80%). The tumor suppressor gene encodes a 312 kDa protein, the function of which includes regulation of cell adhesion, cell migration, microtubular assembly, and signal transduction. At present, the only well-defined molecular intervention of the APC protein involves the WNT pathway, where it functions as a regulator of the β catenin levels. Unphosphorylated β catenin participates in the transduction pathway that leads to the coactivation of several target genes, among them *c-myc* and *cyclin D1* (discussed in Chapter 22). Consequently, the biallelic mutation of the *apc* gene in a somatic cell is most likely to be responsible for clonal expansion of a cell carrying the mutants. However, as pointed out by Fodde (15), the role of APC is not limited to signal transduction in the WNT pathway. It also links E-cadherin and α-catenin, two molecules connected with the adherens junctions. Thus, APC could also modulate cell adhesion by regulating the intracellular distribution of α-catenin, a control mechanism that may be disrupted by *apc* mutations. APC's function is also associated with that of microtubules but this is independent of *apc*'s role in the WNT pathway and also involves APC's C-terminus. The consequences of interruption of this function are not yet clear, but it is possible that they may contribute to chromosomal instability (CIN) (15).

## *K-ras* Gene

*K-ras* is mutated in at least 50% of cancers of the CRC. Mutated *K-ras* constitutively activates downstream pathways (e.g., mitogen activated protein kinase [MAPK]) and generates a mitogenic response. The mutations of *K-ras* are believed to coincide with the passage from early adenoma to intermediate and possibly late adenoma. The mitogenic stimulus of mutated *K-ras* is dominant and only the mutation of one allele suffices.

## The *smad* Genes

The *SMAD*s are also believed to contribute to active proliferation of adenomas to cancer. *SMAD4* and *SMAD2* are components of the transduction pathway of transforming growth factor-β (TGF-β). TGF-β is a perplexing gene product that under the appropriate circumstance can stimulate cell growth and arrest cell differentiation and apoptosis. Its type II receptor is itself susceptible to frameshift mutations in mismatch repair-deficient cells. However, TGF-β's molecular contribution to cancer progression, with the exception of cell proliferation, is not yet clear. In contrast to what is observed for other genes involved in CRC progression, the temporal association of *smad*s genes with cancer progression can only be surmised. *smad4* (18q21-1) deletions or LOH are seen in cancers of the pancreas, while deletion and point mutations of *smad4* and *smad2* have been detected in colorectal cancers. These mutations are suspected to be associated with poor prognosis and to be a predictor of metastasis (16).

### *p53*

*p53* (or *tp53*) is mutated in many cancers. The mutation is associated with the abrogation of $G_1$/S and possibly other checkpoints. Cells harboring a damaged DNA proliferate and escape apoptosis.[1] Cells harboring a *p53* mutation grow because of their inability to arrest the cell cycle (even if their DNA is damaged) and also because they are resistant to apoptosis. As a result, the cell population at the core of the tumor becomes hypoxic, an environment that is believed to favor the proliferation of *p53* mutated cells, leading to the predominance of a *p53* mutated cell population in the tumor, thereby favoring dysplasia and more uncontrolled proliferation. It cannot be ruled out that by allowing the escape at the $G_1$/S checkpoint of cells harboring a damaged DNA, the *p53* mutated cells may contribute to the genesis of a clone with genomic instability.

Germline mutations of mismatch repair genes (*mlh1* and *msh2,* chromosomes 3p21 and 2p14-16, respectively) are associated with hereditary non-polyposis colonic cancer and a high risk of CRC (~80%). The cancer risk is not limited to CRC because the risk also increases for cancers of the stomach, small intestine, ovaries, liver, kidney, and ureter. Moreover, somatic mutations of HNPCC have also been reported in sporadic CRC. While the germ mutation is an initial event in HNPCC, the somatic mutations are believed to occur late in the cancer progression phase in most cases of sporadic CRC. In both cases, mutations of the mismatch repair molecules cause microsatellite instability. It has been suggested that mismatch repair defects may also initiate the oncogenic process in some CRCs (13–15).

The potential role of epigenetic methylation in the pathogenesis of colorectal cancer needs to be considered. Close to half of the human genes contain 0.5 to 2 kB regions of CpG islands (rich in cytosine and guanine). Methylation of these regions in the 5′ sequence of the gene silences transcription (17–20). Methylation of the gene's (CpG) islands serves to regulate gene expression in normal cells. For example, during the progress of cell division the genes required for passage into the next phase of the cycle (e.g., S) have demethylated promoters. In contrast, the promoters of the genes required in a previous phase (e.g., $G_1$) are methylated and thereby silenced when they are no longer needed for progress into a next phase. Consequently, disturbed methylation can disrupt the orderly process of cell division (21). Many genes that contribute to the pathogenesis of colorectal cancer can be mutated by epigenetic methylation. They include *APC, K-ras, P53, P16, MHL-1, CDKN2A* (cyclin-dependent kinase inhibitor 2A), *MGMT* (0-6 methyl guanine-DNA methyl transferase) (22).

In addition, the expression of other genes may become deregulated by mutations or be disturbed by crosstalk between the transducing pathways, and thereby help the evolution of an early well differentiated cancer into a florid one that is invasive, poorly differentiated, and metastasizing. For example, at some stage the growing tumor needs to be supported by an appropriate vasculature. Angiogenic factors are released and stim-

[1]In the rare hereditary disease, Li-Fraumeni syndrome, germline mutations of *p53* are responsible for the disease that is associated with a variety of cancers occurring at a young age (childhood osteosarcoma; soft tissue sarcomas; leukemias; and breast, brain, and adrenocortical cancers), but are seldom found in colorectal cancers.

ulate vascular growth. The cancer may then invade surrounding tissue, possibly with the help of proteases, and also find its way to lymph nodes, liver, and bones by mechanisms not entirely known. These important events are not reviewed, but it should be remembered that they are largely responsible for drug resistance and failure of surgery, radiotherapy, and chemotherapy of advanced cancers.

## Interaction Between Genes and Environment in Colorectal Cancer

Geographic distributions and epidemiologic investigations have clearly established an association between colorectal cancers that is independent of age, the major risk factor for most adult cancers. The incidence of cancer of the colon rises in the offspring of persons who migrate from a low risk to a high risk region. This observation implicates the environment as a contributor of cancer of the colorectum. In 1995, Spjut and Pratt showed that 3′-2′-dimethyl-4-aminodiophenyl can induce colon cancers in rats (23). In humans, numerous environmental factors have been incriminated in the incidence of cancer of the colorectum. Probably the most prevalent is diet. High consumption of red meat is a major suspect. Cooking red meat at high temperatures generates heterocyclic amines, which are metabolized by cytochrome P450 sulfotransferases and N acetyltransferases that generate bioactive carcinogenic compounds. The molecular pathway to carcinogenesis starts with a procarcinogen that is metabolically converted to an active carcinogen (that may also be a mutagen). The mixed function oxidases are among the enzymes involved in the process (24).

The vast research on the intermediate metabolism of procarcinogens and carcinogens performed since the 1950s is not discussed, yet consider the fate of polycyclic aromatic hydrocarbons as an illustration of their significance:

- Cytochromes P450 activate aromatic polycyclic hydrocarbons by N- or C-oxidation.
- There are more than 40 isoforms of the enzyme.
- The cytochromes P450 have a relatively broad specificity. They bind to and modify various substrates, and some are activated and others deactivated. Aromatic polycyclic hydrocarbons are mostly activated, but they can also be inactivated. Yet deactivation of polycyclic hydrocarbons is minor.
- The enzymes involved may be competition for substrates.
- Three isoforms of inducible epoxide hydrolases exist that are capable of inactivating polycyclic aromatic hydrocarbon.

In summary, the activation or deactivation of polycyclic hydrocarbons does not only depend on the redundancy of the substrates (e.g., alcohol and a carcinogen), but also on the repertoire or cytochromes P450, their catalytic specificity, and their genetic control. The enzymatic activation and deactivation of several other classes of carcinogens include aromatic amines, aflotoxin, heterocyclic amines, nitrosamines, and benzene, which are regulated by the environment and genetics in a pattern not identified but similar to the activation and inactivation of aromatic polycyclic hydrocarbons (24, 25).

Fearon, Hamilton, and Vogelstein (26) proposed a model for CRC based on the correlation of the genetic findings with well-defined stages in the evolution of CRC. The phases unfold in a slow, multistep progression that emerges from the normal epithelium into an invading and metastasizing cancer over a period of 20 to 40 years. The gross and histologic properties of each phase were carefully defined starting with the normal epithelium, followed by the dysplastic ACF, the early adenoma, the intermediate adenoma, the late adenoma and carcinoma, and metastatic cancer. These observations were correlated with the incidence of mutations or LOH of various gene products: APC, K-RAS, P53, SMAD4 and SMAD2, HNPCC, P53, and others. Mutations of the *apc* gene (more rarely of the β catenin gene) are believed to initiate the sequence by stimulating cell proliferation, thereby causing the appearance of small adenomas (less than 1 cm). A mutation of the *hpncc* gene may be responsible for initiation in approximately 13% of the early adenomas. Mutations of *K-ras* (~50% mutations) cause further clonal expansion and the passage from small to intermediate adenoma (larger than 1 cm but without carcinomatous transformation). *smad4* and *smad2* are suspected to contribute to the passage of intermediate to late adenoma (greater than 1 cm with signs of carcinomatous transformation). *p53* LOH is associated with more clonal expansion and dysplasia that ultimately evolves into cancer because of genomic instability. This instability, however, can appear at anytime and last during the entire evolution process from normal epithelium to cancer (**FIGURE 24-1**) (13, 14, and 26).

Two patterns of genomic instability contribute later to the cancer transformation process: microsatellite instability (MSI) and chromosomal instability (CIN). Mutations of *hpncc* are in part responsible for MSI. However, other genetic insults, possibly leading to secondary damage of the chromosomes, are believed to be needed to bring about the development of an early epithelial cancer into one that is invasive and metastasizing (**FIGURE 24-1**).

This elegant model for sporadic colorectal cancer in which the environment and genes interact to cause colorectal cancer is likely to apply for many (if not most) epithelial cancers (*mutatis mutandis*). Yet there are unanswered questions concerning the exact role of genes. For example, why would the *APC* gene molecule of the colorectal cells be a preferred first target for biallelic mutations, and which molecular changes are responsible for hypermutability?

The *APC* gene contains a mutation cluster region (MCR) between codons 1286 and 1513 (27). Ninety-nine percent of the mutations are either nonsense or frameshift mutations that yield small deletions, small insertions, or point mutations that often result in truncation of the APC protein; the truncated mutations occur in the MCR in 60% of the somatic mutations. Development of LOH constitutes the second somatic hit. On the basis of what is currently known, the biallelic loss of APC function can explain the proliferation of cell with survival advantages that also include resistance apoptosis. However, cell proliferation and loss of apoptosis may make a tumor (e.g., an early adenoma) but not necessarily a cancer.

The change that marks the passage from the normal cell population to cancer cell is genomic instability (microsatellite) (28–31) or chromosomal instability. This can be introduced by an early mutation in the mismatch repair gene, gene silencing by methylation, or chromosomal instability. However, Fodde proposes that the biallelic mutation of APC could be responsible for both cell proliferation and CIN (15).

Could it be possible that this occurs because of a disruption of the microtubular assembly? Interactions between genes and environment in the pathogenesis of sporadic cancers include exposure to environmental chemicals (procarcinogens) through ingestion or inhalation, followed by their metabolic conversion to reactive chemical (carcinogens) that bind to proteins and DNA. Special genes control the expression in the phenotype of the enzymes and regulatory molecules involved in the conversion of the procarcinogen into an active carcinogen.

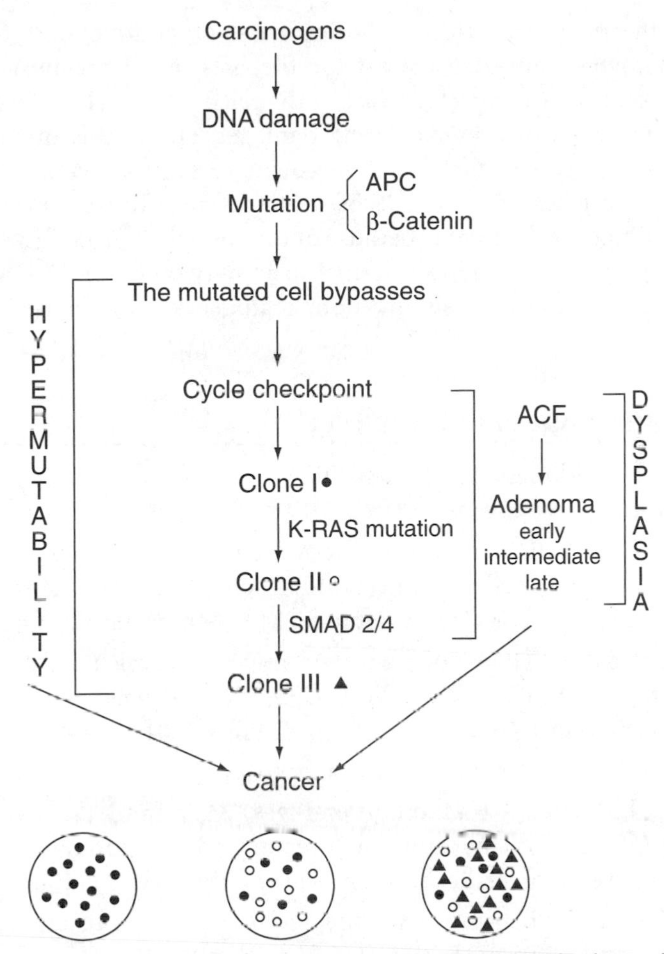


**FIGURE 24-1** Initiation and progression of colorectal cancer. Illustration of the role of genes in the progression of colorectal cancer (13, 14, 26). A mutation of the gene that encodes the APC protein initiates the process of uncontrolled cell proliferation. The process is activated by mutations in *K-ras* and *smad2*. A *p53* mutation triggers the start of a new clone of cells that resist apoptosis even after their DNA is damaged. Cellular dysplasia appears at the adenoma stage. Cellular hypermutability can occur at any time during the progression of the cancer, but it occurs mainly in late adenomas. Two molecular mechanisms contribute to hypermutability: mutations of special genes such as genes involved in DNA repair (e.g., *hpncc* and *p53*) and genes involved in maintaining chromosomal stability (the *apc* gene?). During the succession of the mutations, new clones appear with some of them having survival advantages over the other cell populations. The cells of the most aggressive clones overcome the cell populations of the least aggressive clones. (Modified after Kinzler and Vogelstein [13].)

It is generally assumed that in most cases of sporadic cancer DNA damage constitutes the primary molecular insult. Because of the efficiency of the different DNA repair pathways often functioning in cooperation, most of the DNA damage is repaired. The efficiency of the DNA repair mechanisms depends on both the constitutive expression of a multitude of proteins (enzymes, activators, and inhibitors) and their renewal capacities, which requires a healthy panoply of genes. The DNA repair machinery is not readily overwhelmed; still, it is not entirely foolproof as it can fail either because it is overwhelmed by extensive damage or because of defective genes. When it fails, the cell rarely escapes a form of programmed cell death (apoptosis). But when a cell escapes apoptosis and is still able to respond to growth stimuli (hormones, growth factors) or—worse—has learned to make growth its constitutive status (mutations of the *apc*, *β-catenin*, and the *ras* genes), thus excluding most other functions (differentiation, apoptosis), it becomes immortal at least for the lifespan of the host. However, the guardians of the genome are expected to arrest the cell cycle and close checkpoints ($G_1$/S or $G_2$/M) to the passage of cells harboring damaged DNA. This often fails through mutations and inactivation of those very guardians (e.g., *p53*, reviewed in Chapter 9), thereby opening the gates to a cascade of mutated constitutively proliferating cells. Clones of hypermutated cells emerge because of microsatellite instability (e.g., mutation of missmatch repair genes) or chromosomal instability with survival advantages, cell proliferation, resistance to apoptosis, invasion, angiogenesis, and metastasis.

## References and Suggested Reading


1. Ronan, A., and Glikman, B. Human DNA repair genes. *Environ Mol Mutagen.* 2001;37:241–83.
2. Andrew, S.E., Glazer, P.M., and Jirik, F.R. Mutagenesis and tumor development in DNA mismatch repair-deficient mice. In: Ehrlich, M., ed. *DNA Alterations in Cancer: Genetic and Epigenetic Changes.* Natick, MA: Eaton Publishing; 2000:177–89.
3. Flores, C.C. Repair of DNA double-strand breaks and mismatches in *Drosophila.* In: Nickoloff, J.A., and Hoekstra, M.F., eds. *DNA Damage and Repair. Advances from Phage to Humans.* Totowa, NJ: Humana Press 2001:173–206.
4. Lawrence, L. A mutator phenotype in cancer. *Cancer Res.* 2001;61:3230–9.
5. Yamamoto, H., Weber, T.K., Rodriguez-Bigas, M.A., and Perucho, M. Somatic frameshift mutations in DNA mismatch repair and proapoptosis genes in hereditary nonpolyposis colorectal cancer. *Cancer Res.* 1998;58:997–1003.
6. Loeb, L.A. Mutator phenotype may be required for multistage carcinogenesis. (Review). *Cancer Res.* 1991;51:3075–9.
7. Vasen, H.F., Mecklin, J.P., Khan, P.M., and Lynch, H.T. The International Collaborative Group on Hereditary Non-Polyposis Colorectal Cancer (ICG-HNPCC). *Dis Colon Rectum.* 1991;34:424–5.
8. Boland, C.R. Hereditary nonpolyposis colorectal cancer (HNPCC). In: Scriver, C.R., Beaudet, A.L., Sly, W.S., and Valle, D., eds. *The Metabolic and Molecular Bases of Inherited Disease.* New York, NY: McGraw-Hill Medical Publishing Division; 2001:769–83.
9. Fearnhead, N.S., Wilding, J.L., and Bodmer, W.F. Genetics of colorectal cancer: hereditary aspects and overview of colorectal tumorigenesis. *Br Med Bull.* 2002;64:27-43.
10. Boland, C.R., Thibodeau, S.N., Hamilton, S.R., Sidransky, D., Eshleman, J.R., Burt, R.W., Meltzer, S.J., Rodriguez-Bigas, M.A., Fodde, R., Ranzani, G.N., and Srivastava, S. A National Cancer Institute Workshop on Microsatellite Instability for cancer detection and familial predisposition: development of international criteria for the determination of microsatellite instability in colorectal cancer. *Cancer Res.* 1998;58:5248–57.
11. Lipton, L.R., Johnson, V., Cummings, C., Fisher, S., Risby, P, Eftekhar Sadat, A.T., Cranston, T., Izatt, L., Sasieni, P., Hodgson, S.V., Thomas, H.J., and Tomlinson, I.P.

Refining the Amsterdam Criteria and Bethesda Guidelines: testing algorithms for the prediction of mismatch repair mutation status in the familial cancer clinic. *J Clin Oncol.* 2004;22:4934–43.

12. Bird, R.P., McLellan, E.A., and Bruce, W.R. Aberrant crypts, putative precancerous lesions, in the study of the role of diet in the aetiology of colon cancer. *Cancer Surv.* 1989;8:189–200.
13. Kinzler, K.W., and Vogelstein, B. Lessons from hereditary colorectal cancer. *Cell* 1996;87:159–70.
14. Kinzler, K.W., and Vogelstin, B. Colorectal tumors. In: Scriver, C.R., Beaudet, A.L., Sly, W.S., Valle, D., eds. *The Metabolic and Molecular Bases of Inherited Disease.* New York, NY: McGraw-Hill Medical Publishing Division; 2001:1033–61.
15. Fodde, R. The APC gene in colorectal cancer. (Review). *Eur J Cancer.* 2002; 38:867–71.
16. Reymond, M.A., Dworek, O., Remke, S., Hohenberg, W., Kirchner, T., and Kockerling, F. DCC protein as a predictor of distant metastases after curative surgery for rectal cancer. *Dis Colon Cancer.* 1998;41:755–60.
17. Esteller, M., Sparks, A., Toyota, M., Sanchez-Cespcdes, M., Capella, G., Peinado, M.A., Gonzalez, S., Tarafa, G., Sidransky, D., Meltzer, S.J., Baylin, S.B., and Herman, J.G. Analysis of adenomatous polyposis coli promoter hypermethylation in human cancer. *Cancer Res.* 2000;60:4366–71.
18. Jass, J.R., Young, J., and Leggett, B.A. Evolution of colorectal cancer: change of pace and change of direction. *Gastroenterol Hepatol.* 2002;17:17–26.
19. Baylin, S.B., Herman, J., Gaff, J.R., Vertino, P.M., and Issa, J.P. Alterations in DNA methylation: a fundamental aspect of neoplasia. *Adv Cancer Res.* 1998;72:141–96.
20. Baylin, S.B., and Herman, J.G. DNA hypermethylation in tumorigenesis: epigenetics joins genetics. *Trends Genet.* 2000;16:168–74.
21. Toyota, M., Ohe-Toyota, M., Ahuja, N., and Issa, J.P. Distinct genetic profiles in colorectal tumors with or without the CpG island methylator phenotype. *Proc Natl Acad Sci USA.* 2000;97:710–5.
22. Esteller, M., Tortola, S., Toyota, M., Capella, G., Peinado, M.A., Baylin, S.B., and Herman, J.G. Hypermethylation-associated inactivation of p14 (ARF) is independent of p16(INK4a) methylation and p53 mutation status. *Cancer Res.* 2000; 60:129–33.
23. Spjut, H.J., and Spratt, J.J. Endemic and morphologic similarities existing between spontaneous colon neoplasms in man and 3′,2′-dimethyl 4 aminobiophenyl induced colonic neoplasms in rats. *Ann Surg.* 1965;161:309–24.
24. Wogan, G.N., Hecht, S.S., Felton, J.S., Conney, A.H., and Loeb, L.A. Environmental and chemical carcinogenesis. *Semin Cancer Biol.* 2004;14:473–86.
25. Ingelman-Sundberg, M. Polymorphism of cytochrome P450 and xenobiotic toxicity. (Review). *Toxicology.* 2002;181–2:447–52.
26. Fearon, E.R., Hamilton, S.R., and Vogelstein, B. Clonal analysis of human colorectal tumors. Science. 1987;238:193–7.
27. Miyoshi, Y., Nagase, H., Ando, H., Horii, A., Ichii, S., Nakatsuru, S., Aoki, T., Miki, Y., Mori, T., and Nakamura, Y. Somatic mutations of the APC gene in colorectal tumors: mutation cluster region in the APC gene. *Hum Mol Genet.* 1992;1:229–33.

28. Jaob, S., and Praz, F. DNA mismatch repair defects: role in colorectal cancinogenesis. *Biochimie.* 2002;84:27–47.

29. Rampino, N., Yamamoto, H., Ionov, Y., Li, Y., Sawai, H., Reed, J.C., and Perucho, M. Somatic frameshift mutations in the BAX gene in colon cancers of the microsatellite mutator phenotype. *Science.* 1997; 275:967–9.

30. Lu, S.L., Kawabata, M., Imamura, T., Akiyama, Y., Nomizu, T., Miyazono, K., and Yuasa, Y. HNPCC associated with germline mutation in the TGF-beta type 11 receptor gene. *Nat Genet.* l998;19:17–8.

31. Garcea G., Sharma R.A., Dennison A., Steward W.P., Gescher A., and Berry D.P. Molecular biomarkers of colorectal carcinogenesis and their role in surveillance and early intervention. *Eur J Cancer.* 2003;39:1041–52.

CHAPTER

# 25

# ROLE OF FANCONI ANEMIA, BREAST CANCER, AND BLOOM SYNDROME GENES IN ONCOGENESIS

Fanconi anemia (FA) was first described in 1927 by the Swiss physician Guido Fanconi (1892–1979). The syndrome is transmitted as an autosomal recessive disorder. Patients usually present with congenital anomalies that vary widely from case to case, a progressive bone marrow depression, and cancer predisposition, mostly myelogenous leukemia (1).

Chromosomal breakage (clastogenesis) is a characteristic common denominator in all cases and is therefore the diagnostic component of choice. Indeed, the phenotypic manifestations of the disorder are so variable that they partially overlap with other genetic or nongenetic conditions, making a diagnosis based on clinical symptoms alone impractical. The variability also explains the difficulties encountered in determining, until recently, the exact incidence of FA. The number of carriers is estimated to be 1 in 300, but this number is likely to be too low. The determination of oversensitivity to crosslinking by exposure of cells of pre-anemic and aplastic anemia patients to diepoxybutane hopefully will provide a more realistic estimate of the carriers' incidence (2). This chapter focuses on the genetics of FA and its role in DNA repair and cancer.

## FA Genes and Gene Products

Eight complementation groups of FA have been identified (A, B, C, D1, D2, E, F, G) by fusing somatic cells obtained from FA patients (**TABLE 25-1**). Cells of patients belonging to the same complementation group remain sensitive to crosslinking agents (e.g., mitomycin C); in contrast, cells of patients of different complementation groups

**Table 25-1 FA genes and their products**

| Gene Products | % of FAs | Locus | kDa | Amino Acid |
|---|---|---|---|---|
| FANCA | 10–66 | 16p24.3 | 163 | 1455 |
| FANCB | 4 | 13q-12-3 | 380 | |
| FANCC | 12–13 | 9p22.3 | 63 | 558 |
| FANCD1 | 4 | 13q12-13 | 380 | |
| FANCD2 | | 3p25.3 | S,155<br>L,162 | 1451 |
| FANCE | 12 | 6p21-22 | | 5536 |
| FANCF | Rare | 11p15 | 42 | 374 |
| FANCG | Rare | 9p13 | 68 | |

FANCB = BRCA2; FANCG-XRC09. Refer to Pichierri et al. (7).

become resistant to crosslinking agents. These observations led to identification and ultimately cloning of seven genes (Table 25-1) dispersed throughout the genome. The most frequent complementation groups represented in patients are A (~65%), C (~15%), and G (~10%). Many mutations have been detected in each complementation group; however, some mutations appear more frequently in some populations. For example, FANCC IVS4 + 4A → T is found in Ashkenazi Jews and there are FANCA mutations in Africaners.

## The FANC-BRCA Pathway

Except for FANCB, and FANCD2, and FANCG, the amino acid sequences of the proteins encoded by the *FANC* genes have no known homologous counterparts. Although FANC proteins have been found in the cytoplasm, they function in the nucleus where they form complexes with themselves and with other proteins. Thus, FANC proteins are involved in interactions with ATM, BRCA1 and BRCA2, and the molecules mutated in the Nÿjmegen breakage syndrome (NBS). FANCA, FANCC, FANCF, FANCG, and FANCE form a stable nuclear complex in normal cells during cell replication or after DNA damage. FANCD2 exists in two forms: a primary translation product, FANCD2S (short) and a longer form, FANCD2L, which is monoubiquitinated at lysine 561 (3). The complex contributes to monoubiquitination of the lysine 561 of FANCD2. The activated FANCD2 is translocated to the nucleus where it assembles at DNA repair foci with BCRA1 and RAD51. The DNA binding capacity of the complex is enhanced by exposure of the cells to crosslinking agents (4). Whether the multimolecular complex senses or repairs the DNA damage, or it serves only to stabilize the chromosome is not yet known. It is likely that other yet unknown proteins might also join the complex.

Monoubiquitination at K561 is specifically required for the formation of foci, but it is not known whether phosphorylation prior to ubiquination also occurs. Neither is it known whether the pathway is regulated by deubiquitination (**FIGURE 25-1**) (5, 6).

## FANC Proteins and ATM

An interaction between the ATM and the FANC molecules was suspected because of the radiosensitivity observed in FANCD2 complementation cells, and only in those

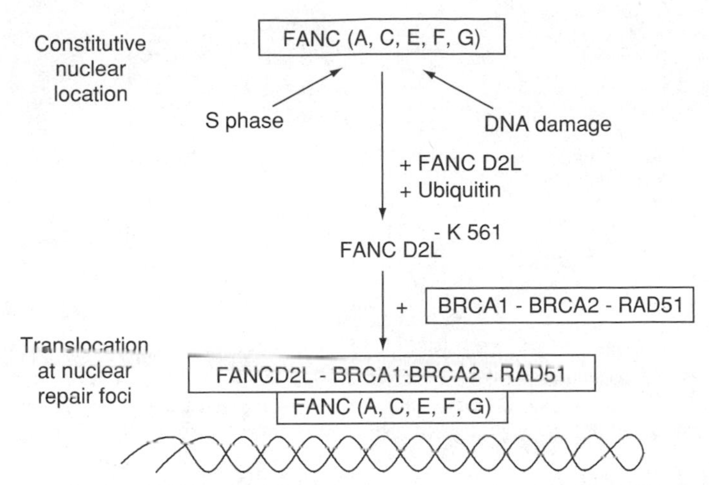


**FIGURE 25-1** The contribution of FANC and BRCA gene products to nuclear foci. The FANC proteins (A, C, E, F, G) constitute a stable nuclear complex. During the S phase or after DNA damage, FANCD2C is ubiquitinated at K561. The activated FANCD2L binds the BRCA1–BRCA2–RAD51 complex, which is joined by the constitutive FANC (A, C, E, F, G). The multiprotein assemblage is helived to contribute to the cell cycle arrest and to participate in crosslink repair.

cells. ATM phosphorylates FANCD2 at serine 222, an event that is believed to arrest the cycle at the $G_1$/S checkpoint. The phosphorylation of FANCD2 by ATM is not related to its ubiquitination, indicating that the FANC protein response to ionizing radiations separates it from other types of DNA damage (**FIGURE 25-2**).

## FANC Proteins and NSB1

Cells obtained from patients afflicted with NBS are oversensitive to mitomycin C, but to a lesser extent than cells obtained from FA patients. Also, one patient harboring biallelic mutations in the *nbs1* gene and presenting with an FA phenotype was reported. Therefore, it has been suggested that the FANC and the NBS pathways are connected. FANCD2L (ubiquitinated at K561) and NBS proteins associate in nuclear foci after mitomycin challenge. The formation of the complex requires MRE11. After exposure of DNA to crosslinking agents, ATM phosphorylates NBS1, which is required for the S222 phosphorylation of FANCD2L (ubiquitinated at K561) (**FIGURE 25-3**).

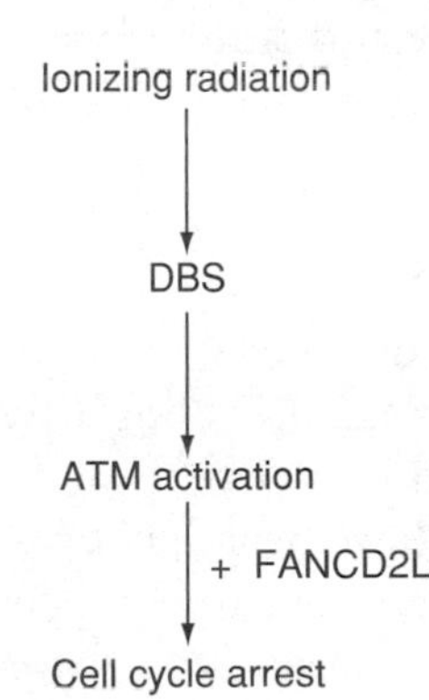


**FIGURE 25-2** The FANC–ATM connection. The double strand breaks caused by ionizing radiation lead to the activation of ATM, which in the presence of FANCD2 causes cell cycle arrest.

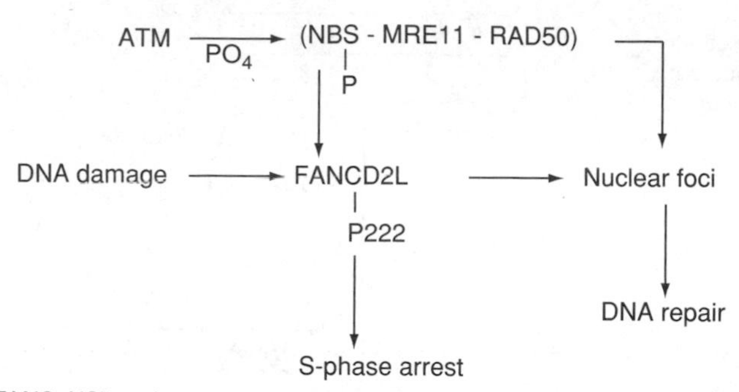


**FIGURE 25-3** The FANC–NSB pathway. After exposure to mitomycin, ATM activates the NSB protein, which is part of the NSB–MER11–RAD50 complex. For the activation of NSB, the phosphorylation of NSB is required for the phosphorylation of FAND2L at serine 222. Both the NSB–MER11–RAD50 and the phosphorylated FANCD2L join the nuclear foci from where they are called at the DNA damage sites for repair by homologous recombination.

In summary, FANC proteins are at the crossroads of three repair pathways: double strand breaks, interstrand crosslinks, and recombination repair. It can be anticipated that as more is learned about the intricacies of the interactions of the FANC proteins among themselves and with other proteins (such as ATM, BRCA1, BRCA2, NBS, and possibly with other, as yet unknown proteins), a more precise understanding of the interrelations between different DNA repair pathways may be gained. Hopefully, this will help to clarify the molecular response to excessive exposure from genotoxic agents (iatrogenic or environmental), and allow a better understanding of their role in carcinogenesis (7, 8).

## Fanconi Anemia and Cancer

The Fanconi anemia syndrome is associated with cancer susceptibility, a likely consequence of chromosomal instability. The products of the *FANC* genes function in cooperation with other proteins; BRCA1 and BRCA2 are prominent among them. The large protein complexes secure the repair of DNA damage, including double strand breaks and crosslinks. Although the condition is rare (1 in 300,000 live births), it is significant that FA patients are predisposed to leukemia and various types of solid tumors.

Moreover, if leukemias could be excluded, it is estimated that the cumulative risk for other cancers in FA patients is on the order of 76% by the age of 45 (9), namely, cancers of the upper respiratory and gastroesophageal tracts and of the gynecological system. The most common sites are the vulva, esophagus, and head and neck (10). These cancers are usually more aggressive and occur at a younger age than in the general population. The failure(s) in DNA repair that causes the chromosomal instability is discussed below (11).

## Fanconi Anemia and Apoptosis

Although this discussion focuses on cancer, the occurrence of severe anemia, pancytopenia, and congenital anomalies in FA cannot be ignored. A disturbance in the regulation of apoptosis could explain both aspects of the syndrome. Studies on a potential deregulation of P53 function have been inconclusive. However, when the product of the *FANCC* gene is defective, the Fas pathway is more active in these Fanconi anemia cells than in other complementation types. Moreover, constitutive expression of FANCC c-DNA protects mice and human cell lines from apoptosis caused by growth factor withdrawal. Mutations of *FANCC* raises the induction of interferon-gamma (IFN-γ), and IFN-γ sensitizes hematopoietic stem cells to Fas and tumor necrosis factor-alpha (TNF-α) and consequently to apoptosis in $FANCC^{-/-}$ mice (12).

*FANCC* mutations may not be the only agents to induce apoptosis in FA; *FANCA* and *FANCG* have also been implicated (13). The cooperation of several *FANC* genes in modulation of apoptosis could explain why the phenomenon is observed in all complementations (11).

## Fanconi Anemia and Telomeres

Telomeres protect chromosomal ends and thereby prevent unwanted fusion and they play a critical role in maintaining the integrity of the genome and the individuality of each chromosome. Telomeres are protected from end fusion by the TRF2 protein. Telomeres of repeatedly dividing cells shorten because polymerases are unable to replicate the terminal sequence of the lagging strands. This telomeric shortening is associated with senescence, aging, and anoxia. The telomere of lymphocytes collected from patients with aplastic anemia and myelodysplastic syndrome are shorter (14). Callen and associates studied the telomere status in FA A cells and demonstrated increased chromosome end fusion independent of TRF2 (15). Moreover, they found that the telomeric signals are increased in FA cells as well as the extrachromosomal telomeric DNA. Thus, despite the fact that telomeres are shorter in FA cells than in controls and the degree of the shortening is similar in both arms of the chromosome, extrachromosomal telomeric DNA continues to be synthesized. These findings indicate that in addition to shortening caused by cell replication, breakage of the arrays of TTAGGG sequence causes telomeric shortening in FA cells (15).

## Molecular Functions of FA Proteins

Because the *FANC* genes and their products have no known homologues in vertebrates or invertebrates, it has been difficult to identify their exact molecular function. Also, the lack of homology with other proteins suggests that FANC proteins appeared late in evolution and function quite differently from the more traditional molecular repair system. However, the discovery, identification, and cloning of *FANC D2* shed some new light on the function of FANC proteins. The gene's locus was identified as a 200 kb region in the short arm of chromosome 3 (3p25.3). It encodes a nuclear protein (1451 amino acids).

The major difference between *FANCD2* and other FANC proteins is that its amino acid sequence is highly conserved among worms (*Caerhabditis elegans*), insects (*Drosophila*), and even in a plant (*Arabidopsis*) (3). The *FANCD2* sequence also includes a highly conserved amino acid, lysine 561, which is monoubiquitinated that contributes to its relocation to nuclear foci within the nucleus after complexing with the FANC (A, C, E, G, F) complex.

The phosphorylation of FANCD2 at serine 222 by ATM suggests that it contributes to the $G_1$/S checkpoint arrest, thereby providing an opportunity for the cell to repair its DNA. However, there is also evidence that the FANC pathway cooperates with the BRCA pathway at an important stage of interstrand crosslink repair: the homologous recombination phase. This function will be discussed after considering the roles of BRCA in cancer.

## *brca-1* and *brca-2*

The *brca-1* gene located on chromosome 17 (17q12-q21) and the *brca-2* gene located on chromosome 13 (13q12q13) are two genes whose mutations are associated with hereditary breast cancers. The genes were discovered at the end of the 20th century. Mutations of the *brca-1* gene were first believed to be responsible for 45% of familial breast cancer, but later studies showed that this estimate was too high, and it is now believed to be rather on the order of 20% to 30%. The genes are also responsible for 10% to 20% of familial ovarian cancer. *brca-1* is a tumor suppressor gene transmitted as an autosomal dominant Mendelian trait with high penetrance. The lifetime risk for having developed breast cancer for female carriers aged between 55 and 85 is 87% and 40% to 60% for ovarian cancer. Familial cancers appear at an earlier age, are often more aggressive, and are estrogen and progesterone negative. *brca-1* carriers also have a greater risk for bilateral breast cancer (cumulative risk of 65% by the age of 70).

The *brca-1* gene also modulates the risk for other cancer types. The relative risk for cancer of the prostate is 3.3 in male carriers. An excess of cancers of the colon in both males and female carriers has been reported, but more extensive investigations are needed. Male breast cancer is rare in carriers with a germline mutation of *brca-1* and does not exceed the sporadic incidence (**TABLE 25-2**).

### *brca-1*

#### *brca-1* Mutations and Cancer of the Breast

At first, only a few mutations were associated with breast cancer (~18), but it soon became obvious that this number was much too low. Since their first identification, more than 500 sequence variations have been reported; they include frameshift, missense mutations, and splice site variations. Many of the mutations are recurrent and a few have been observed only once. The two most common are 185delAG in exon 2 and 5382insC in exon 20 (8 and 12 in Ashkenazi Jews and 1 in 1500 in Caucasians). It is significant that as many as 21% of Jewish women who develop cancer of the breast before age 40 carry the 185delAG mutation (16–18).

**Table 25-2 The *brca-1* and *brca-2* chromosomes and products**

| | *brca-1* | *brca-2* |
|---|---|---|
| Chromosome | 17q12q21 | 13q12-q13 |
| Length of gene | ?? kb | 115 kb |
| Genomic DNA | 80 kb | 70 kb |
| Number of exons | 24 (22 coding) | 262 or 27 |
| Exon 11 | 3.5 kb | 4.8 kb |
| MRNA | 7.8 kb | 11.4 kb<br>no splice variants |
| Protein | n220 kb | 348 kDa |
| Amino acids | 1863 | 33418 |

## BRCA1 Protein

The *brca-1* gene encodes a 1836-amino-acid protein (220 kDa), a nuclear phosphoprotein with little homology[1] with other known proteins except for a RING finger, other Zn finger domains, and the BRCT sequences. The BRCT sequences mediate protein-protein interaction and DNA binding. BRCT is often found in DNA repair proteins. Two repeats of the BRCT are located in the C terminal region of the protein (amino acids 1663–1736 and amino acids 1760–1855). The BRCT domain interacts with RNA helicase A, contributes to the gene's transactivation, and links BRCA1 to RNA polymerase II. The ring finger located in the N terminal (amino acid 20-68) interacts with two proteins, BARDI and BAP1. BARDI contains a Zn finger domain and an α BRCT sequence. During the S phase, it is found with BRCA1 in nuclear foci, and also interacts with RAD51 and BRCA2. The BRCA1 activates protein 1 (BAP1), a ubiquitin hydrolase that enhances BCRA1 growth suppression properties. A sequence (amino acid 224–304) interacts with P53 and stimulates P21 expression. RAD51 binds to BRCA1 (amino acid 758–1064) downstream from the binding site of P53. Between these two binding sites there are two nuclear localization signals (amino acids 500–508 and amino acids 609–615).

## RING Domain Structure of the N-Terminal

The RING motif (~50 amino acid) of BRCA-1 is ensconced in the RING domain (amino acids 8–90). The RING motif contains a 8 amino acid Cys and His sequence (Cys3HisCys4) that interacts with two Zn atoms. The RING motif is linked to two antiparallel α helices to form the RING domain. The RING domain of BRCA1 interacts with a similar RING domain of BARD1 to form a four helix bundle from which the two RING motifs emerge. The RING domains of both molecules are highly conserved. The formation of the BRCA1-BARD1 complex generates an enzyme, an E3 ubiquitin ligase, for which several substrates have been suspected (histone, P53, FANCD2 centrosomal proteins, and tubulin) (19). A new substrate for the ubiquitin ligase, namely the hyperphosphorylated (ser 5) Rpb1(20) subunit of RNA polymerase II (21, 22), was recently discovered. The finding suggests that a link may exist between the BRCA1-BARD1 complex and transcription coupled repair through the degradation of the RNAPII operating in regions close to the adducts or double-strand breaks (21).

The RING domain also includes a nuclear localization domain and a nuclear export domain, which when operating simultaneously could cancel each other out. However, the heterodimerization with BARD1 abolishes the nuclear localization function. Finally, BRCA1 can also bind to the Ub hydrolase BAP1.

It is notable that some cancer-predisposing missense mutations are associated with the RING motif, including C61G, which disrupts the peptide folding. However, all of the detected mutations are not necessarily as structurally disturbing (16). The similarity of the RING domain to zinc fingers led to the assumption that it functions at the actual DNA binding site of the BRCA1, but this proved not to be the case. BRCA1 possess a transactivation domain in the region of the second BRCT, and it interacts with BARD1, which is believed to be a subunit of the transcriptional complex. The promoters of *p21, mdm2,* and *bax* genes are coactivated by BRCA1 and P53. The very structure and properties of BRACI suggest possible roles in cell biology: transcription, gene expression, growth suppression, and DNA repair.

[1]All evidence suggests that BRCA1 is critical to genomic integrity. Why then is it poorly conserved through evolution? Homologues exist in mouse, rat, chicken, and dog. They include conserved sequences in the long exon 11 even in *Xenopus. C. elegans* possesses BRCA1, BRCT, and BARD1 ortologs, which are believed to function in DNA repair.

### C-Terminal Domain of BRCA1

The C-terminal of BRCA1 contains a binding domain for two BRCT repeats. These are conserved sequences frequently found in molecules involved in DNA repair. The structure of the repeat is relatively simple: at the center are four parallel $\beta$ strands and at the flanks are $\alpha$ helices (22). Dimers of BRCT are packed against each other and a few missense mutations have been uncovered at the site of contact. For example, the A1708E mutation destabilizes the folding of BRCT repeats. BRCT was also suspected to secure the binding between BRCA1 and DNA, but data supporting this notion are incomplete.

### BRCA1 and DNA Damage

BRCA1 mutated cells are highly sensitive to DNA damage. $BRCA1^{-/-}$ cells are about five times more sensitive to IR and $H_2O_2$ than $BRCA1^{+/+}$ cells. They are also hypersensitive to UV light, but much less sensitive to X- or $\gamma$-rays (21, 23). BRCA2 defective cells are also hypersensitive to DNA damage. BRCA1 binds to RAD51, a molecule, which in *Saccharomyces cerevisiae* is required for soluble strand break (DSB) repair and homologous recombination. Rad 51 mutants exposed to ionizing radiation (IR) are sensitive to DSB and interstrand crosslinks (ICL). The vertebrate homologue of Rad 51, including the human, participates in the formation of nucleoprotein-filaments in vitro. Its activity is DNA dependent, as is the *Escherichia coli* recombinational strand transferase to which it is related. RAD1 has strand transfer activity in vitro and its expression is cell cycle-dependent (from late $G_1$ through the end of $G_2$). Finally, like BRCA1, RAD51 is located in nuclear foci that appear in the premeiotic S phase and, after exposure to ionizing radiation, are translocated at the sites of DNA damage where they remain even during interphase.[2]

These observations show that RAD51 is required for DNA repair, genome stability, and cell growth. In humans, RAD51 overlaps with BRCA1 nuclear foci and the latter are dispersed to relocalize with DNA sites to which PCNA is bound, a further indication that RAD51 contributes to DNA repair in humans. As discussed below, BRCA2 is believed to contribute to the amalgamation of BRCA1 and RAD51 by interaction of the two proteins at sites involving amino acids 3191–3232 (36 amino acids at a minimum) in the BRCA1 and the highly conserved aminoterminal (amino acids 1–43) sequence of RAD51. The C-terminal of RAD51 (amino acids 98–339) has also been shown to interact with six of the BRC repeats of BRCA2. Thus, it appears that BRCA1, RAD51, and BRCA2 cooperate. Antibodies raised against different epitopes of either BRCA1 or BRCA2 coprecipitate the two proteins as well as RAD1. The association of BRCA1, BARD1, BRCA2, and RAD51, their colocalization in nuclear foci and their translocation to DNA repair sites, all vouch for a coordinate role of these proteins in DNA repair. The interaction of BRCA1 with other proteins involved in DNA repair, in particular with the RAD50-MREII-NBS1, which is a complex discussed in Nijmegen breakage syndrome and in the FA pathway, further support the notion of BRCA1 involvement in DNA repair.

### *brca-1* Is a Tumor Suppressor Gene

Loss of heterozygosity (LOH) suggested, even before the gene was isolated, that *brca-1* is a tumor suppressor gene. The effects of *brca-1* on growth and development were

[2] $RAD51^{-/-}$ embryos are oversensitive to ionizing radiation; their cell proliferation is impaired, and they show chromosomal loss and apoptosis.

investigated in mouse embryo and explanted blastocyst $BRCA1^{-/-}$. The ablation of BRCA1 caused a marked decrease in the rate of growth associated with a considerable overexpression of P21. However, control of the checkpoint only partly explains the growth impairment, because P21- or P53-null mutations cannot rescue the growth deficit completely (21, 24, 25).

Exposure of human breast epithelial cells to antisense BRCA1 accelerates their growth. Antisense BRCA1 has no such effect on non-mammary epithelial cells.[3] The growth of breast and ovarian cancer cells is restrained by wtBRCA1 overexpression, but other malignant cell types are not affected. Retroviral transfer of wild-type BRCA1 inhibits the growth of MCF7 tumors in nude mice (25). Thus, it appears that BRCA1 is needed for development, during which time it stimulates growth. It also acts as a tumor suppressor for at least mammary and ovarian cancers. The molecular interactions that restrict BRCA1 attack to mammary and ovarian cells remain unknown.

## BRCA1 and the Centrosome

Duplication of centrosomes is strictly regulated. A bipolar spindle is an absolute requirement for equal distribution of chromosomes during mitosis. Multiple centrosomes give rise to a multipolar spindle that pulls chromosomes in several directions. There are one or two centrosomes per cell. Duplication occurs during the $G_1$ stage of the cell cycle and lasts through the S phase. Multiple centrosomes are associated with cancer and are often seen in highly malignant cancers (i.e., of the breast). BRCA1 is believed to arrest the cell cycle in $G_1$/S, and $G_2$/S failure at the $G_2$/M checkpoint could be responsible in part for the multiplicity of centrosomes and genomic instability, but it also cannot be ruled out that BRCA1 regulates the number of centrosomes by transactivation of genes involved in their construction. However, centrosomal multiplication is not unique to cancer of the breast and may be a cause for loss of integrity of the genome (26).

## *brca-2*

The basic properties of the *brca-2* gene and protein are summarized in Table 25-2. The BRCA2 protein is large, with no sequence similarity to any other known protein including BRCA1, although it is moderately conserved among mammalian species. The molecule contains some characteristic domains that need mentioning. In its higher N-terminal is a transcription activation domain encoded by exon 3 of the gene. Eight BCRT repeats are located downstream in the center of the molecule, six of which contribute to the binding of RAD51. The product of exon 27 at the end of the COOH terminal possesses an additional binding site for RAD51 and two nuclear localization signals (amino acids 3263–3269 and 3381–3325) (17).

More than 250 mutations have been detected in *brca-2*, 70% of which are truncations and 30% missense mutations. The gene contains no hotspots for mutations. None of the mutations have revealed clues that explain the function of the gene. This is not surprising, *brca-2* is a very large gene that encodes a protein whose sequences for the most part are not shared by any other known protein and, as mentioned earlier, it contains no hotspots for mutations. Consequently, whether those mutations represent non-functional polymorphism, benign variant, or are disease associated remains unclear, except for the following. The 999del5 first observed in Finland is apparently associated with cancers (8.5% of breast, 7.9% of ovarian, 2.7% of prostate) appearing

[3]The findings made in mouse embryo conflicts with those made in human cells.

in patients younger than 65 years of age. *brca-2* mutations have been reported in the germline in a few patients with pancreatic cancer (7.3%). The 6174delT mutation was found in association with breast cancer in Icelandic patients (7% in patients under age 40 and 5% in patients over 50). Simultaneous mutations in *brca-1* and *brca-2* have been reported in sporadic cancer.

In summary, the extent of the contribution of *brca-2* to the incidence of breast cancer remains unanswered. A fair estimate is that 10 to 20% high risk mutations of the *brca-2* gene are present in families where breast and ovarian cancer is present (27).

## Structure of the BRC Complex in *brca-2*

Yang and associates investigated the structure of the C terminal of *brca-2* associated with DSS1 (28). DSS1 is a 70 amino acid protein that is highly acidic and found in mice and humans. It is the product of one of three genes believed to contribute, when mutated, to the split hand/foot disease, a condition in which some fingers are missing while the remaining ones fuse. The wild-type gene is encoded by a locus on chromosome 7q21.3-q22.1 and is believed to play a role in growth control (29). DSS1 is conserved from yeast to mammals and was found to interact with BRCA2 (30) (**FIGURE 25-4**).

The BRCA2–DSS1 complex is made of a platform composed of four globular structures from which two helices emerge to form a tower-like projection. Three of the globular peptides that form the base in the C-terminal region possess an oligonucleotide binding fold (OB folds: OB1, OB2, OB3) similar to those often found in proteins that bind single stranded DNA, for example the replication protein A (RPA). The sequence of the fourth globular domain located in the N-terminal region mainly extends into $\alpha$ helical structures. The two helices of the tower stand between OB2 and OB3 folds and support a triple helical bundle that binds double stranded DNA (28).

The significance of the binding of DSS1 to BRCA2 resides in the similarities of DSS1's properties with those of oligonucleotides. Therefore, the molecule (37% acidic

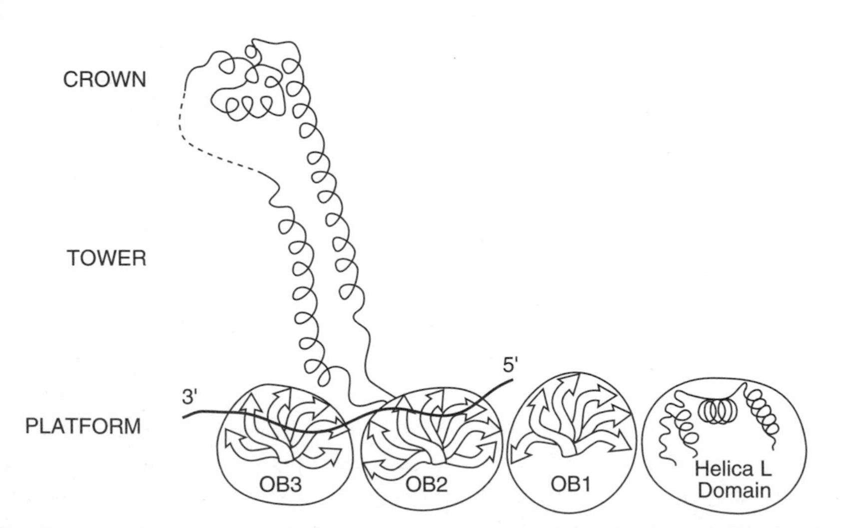


**FIGURE 25-4** Schematic representation of the structure of the C terminal domain of *brca-2*. The domain is divided into three components: the platform, the tower, and the crown (see text). The platform is composed of three OB folds (from left to right OB1, OB2, and OB3) and the alpha-helical arrangement to the left (toward the N terminal). DSS1 (not shown) wraps around and even penetrates OB1 and OB2. Single-stranded DNA is attached to DNA binding sites of OB1 and OB2. Double-stranded DNA is believed to bind to the triple helical bundle of the crown. (Adapted after Yang, H., et al. [21] and Jasin, M. [30].)

residues, 13% aromatic residues) could modulate the binding of DNA to the BRCA2. It is worth noting that only the highly conserved sequences of DSS1 are involved in the binding and, moreover, two of the BRCA2 residues involved in the DSS1 BCRA2 ($Al^{2564}$ and $Arg^{2580}$) are mutated in cancer. Single strand DNA binds to OB folds 3 and 2; double stranded DNA binds to the three helical bundles. Four of the frequently mutated residues of BRCA2 are located in the tower and others are observed in the OB domain.

Using an in vitro system, it was further established that the BRCA2–COOH terminal stimulates RAD51-mediated recombination (28). Although the significance of these elegant studies can still not be fully appreciated, these investigations have shed light on the role of BRCA2 in recombination and begin to provide clues about the nature of its function in cancer.

BRCA2's kinetics are similar to that of BCRA1. BRCA2 begins to rise during the $G_1$/S phase, reaches its peak at the end of $G_1$/S, and persists during the $G_2$/M checkpoint. This suggests that BRCA2 and BRCA1 operate together. BRCA2 binds RAD-1 and therefore is suspected of operating during transcription in cooperation with BRCA1.

## *BRCA1, BCRA2,* and DNA Repair

Homologous recombination (HR) is essential for restoration of the integral DNA sequence after its damage caused by either cell replication or exogenous agents. HR is an accurate process in contrast to non-homologous end joining (NHEJ; see Chapters 10 and 11), which fails to secure homologous strand restoration. The role of *BRCA1* and *BRCA2* in these forms of repair has been investigated in mutant mice and in *BRCA1* and *BRCA2* mutant cell lines (reviewed in 19 and 31).

The interaction of *BRCA1* and *BRCA2* with RAD51 suggests that both genes participate in HR repair; indeed, the homologue of RAD51 has been clearly demonstrated to function in double strand breaks and HR repair in yeast. The association between *BRCA1* and *BRCA2* and RAD51 has been demonstrated in nuclear foci in vivo by coimmunoprecipitation. The foci disperse and are translocated at the DNA repair site after exposure of the cells to ionizing radiation (IR).

Further evidence for the participation of *BRCA1* and *BRCA2* in HR repair was obtained by using mutant embryos. $BRCA1^{-/-}$ embryos have a shorter lifespan, are oversensitive to DNA damaging agents, and show chromosomal instability, a phenotype resembling embryos defective in two RAD51 related genes *xrcc-1* and *xrcc-2*. Mouse embryonic cells (ES) in which the BRCA1 carries a substrate for the I-SceI endonuclease are four to five times more sensitive to mitomycin C and ionizing radiations, and manifest chromosomal instability that can be restored to normal by correcting the *BRCA1* mutation (32).

*BCRA2* mutant embryos are oversensitive to IR and cell proliferation is arrested (32). Similarly, the association and colocalization of *BRCA2* and RAD51 lead to the assumption that they also contribute to homologous recombination from repair. Deletion of the BRC repeats that are involved in RAD51 binding causes hypersensitivity to IR, methyl methanesulphonate (MMS), and cisplatin, and loss of the $G_2$/M checkpoint. Cells harboring two mutated *BRCA2* alleles have a low response to genotoxic agents that can be corrected by complementation (34, 35).

Capan-1 cells are interesting; they are derived from a pancreatic cancer and harbor the BRCA2 6174delT mutation, which has been observed in Ashkenazi Jewish families with hereditary cancers. The Capan-1 cells are sensitive to MMS, but the sensitivity can be eliminated by introducing a wild-type *BRCA2* gene. However, the introduction of a *BRCA2* gene in which BRCT motif(s) is deleted or mutated has no effect, again implicating the BCRA-2RAD51 connection in the double strand break and homologous recombination repairs (36).

In summary, in addition to the observations made with *brca-1*, the association of RAD51, *brca-2*, and embryonic or cell line mutations of the gene strongly support a role for both BRCAs in homologous recombination repair of double strand breaks.

## Cell Cycle Checkpoints, Chromosomal Integrity, and *brca* Genes

There is no convincing evidence that the $G_1/S$ checkpoint is affected by either *BRCA*. The checkpoint operates normally in allelic mutants.

Both *brca-1* and *brca-2* have been implicated in other checkpoint controls, thereby arresting the cycle in the face of DNA damage and giving the cell time to either repair its DNA or eliminate itself by apoptosis. The mitotic rate of MEF cells harboring normal *brca-1* that are exposed to $\gamma$ radiation is markedly reduced, but targeted mutation (a deletion in exon 11) of the same cells fails to impair cell division after the same exposure to $\gamma$ radiations, because of failure of the $G_2/M$ checkpoint (37). Chen and associates have disrupted the BRCT complex between the wild-type *brca-2* and RAD51 in breast cancer cells in which four BRCT motifs were overexpressed. The disruption led to a dominant negative response, characterized by radiation hypersensitivity and loss of the $G_2/M$ checkpoint (38).

Mice have 40 chromosomes. Wild-type embryonic cells lines have approximately 3% abnormal chromosomes (smaller or larger than normal). In embryonic cells in which targeted deletions have been introduced in BRCA $I^{11-/-}$ and P53 ($P53^{-/-}$), the incidence of chromosomal anomalies is 10 times greater (30% of metaphases) (39). The changes not only affected the number of chromosomes, but they often were dicentric or translocated. Multiple centrosomes are also observed often, causing multipolar spindles to appear. Similar findings were made in BRCA $I^{11-/-}$ MEF cells (34). *BRCA1* is believed to directly contribute to the chromosomal stability by associating with centrosomes during mitosis. Chromosomal instability also has been observed in mice cells harboring *BRCA2* mutations (exon 11 deletions). MEF cells obtained from embryos carrying the mutation show chromatid breaks and exchange in ~40% of metaphases. Chromosomal instability has similarly been observed with a deletion of exon 27; finally, an exon that contributes to the binding of RAD51 is also associated with chromosomal instability (40).

Structural or numerical alterations of chromosomes are not the only cause of chromosomal instability because they also can be caused by loss of telomeres, amplifications of chromosomes and centrosomes, or incorrect segregation. These alterations have been observed in *brca-1* and *brca-2* mice (41, 42). At present it is not known how the severity in human chromosomal instability relates to that observed in mice.

In summary, *brca-1* and *brca-2* are involved in cell cycle checkpoint control, transcriptional transactivation and DNA repair. Mutations that affect either of these functions could lead to increased cell proliferation combined with chromosomal instability events that are associated with cancer. However, the known mutations, multiple as they are, cannot initiate cancer, but only facilitate the process. It is also possible that among the many mutations already catalogued, some might qualify as true cancer initiators (see 43 and 44 for reviews).

The complete molecular mechanisms by which the events are effectuated in the wild-type and altered in the mutated cell remain to be uncovered. The findings of important interactions between the FANC and the BRCA1 and BRCA2 pathways will certainly help to clarify the steps involved in the carcinogenesis of breast and ovarian and possibly of other cancer types. Finally, the factors that determine organ specificity (breast and ovary) of the developing cancers in mice and humans remains to be explained (17).

## Bloom Syndrome

Bloom syndrome (BS) is a very rare genetic syndrome (220 cases reported since 1975), but like many rare conditions, BS not only has taught scientists a great deal about the function of the wild-type gene, it also challenged our understanding of genetic and sporadic cancer. Bloom syndrome is a hereditary disease transmitted in an autosomal recessive pattern. The gene is located on chromosome 15q26.1 and it encodes a protein made of 1417 amino acids. The cDNA sequence contains the code for a homologue of the RegQ helicases. This DNA helicase and its homologues are Mg- and adenosine 5′-triphosphate (ATP)-dependent enzymes that catalyze the unwinding of the helix during DNA replication or repair, recombination, and RNA transcription (45). The enzyme binds to one strand and moves on the DNA in a 3′ → 5′ direction. The prototype is the *E. coli* RecQ, which contributes to the correction of illegitimate recombination (46). Helicases are found in prokaryotes and eukaryotes from yeast to humans, but while lower eukaryotes contain only one or two such enzymes, higher eukaryotes contain more (humans have five). Yeast contains only one helicase, SGS1 (slow growth suppressor), which physically interacts with topoisomerase III. Mutations of both SGS1 and topoisomerase III cause abnormally frequent recombinations.

Despite their homology, RecQ and SGS1 differ by their size (610 vs. 1447 amino acids) and by the function of their N- and C-terminals. The RecQ helicase function is located in the N terminal (~300 amino acids). Most of the function of the C-terminal is still unknown. Both N- and C-terminals are highly positively charged. In contrast, in the SGS1, the helicase activity (650 amino acids) is located in the C-terminal and the N-terminal is highly negatively charged. Five RecQ helicases have been found in humans: BLM, WRH, RECQ4, RECQL, and RECQ5β. Mutations of the first three are associated with cancer-prone hereditary diseases: Bloom, Werner, and Rothmund-Thompson syndromes respectively. This chapter focuses on the Bloom syndrome. The BLM protein is a relatively large protein whose sequence includes several conserved motifs.

### Bloom Helicase

The N terminal in Bloom helicase contains the acidic region and the highly conserved helicase domain is located in the center of the molecule. It is flanked by an acidic region and the C-terminal that contains the RecQC-terminal homologues region, and further downstream two nuclear localization signals (**FIGURE 25-5**).

Mutated BLM cells reveal numerous chromosomal anomalies: breaks, gaps, and rearrangement. These anomalies result from excessive sister chromatid exchanges (SCE) or chromosomal exchanges. Such lesions reflect hypermutability and explain (at least in part) the various developmental defects and the predisposition to cancer in BS patients. While less than 10 SCEs are seen in non-BS cells during metaphase, this number is 10 times greater for cells derived from BS patients. Mutations of the *BLM* gene are ultimately responsible for the high SCE.

Although rare, BS can occur in most (if not all) populations; its incidence occurs at relatively high levels in Ashkenazi Jews (1 in 107 persons among Ashkenazi Jews of New

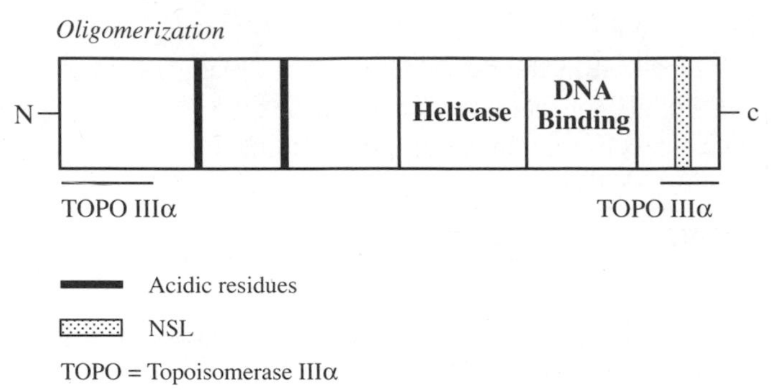

**FIGURE 25-5** The molecular associations of the BLM gene. TOPO III$^{\alpha}$ = topoisomerase III$^{\alpha}$.

York City). Most patients with BLM have null alleles and, rarely, missense mutations. A 6 bp deletion and a 7 bp insertion in exon 10 of the gene have been found to be associated with Bloom syndrome in Ashkenazi Jews.

The exact nature of the substrate of the BLM helicase is not known. In general, the enzyme does not bind to distorted blunt ended double stranded DNA; however, it can bind double stranded DNA when the latter is part of a replication fork or a single strand 3 tail (47). In vitro the helicase targets Holliday junctions.[4] The helicase unwinds a variety of substrates in vitro including hairpins and G-quadruplex DNA,[5] G-rich strands of ribosomal and telomeric DNA. Preferred substrates for other RecQ helicases are replication forks stalled by DNA adducts or other types of lesions. Collapsed forks occur rather frequently during replication.

The alterations observed in vivo in the cells of patients with Bloom syndrome or in viable BLM null mice ($BLM^{-/-}$) include (48) slower growth than their normal counterpart and increased sister chromatid exchange. Biochemical studies reveal that the cells contain abnormal replication intermediates, and show delayed replication fork progress and other signs of interference with growth and illegitimate recombination, for example, elevated occurrence of quadriradial chromosomes that result from incomplete recombination between homologous chromosomes.

Jansack et al. purified the core of the BLM protein (amino acids 642–1290) because of its homology with the *E. coli* Q helicase, another suppressor of illegitimate recombination (49). The monomer binds to single stranded DNA and possess ATPase and helicase functions. The BLM core can substitute, at least in part, for the *E. coli* helicase when introduced in RecQ devoid bacteria; it thereby reduces the occurrence of illegitimate recombination. Moreover, mutations of the core markedly reduce the catalytic function of both the ATPase and the helicase. In summary, the BLM helicase slows growth, probably by cycle delay or arrest, and ultimately prevents illegitimate recombination. BLM interacts with several proteins involved in the maintenance of the integrity of the genome.

---

[4]The finding that the BLM protein fosters branch migration in Holliday junctions further supports a role for it in recombination.

[5]Tetraplex or quadruplex DNA results from the folding of guanine-rich repeat sequences in a non-Watson Crick structure that generates a DNA region made of four parallel separate strands held by hydrogen bonds and van der Waals forces.

P53 and BLM need to interact for resolution of stalled replication forks (50, 51). The BLM protein physically interacts with topoisomerase IIIα and the interaction is required for the cooperative activities of the helicase and the topoisomerase (Figure 25-4) (52, 53). Fibroblasts whose DNA has been damaged during the S phase form foci containing BLM and γH2AX at the site of the replication fork damaged with strand breaks. BLM is required for BRCA1-NBSI function. Cells that assemble the BRCA-NBSI in the absence of BLM are eliminated in a P53-dependent fashion (54). The mismatch repair enzyme, hMSH2/6, coimmunoprecipitates with BML, P53, and RAD51 (55, 56). The mismatch repair enzymes are believed to contribute to homologous recombination repair. BLM complexes with the telomere repeat protein TRF 2 in cells that engage in the alternative lengthening of telomere (ALT), a pathway that involves recombination. The formation of TRF 2/BLM complexes require that BLM harbors a functional helicase. The finding suggests that BLM may contribute to telomeric lengthening in ALT cells (57). In conclusion, it appears that the Bloom helicase plays a critical role in homologous recombination repair of the genome and in telomere lengthening.

## BLM and Cell Cycle Checkpoint

Although there is evidence that BLM[6] affects several cell cycle checkpoints, the data are incomplete. The cell levels of BLM during the cell cycle argue in favor of a functional role during the cycle. BLM rises sharply during S phase, plateaus during the $G_2$, and drops during the next $G_1$ (58). BLM is found in the nucleus as part of the BRCA1-associated genome surveillance complex (BASC) in which several repair proteins are found, including ATM (59). These observations led Ababou and collaborators to investigate the functional relationship between ATM and BLM after DNA damage caused by ionizing radiation (57). They showed that the P53 response and the function of the $G_1$/S checkpoint were normal in primary BS fibroblasts exposed to γ rays. In contrast, there was a reproducible defect at the $G_2$/M checkpoint that was manifested by more cells escaping $G_2$/M to enter mitosis and by a rise in SCEs in BS cells compared to wild-type cells. Finally, it was found that ATM phosphorylates BLM probably at one of the three phosphorylation consensus motifs identified in the molecule (60, 61).

Phosphorylation of BLM is linked to a molecular interaction between BLM and ATM.

However, if the role of BLM is to prevent SCE by promoting branch migration of Holliday junctions, and if BLM phosphorylation is an absolute requirement for such functions, then it can be expected that cells from AT patients would present an increased incidence of SCE, yet this is not the case. Therefore, Ababou et al. (60) proposed that the BLM is phosphorylated in an ATM-dependent manner at the site of a type of DNA damage other than potential SCEs, perhaps double strand breaks. Their study suggested that BLM responds to ionizing radiation in a pathway in which ATM functions upstream. Whatever the role of ATM in BLM phosphorylation, it is not incompatible with the multiple functions of BLM during the cell cycle. For example, its association with BASC (56), its ATM-dependent association with PML bodies after γ irradiation, and its P53-dependent cleavage during apoptosis (62, 63) and, moreover, the phosphorylation of BLM by other kinases cannot be excluded.

[6] BLM is found in the nucleolus mainly during the S phase. After exposure to genotoxins or radiation, BLM is translocated to the nucleus in the nuclear foci, where it participates in a rich complex of proteins: the BASC, most of which contribute to DNA repair.

BLM associates with the polymorpholeukocyte (PML) protein in PODS and with several families of DNA repair proteins in the nucleus. PODS contain the protein functionally modified by chromosomal translocation of its encoding gene in acute promyelocytic leukemia. Although many have been proposed, including protein storage turnover, etc., the exact function of these bodies is unclear. In the case of the BLM-PML association, the complex is believed to be translocated at the site of single strand breaks.

## BLM is a Partner of the BASC Protein Complex

The number of SCEs in UVC-exposed BS cells is greater than in wild-type cells similarly exposed. Moreover, BLM-deficient cells treated with UVC also allow cells to bypass the $G_2$/M checkpoint, yet the number of dividing cells is decreased, most likely because of the slowing down of the progress of the replication fork. Thus, $G_2$/M escape is observed in BS cells with both $\gamma$ and UVC radiations, an intriguing observation because the DNA damage inflicted by these two agents is very different.

It is established that the phosphorylated BLM maintains the helicase's ability to interact with topoisomerase III$\alpha$. Phosphorylated BLM also exists in a more soluble form (extractable with detergent) and does not associate with proteins bound to chromatin. However, exposure of cells to ionizing radiation causes BLM dephosphorylation, and the dephosphorylated BLM moves to the nuclear fraction (not dissolvable by detergent). On the basis of these findings, Dutertre and associates proposed that BLM phosphorylation is linked to its role in mitosis (46). Their study demonstrated that cdc2 is the functional kinase, suggesting that phosphorylation of BLM in mitotic cells provides them with stores of soluble BLM that can rapidly be mobilized to join the nuclear matrix soon after dephosphorylation by ionizing radiations. In the matrix, BLM joins other components of BASC and becomes ready to function in DNA repair and prevent SCE.

## Symptoms in Bloom Syndrome

The two major manifestations of BS are the small, albeit proportionate, body size and a very marked predisposition to cancer (a higher incidence of any sporadic cancer type at the sites that are observed in the general population). There are several other features that can be present and that may help to distinguish Bloom syndrome from other conditions associated with short stature. They include a typical abnormal facies, photosensitivity (of faces, dorsal hands, and forearms), immunodeficiency, diabetes, infertility in males (due to failing spermatogenesis), reduced fertility in females, and often a low IQ.

The cancers observed in BS patients include carcinomas, mesenchymal tumors, and acute myelogenous leukemia. They occur at a much younger age (average age of first cancer diagnosis is 24 years). These cancers are multiple and often more aggressive than their counterpart in the general population (60).

The Bloom syndrome stands out for being less discriminatory with respect to the type of cancer risk associated with the BLM mutation. A hundred cancers were reported in 71 patients (among 168 patients registered) in the BS registry. The diagnosis of these cancers included 21 leukemias, 23 lymphomas, 5 children's cancers (1 medulloblastoma, 2 Wilms tumors, and 2 osteosarcomas). In addition, 51 cancers commonly found in the general population were also detected: 8 of the skin, 10 of the head and neck, 19 of the gastrointestinal tract, 7 of the breast, 5 of the uterus, 1 of the lung, and 1 metastatic (primary unknown). Most of the cancers (93%) were detected before the age of 40 (see reference 64 for details). This unfortunate experiment of nature is unique, but science would fail the Bloom syndrome's victims and the general population if it did not seek to uncover the details of the molecular genetic alterations responsible for the types and distribution of cancers (61). The contribution of environmental factors to the cancer risk associated with the BS is unclear, but even if environmental factors add to the cancer risk associated with the BS, they do not explain the high incidence, early age of onset, or the distribution and variety of the cancers that afflict the Bloom patient. Cooperation between environmental and genetic factors is not excluded, however.

## References and Suggested Reading


1. Wong, J.C., and Buchwald, M. Disease model: Fanconi anemia. *Trends Mol Med.* 2002;8:139–42.
2. Auerbach, A.D., Buchwald, M., and Joenje, H. Fanconi anemia. In: Scriver, C.R., Beaudet, A.L., Sly, W.S., and Valle, D., eds. *The Metabolic and Molecular Bases of Inherited Disease,* 8th ed. New York, NY: McGraw-Hill; 2001:758–68.
3. Joenje, H., and Arwert, F. Connecting Fanconi anemia to BRCA1. *Nat Med.* 2001;7:406–7.
4. Hoatlin, M., and Reuter, T. Foci on Fanconi. *Trends Mol Med.* 2001;7:237–9.
5. Folias, A., Matkovic, M., Bruun, D., Reid, S., Hejna, J., Grompe, M., D'Andrea, A., and Moses, R. BRCA1 interacts directly with the Fanconi anemia protein FANCA. *Hum Mol Gen.* 2002;11:2591–7.
6. Taniguchi, T., Garcia-Higuera, I., Andreassen, P.R., Gregory, R.C., Grompe, M., and D'Andrea, A.D. S-phase-specific interaction of the Fanconi anemia protein. FANCD2, with BRCA1 and RAD51. *Blood.* 100:2414–20.
7. Pichierri, P., Averbeck, D., and Rosselli, F. DNA cross-link-dependent RAD50/MRE11/NBS1 subnuclear assembly requires the Fanconi anemia C protein. *Hum Mol Genet.* 2002;11:2531–46.
8. D'Andrea, A.D., and Grompe, M. The Fanconi anaemia/BRCA pathway. *Nat Rev Cancer.* 2003;3:23–34.
9. Alter, B.P., Greene, M.H., Velazquez, I., and Rosenberg, P.S. Cancer in Fanconi anemia. *Blood.* 2003;101:2072.
10. Rosenberg, P.S., Greene, M.H., and Alter, B.P. Cancer incidence in persons with Fanconi anemia. *Blood.* 2003;101:822–6.
11. Bogliolo, M., Cabre, O.,Callen, E., Castillo, V., Creus, A., Marcos, R., and Surralles, J. The Fanconi anemia genome stability and tumor suppressor network. *Mutagenesis.* 2002;17:529–38.
12. Ratburn, R.K., Christianson, T.A., Faulkner, G.R., Jones, G., Keeble, W., O'Dwyer, M., and Bagby, G.C. Interferon-gamma-induced apoptotic responses of Fanconi anemia group C hematopoietic progenitor cells involve caspase 8-dependent activation of caspase 3 family members. *Blood.* 2000;96:4204–11.
13. Futaki, M., Igarashi, T., Watanabe, S., Kajigaya, S., Tatsuguch, J.M., Wang, J., and Liu, J.M. The FANCG Fanconi anemia protein interacts with CYP2E1: possible role in protection against oxidative DNA damage. *Carcinogenesis.* 2002;23:67–72.

14. Brummendorf, T.H., Maciejewski, J.P., Mak, J., Young, N.S., and Lansdorp, P.M. Telomere length in leukocyte subpopulations of patients with aplastic anemia. *Blood.* 2001;97(4):895–900.
15. Callen, E., Samper, E., Ramirez, M.J., Creus, A., Marcos, R., Ortega, J.J., Olive, T., Badell, I., Blasco, M.A., and Surralles, J. Breaks at telomeres and TRF2-independent end fusions in Fanconi anemia. *Hum Mol Genet.* 2002;11:439–44.
16. FitzGerald, M.G., MacDonald, D.J., Krainer, M., Hoover, I., O'Neil, E., Unsal, H., Silva-Arrieto, S., Finkelstein, D.M., Beer-Romero, P., Englert, C., Sgroi, D.C., Smith, B.L., Younger, J.W., Garber, J.E., Duda, R.B., Mayzel, K.A., Isselbacher, K.J., Friend, S.H., and Haber, D.A. Germ-line BRCA1 mutations in Jewish and non-Jewish women with early-onset breast cancer. *N Engl J Med.* 1996;334:143–9.
17. Couch, F., and Weber, B.L. Breast cancer. In: Scriver, C.R., Beaudet, A.L., Sly, W.S., and Valle, D., eds. *The Metabolic and Molecular Bases of Inherited Disease.* New York, NY: McGraw Hill; 2000:999–1032.
18. Rebbeck, T.R. Inherited susceptibility to breast cancer in women: High-penetrance and low-penetrance genes. In: Ehrlich, M., ed. *DNA Alterations in Cancer: Genetic and Epigenetic Changes.* Natick, MA: Eaton Publishing; 2000:253–72.
19. Jasin, M. Homologous repair of DNA damage and tumorigenesis: the BRCA connection. (Review). *Oncogene.* 2002;21:8981–93.
20. Choudhury, A.D., Xu, H., Modi, A.P., Zhang, W., Ludwig, T., and Baer, R. Hyperphosphorylation of the BARD1 tumor suppressor in mitotic cells. *J Biol Chem.* 2005;280:24669–79.
21. Starita, L.M., Horwitz, A.A., Keogh, M.C., Ishioka, C., Parvin, J.D., and Chiba, N. BRCA1/BARD1 ubiquitinate phosphorylated RNA polymerase II. *J Biol Chem.* 2005;200:24498–505.
22. Joo, W.S., Jeffrey, P.D., Cantor, S.B., Finnin, M.S., Livingston, D.M., and Pavletich, N.P. Structure of the 53BP1 BRCT region bound to p53 and its comparison to the Brca1 BRCT structure. *Genes Dev.* 2002;16:583–93.
23. Gowen, L.C., Avrutskaya, A.V., Latour, A.M., Koller, B.H., and Leadon, S.A. BRCA1 required for transcription-coupled repair of oxidative DNA damage. *Science.* 1998;1281:1009–12.
24. Cressman, V.L., Backlund, D.C., Hicks, E.M., Gowen, L.C., Godfrey, V., and Koller, B.H. Mammary tumor formation in p53- and BRCA1-deficient mice. *Cell Growth Differ.* 1999;10:1–10.
25. Holt, J.T., Thompson, M.E., Szabo, C., Robinson-Benion, C., Arteaga, C.L., King, M.C., and Jensen, R.A. Growth retardation and tumour inhibition by BRCA1. *Nat Genet.* 1996;12:298–302.
26. Nigg, E.A. Centrosome aberrations: cause or consequence of cancer progression? *Nat Rev Cancer.* 2002;2:815–25.
27. Schubert, E.L., Lee, M.K., Mefford, H.C., Argonza, R.H., Morrow, J.E., Hull, J., Dann, J.L., and King, M.C. BRCA2 in American families with four or more cases of breast or ovarian cancer: recurrent and novel mutations, variable expression, penetrance, and the possibility of families whose cancer is not attributable to BRCA1 or BRCA2. *Am J Hum Genet.* 1997;60:1031–40.
28. Yang, H., Jeffrey, P.D., Miller, J., Kinnucan, E., Sun, Y., Thoma, N.H., Zheng, N., Chen, P.L., Lee, W.H., and Pavletich, N.P. BRCA2 function in DNA binding and

recombination from a BRCA2-DSS1-ssDNA structure. *Science.* 2002;297: 1837–48.

29. Marston, N.J., Richards, W.J., Hughes, D., Bertwistle, D., Marshall, C.J., and Ashworth, A. Interaction between the product of the breast cancer susceptibility gene BRCA2 and DSS1, a protein functionally conserved from yeast to mammals. *Mol Cell Biol.* 1999;19:4633–42.

30. Jasin, M. Homologous repair of DNA damage and tumorgenesis: the BRCA connection. *Oncogene.* 2002;21;8981–93 (review).

31. Powell, S.N., and Kachnic, L.A. Roles of BRCA1 and BRCA2 in homologous recombination, DNA replication fidelity and the cellular response to ionizing radiation. (Review). *Oncogene.* 2003;22:5784–91.

32. Moynahan, M.E., Cui, T.Y., and Jasin, M. Homology-directed DNA repair, mitomycin-c resistance, and chromosome stability is restored with correction of a Brca1 mutation. *Cancer Res.* 2001;61:4842–50.

33. McAllister, K.A., Hauden-Strano, A., Hagevik, S., Brownlee, H.A., Collins, N.K., Futreal, P.A., Bennett, L.M., and Wiseman, R.W. Characterization of the rat and mouse homologues of the BRCA2 breast cancer susceptibility gene. *Cancer Res.* 1997;57:3121–5.

34. Xia, F., Taghian, D.G., DeFrank, J.S., Zeng, Z.C., Willers, H., Iliakis, G., and Powell, S.N. Deficiency of human BRCA2 leads to impaired homologous recombination but maintains normal nonhomologous end joining. *Proc Natl Acad Sci USA.* 2001;98:8644–9.

35. Kraakman-van der Zwet, M., van Lange, R.E., Essers, J., van Duijn-Goedhart, A., Wiggers, I., Swaminathan, S., van Buul, P.P., Errami, A., Tan, R.T., Jaspers, N.G., Sharan, S.K., Kanaar, R., and Zdzienicka, M.Z. Brca2 (XRCC11) deficiency results in radioresistant DNA synthesis and a higher frequency of spontaneous deletions. *Mol Cell Biol.* 2002;22:669–79.

36. Chen, P.L., Chen, C.F., Chen, Y., Xiao, J., Sharp, Z.D., and Lee, W.H. The BRC repeats in BRCA2 are critical for RAD51 binding and resistance to methyl methanesulfonate treatment. *Proc Natl Acad Sci USA.* 1998;95:5287–92.

37. Xu, X., Weaver, Z., Linke, S.P., Li, C., Gotay, J., Wang, X.W., Harris, C.C., Ried, T., and Deng, C.X. Centrosome amplification and a defective $G_2$-M cell cycle checkpoint induce genetic instability in BRCA1 exon 11 isoform-deficient cells. *Mol Cell.* 1999;3:389–95.

38. Chen, C.F., Chen, P.L., Zhong, Q., Sharp, Z.D., and Lee, W.H. Expression of BRC repeats in breast cancer cells disrupts the BRCA2-Rad51 complex and leads to radiation hypersensitivity and loss of G(2)/M checkpoint control. *J Biol Chem.* 1999;274:32931–5.

39. Shen, S.X., Weaver, Z., Xu, X., Li, C., Weinstein, M., Chen, L., Guan, X.Y., Ried, T., and Deng, C.X. A targeted disruption of the murine Brca1 gene causes gamma-irradiation hypersensitivity and genetic instability. *Oncogene.* 1998;17:3115–24.

40. Patel, K.J., Yu, V.P., Lee, H., Corcoran, A., Thistlethwaite, F.C., Evans, M.J., Colledge, W.H., Friedman, L.S., Ponder, B.A., and Venkitaraman, A.R. Involvement of Brca2 in DNA repair. *Mol Cell.* 1998;1:347–57.

41. Tutt, A., Gabriel, A., Bertwistle, D., Connor, F., Paterson, H., Peacock, J., Ross, G., and Ashworth, A. Absence of Brca2 causes genome instability by chromosome breakage and loss associated with centrosome amplification. *Curr Biol.* 1999; 9:1107–10.

42. Ban, S., Shinohara, T., Hirai, Y., Moritaku, Y., Cologne, J.B., and MacPhee, D.G. Chromosomal instability in BRCA1- or BRCA2-defective human cancer cells detected by spontaneous micronucleus assay. *Mutat Res.* 2001;474:15–23.
43. Thompson, L.H., and Schild, D. Recombinational DNA repair and human disease. *Mutat Res.* 2002;509:49–78.
44. Brenneman, M.A. BRCA1 and BRCA2 in DNA repair and genome stability. In: Nickoloff, J.A., and Hoekstra, M.F., eds. *DNA Damage and Repair.* Totowa, NJ: Humana Press; 2000:237–68.
45. Soultanas, P., and Wigley, D.B. Unwinding the "Gordon knot" of helicase action. *Trends Biochem Sci.* 2001;26:47–54.
46. Hanada, K., Ukita, T., Kohno, Y., Saito, K., Kato, J., and Ikeda, H. RecQ DNA helicase is a suppressor of illegitimate recombination in *Escherichia coli. Proc Natl Acad Sci USA.* 94:3860–5.
47. Karow, J.K., Constantinou, A., Li, J.L., West, S.C., and Hickson, I.D. The Bloom's syndrome gene product promotes branch migration of Holliday junctions. *Proc Natl Acad Sci USA.* 2000;97:6504–8.
48. Chester, N., Kuo, F., Kozak, C., O'Hara, C.D., and Leder, P. Stage-specific apoptosis, developmental delay, and embryonic lethality in mice homozygous for a targeted disruption in the murine Bloom's syndrome gene. *Genes Dev.* 1998;12:3382–93.
49. Jansack, P., Garcia, P.L., Hamburger, F., Makuta, Y., Shiraishi, K., Imai, Y., Ikeda, H., and Bickle, T.A. Characterization and mutational analysis of the RecQ core of the Bloom syndrome protein. *J Mol Biol.* 2003;330:29–42.
50. Sengupta, S., Linke, S.P., Perdeux, R., Yang, Q., Farnsworth, J., Garfield, S.H., Valerie, K., Shay, J.W., Ellis, N.A., Wasylyk, B., and Harris, C.C. BLM helicase-dependent transport of p53 to sites of stalled DNA replication forks modulates homologous recombination. *EMBO J.* 2003;22:1210–22.
51. Yang, Q., Zhang, R., Wang, X.W., Spillare, E.A., Linke, S.P., Subramanian, D., Griffith, J.D., Li, J.L., Hickson, I.D., Shen, J.C., Loeb, L.A., Mazur, S.J., Appella, E., Brosh, R.M. Jr., Karmakar, P., Bohr, V.A., and Harris, C.C. The processing of Holliday junctions by BLM and WRN licases is regulated by p53. *J Biol Chem.* 2002;270:1980–7.
52. Wu, L., and Hickson, I.D. The Bloom's syndrome helicase stimulates the activity of human topoisomerase IIIalpha. *Nucl Acids Res.* 2002;30:4823–9.
53. Hickson, I.D., Davies, S.L., Li, J.L., Levitt, N.C., Mohaghegh, P., North, P.S., and Wu, L. Role of the Bloom's syndrome helicase in maintenance of genome stability. (Review). *Biochem Soc Trans.* 2001;29:201–4.
54. Davalos, A.R., and Campisi, J. Bloom syndrome cells undergo p53-dependent apoptosis and delayed assembly of BRCA1 and NBS1 rapir complexes at stalled replication forks. *J Cell Biol.* 2003;162:1197–209.
55. Pedrazzi, G., Bachrati, C.Z., Selak, N., Studer, I., Petkovic, M., Hickson, I.D., Jiricny, J., and Stagljar, I. The Bloom's syndrome helicase interacts directly with the human DNA mismatch repair protein hMSH6. *Biol Chem.* 2003;384:1155–64.
56. Yang, Q., Zhang, R., Wang, X.W., Linke, S.P., Sengupta, S., Hickson, I.D., Pedrazzi, G., Perrera, C., Stagljar, I., Littman, S.J., Modrich, P., and Harris, C.C. The mismatch DNA repair heterodimer, hMSH2/6, regulates BLM helicase. *Oncogene.* 2004;23:3749–56.

**57.** Stavropoulos, D.J., Bradshaw, P.S., Li, X., Pasic, I., Truong, K., Ikura, M., Ungrin, M., and Meyn, M.S. The Bloom syndrome helicase BLM interacts with TRF2 in ALT cells and promotes telomeric DNA synthesis. *Hum Mol Genet.* 2002;11: 3134–44.

**58.** Duterte, S., Ababou, M., Onclercq, R., Delic, J., Chatton, B., Jaulin, C., and Amor-Gueret, M. Cell cycle regulation of the endogenous wild type Bloom's syndrome DNA helicase. *Oncogene.* 2000;19:2731–8.

**59.** Wang, Y., Cortez, D., Yazdi, P., Neff, N., Elledge, S.J., and Qin, J. BASC, a super complex of BRCA1-associated proteins involved in the recognition and repair of aberrant DNA structures. *Genes Dev.* 2000;14:927–39.

**60.** Ababou, M., Duterte, S., Lecluse, Y., Onclercq, R., Chatton, B., and Amor-Gueret, M. ATM-dependent phosphorylation and accumulation of endogenous BLM protein in response to ionizing radiation. *Oncogene.* 2000;19:5955–63.

**61.** Beamish, H., Kedar, P., Kaneko, H., Chen, P., Fukao, T., Peng, C., Beresten, S., Gueven, N., Purdie, D., Lees-Miller, S., Ellis, N., Kondo, N., and Lavin, M.F. Functional link between BLM defective in Bloom's syndrome and the ataxia-telangiectasia-mutated protein, ATM. *J Biol Chem.* 2002;277:30515–23.

**62.** Bischof, O., Galande, S., Farzaneh, F., Kohwi-Shigematsu, T., and Campisi, J. Selective cleavage of BLM, the Bloom Syndrome protein, during apoptotic cell death. *J Biol Chem.* 2001;276:12068–75.

**63.** Hickson, I.D. RecQ helicases: caretakers of the genome. *Nat Rev Cancer.* 2003; 3:169–78.

**64.** German, J. Bloom's syndrome. XX. The first 100 cancers. *Cancer Genet Cytogenet.* 1997:93:100–6.

CHAPTER

# 26 CANCER AND APOPTOSIS

Genetic, metabolic, and proliferative homeostasis are major guarantors of the functional and structural harmony of tissues, organs, and ultimately the entire organisms. Loss of proliferative homeostasis (the balance between cell division and apoptosis) or of genetic integrity (through germ or somatic mutations or both) followed by a hypermutability phenotype generates cancer. Growth is the most striking manifestation on physical examination, imaging (x-rays, computed tomography [CT], nuclear magnetic resonance [NMR], ultrasounds), or under the microscope. The cells not only proliferate, but also refuse to die. Faced with these enormous survival advantages, a logical therapy is to kill all the cancer cells without mercy, but without creating a wasteland of the surrounding tissue. Surgery, radiation therapy, and chemotherapy, and in some cases immunotherapy are used to accomplish this end. These techniques are often successful, but they can also fail because the cancer cell survival advantages allow them to grow beyond their normal confines through invasion and metastasis, thereby causing recurrences. Even cancer cells exposed to radiation or chemotherapeutic agents survive by hypoxia or drug resistance mechanisms, thus escaping cell death.

Assuming that mechanisms of drug resistance are excluded, how do cells that may be derived from tissues highly sensitive to ionizing radiation or genotoxic agents (e.g., lymphomas or germ cell tumors) escape apoptosis?[1]

Many molecules are involved in apoptosis, including: ligands, receptors, adaptor proteins with or without death domains, multiple regulators, traditional signal transducing pathways, permeability transition pore complex transcription activators or receptors, and ultimately activation of proapoptotic and inactivation of antiapoptotic molecules. In principle, interference with apoptosis could occur at any one or several of these steps; however, to facilitate this discussion, interference with antiapoptotic proteins (e.g., Bcl2, $Bclx_L$, etc.) are considered separately. This interference may occur at various levels, including mutation of their genes and their transcripts (including splicing variants, posttranslational events, appearance or disappearance of inhibitors, protein degradation, among others). Moreover, the mechanisms of molecular interference with apoptosis may vary from cancer to cancer and possibly from cell to cell within a cancer. Finally, the measure of overexpression or underexpression of any one

[1]Although drug resistance mechanisms are a curse for the therapist, whether it be malaria or cancer, and the molecular mechanisms of the process have evolved from parasites to humans, they are not discussed in this chapter.

molecule may be of little significance because of the frequent redundancy of pro- and antiapoptotic molecules.[2]

In situ determination of the rate at which cells undergo apoptosis is difficult for at least two reasons: 1) once started, death and engulfment of the corpse are quick and, therefore, it is not easy to detect early stages of apoptosis; 2) the late stages of apoptosis may be confusing and difficult (if not impossible) to diagnose morphologically. In addition, labeling of DNA fragments (TUNEL) in situ is controversial because it does not always distinguish early apoptotic from early necrotic cells (1). A monoclonal antibody directed at single stranded DNA is believed to be a specific marker for apoptotic cells (2, 3). However, the fact is that the apoptotic index (AI) contributes to defining the chances of recurrence-free survival into two categories. A low AI is associated with short recurrence-free survival in some cancers, such as non-Hodgkin lymphoma). In contrast, a high AI has been reported in other cancers, such as breast cancer (3, 4). These findings are not necessarily conflicting. The first category usually includes rapidly growing cancers and advanced cancers of the gastrointestinal (GI) tract (e.g., lymphoma and high grade neurological tumors) and the second category includes relatively slow growing cancers (e.g., hepatocarcinoma, prostate, and ependymoma). Several hypothetical explanations have been proposed, but only more investigation will provide the final answers to these intriguing findings.

The association of cell proliferation and cancer is so consistent that cell growth is often considered to be the *primum movens* of cancer. The question remains, does the association of the two events link them in a cause-effect relationship? Consider the evolution of lesions of the breast associated with growth fibroadenoma, fibrocystic changes, and carcinoma.

All three depart from the homeostatic state of the normal breast, which maintains a careful balance between cell proliferation and cell death (5). While fibroadenoma is not associated with cancer, fibrocystic disease is associated with it. The difference must reside in the appearance of a new clone of cells that proliferates with substantial survival advantages, most likely as a result of somatic mutations that not only favor growth over cell death, but also invasion and metastasis (probably as a consequence of hypermutability).

## TNF Family of Receptors and Ligands and Apoptosis in Cancer

Disease, including cancers, has been associated with mutations in germ cells or somatic cells, transcription variations, or posttranslational events involving some tumor necrosis factor (TNF) receptors or ligands. Since induction of apoptosis is a critical biologic function of these receptors, the targeted manipulation of some of the steps in a pathway may assist cancer treatment.

The interaction between TNF family of receptors ($n = 26$) and their ligands ($n = 18$) constitutes one among many other steps in apoptosis. The receptors are type I proteins with a transmembrane portion that extends into an extracellular N-terminal and an intracellular C-terminal. There is little homology in the ligand binding sequence, but there is 25% to 30% homology in the intracellular sequence. Specificity resides with the

[2]Such statements may at first glance be discouraging. Fortunately, the cells have a preference for securing survival and the identification of the latter aids in cancer prognosis and therapy. This is often the case, but it is encouraging that the relationship between overexpression of Bcl2, although a negative prognostic indicator in many cancers, has occasionally been reported as a positive prognosticator and vice versa for the proapoptotic Bax.

ligand binding sequences and in the intracellular portion of the receptors, which share similar protein-protein interactions that allow the buildup of the signal transducing pathways. The transducing pathways include scaffolds built by TNFR1-associated factor (TRAF), TNFR1-associated death domain (TRADD), receptor-interacting protein (RIP) proteins, and proteins containing death domain sequences that interact with Fas-associated death domain (FADD) and RIP-associated ICH1/CED3 homologous protein with death domain (RAIDD; discussed in Chapter 3). The scaffolds connect with various signal transducing pathways including mitogen-activated protein kinases (MAPKs), Akt, Jun, and NF-κβ (6, 7).

Mutations of TNF receptor 1 (TNFR1) have not been associated with cancer, but with an inflammatory syndrome that is dominantly inherited: the inherited periodic fever syndrome (8).

Spontaneous mutations of the FAS receptor are linked in mice and humans with extensive lymphoproliferation (which causes severe lymph node, spleen, and enlargement) coupled to autoimmune manifestations that vary with the genetic mosaic of the victim. The condition is referred to as autoimmune lymphoproliferative syndrome, or ALPS type I, or Canal Smith syndrome. In humans, the disease appears in childhood and causes lymphoproliferation and autoimmune disease (hyperimmunoglobulinemia). Of interest is the finding that the lymphocytes of ALPS patients resist cell death (9). Fas mutations have also been detected in a variety of human cancers: plasmocytomas in B and T cell tumors, including the Reed-Sternberg cells in Hodgkin disease, in bladder, colon, gastric cancers with mucous membrane pemphigoid (MMP) features, and in hepatoblastoma. These findings usually cover only a small number of cases, but that does not mean that they are insignificant; indeed, mutations were detected in twelve among 43 patients with bladder cancers (of the 12 mutations 10 involved a G → A transition at codon 251 in the dead domain) (10). In conclusion, in addition to causing autoimmune responses, FAS mutations are associated with cell proliferation and resistance to cell death. Therefore, their detection could well assist in the design of targeted molecular therapies at the level of the receptor, ligand, or transducing pathway, but whether they have a chance of success in face of the high redundancy of receptors and ligands remains to be proved.

## TRAIL, Apoptosis, and Cancer

Among the TNF receptors, few are as intriguing as TRAIL (tumor necrosis factor–related apoptosis-induced ligand) because of its capacity to induce apoptosis in cancer cells selectively.[3] These special properties of TRAIL have led to several therapeutic trials in which it is used in combination with chemotherapy and radiotherapy in the treatment of a variety of cancers, such as colon, pancreas, breast, prostate, thyroid, kidney, bladder, lung and brain, skin, and multiple myeloma. Many of these cancers do not readily respond to traditional therapies and, therefore, it is hoped that the addition of TRAIL to the polytherapy schedule will help to cure the patient.

The major object of this brief discussion is the role of TRAIL receptors and ligands in apoptosis, particularly as it pertains to cancer cells. The TRAIL receptor family constitutes a subfamily of the TNF family of receptors. There are four known TRAIL receptors: TRAIL R1, R2, R3, and R4. TRAIL R1 and R2 are type I transmembrane

[3]Hepatotoxicity has been reported.

receptors that can induce apoptosis using transduction signals similar to those used by FAS (11). Indeed, it appears that TRAIL may signal through FADD and caspase 8, which in turn triggers the caspase cascade (12). Neither R3 nor R4 induce apoptosis. R3 has no cytoplasmic tail and consequently no dead domain; R4 has a short cytoplasmic tail, an incomplete death domain, and is able to activate NF-κβ (13).

All TRAIL receptors are broadly but variably expressed in many normal tissues and, thus, proapoptotic and "decoy" TRAIL receptors often coexist in the same tissue. Their ratio is believed to determine the level of sensitivity to apoptosis. However, the correlation between sensitivity to apoptosis and TRAIL expression is not impressive, even in tumor cells lines (14). Because of this lack of correlation, it was suggested that other factors must contribute to the selective sensitivity of cancer cells. FLIP (FLICE inhibitor protein) is among the suspects. FLIP is recruited by the death receptor signal complexes and inhibits caspases. In animals, FLIP contributes to regulation of T cell and heart development and in humans it is suspected to contribute to several neurodegenerative diseases (15, 16). However, the existence of other as yet unidentified factors cannot be excluded.

The gaps in understanding the functions of TRAIL receptors are further complicated by the various roles of different signal transducing pathways that regulate the cell's response to the activation of the TRAIL receptors. Among these pathways are the Akt and the NF-κβ signal transducing pathways. Akt phosphorylates and inactivates several proapoptotic agents: BAD, caspase 9, and the forkhead transcription factors AFK and FKHR. However, it also activates NF-κβ, a transcription factor that favors cellular proliferation over apoptosis. It seems that NF-κβ also up-regulates the death receptor TRAIL R2 but not R4, and thereby shifts the balance toward apoptosis (17).

In summary, encouraging experiments in vitro and in vivo and clinical trials (18) give hope that a better understanding and quantification of the competing regulatory mechanisms of the TRAIL receptors will lead to improvements in cancer therapy. Of interest are some trials in which monoclonal antibodies targeted to TRAIL receptors were used.

## Bcl-2 Family and Apoptosis in Cancer

The Bcl-2 is a complicated family of opposites as it houses antiapoptotic and proapoptotic molecules that manifest individual functional behaviors (refer to Chapter 5). The antiapoptotic pro-life molecules include the founder Bcl-2 and its relatives, $Bclx_L$, Mcl1, and Bclw. The proapoptotic members of the family fall into two groups: the $BH_1$-$BH_3$ group (e.g., Bax, Bak, Bod, etc.), and the $BH_3$-only group (e.g., Bid, Bim, Bimf, Noxa, and Puma). Whether pro-life or pro-death, each of these molecules in its own way regulates the ratio of life and death of the cell's population in a special organ at special times, and therefore are responsible for homeostasis. In cancer cells the balance tips toward survival.

In the apoptotic pathway, the pro- and antiapoptotic proteins function between the guardians of genomic integrity (e.g., P53) and the upstream caspases. The Bcl-2 family is not usually much affected by the TNF or even the FAS ligand death signals, but it is moved into cooperation or antagonism when faced with cytokinine deprivation, exposures to ionizing or ultraviolet (UV) radiation, and a variety of endogenous or exogenous genotoxic agents. The imbalance between anti- and proapoptotic molecules triggers cytochrome C release (and other proteins of the intermatrix membrane and activation of the caspase cascade. The antiapoptotic molecules tether some of the traditional proapoptotic molecules such as Bax, Bak, and Bod, which are three homology domain-only proteins.

The pathway used for apoptosis by the "$BH_3$-only" members of the family is more variable. For example, Bid responds to caspase 8 and is therefore a downstream death effector (19) or to granzyme B released in dying T cells (20). Many types of DNA damage cause P53 up-regulation and apoptosis following release and activation of Puma and Noxa; however it is not ruled out that other guardians of the genome (P63, P73), or other $BH_3$ molecules may act cooperatively or independently in the process (21). Taxol and vincristine, two drugs that damage the microtubule, activate Bim's and Bod's chemical attack on the actin motor complex myosin V in a way reminiscent of the activation of proapoptotic Bmf by cytocatalazin (22).

Although an honest desire to understand the intricacies of molecular mechanisms alone justifies research, the motivation for biological research is often driven by more distant goals: to kill the cancer cells that have escaped homeostasis. Faced with the complexity of the equilibrium between the proapoptotic and the antiapoptotic molecules, the task appears Herculean. Can the available observations help our understanding of the intricacies of the regulation of this equilibrium? Consider the function of Bcl-2, which inhibits apoptosis by binding proapoptotic molecules. For example, Bax and Bcl-2 prevent apoptosis by stopping the release of cytochrome C from mitochondria.

Consequently, it can be expected that Bcl-$2^{-/-}$ animals will be oversensitive to apoptosis, assuming that there is no overlap between Bcl-2 and other antiapoptotic proteins in the tissue, organ, or entire organism. Indeed, Bcl-$2^{-/-}$ mice do not live beyond a few days because of massive apoptosis of lymphocytes, intestinal epithelial cells, and neuronal cells. Their kidneys are malformed (polycystic kidneys) and their fur remains hypopigmented. Moreover, in humans the translocation of the Bcl-2 locus causes lymphoma or lymphocytic leukemia because the rearrangement is associated with overexpression of the *bcl-2* gene and prevention of apoptosis. The *bcl-2* translocation is not sufficient, however, to cause neoplastic transformation (cancer). Other genetic changes are required, as is illustrated by Bcl-2/IgH transgenic mice, which after developing severe lymphoid hyperplasia need a second genetic defect, namely, rearrangement of the *c-myc* oncogene, for neoplastic transformation (23).

In hereditary non-polyposis colorectal cancer (HNPCC), the mismatch repair defect is associated with microsatellite mutator phenotype causing a multitude of gene mutations, among them that of *bax*. The *bax* gene mutations have been detected in 50% of HNPCC (24). The locus for the frameshift mutation is a 8 deoxyguanosines sequence G(8) located in the third exon, codons 38–41. There is an intriguing inverse correlation between *bax* and *p53* mutations in HPNCC with microsatellite mutation phenotype. When *bax* mutations are present in tumors that harbor mismatched repair gene mutations, *p53* mutations are absent in those tumors. *Bax* mutations involving the G(8) tracts have also been detected in approximately 20% of hematopoietic tumors.

Bax transcription is P53-dependent. When both *p53* and *bax* are present, *bax* contributes to the apoptotic response to damage. A frameshift mutation of *bax* is most likely to abrogate this function. However, it appears that the wild-type *p53* can induce the expression of an ineffective mutated *bax* and thereby contributes to the failure of the cells to enter apoptosis (25).

Bcl-2, a major protector against apoptosis, under the proper circumstances can function as an oncogene and contribute to the progression of hypermutable clones. This occurs, for example, when NIH 3T3 cells are transfected with *bcl-2* and injected into male mice, or when clones of human breast cancer cell lines transfected with *bcl-2* endow these cells with metastatic capacity (26).

In vitro experiments, observations in knockout mice, and mutations of members of the Bcl-2 family all suggest that the activation of antiapoptotic or inactivation of the proapoptotic members of the Bcl-2 family may contribute to apoptosis resistance observed in cancer. Therefore, it is logical to attempt to restore apoptosis by inhibiting Bcl-2 by various means, for example:

- By targeting Bcl-2 with antisense nucleotides or single chain antibodies (e.g., in non-Hodgkin lymphoma) (27, 28).
- By supplying cells (with nonfunctional proapoptotic proteins) with $BH_3$-only peptides that block the interaction between Bak or Bax with Bcl-2 (29).
- By direct introduction of the proapoptotic genes coupled with an appropriate vector (30).

## Transcription Factors that Affect Apoptosis

### The Rb/E2F Complex, Apoptosis, and Cancer

The Rb/E2F1 cell cycle regulatory complex is a frequent target in human cancer, and it is even suspected that its deregulation contributes to the evolution of almost every cancer. Retinoblastoma (Rb) mutations have been discussed. This section focuses on the relevance of E2F protein's deregulation and its effect on apoptosis in cancer cells. The deregulation of E2Fs cannot be disassociated from that of the Rb family of proteins (**FIGURE 26-1**).

The E2F family is composed of six members. Each forms dimers with one of two other peptides: DP1 and DP2. Each possess a DNA-binding site and E2Fs (1, 2, 3, 4, and 5) also have a transactivation site and a pocket protein binding site. E2F6, the function

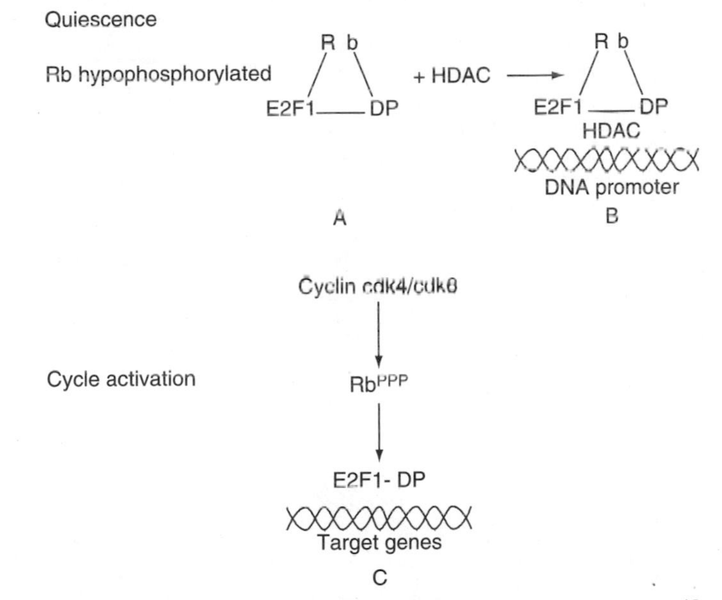


**FIGURE 26-1** Transactivation of target genes by the Rb-E2F pathway. (A) During quiescence ($G_0$ to $G_1$ phase), Rb is hypophosphorylated and forms a complex with E2F, 19 amino acids of the E2F transactivation domain bind to the E2F binding site of the promoter, causing passive repression of the target gene. (B) In a second step, Rb recruits histone deacetylase and histone methyltransferase, S WI/ SNF, and polycomb proteins. Nucleosome condensation (packaging) is induced and causes active repression of target genes. (C) When mitogens activate cyclin D, cdk4, and cdk6, or cyclin E and cdk2, Rb becomes hyperphosphorylated. It separates from E2F/DP, and allows target gene activation.

of which is poorly understood, has neither a transactivation nor a pocket protein binding site (**FIGURE 26-2**).

The pocket protein family (Rb, p130, and p107) forms complexes with various E2Fs—E2F1, E2F2, E2F3—each are known to bind to Rb. E2F4 complexes to Rb, p130, and p107. E2F5 binds to the pocket protein 130, and the complex is believed to repress E2F target genes (31). E2F6 cannot bind pocket proteins.

E2F complexes play a critical role in the cell cycle. The different E2Fs interact with the promoter sites of a panoply of genes and act to either transactivate or repress the gene. Currently, none of the genes that are regulated by E2F have been identified, except those that have encoded kinases, cyclins, and enzymes involved in DNA synthesis (e.g., cdc2, cdc25a, cyclin E, DHFR, thymidine kinase, and DNA polymerase $\alpha$). When overexpressed in quiescent cells, E2F1, E2F2, and E2F3 induce their passage from the $G_0$ phase to the $G_1$/S checkpoint and, therefore, can function as oncogenes and transform cells (32). Note that E2F4 and E2F5 lack a nuclear localization signal and are unable to enter the nucleus unassisted. E2F1/Rb is the most studied among the E2F complexes.

Hyperphosphorylation of Rb dissociates it from E2F, enabling E2F to bind and transactivate promoters of the genes that drive the cycle. The cyclin kinases are associated with growth inhibitory signals (P53/P21/P16, and transforming growth factor-$\beta$ [TGF-$\beta$]). It is logical to anticipate that deregulation of the cycle may lead to cell proliferation. Deregulation could occur in at least three ways:

1. Amplification of the cyclin D or of kinase cyclin D/cdk4 and cdk6, which hyperphosphorylate more Rb and release more E2F for transcription of E2F target genes.
2. Loss of P16, an inhibitor of cdk4/cdk6.
3. Loss or degradation of Rb through ubiquination and proteosomal digestion, as occurs with the expression of the HPV E1 oncoprotein (**FIGURE 26-3**).

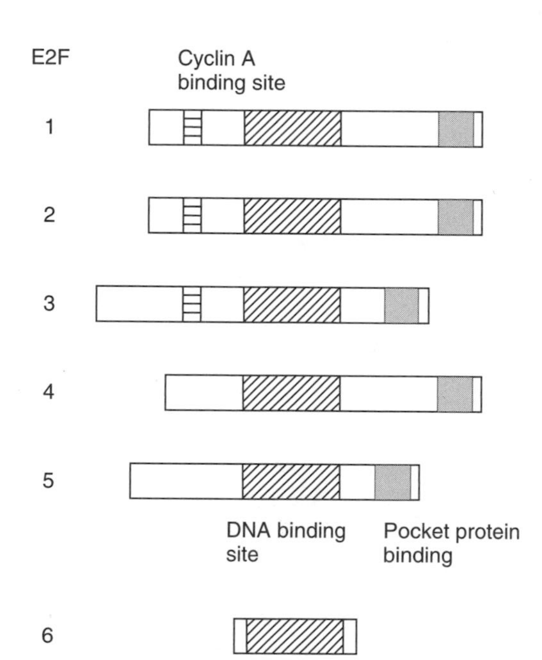


**FIGURE 26-2** Schematic representation of the relative positions of the cyclin kinase binding, the DNA binding, and the pocket binding sites in the E2F1, E2F2, E2F3, E2F4, E2F5, and E2F6 proteins. Rb binds to E2F1, E2F2, E2F3, proteins P107 and P130 bind to E2F4, and P130 binds to E2F5. E2F6 has no pocket binding site.

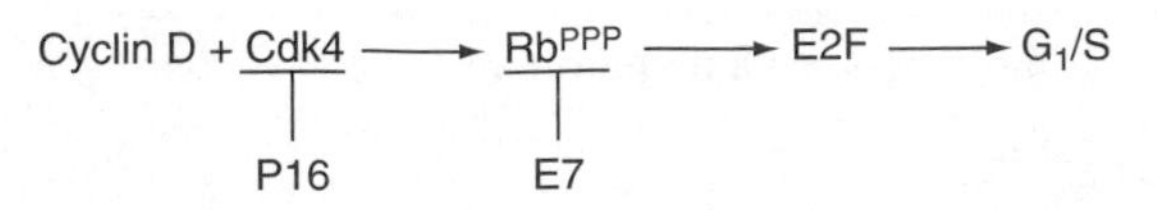


**FIGURE 26-3** Simplified illustration of the pathway of activation of the $G_1$/S transition. Rb hyperphosphorylation frees the E2F transcription factor for activation of the genes that encode the proteins required for DNA synthesis. The pathway is modulated at least in part by the cyclin inhibitor's p16. Activation of the pathway may result from cyclin amplification, loss of inhibitors (p16), or Rb inactivation by the E7 viral oncoprotein.

Early available knowledge indicated that the main function of the E2F protein family was to activate genes that engage and thereby drive the cell cycle. It followed that E2F upstream deregulation was assumed to lead to uncontrolled cell proliferation only. Therefore, it was anticipated that the lack of the E2F1 oncogene would be associated with a low incidence of tumors or would suppress existing tumors. However, in 1996 Field et al. showed that mice defective in E2F1 present an excess of T cells because they fail to enter apoptosis (33), and Yamasaki and colleagues demonstrated that tumors are induced in mice that lacked E2F1 because apoptosis failed (34). These and other experiments reviewed by Phillips and Vousden (35) led to the notion that E2F1 induces apoptosis during normal development.

Although E2F3 and E2F2 have also been implicated in induction of apoptosis, E2F1 seems to be the principal actor in the process. As pointed out by Phillip and Vousden (35), these findings attribute a bipolar behavioral pattern of E2F1, as it acts both as an oncogene and a tumor suppressor gene. This notion may appear paradoxical at first, but it is not unreasonable for a single system to be able to regulate, at the appropriate time, cell proliferation and cell death, two critical components of homeostasis of the cell population. However, the balance between promotion of cell proliferation and oncogenic function may vary with the tumor type (36).

## Mechanism of E2F1 Apoptosis

The addition of growth factors inhibits apoptosis induced by E2F1, as is the case for Myc induced apoptosis. The potential mechanisms that trigger E2F1 apoptosis can be divided into three groups: (a) inhibition of antiapoptotic agents (the NF-κβ and JNK/SAPK signal pathways, or IAPs and Smac diablo; see Chapters 3 and 7); (b) P53 dependent processes; and (c) P53-independent processes. NF-κβ transactivates several antiapoptotic genes (37) and this signal is interrupted by E2F1 (38, 39), possibly by degradation of the signaling complex.

## NF-κB, Apoptosis, and Cancer

Nuclear factor-κBs (members of the REL family of proteins) function as ubiquitous transcription factors. They are activated by a multitude of signals that include cytokines (e.g., interleukin, TNF-α), bacterial or viral products (lipopolysaccharide [LPS]), growth factors, infectious agents, and several proapoptotic agents (oxidative stress, genotoxins, γ-rays). They play a central role in immune reactions, inflammation, cell proliferation, cell survival, and apoptosis. In brief, NF-κBs are transcription factors that are activated by a variety (sometimes contradictory) of signals and, therefore, it is expected that specificity of the ultimate response will require strict control of the transducing factors.

Currently there are five known members of the NF-κB family (**TABLE 26-1**).

Three Rels are generated as transcriptionally functional proteins; activation of P100 and P105/50 requires processing of the large polypeptide (p100 and p105) to a smaller

**Table 26-1 Five members of the NF-κB family**

| | |
|---|---|
| Rel A or p65 | |
| Rel B | |
| C-Rel | |
| P100/52 | NF-κB2 |
| P105/50 | NF-κB1 |

one (p52 and p50), respectively. Each protein contains a Rel homology domain that is required for DNA binding, dimerization and association with the NF-κB inhibitor(s). The traditional NF-κB is a heterodimer of Rel A–NF-κB1.

Nuclear factor-κBs are found in the cytoplasm in the form of a homodimer or heterodimer. They are in the cytoplasm because their nuclear localization signal, present in the C terminus of the molecule near the Rel, is masked by the inhibitors IKK1 and IKK2 (or IKKα and IKKB).

Activation of NF-κB requires that it be freed from the inhibitor. This is secured by phosphorylation of the inhibitor by members of the MEKK related family of kinases (e.g., NF-κB inducing kinase [NKK] and MEKK1), followed by its ubiquination by an E3 ubiquitin ligase and 26 proteosomal degradations. IKK1 is phosphorylated at serines 32 and 36, and IKK2 at serines 19 and 23 (40–42).

Activation of TNF-α via TRAF2 activates NF-κB, a survival signal that interferes with apoptosis. TRAF1 and TRAF2 contribute to phosphorylation of IKKB–NF-κB inhibitors and thereby to the activational NF-κB (**FIGURE 26-4**).

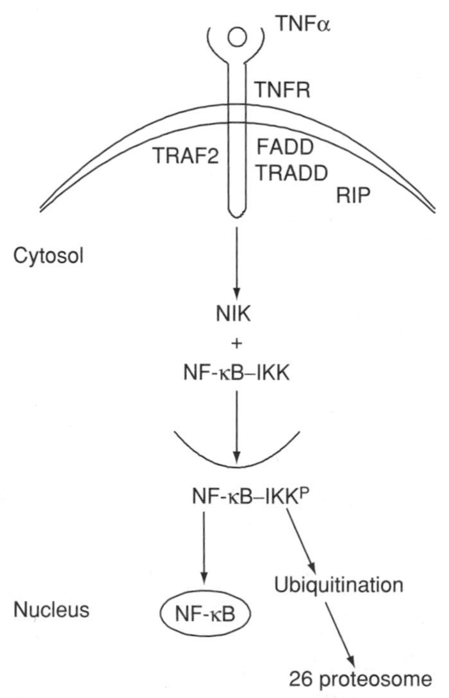


**FIGURE 26-4** Activation of NF-κB. NF-κB bound to its inhibitor (NF-κB–IKK) is activated by phosphorylation of IKK. Phosphorylated IKK separates from NF-κB and is further ubiquitinated and degraded.

We have known since the early 1990s the association of NF-κB and cancer. NF-κB2 is translocated in Hodgkin disease and cutaneous lymphomas; IκB-A also can be mutated in Hodgkin disease (43–45). Since then, several other associations of NF-κB and cancer have been reported. Of course, this is not surprising since the NF-κB engages a survival pathway, in part through stimulation of proapoptotic agents, inhibitors of apoptosis (IAPs), Bcl-2, TNF-associated factor 2 (TRAF2), and others.

Then how does E2F1 secure apoptosis in face of an active antiapoptotic pathway? E2F1 interferes with the antiapoptotic pathway, which is at least in part triggered by TRAF2, by down-regulating TRAF2 (38). The transactivation domain of the E2F1 molecule is apparently not necessary for TRAF2 down-regulation. The significance of this elegant mechanism of feedback control of antiapoptotic steps needs further studies, because the extent to which E2F1 inhibits the antiapoptotic process in vivo is still unknown.

At least two other antiapoptotic proteins have been shown to be targeted by E2F. E2F activates Apaf-1 and indirectly activates the apoptosome (Apaf-1, cytochrome C, and caspase 9), thereby stimulating apoptosis (**FIGURE 26-5**). E2F is also believed to inhibit Mcl-1, a member of the Bcl-2 family, through transcription repression (46, 47). Finally, E2F can also activate apoptosis by stabilizing P53. E2F activates the ARF promoter (directly or indirectly by activating the death associated protein [DAP]), which leads to a rise in the levels of the P14 ARF protein, inhibiting hMDM2 (human homologue of mouse double minute 2) and thereby preventing P53 degradation (48). It has also been proposed that P73 could titrate MDM2 by binding to it (49). As a result, MDM2 is not degraded, but the P73[4] gene targets (some of which are also P53 targets) can be activated, thus enabling apoptosis.

In summary, the Rb-E2F pathway is primarily a cycle-inducing pathway that controls its own performance by also inducing apoptosis when needed for the maintenance of homeostasis (Figure 26-5) (reviewed in 38, 41, 51).

## C-myc, Apoptosis and Cancer

C-myc functions as an oncogene and is deregulated by translocation in Burkitt lymphoma. Its functions are frequently disturbed in many other cancer types by various means: through increased copies in prostate carcinoma; increased expression and amplification in small cell lung cancer; increased expression in colon cancer through (APC?) and increased expression through translation and stabilization of the protein in breast cancer (52 and references therein). The function of the wild-type gene explains the contribution of the deregulated gene to cancer. C-myc is needed during development of the embryo and throughout life after birth. Its expression is minimal in quiescent cells, but it rises soon after the cell's exposure to growth factors (e.g., catenin through the WNT pathway) and engages the cell to enter the cycle. This is achieved by activating cyclins and Cdks (cyclin D and Cdk4, and cyclin E and Cdk2), and by sequestering and degrading cyclin inhibitors (P27, P15, and P21). C-myc also plays a critical role in initiating differentiation by modulating the interaction of Myc, Max, and Mad in the Myc/Max/Mad protein network. In brief, the down-regulation of *c-myc* interrupts the expression of growth promoting target genes and new genes involved in differentiation are activated (52).

[4]P73 is seldom mutated in cancer; however, gene hypermethylation and silencing has been reported in Hodgkin disease (50).

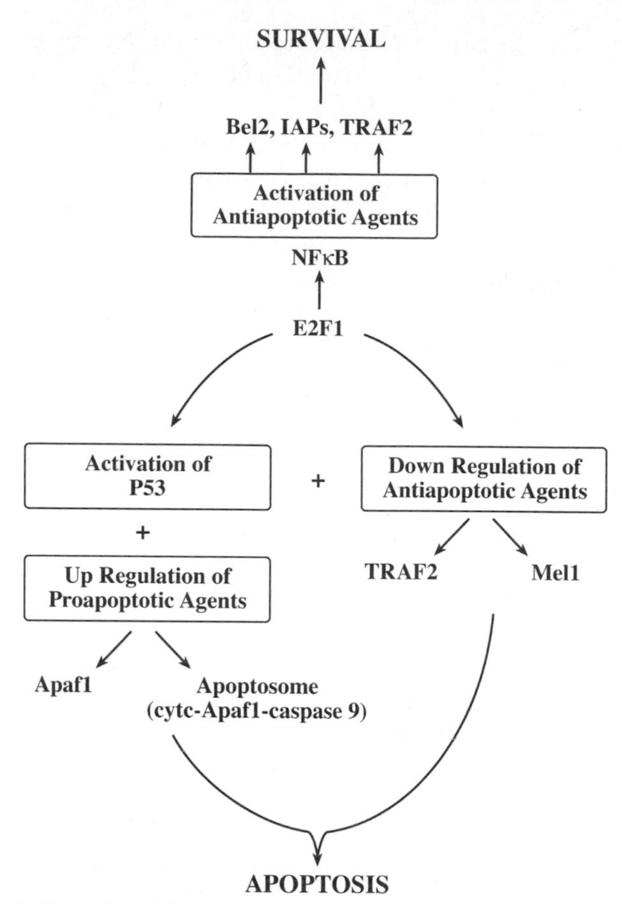


**FIGURE 26-5** Schematic illustration of the potential pathway to apoptosis engaged by E2F1. NF-κB engages the survival pathway by upregulating proapoptotic agents, among which are Bcl-2, IAPs, and TRAF2. The switch to the apoptotic pathway involves downregulation of the antiapoptotic agents TRAF2 and possibly Mcl1, and upregulation of the proapoptotic agents Apaf1 and the apoptosome.

Myc also contributes to apoptosis. This again is not surprising since cell cycle and the apoptotic pathways are coupled. It also makes sense because inadvertent overgrowth automatically should be overcompensated by programmed cell death or vice versa to maintain homeostasis. Many if not all agents that trigger apoptosis are believed to engage Myc's stimulation of apoptosis. The mitochondrial membrane protein Bax is activated, the mitochondrial membrane permeability is raised, cytochrome C is released, the apoptosome (Apaf1, cytochrome C, and procaspase 9) is assembled, and the caspase cascade is unleashed. To sum up, Myc contributes to cell proliferation, cell differentiation, and cell death at all stages of embryonic and adult life.

Is the balance of these very different effects the same in all organs and all tissues? In the suprabasal layer of the skin of transgenic mice, Myc causes cell proliferation and the proapoptotic effect is minimal; the net result is epidermal hyperplasia. Deactivation of Myc arrests the cell proliferation and keratinocyte differentiation; in contrast, in pancreatic β cell Myc stimulation causes apoptosis that leads to diabetes.

Reversal of Myc activation in mice carrying a Myc induced mammary adenocarcinoma causes regression of the cancer in most animals; those that failed had acquired a genetic defect in the Ras pathway (53). In transgenic mice Myc can prevent apoptosis in β cell tumors by inducing $Bclx_L$. The tumors are caused in part by proliferation of the vasculature. Deactivation of Myc causes the vasculature to collapse and the tumor regresses (54).

## C-myc Activation of P53 and Proapoptotic Proteins

Bax is a proapoptotic protein that may be activated by Myc, at least in some cases by Myc's activation of P53. Moreover, P53 induces in addition to Bax: Noxa, Puma, and P53 A1P1.

However, Myc can also act downstream of P53 by inducing $p19^{ARF}$, which inhibits MDM2 and thereby protects P53 from degradation, or upstream of P53 by inducing death receptor ligands such as TNF, FAS, or TRAIL (55, 56).

## Survivin and Apoptosis in Cancer

Survivin is a promising but also an elusive molecule. It is promising because of its potential as a target for cancer therapy, and elusive in its structure, its distribution, and its function. Survivin is the smallest molecule among the mammalian IAP family (16.5 kDa) (see Chapter 8). The gene is located at the telomeric site of chromosome 17 (17q25) in humans; it is composed of 14.7 kb and encodes a polypeptide 142 amino acids long. The gene includes four exons and three entrons, a TATA less promoter with GC-rich regions preceding the open reading frame. The molecule has little in common with the other members of the IAP family. It contains only one Cys and a Bir repeat (~70 amino acids) but no RING finger motif. The wild-type encoded protein is composed of an N-terminal, a BIR domain, and a C-terminal containing an α helix. In addition, two splicing variants exists: survivin 2B (165 amino acids) includes an alternate for exon 2 and survivin ΔEx3 in which exon 3 is deleted (137 amino acids), thereby generating a frameshift in the sequence. Moreover, its COOH terminal sequence differs in that it contains a $BH_2$ domain and a mitochondrial localization domain, which may be significant for the molecule's function. Survivin forms a dimer. The monomer interphases assume a bow type shape. The bow structure reveals clues as to the antiapoptotic function of survivin, dimerization, and the presence of an extended negatively-charged surface around asparagine 71 are required for the antiapoptotic function (57). The distribution of survivin is linked to the evolution of the cell cycle. Survivin is practically undetectable in differentiated cells. In contrast, in cells engaged in replication, such as fetal cells and lymphocytes or endothelial cells, survivin expression is cell cycle dependent (58). However, the marked rise of survivin expression in cancer cells compared to normal adult cells is what stirs the unique interest in the molecule. The controversies about survivin's function nevertheless have obscured its potential role as a target for therapy (59).

Survivin expression has been associated with the $G_2$/M phase of the cycle and is believed to contribute to the regulation of that checkpoint (58). A role in regulation of the $G_1$/S checkpoint is also suggested (60). Of no less significance is the fact that the association of survivin with the cell cycle cannot be dissociated from the fate of the chromosomes and their proteins. Some special chromosomal proteins, referred to as passenger chromosomal proteins, are believed to play a critical role in coordinating cytoskeletal functions and chromosomes from the prophase to telophase. During interphase these proteins are distributed along the length of the chromosomes. During metaphase they concentrate in the centromere, and during the passage from metaphase to telophase, they promptly move to the central region of the spindle. Finally, during cytokinesis they are found in the intracellular bridge flanking the midbody. These orchestrated movements led to the notion that the "passenger proteins" are critical to anaphase and telophase. Among these proteins are the Aurora family, inner centromere protein (INCENP), and survivin (61, 62). Aurora B–INCENP-B–Survivin form a complex. The properties of the Aurora family of proteins are not discussed in this chapter; but it must be pointed out they are special kinases that regulate the interaction between kinetochore, centrosome, and cytoskeleton (63). Aurora B is part of a special family of kinases that phosphorylate histone H3. INCENP directs the kinase to the appropriate target and survivin is believed to be part of the complex. All three proteins are likely to act in concert at various stages of mitosis. Indeed, Wheatley and associates have shown that survivin interacts directly with the other members of the complex (64). These findings provide some evidence on the molecular association of survivin, its movement, and its potential role during the cell cycle. Of course, they do not rule out that other potential molecular combinations contribute to these events. Yet, they do little to explain the most intriguing properties of survivin, its overexpression in cancer and its antiapoptotic effect. Cancer overexpression of survivin is impressive both by its ubiquity and its levels. Survivin is very low in primary cell lines but very high in tumor cell lines. Survivin is overexpressed in most cancers studied (65). Its expression is markedly elevated in all common cancers, compared to its levels in the normal adult tissue from which they are derived or in the tissue that surround them. This is not only true of for the wild-type survivin, but also for the alternative spliced survivins 2 B and ΔEx3 (66).

All evidence shows that overexpression of survivin in cancer cells is not simply proportional to the degree of cell proliferation, because it significantly exceeds what is anticipated from cell proliferation alone. Therefore, it is suggested that overexpression is not linked to a single phase but persists through all phases of the cell cycle, because of deregulation of the gene through amplification of the 17q25 locus or selective demethylation (67). For example, mutation of the *apc* gene and stabilization of the WNT-β-catenin pathway (see Chapter 22) causes up-regulation of survivin in colorectal cancer.

Overexpression of survivin is known to be associated with activation of cell division, but is this the only mechanism of cell protection associated with survivin, a molecule of the IAP family, that contains a single BIR repeat? Of course, from the start, survivin was suspected to possess antiapoptotic properties. Research since 1997 has generated an extensive literature confirming the antiapoptotic properties of survivin; Li provides an excellent review on the topic (65).

Our understanding of the molecular steps involved in the prevention of apoptosis by survivin is incomplete. A hypothetical pathway has been proposed. The antiapoptotic effect is linked to the progress of mitosis, principally during metaphase. At that stage, the cyclin B1-cdc2 phosphorylates survivin at threonine 434. This phosphorylation of survivin enables it to bind caspase 9,[5] thereby preventing the unleashing of the caspase cascade in the extrinsic pathway (68). It appears that crosstalk between extrinsic and intrinsic pathways can lead to further interference with apoptosis, by cytochrome C release and activation of the apoptosome (69).

Although a great deal remains to be learned about the role of survivin in the cell cycle and apoptosis during embryonic development and cancer, expression of survivin

[5]The mechanism(s) of the inhibition of caspase is unknown.

in cancer is associated with poor prognosis. These cancers in which survivin is overexpressed progress more rapidly, the time that elapses between remission and recurrence is short, and the tumor resists traditional treatment. Moreover, cells expressing high levels of survivin are resistant to radio- and chemotherapy (70).

A molecule that is overexpressed in cancer, contributes to cell proliferation, prevents apoptosis, and is absent in normal adult tissues is bound to be a preferred target for cancer therapy. It can be attacked at different phases of its appearance in the cell or of its function. Thus, the targets may include its gene, its mRNA, the protein product itself, the survivin transducing pathway, and its regulatory pathways. Hopefully, suppression of survivin functions could slow down the progress of the cancer cells without damaging the surrounding cells, since the expression of survivin is absent or minimal in the surrounding healthy cells.

Methods used to arrest cancer cells are only mentioned:

- Interference with expression by transfection of antisense nucleotides, ribosomes or dominant negative mutants show promise. They slow cancer growth and induce apoptosis.
- Inhibitors of cdks, among them cdc2.

### Survivin-Derived Peptide Immunotherapy

Generation of antigen-specific cytolytic T cells against survivin peptides abolishes survivin's function in cells in vivo and in vitro. The approach is based on the notion that survivin is a potential tumor-associated antigen (TAA). Tumor specific antigens have been sought in several cancers, among which are melanoma and the testis. The most investigated are probably the melanocytic-specific proteins. Peptides derived from these antigens are used to induce specific cytotoxic T cells in the hope of generating effective agents against melanosarcomas, but they are only attacking that single type of cancer. Being more generally found in cancer, survivin could be the source of an effective vaccine to many if not all cancers. CD8 + naïve T cells can be primed when exposed to survivin-derived antigen presenting cells (e.g., monocytes-derived dendritic cells or cancer cells) (reviewed in 65, 71). There are several ways by which a TAA may be presented to a T cell, but only two are mentioned here: direct presentation by tumor cells or indirect transfer by antigen presenting cells (APCs). Survivin-activated CD8 + T cells have been detected in sentinel metastatic lymph nodes. Of significance is the observation that different cancers appear to present similar sets of survivin epipitopes.

## References and Suggested Reading


1. Grasl-Kraupp, B., Ruttkay-Nedecky, B., Koudelka, H., Bukowska, K., Bursch, W., and Schulte-Hermann, R. In situ detection of fragmented DNA (TUNEL assay) fails to discriminate among apoptosis, necrosis, and autolytic cell death: a cautionary note. *Hepatology.* 1995;21:1465–8.
2. Frankfurt, O.S., Robb, J.A., Sugarbaker, E.V., and Villa, L. Apoptosis in breast carcinomas detected with monoclonal antibody to single-stranded DNA: relation to bcl-2 expression, hormone receptors, and lymph node metastases. *Clin Cancer Res.* 1997;3:465–71.
3. Konstantinidou, A.E., Korkolopoulou, P., and Patsouris, E. Apoptotic markers for tumor recurrences: a minireview. *Apoptosis.* 2002;7:461–70.

4. Korkolopoulou, P.A., Konstantinidou, A.E., Patsouris, E.S., Christodoulou, P.N., Thomas-Tsagli, E.A., and Davaris, P.S. Detection of apoptotic cells in archival tissue from diffuse astrocytomas using a monoclonal antibody to single-stranded DNA. *J Pathol.* 2001;193:377–82.
5. Allan, D.J., Howell, A., Roberts, S.A., Williams, G.T., Watson, R.J., Coyne, J.D., Clarke, R.B., Laidlaw, I.J., and Potten, C.S. Reduction in apoptosis relative to mitosis in histologically normal epithelium accompanies fibrocystic change and carcinoma of the premenopausal human breast. *J Pathol.* 1992;167:25–32.
6. Younes, A., and Kadin, M.E. Emerging applications of the tumor necrosis factor family of ligands and receptors in cancer therapy. *J Clin Oncol.* 2003;21:3526–34.
7. Naismith, J.H., and Sprang, S.R. Modularity of the TNF receptor family. *Trends Biochem Sci.* 1998;23:74–9.
8. McDermott, M.F., Aksentijevich, I., Galon, J., McDermott, E.M., Ogunkolade, B.W., Centola, M., Mansfield, E., Gadina, M., Karenko, L., Pettersson, T., McCarthy, J., Frucht, D.M., Aringer, M., Torosyan, Y., Teppo, A.M., Wilson, M., Karaarslan, H.M., Wan, Y., Todd, I., Wood, G., Schlimgen, R., Kumarajeewa, T.R., Cooper, S.M., Vella, J.P., Kastner, D.L., et al. Germline mutations in the extracellular domains of the 55 kDa TNF receptor, TNFR1, define a family of dominantly inherited autoinflammatory syndromes. *Cell.* 1999;97:133–44.
9. Straus, S.E., Sneller, M., Lenardo, M.J., Puck, J.M., and Strober, W. An inherited disorder of lymphocyte apoptosis: the autoimmune lymphoproliferative syndrome. *Ann Intern Med.* 1999;130:591–601.
10. Mullauer, L., Gruber, P., Sebinger, D., Buch, J., Wohlfart, S., and Chott, A. Mutations in apoptosis genes: a pathogenetic factor for human disease. *Mutat Res.* 2001;488:211–31.
11. Kischkel, F.C., Lawrence, D.A., Chuntharapai, A., Schow, P., Kim, K.J., and Ashkenazi, A. Apo2L/TRAIL-dependent recruitment of endogenous FADD and caspase-B to death receptors 4 and 5. *Immunity.* 2000;12:611–20.
12. Bodmer, J.L., Holler, N., Reynard, S., Vinciguerra, P., Schneider, P., Juo, P., Blenis, J., and Tschopp, J. TRAIL receptor-2 signals apoptosis through FADD and caspase-8. *Nat Cell Biol.* 2000;2:241–3.
13. Degli-Esposti, M.A., Dougall, W.C., Smolak, P.J., Waugh, J.Y., Smith, C.A., and Goodwin, R.G. The novel receptor TRAIL-R4 induces NF-kappa B and protects against TRAIL-mediated apoptosis, yet retains an incomplete death domain. *Immunity.* 1997;7:813–20.
14. Griffith, T.S., and Lynch, D.H. TRAIL: a molecule with multiple receptors and control mechanisms. *Curr Opin Immunol.* 1998;10:559–63.
15. Micheau, O. Cellular FLICE-inhibitory protein: an attractive therapeutic target? *Expert Opin Ther Targets.* 2003;7:559–73.
16. Griffith, T.S., Rauch, C.T., Smolak, P.J., Waugh, J.Y., Boiani, N., Lynch, D.H., Smith, C.A., Goodwin, R.G., and Kubin, M.Z. Functional analysis of TRAIL receptors using monoclonal antibodies. *J Immunol.* 1999;162:2597–605.
17. Shetty, S., Gladden, J.B., Henson, E.S., Hu, X., Villanueva, J., Haney, N., and Gibson, S.B. Tumor necrosis factor-related apoptosis inducing ligand (TRAIL) up-regulates death receptor 5 (DR5) mediated by NFkappaB activation in epithelial derived cell lines. *Apoptosis.* 2002;7:413–20.

18. Ashkenazi, A., Pai, R.C., Fong, S., Leung, S., Lawrence, D.A., Marsters, S.A., Blackie, C., Chang, L., McMurtrey, A.E., Hebert, A., DeForge, L., Koumenis, I.L., Lewis, D., Harris, L., Bussiere, J., Koeppen, H., Shahrokh, Z., and Schwall, R.H. Safety and antitumor activity of recombinant soluble Apo2 ligand. *J Clin Invest.* 1999; 104: 155–62.
19. Zhang, X.D., Zhang, X.Y., Gray, C.P., Nguyen, T., and Hersey, P. Tumor necrosis factor-related apoptosis-inducing ligand-induced apoptosis of human melanoma is regulated by smac/DIABLO release from mitochondria. *Cancer Res.* 2001; 61:7339–48.
20. Sutton, V.R., Davis, J.E., Cancilla, M., Johnstone, R.W., Ruefli, A.A., Sedelies, K., Browne, K.A., and Trapani, J.A. Initiation of apoptosis by granzymes B requires direct cleavage of bid, but not direct granzymes B-mediated caspase activation. *J Exp Med.* 2000;192:1403–14.
21. Flores, E.R., Tsai, K.Y., Crowley, D., Sengupta, S., Yang, A., McKeon, F., and Jacks, T. p63 and p73 are required for p53-dependent apoptosis in response to DNA damage. *Nature.* 2002;416:560–4.
22. Puthalakath, H., Villunger, A., O'Reilly, L.A., Beaumont, J.G., Coultas, L., Cheney, R.E., Huang, D.C., and Strasser, A. Bmf: a proapoptotic BH3-only protein regulated by interaction with the myosin V actin motor complex, activated by anoikis. *Science.* 2001;293:1829–32.
23. Adams, J.M., and Cory, S. The Bcl-2 protein family: arbiters of cell survival. *Science.* 1998;281:1322–6.
24. Rampino, N., Yamamoto, H., Ionov, Y., Li, Y., Sawai, H., Reed, J.C., and Perucho, M. Somatic frameshift mutations in the BAX gene in colon cancers of the microsatellite mutator phenotype. *Science.* 1997;275:967–9.
25. McCurrach, M.E., Connor, T.M., Knudson, C.M., Korsmeyer, S.J., and Lowe, S.W. Bax-deficiency promotes drug resistance and oncogenic transformation by attenuating p53-dependent apoptosis. *Proc Natl Acad Sci USA.* 1997;94:2345–9.
26. Del Bufalo, D., Biroccio, A., Leonetti, C., and Zupi, G. Bcl-2 overexpression enhances the metastatic potential of a human breast cancer line. *FASEB J.* 1997; 11:947–53.
27. Webb, A., Cunningham, D., Cotter, F., Clarke, P.A., di Stefano, F., Ross, P., Corbo, M., and Dziewanowska, Z. BCL-2 antisense therapy in patients with non-Hodgkin lymphoma. *Lancet.* 1997;349:1137–41.
28. Banerjee, D. Technology evaluation: G-3139. *Curr Opin Mol Ther.* 1999;1:404–8.
29. Finnegan, N.M., Curtin, J.F., Prevost, G., Morgan, B., and Cotter, T.G. Induction of apoptosis in prostate carcinoma cells by BH3 peptides which inhibit Bak/Bcl-2 interactions. *Br J Cancer.* 2001;85:115–21.
30. Kagawa, S., Pearson, S.A., Ji, L., Xu, K., McDonnell, T.J., Swisher, S.G., Roth, J.A., and Fang, B. A binary adenoviral vector system for expressing high levels of the proapoptotic gene bax. *Gene Ther.* 2000;7:75–9.
31. Dyson, N. The regulation of E2F by pRB-family proteins. *Gene Dev.* 1998;12: 2245–62.
32. Xu, G., Livingston, D.M., and Krek, W. Multiple members of the E2F transcription factor family are the products of oncogenes. *Proc Natl Acad Sci USA.* 1995; 92:1357–61.

33. Field, S.J., Tsai, F.Y., Kuo, F., Zubiaga, A.M., Kaelin, W.G. Jr., Livingston, D.M., Orkin, S.H., and Greenberg, M.E. E2F-1 functions in mice to promote apoptosis and suppress proliferation. *Cell.* 1996;85:549–61.

34. Yamasaki, L., Jacks, T., Bronson, L., Goillot, E., Harlow, E., and Dyson, N.J. Tumor induction and tissue atrophy in mice lacking E2F-1. *Cell.* 1996;85:537–48.

35. Phillips, A.C., and Vousden, K.H. E2F-1 induced apoptosis. *Apoptosis.* 2001; 6:173–82.

36. Yamasaki, L., Bronson, R., Williams, B.O., Dyson, N.J., Harlow, E., and Jacks, T. Loss of E2F-1 reduces tumorigenesis and extends the lifespan of Rb1(+/–) mice. *Nat Genet.* 1998;18:360–4.

37. Nip, J., Strom, D.K., Fee, B.E., Zambetti, G., Cleveland, J.L., and Hiebert, S.W. E2F-1 cooperates with topoisomerases II inhibition and DNA damage to selectively augment p53-independent apoptosis. *Mol Cell Biol.* 1997;17:1049–56.

38. Phillips, A.C., Ernst, M.K., Bates, S., Rice, N.R., and Vousden, K.H. E2F-1 potentiates cell death by blocking antiapoptotic signaling pathways. *Mol Cell.* 1999; 4:771–81.

39. Tanaka, H., Matsumura, I., Ezoe, S., Satoh, Y., Sakamaki, T., Albanese, C., Machii, T., Pestell, R.G., and Kanakura, Y. E2F1 and c-Myc potentiate apoptosis through inhibition of NF-kappa B activity that facilitates MnSOD-mediated ROS elimination. *Mol Cell.* 2002;9:1017–29.

40. Hersko, A., and Ciechanover, A. The ubiquitin system. *Ann Rev Biochem.* 1998;67: 169–78.

41. Mercurio, F., and Manning, A.M. Regulation of NF-κB function: novel molecular targets for pharmacological intervention. In: Gutkind, J.S., ed. *Signaling Networks and Cell Cycle Control: The Molecular Basis of Cancer and Other Diseases.* Washington, DC: National Institutes of Health. Totawa, NJ: Humana Press; 2000:429–38.

42. Israel, A. The IKK complex: an integrator of all signals that activate NF-kappa B? *Trends Cell Biol.* 2000;10:129–33.

43. Neri, A., Chang, C.C., Lombardi, L., Salina, M., Corradini, P., Maiolo, A.T., Chaganti, R.S., and Dalla-Favera, R. B cell lymphoma-associated chromosomal translocation involves candidate oncogene lyt-10, homologous to NF-kappa B p50. *Cell.* 1991;67:1075–87.

44. Fracchiolla, N.S., Lombardi, L., Salina, M., Migliazza, A., Baldini, L., Berti, E., Cro, L., Polli, E., Maiolo, A.T., and Neri, A. Structural alterations of the NK-kappa B transcription factor lyt-10 in lymphoid malignancies. *Oncogene.* 1993;8:2839–45.

45. Wood, K.M., Roff, M., and Hay, R.T. Defective I kappa B alpha in Hodgkin cell lines with constitutively active NF-kappa B. *Oncogene.* 1998;16:2131–9.

46. Furukawa, Y., Nishimura, N., Furukawa, Y., Satoh, M., Endo, H., Iwase, S., Yamada, H., Matsuda, M., Kano, Y., and Nakamura, M. Apaf-1 is a mediator of E2F-1-induced apoptosis. *J Biol Chem.* 2002;277:39760–8.

47. Croxton, R., Ma, Y., Song, L., Haura, E.B., and Cress, W.D. Direct repression of the Mcl-1 promoter by E2F1. *Oncogene.* 2002;21:1359–69.

48. Pomerantz, J., Schreiber-Agus, N., Liegeois, N.J., Silverman, A., Alland, L., Chin, L., Potes, J., Chen, K., Orlow, I., Lee, H.W., Cordon-Cardo, C., and DePinho, R.A. The

Ink4a tumor suppressor gene product, p19Arf, interacts with MDM2 and neutralizes MDM2's inhibition of p53. *Cell.* 1998;92:713–23.

49. Balint, E., Bates, S., and Vousden, K.H. Mdm2 binds p73 alpha without targeting degradation. *Oncogene.* 1999;18:3923–9.

50. Martinez-Delgado, B., Melendez, B., Cuadros, J., Jose Garcia, M., Nomdedeu, J., Rivas, C., Fernandez-Piqueres, J., and Benitez, J. Frequent inactivation of the p73 gene by abnormal methylation or LOH in non-Hodgkin's lymphomas. *Int J Cancer.* 2002;102:15–9.

51. O'Connor, S.L., Briones, F., Chari, N.S., Cho, S.H., Hamm, R.L., Kadowaki, Y., Lee, S., Spurgers, K.B., and McDonnell, T.J. Apoptosis and cancer: pathogenic and therapeutic implications. In: Yin, X-M., and Dong, Z. *Essentials of Apoptosis: A Guide for Basic and Clinical Research.* Totowa, NJ: Humana Press; 2003:177–200.

52. James, L., and Eisenman, R.N. Myc and Mad bHLHZ domains possess identical DNA-binding specificities but only partially overlapping functions in vivo. *Proc Natl Acad Sci USA.* 2002;99:10429–34.

53. O'Connell, B.C., Cheung, A.F., Simkevich, C.P., Tam, W., Ren, X., Mateyak, M.K., and Sedivy, J.M. A large scale genetic analysis of c-Myc-regulated gene expression patterns. *J Biol Chem.* 2003;278:12563–73.

54. Pelengaris, S., Khan, M., and Evan, G.I. Suppression of Myc-induced apoptosis in beta cells exposes multiple oncogenic properties of Myc and triggers carcinogenic progression. *Cell.* 2002;109:321–34.

55. Kasibhatla, S., Beere, H.M., Brunner, T., Echeverri, F., and Green, D.R. A non-comical DNA-binding element mediates the response of the Fas-ligand promoter to c-Myc. *Curr Biol.* 2000;10:1205–8.

56. Brunner, T., Kasibhatla, S., Pinkoshi, M.J., Frutschi, C., Yoo, N.J., Echeverri, F., Mahboubi, A., and Green, D.R. Expression of Fas ligand in activated T cells is regulated by c-Myc. *J Biol Chem.* 2000;275:9767–72.

57. Muchmore, S.W., Chen, J., Jakob, C., Zakula, D., Matayoshi, E.D., Wu, W., Zhang, H., Li, F., Ng, S.C., and Altieri, D.C. Crystal structure and mutagenic analysis of the inhibitor-of-apoptosis protein survivin. *Mol Cell.* 2000;6:173–82.

58. Li, F., Ambrosini, G., Chu, E.Y., Plescia, J., Tognin, S., Marchisio, P.C., and Altieri, D.C. Control of apoptosis and mitotic spindle checkpoint by survivin. *Nature.* 1998;396:580–4.

59. Reed, J.C. The Survivin saga goes in vivo. *J Clin Invest.* 2001;108:965–9.

60. Suzuki, A., Hayashida, M., Ito, T., Kawano, H., Nakano, T., Miura, M., Akahane, K., and Shiraki, K. Survivin initiates cell cycle entry by the competitive interaction with Cdk4/p16(INK4a) and Cdk2/cyclin E complex activation. *Oncogene.* 2000;19:3225–34.

61. Adams, R.R., Carmena, M., and Earnshaw, W.C. Chromosomal passengers and the (aurora) ABCs of mitosis. *Trends Cell Biol.* 2001;11:49–54.

62. Terada, Y. Role of chromosomal passenger complex in chromosome segregation and cytokinesis. *Cell Struct Func.* 2001;26:653–7.

63. Carmena, M., and Earnshaw, W.C. The cellular geography of aurora kinases. *Nat Rev Cell Biol.* 2003;4:842–54.

64. Wheatley, S.P., Carvalho, A., Vagnarelli, P., and Earnshaw, W.C. INCENP is required for proper targeting of survivin to the centromeres and the anaphase spindle during mitosis. *Curr Biol.* 2001;11:886–90.

65. Li, F. Survivin study: what is the next wave? *J Cell Physiol.* 2003;197:8–29.

66. Hirohashi, Y., Torigoe, T., Maeda, A., Nabeta, Y., Kamiguchi, K., Sato, T., Yoda, J., Ikeda, H., Hirata, K., Yamanaka, N., and Sato, N. An HLA-A24-restricted cytotoxic T lymphocyte epitope of a tumor-associated protein, survivin. *Clin Cancer Res.* 2002; 8:1731–9.

67. Islam, A., Kageyama, H., Takada, N., Kawamoto, T., Takayasu, H., Isogai, E., Ohira, M., Hashizume, K., Kobayashi, H., Kaneko, Y., and Nakagawara, A. High expression of survivin, mapped to 17q25, is significantly associated with poor prognostic factors and promotes cell survival in human neuroblastoma. *Oncogene.* 2000; 19:617–23.

68. Verdecia, M.A., Huang, H., Dutil, E., Kaiser, D.A., Hunter, T., and Noel, J.P. Structure of the human anti-apoptotic protein survivin reveals a dimeric arrangement. *Nat Struct Biol.* 2000;7:602–8.

69. Shankar, S.L., Mani, S., O'Guin, K.N., Kandimalla, E.R., Agrawal, S., and Shafit-Zagardo, B. Survivin inhibition induces human neural tumor cell death through caspase-independent and -dependent pathways. *J Neurochem.* 2001;79:426–36.

70. Zaffaroni, N., and Daidone, M.G. Survivin expression and resistance to anticancer treatments: perspectives for new therapeutic interventions. *Drug Resist Updat.* 2002;5:65–72.

71. Yamamoto, T., and Tanigawa, N. The role of survivin as a new target of diagnosis and treatment in human cancer. *Med Electron Microsc.* 2001;34:207–12.

AFTERWORD

# MOLECULAR PATHOGENESIS OF CANCER

After fertilization, the two haploid gametes combine, divide, and generate diploid progenitor cells that grow into an embryo through repeated cell divisions. The ultimate molding of the original multicellular precursor into a fetus, infant, and adult requires a measured blend of cell division and cell death. The morphology of these events was described at the end of the nineteenth and in the early twentieth centuries. Less than a year before the end of World War II, Avery, MacLeod, and McCarty (1) linked genes to DNA by transferring of genetic information with the pure DNA of one pneumococcus strain with a smooth coat to another pneumococcus strain with a rough coat. The transfer caused a smooth coat phenotype to appear in the rough pneumococci and their succeeding generations. This experiment established the DNA predominance over proteins in transferring genetic information.

The description of the DNA structure by Watson and Crick (2) opened a new era in the science of life. It made it possible for Kornberg (3), Nirenberg et al. (4), Khorana et al., and Jacob and Monod (6) to imagine and perform experiments that ultimately revealed:

- The nature and the sequence of action of the molecules (polymerase, polynucleotide ligase, thymidylic kinase, etc.) involved in DNA synthesis.
- The codes and pathways for transcription of DNA into mRNA, and the translation of messenger RNA into proteins.

The sum of these findings is at the source of our understanding of the molecular processes involved in the prokaryotic and eukaryotic life. Investigations of cell replication from bacteria to mammals soon uncovered molecules involved in various stages of the cell cycle.

A balance between life and death is essential to homeostasis. Although it was known that cellular life and death constantly blend during the entire life span of metazoans, from tadpoles to humans, little if anything was known about this mode of death, except that it differs from necrosis and is likely to be genetically programmed. The studies of Kerr stimulated interest in what was named apoptosis (7, 8). Apoptosis is genetically controlled programmed cell death that occurs not only during embryonic life but also persists through the entire life span of the animal in health and disease. Apoptosis is indispensable to homeostasis in constantly replicating organs (e.g., the skin, the digestive, respiratory and urinary tracts, bone marrow), in occasionally replicating parenchyma (e.g., liver, pancreas, and kidney), in restoration of lost tissues (wound healing), and in adaptive responses to hormones and cell injuries (Chapters 1 and 3).

Whenever the delicate molecular mechanisms of either induction or transduction leading to programmed cell death are disturbed in the young or the old, they cause unwanted apoptosis in degenerative diseases or loss of apoptosis in cancer.

The molecular processes of apoptosis are so critical to metazoan life that it is not surprising that they remained highly conserved during evolution. The molecular pathways of the induction, the transduction of the death signal, the cells' execution, and the disposal of the remains were most elegantly unraveled in a humble worm (9), the *C. elegans*, and further extended, with subtle variations, from flies to mammals including humans (Chapters 2–8).

In essence, the harmonious development of the fetus and the child, the maintenance of the size and structural harmony of every part of the body largely depends upon securing proper balance between cellular proliferation and programmed cell death (apoptosis).

## Cellular Proliferation and Cancer

In cancers examined under the microscope, the presence of abnormal mitosis is striking and there appears to be a correlation between the incidence of abnormal mitosis and the aggressiveness of the cancer. For these reasons, uncontrolled cell proliferation became the hallmark of cancer and as a result, cell multiplication became the major target of cancer therapy, by surgery and radiation first and later by various antimetabolites (10).

### Genome Integrity is Essential to Correct Cell Replication

The clarification of the molecular mechanism of DNA synthesis that started in bacteria was soon resolved, at least in part, in mammalian cells (e.g., regenerating liver and cells in culture). However, the molecular sequences that engage and keep the steps of the cycle moving through the $G_2$ stage, anaphase, metaphase, and telophase only began to be discovered during the latter half of the twentieth century (11–13). The cell cycle is initiated by changes in nutrients and the association of ligands (e.g., growth factors) with intramembrane cytoplasmic or nuclear receptors. Different pathways (NF-κB, c-jun, PIK3, SMADs, etc.) transduce the emerging signal from the cell membrane to the genome. The latter then choreographs the molecular events required for each phase of the cycle. These events are many. For instance, in $G_1$ phase of the cell cycle, they include Rb phosphorylation, cyclins and cyclin kinases activation, elaboration of various regulatory proteins (other kinase inhibitors, e.g., p21, p16, etc.), and phosphatases (e.g., PTEN, etc.). Many of these proteins are synthesized de novo at the appropriate time and others are constitutive; they may be activated or deactivated depending on the protein's function in the cycle. Phosphorylation and acetylation or dephosphorylation and deacetylation are often involved in the activation or the deactivation processes. When the functions of the proteins are completed, they are cancelled either by protein inhibitors or by protein degradation (14) through the ubiquitin–proteosomal pathway (Chapters 3, 14, 17, 19, 20, and 21).

Integrity of the genome is not only indispensable for the correct performance of each step of the cycle but also for the daughter cells functions, including their capacity to give rise to a genetically correct progeny. This goal can only be reached if the genomic blue print is copied without mistakes in the new DNA strand during the S phase and when the new DNA is ensconced into its correct place in the chomosome, and the correct number of chromosomes is transferred, free of breaks, amplifications, translocations, deletions, and point mutations to the daughter cell.

## Damaged DNA Needs to be Repaired

DNA replication is carefully regulated by polymerases, enzymes that not only replicate most of the strands faithfully but are also competent proofreaders. Despite such fidelity, the replication process is not entirely free of occasional errors. DNA may come in contact with reactive oxygen species generated endogenously during metabolism, or exogenously as a result of exposure to ionizing photons or particulate radiations and environmental reactive oxygen species including some chemicals used for cancer therapy. These genotoxic agents cause base alterations, single and double strands breaks, and sometimes crosslinks that must be repaired (Chapter10).

Life on earth would not be what it is, if at an early stage and all during the many millennia of evolution, prokaryotes and eukaryotes had not become well endowed with complex multimolecular pathways for DNA repair. Each pathway for DNA repair is adapted to a special group of lesions: base damage, nucleotide excision repair, strand breaks (single or double), recombination, and crosslink repair. Sometimes these path ways overlap (Chapter 10). The molecular assemblage involved in a given repair pathway may also be modified depending upon the mode of molecular activities in process at the damaged site of the genome. Nucleotide excision repair surveys the resting global genome almost constantly and is usually very efficient (global nucleotide excision repair, GNER). In contrast, when DNA damage persists at a site of the genome engaged in transcription new special molecules are called up to secure the completion of transcription excision repair (TCR).

We have learned a great deal about the molecules (enzymes, activators, Inhibitors, and other modulators) involved in DNA repair, but the steps that precede and follow the process, especially in dividing cells, are not always clear. Among the most intriguing questions are:

- What keeps damaged DNA from being replicated and transferred to daughter cells? It is known that in some if not in most cases, P53 is at the door of the events (Chapter 9).
- What determines the fate of the cells when the damage is not repaired? We know that the cell has a choice; it may stop dividing and senesce or die (apoptosis), or it may continue to divide and either turn into a cancer or engage into a suicidal sequence of mitosis that explodes into a mitotic catastrophe (15).

Ideally, DNA damage must be repaired before DNA replication takes place. Therefore, the presence of the damage must be sensed and the cell cycle arrested. P53, the "guardian of the genome," arrests the cell cycle, and still the signal emerging at the DNA damage site needs to be first transferred to P53 (16). A family of phosphokinases, among which the product of the *ATM* gene is prominent, is known to phosphorylate and thereby stabilize and activate the transcription factor, P53. Assuming that such a signal is sufficient, it will cause P53 to induce the transcription of more than 30 genes, among them *p21*, a universal cyclin kinase inhibitor, and/or GADD45, a more specific cyclin kinase inhibitor that operates at the $G_1$/S and the $G_2$/S checkpoints. Similar molecular events participate at the intra S and mitotic checkpoints (Chapter 13).

The temporary cell cycle arrest is believed to be necessary for DNA repair to take place, and it seems logical that prior to replication a complete survey of the genome may require such an arrest. In any event, proteins involved in NER, double strand breaks, recombination, and crosslinks repair form molecular complexes found in special nuclear structures that are detectable with the electron microscope, named nuclear filaments. The proteins of the nuclear filaments are translocated to the site of DNA damage for repair, yet little is known about the signal that stimulates the move.

These molecules found in the nuclear filaments have been identified. They form a complex referred to as BASC, which includes many proteins that are all involved in DNA repair. When mutated, these proteins have also been associated with various types of hereditary cancers, including: the Fanconi anemia proteins, the proteins mutated in breast cancer (BRCA 1-2), the protein mutated in the Nijmegen breakage syndrome, the Bloom protein, and proteins involved in mismatch repair (Chapter 25). Several mechanisms have been proposed that trigger apoptosis after P53 activation, but the extent and the exact nature of their participation in specific in vivo situations remains unclear. Once the DNA repair is complete and correct replication can resume, the question of whether this resumption is automatic or requires signals still to be identified, is unknown. However, some cells harboring unrepaired DNA may evade the cycle's checkpoints (17, 18). If one of the mutated cells replicates and generates a clone with survival advantages, it thereby becomes the forebearer of a multicellular cancer.

## Anatomic Pathology of Cancer

Pathologists have studied the subtleties of the anatomy, histology, and pathophysiology of cancer and, thereby, have provided important clues for cancer diagnosis and its response to surgical, radiation, and chemical therapy. These important contributions are not discussed in this book.

Otto Warburg (19, 20) opened a new era of cancer research in the decade preceding World War II. The German biochemist discovered anomalies in the energy supplies of the cancer cell that had a preference for aerobic glycolysis. Such preference is a manifestation of a disturbance in gene expression. The significance of Warburg's findings was later revised, yet his findings had a lasting benefit by stimulating the investigators' interest in the biochemistry of the cancer cell. After peace returned to Europe, many biochemists such as Weinhouse, Greenstein, and Potter among others returned to the study of cancer intermediary metabolism (21–23), the discovery of the purine's and pyrimidine's intermediary metabolism (24), and of the molecular pathways involved in the transcription of DNA into messenger RNA and the translation of mRNA into protein (25) soon followed the discovery of the DNA structure (2).

When these basic facts were established, it became possible to search for or design chemicals that interfere with critical steps of the macromolecular biosynthesis of nucleic acids and proteins and, thereby, arrest or slow the rate of cell proliferation. There is now a long list of chemotherapeutic agents that when used alone or in combination are effective in different types of cancers. The association of chemotherapy with radiotherapy or with surgery has prolonged lives and often cured cancers, yet these agents also reach normal cells, which limits their use (10).

## A Genomic Lesion Initiates the Progenitor Cancer Cell

The need for developing therapies that selectively target cancer cells, without causing any or only minimal damage to normal cells, encouraged research on the nature of the primary molecular injury of cancer. Hopes for the future rests in part with discoveries on the succession of the genetic alterations associated with cancer.

Research in prokaryotes and eukaryotes in vitro and vivo has laid the foundations for such an approach. Clinical observations, experiments in cells in vitro, in animals and patients, and epidemiological studies over the years have established a panoply of facts that yielded clues as to the nature of the primary injury in cancer (Chapter 17). The notion that a DNA lesion initiates a progenitor cell that gives birth to a clone of cancer cells with survival advantages was deduced from three sets of observations:

1. Agents that damage DNA cause cancer.
2. Carcinogens are mutagens in bacteria and cause transformation in vitro.
3. The cancer phenotype is transferred from one generation of cells to another.

Although it was established long ago that ultraviolet and ionizing radiation are mutagenic and carcinogenic in animals and humans (26, 27), it proved difficult, at first, to establish that chemical carcinogens are mutagenic for bacteria and cause transformation and immortalization of cells in culture (Chapter 17). These problems have been resolved in part because of the ingenious experiments of Ames and Gold (28).

The initiated cell escapes DNA repair, replicates, and installs a mutation in the daughter cells.

To mutate, a cell harboring damaged DNA must escape DNA repair. To generate a clone of cancer cells, the mutation must secure survival advantages and the cell must replicate. To replicate, the mutated cell must deceive the cell cycle checkpoints.

The best evidence for an association between DNA damage and cancer comes from experiments of nature. Several gene mutations that encode proteins normally involved in DNA repair have been identified.

The association of skin cancer with *xeroderma pigmentosum*, a disease that results from a defect in nucleotide excision repair of the thymine dimers caused by UV damage, illustrated the existence of a link between DNA repair and cancer (Chapter 10).

Investigations of the molecular pathogenesis of ataxia telangiectasia, the Nijmegen breakage syndrome, Fanconi anemia, Bloom syndrome, and hereditary nonpolyposis colon cancer further demonstrated that the failure of repairing double strand breaks, recombination, crosslink, and mismatch damage is associated with cancer predisposition. Each of these diseases hosts a mutated gene. In each case, the wild type gene encodes a protein contributing to double strand breaks and/or recombination repair. The *BRCA1* and *BRCA2* genes that are associated with predisposition to breast cancers also participate in these complex pathways for DNA repair (Chapters 24 and 25) (29–36).

Each of these genetic conditions is associated with an increased cancer risk. In some, only one type of cells becomes cancerous and the risk for other types of cancer is still in doubt. For example, in ataxia telangiectasia the major risk is for T cell lymphoma (ALL), but a rise in the incidence of breast cancer may also exist, even in the heterozygote (Chapter 25). Other hereditary diseases with cancer predisposition are associated with more than one type of cancer. Patients with Fanconi anemia are at a major risk for developing acute myelocytic lymphoma (AML) by the age of forty. Despite the anticipated early death from various hematological causes (80% death by the age of 40), the risk for nonhematological cancers is markedly elevated among the victims of Fanconi anemia (37). Moreover, the cancers occur at an earlier age than expected in the general population[1] (38, 39).

One of these hereditary diseases, the Bloom syndrome, stands out for being less discriminatory in the type of cancer risk associated with the mutation (Chapter 25) (38, 39).

The contribution of environmental factors to these hereditary cancers is not clear. But even if environmental factors did contribute, they do not explain the high incidence and the early age of onset or the distribution and the variety of these cancers.

## Genes and Sporadic Cancer

Futreal and associates of the Welcome Institute in the UK catalogued the human genes so far reported to be associated with human cancer initiation, promotion, or progression in humans. Their census lists 291 human genes associated with cancer, which is approximately 1% of the total number of human genes. Over 90% of the cancer genes' mutations result from translocations that lead to rearrangement and encoding of a chimeric product. The remaining mutations include base alterations, deletions, or insertions, and amplification or a decrease in copy numbers. Ninety per cent of the mutations occur in somatic cells and the remaining in germ cells. Over 85% of the somatic cell mutations act in a dominant fashion, while close to 95% of the germ cell mutations act in a recessive fashion (40).

If epigenetic alterations are excluded, then genetic damage of the germs or somatic cells contribute to the origin of almost all cancers. In germ cell cancers, the genetics (e.g., mismatch repair of APC mutations) dominate the etiology; however, contributions from environmental agents cannot be ruled out. In sporadic cancer, the environmental factors (e.g., smoking, hepatitis B, asbestos) dominate the etiology. However, the damage caused by the environmental agent affects special genes in somatic cells and the alterations of the global genome only contribute to cancer predisposition and genetic predisposition is often difficult to identify.

In sporadic cancer, the affected somatic progenitor cell harbors a damaged protooncogene or suppressor gene that encodes proteins directly or indirectly involved in:

- Cell replication among them: growth factors (PDGF), protein kinase membrane receptors (EGFR, NEU, and RET), membrane-associated protein kinases (RAS), transcription factors (c-Myc, Forkhead proteins).
- Regulators of transcription factors involved in the cell cycle (Rb-E2F), inhibitors functioning at the cell cycle check points (P21, P27, P16. and P14).

[1]The biostatic and the clinical genetic branches of the National Cancer Institute have reported the cancer risk among patients affected with Fanconi anemia (FA). In these patients, the ratios of observed to the expected cancers (O/E) are 785 for leukemia (AML) and 48 for all solid neoplasms. Prominent among the latter are vulvar (O/E 4317), esophageal (O/E 2362), and head and neck cancers (O/E 706). Although statistically significant, these data are still approximate because some of the patients are still alive; moreover, it is not clear how much the cancer incidence is affected by the type of therapy, which may include stem cell bone marrow transplantation. Yet these findings confirm reports based on the observations of other FA cohorts. FA is definitely associated with an increase in risk for cancer (primarily AML) as well as solid neoplasms, among which cancers of the vulva and the gastrointestinal tract predominate) (for details see reference 37).

- Molecules modulating the signal transducing pathways (APC, regulator of the β catenin pathway, and PTEN).
- The family of the guardians of the genome, their prototype being P53 (mutated in 50% of cancers).
- Adhesion molecules (not discussed).
- The genes involved in nucleotide excision, double strands, and recombination and mismatch repair (Chapters 17–25).

Overexpressed proteins may stimulate cell proliferation (EGF, Rb, cyclin E) or inhibit apoptosis (Bcl-2, Survivin). Underexpressed proteins (P53, DNA repair enzymes) may help mutated genes to bypass cell cycle checkpoints. In either case, the mutated cell is at the origin of a clone of cells with survival advantages. A new mutation appears in an already mutated cell that confers further survival advantages to that single cell. The cell carrying the two mutations proliferates into a new clone that overcomes the preexisting cell population. However, several mutations are needed to make a markedly dysplastic, invasive, and metastasizing cancer. Each new mutation is at the origin of a new clone and ultimately the most aggressive clone dominates the scene[2] (41, 42).

## The Mutator Phenotype

The notion that the progression of cancer requires successive mutations implies that the cancer cell must at some time acquire a "mutator phenotype," as was first suggested by Loeb (43, 44). The minimal number of mutations estimated to be required for generating a florid cancer of the colon is five. A single mutation in a somatic cell is rare and, therefore, it can be anticipated that the establishment of five mutations or more in the same cell will require special events to take place. To date the molecular events involved in the steps that lead to hypermutability in cancer are still not completely understood.

The faithful transfer of the genome from one generation of cells to the next depends on the correct replication of the DNA during the S phase and the transfer of the right number of structurally faultless chromosomes to the daughter cell.

All the requirements for integral genomic transfer are not known, but the contributions of several molecules or molecular complexes involved in maintaining the integrity of the chromosome have already been identified. For instance:

- Molecules involved in DNA synthesis and DNA repair.
- Molecules that control the faithful numeral and structural duplication of chromosomes. This group of molecules includes proteins involved in the structural maintenance of chromosomes, chromosome segregation, and telomere function (45–47) (Chapter 13).

However, other factors can also contribute to a rise in the risk for increased mutations, such as uninterrupted cell replication and polymerases slipping in microsatellite regions of the genome causing mismatches. If the mutations bypass the cell cycle checkpoints and are transmitted to the daughter cells, they may participate in establishing a mutator phenotype. Muations of P53 or of molecules that control its activity (e.g., MDM2) could interfere with the arrest of the cell cycle and thereby help the damaged DNA to bypass the checkpoints.

[2]This succession of new and better clones explains the variability of histological appearance in cancers observed under the microscope, for example, a cell population with levels of differentiation ranging from well to poorly differentiated. The possible occurrence of mutations that accelerate apoptosis senescence, necrosis, or mitotic catastrophe cannot be ruled out.

Cancer has long been known to be associated with numeral or structural chromosomal anomalies. Such chromosomal damage is suspected to be responsible for triggering the mutator phenotype. Too few of the proteins involved in the chromosomal structural maintenance and segregation are known to predict all the potential causes of chromosomal damage. However, consider the following situation in which chromosomal damage leads to mechanical events that contribute to the enhancement of chromosomal instability by the breaking and fusion process, described in 1938 by Barbara McClintock (48). It is a process in which pieces of chromosomes are exchanged among chromosomes. The exposed broken end of one chromatid fuses with the exposed end of another broken chromatid (even a sister chromatid) to form a dicentric chromosome. During anaphase, the two halves of the dicentric chromosome are attracted to the opposite poles of the spindle and form a bridge that breaks under tension. The newly exposed broken pieces repeat the fusion and breaking cycle. The study of the function of telomeres in normal and cancer cells has revealed similar mechanism whereby chromosomal instability may emerge. Telomeres are shortened after each cell replication, but they are repaired by the reverse transcriptase telomerase. Thus, under normal conditions the telomeres keep chromosomes from fusing with other chromosomes (or chromosomal fragments). When the telomerase activity declines, as it does with age, recombination between chromosome ends may lead to dicentric chromosomes that may engage in the fusion breakage process. The capacity of unprotected ends to fuse causes non-reciprocal translocation, amplification and deletions, and may cause either overexpression of gene products or loss of heterozygoty (Chapter 17).

## Genes Influence the Response to Environmental Carcinogens

The etiology of somatic cancer is complicated because it includes both environmental causes and genetic predisposition and it takes a long time for the cancer to appear after exposure to the carcinogens. Moreover, although there is usually a correlation between the type of exposure and the anatomical site of the cancer, it is not necessarily consistent and the histology of the cancer may vary.

The environmental causes of sporadic cancer have been identified through epidemiological studies in humans and experiments in animals. Environmental agents fall into two main classes: radiations and highly reactive chemicals (49, 50).

The response to ionizing radiations is modified by mutations in a germ cell, for example, in Ataxia telangiectasia, the Nijmegen breakage syndrome and the Bloom syndrome.

The chemical carcinogens damage DNA directly or after metabolic activation by cytochrome P450 peroxidase. They can be detoxified by metabolic modification through microsomal flavin containing dehydrogenases, UDP glucuronyl tranferase, epoxide hydrolases, glutathione S-transferase, and N-acetyl transferase. Once a procarcinogen has entered the cell, the incidence of DNA damage depends largely on the balance between activation and detoxification of the metabolite. In reality, this balance is difficult to evaluate for any individual chemical because:

- The substrates, the DNA adducts, are difficult to measure.
- The enzymes that modify the procarcinogen or the carcinogen are often polymorphic (51, 52).
- They have a relatively broad specificity, and some of the enzymes involved may activate the procarcinogen and deactivate the carcinogen.

- The enzymes encounter competition for substrates, for instance, the detoxification of phenobarbital or alcohol may compete with the activation or the detoxification of a carcinogen.
- The cell's capacity for enzyme induction may vary with the carcinogen and the exposed organ (53).

Consider, for instance, the fate of polycyclic aromatic hydrocarbons:

- A cytochrome P-450 activates aromatic hydrocarbon by N- or C-oxidation, but it should be kept in mind that there are more than 40 such enzymes.
- Cytochromes P-450 have a relatively broad specificity. They bind to and modify various substrates, some of which are activated and others deactivated. Polycyclic hydrocarbons are mostly activated but they can also be deactivated, yet the latter activity is minor.
- The enzyme involved may be inducible by common substrates (e.g., alcohol); therefore, there may be competition for substrates.
- Three isoforms of inducible epoxide hydrolases exist that are capable of inactivating polycyclic aromatic hydrocarbons.

The reasoning used in describing the fate of aromatic polycyclic hydrocarbons can also be applied (*mutatis mutandis*) to several other carcinogens including aromatic amines, aflotoxin, heterocyclic amines, nitrosamines and benzene (49, 50).

## DNA Adducts Initiate the Cancer Progenitor Cell

Because there is a significant correlation between the incidence of detectable DNA adducts and the effectiveness of the chemicals causing cancer, DNA is believed to be the major and most critical target of chemical cancinogens (54). The adducts are also believed to cause mutations that initiate the progenitor of the cancer clone (55). These notions need to be reconciled with the realities of human cancers, namely the long latent period that exists between the application of the carcinogen and the clinical manifestations of cancer, the relative rarity of cancer, and the extensive distortion of gene expression observed in the course of cancer progression. It is estimated that the human genome contains $2 \times 10^9$ base pairs. Any part of the genome potentially can be damaged by endogenous DNA damage (56) or by exposure to UV radiation, to ionizing radiations, to reactive free radicals, or to genotoxins introduced in the body by breathing or ingestion. However, several factors can influence the incidence of damage in the genome. For example:

- The rate of binding and the formation of base adducts and carcinogens vary with the chemical or physical proprieties of the carcinogen as well as those of the four bases. Thymine residues are favorite sites for adducts generated by UV radiation (cyclobutane dimmers), and 7-guanine is a favorite site for DNA adduct formation by aflotoxin adducts (56).
- The formation of adducts also depends on the status of the cell cycle at the time of exposure to the carcinogen (e.g. mitosis or interphase).
- Most DNA lesions are repaired; however, the DNA repair capacity of the cell varies with the species (57, 58).
- After deceiving the cell cycle checkpoints, the cells harboring damaged DNA have two choices: die and disappear, or divide and engage in an ephemerous immortality.

In summary, a cell that harbors unrepaired DNA and maintains the capacity to divide forms a clone that possesses survival advantages. Repeated multiplications may by themselves raise the odds for new mutations that further activate the rate of replication, distortions of differentiation (dysplasia), rise in angiogenesis, alterations of cell adhesions, and loss of control of cell movement (invasion and metastasis). If at some point a new mutation adds, for example, inability to enter apoptosis in a cell that already manifests unfettered replication, that cell may generate a new clone of quasi-immortals that will live for the duration of the host's life span unless challenged by immunosuppression, therapy, or hypoxia.

Deception of the cell cycle checkpoints, cell replication, and loss of apoptosis may be associated with a benign growth, but the progression of the benign clone into an uncontrolled cancerous cellular proliferation has additional survival advantages that include dysplasia angiogenesis, invasion, and metastasis, which require new mutations. To sum up, it can be said if a single mutation (in the case of an oncogene or a mutation in one allele) and LOH (in the case of a tumor suppressor gene) can initiate a cancer, then a cascade of somatic mutations is needed for its progression into a full blown cancer with dysplasia, invasion, and metastasis.

The miracle of life is manifested by the survival of a billion cells that proliferate, differentiate, and die in harmony from womb to old age. Life on earth exists because some time, some way, somewhere the inert succeeded in generating DNA, a self-replicating molecule able to be transcribed into RNA language. Messenger RNA contains the blue print for the most versatile molecules, the proteins, that make the multiple forms of life possible. Survival and even programmed cell death require a multitude of molecules to interact in harmony, time, and space like the parts of a chiming clock. In essence, as told to us by Schrödinger (61), the living gained over the inert the ability to renew itself in face of decay by capturing air and food from its immediate environment. The cancer cell is a living cell that through mutation(s) conquers entropy by exploiting its host that gave it birth and food.

This book attempts to present some of what is understood of the complexity of the molecular interactions involved in homeostasis and of the disturbance of that homeostasis in cancer cells. Hopefully the seeds of past research and those of the coming years will blossom into new modes of prevention, early diagnosis, improved prognosis, and targeted therapy.

## References and Recommended Reading


1. Avery, O.T., MacLeod, C. M., and McCarty, M. Studies on the chemical transformation of pneumococcal types. *J Exp Med* 1944;79:137–58.
2. Watson, J.D., and Crick, F.H.C. A structure for deoxyribonucleic acid. *Nature* 1953;171:737–8.
3. Kornberg, A. Pathways of enzymatic synthesis of nucleotides and polynucleotides. In: McElroy, W.D., and Glass, B., eds. *The Chemical Basis of Heredity*. Baltimore: The Johns Hopkins University Press; 1957:579–608.
4. Nirenberg, M., Caskey, T., Marshall, R., Brimacombe, R., Kellogg, D., Doctor, B., Hatfield, D., Levin, J., Rottman, F., Pestka, S., Wilcox, M., and Anderson, F. The RNA code and protein synthesis. *Cold Spring Harbor Symposium Quant Biol* 1966;31:11–24.

5. Khorana, H.G., Buchi, H., Ghosh, H., Gupte, N., Jacob, T.M., Kossel, H., Morgan, R., Narang, S., Ohtsuka, E., and Wells, R.D. Polynucleotide synthesis and genetic code. *Cold Spring Harbor Symposium Quant Biol* 1966;31:39–49.
6. Jacob, F., and Monod, J. Genetic regulatory mechanisms in the synthesis of proteins. *J Molec Biol* 1961;3:318–56.
7. Kerr, JF. Shrinkage necrosis: a distinct mode of cellular death. *J Pathol* 1971;105:13–20.
8. Kerr, JF. History of the events leading to the formulation of the apoptosis concept. Review. *Toxicology* 2002;181–182:471–74.
9. Horvitz, H.R. Worms, life, and death (Nobel lecture). *Chembiochem* 2003;4: 697–711.
10. Ratain, M.J. Pharmacology and cancer therapy, chapt 19. In: deVita, T. Jr., Hellmann, S., and Rosenberg, A.S. *Principles and Practice of Oncology*, 6th ed. Philadelphia: Lippincot Williams and Wilkins; 2001:355–460.
11. Nurse, P.M. Nobel Lecture. Cyclin dependent kinases and cell cycle control. *Biosci Rep* 2002;22:487–99.
12. McGowan, C.H. Regulation of the eukaryotic cell cycle. Review. *Prog Cell Cycle Res* 2003;5:1–4.
13. Harper, J.V., and Brooks, G. The mammalian cell cycle: an overview. (Review). *Meth Mol Biol* 2005;296:11353.
14. Hunt, T. Nobel Lecture. Protein synthesis, proteolysis, and cell cycle transitions. *Biosci Rep* 2002;22:465–86.
15. Castedo, M., Perfettini, J.L., Roumier, T., Andreau, K., Medema, R., and Kroemer, G. Cell death by mitotic catastrophe: a molecular definition. Review. *Oncogene* 2004;23:2825–37.
16. McGowan, C.H., and Russell, P. The DNA damage response: sensing and signaling. Review. *Curr Opin Cell Biol* 2004;16:629–33.
17. Smith, A.P., Gimenez-Abian, J.F., and Clarke, D.J. DNA-damage-independent checkpoints: yeast and higher eukaryotes. Review. *Cell Cycle* 2002;1:16–33.
18. Dash, B.C., and El-Deiry, W. Cell cycle control mechanism that can be disrupted in cancer. In: Shontal A.H., ed. *Checkpoint Control and Cancer*. Totowa, NJ: Humana Press; 2004:99–161.
19. Warburg, O. The prime cause and prevention of cancer. Lecture, Meeting of Nobel Laureates, June 30, 1966, K. Triltsch, Wurzburg, Germany, 1967.
20. Warburg, O. On the origin of cancer cells. *Science* 1956;123:309–14.
21. Weinhouse, S. The Warburg hypothesis fifty years later. *Z Krebsforsch Klin Onkol* 1976;87:115–26.
22. Greenstein, J.P. *Biochemistry of Cancer*, 2nd ed. New York: Academic Press; 1954.
23. Potter, V.R. The biochemical approach to the cancer problem. *Fed Proc* 1958;17:691–7.
24. Hartman S.C., and Buchanan, J.M. Nucleic acids, purines, pyrimidines (nucleotide synthesis). *Annu Rev Biochem* 1959;28:365–410.
25. Metzler DE. Ribosomes and the synthesis of proteins, chapt 29. In: *Biochemistry the Chemical Reaction in Living Cells*, 2nd ed. New York: Academic Press; 2003; 1669–738.

26. Van Lancker, JL. *Molecular and Cellular Mechanisms in Disease*, vol II. New York, Berlin: Springer Verlag; 1976:714–61.
27. Hall, J., and Angele, S. Radiation, DNA damage and cancer. Review. *Mol Med Today* 1999;54:157–64.
28. Ames, B.N., and Gold, L.S. Mitogenesis, mutagenesis and animal cancer tests. In: *Chemically Induced Cell Proliferation. Implication for Risk Assessment.* New York: Wiley; 1991:369:1–20.
29. Andreassen, P.R., D'Andrea, A.D., and Taniguchi T. ATR couples FANCD2 monoubiquitination to the DNA-damage response. *Genes Dev* 2004;18:1958–63.
30. Niedzwiedz, W., Mosedale, G., Johnson, M., Ong, C.Y., Pace, P., and Patel, K.J. The Fanconi anaemia gene FANCC promotes homologous recombination and error-prone DNA repair. *Mol Cell* 2004;15:607–20.
31. Venkitaraman, A.R. Tracing the network connecting BRCA and Fanconi anaemia proteins. Review. *Nat Rev Cancer* 2004;4:266–76.
32. Narod, S.A., and Foulkes, W.D. BRCA1 and BRCA2: 1994 and beyond. Review. *Nat Rev Cancer* 2004;4:665–76.
33. Yoshida, K, and Miki, Y. Role of BRCA1 and BRCA2 as regulators of DNA repair, transcription, and cell cycle in response to DNA damage. Review. *Cancer Sci* 2004;95:866–71.
34. Digweed, M., and Sperling, K. Nijmegen breakage syndrome: clinical manifestation of defective response to DNA double-strand breaks. Review. *DNA Repair (Amst)* 2004;3:1207–17.
35. Kobayashi, J., Antoccia, A., Tauchi, H., Matsuura, S., and Komatsu, K. NBS1 and its functional role in the DNA damage response. Review. *DNA Repair* (*Amst*) 2004; 3:855–61.
36. Kaneko, H., and Kondo, N. Clinical features of Bloom syndrome and function of the causative gene, BLM helicase. Review. *Expert Rev Mol Diagn* 2004;4:393–401.
37. Rosenberg, P.S., Greene, M.H., and Alter, B.P. Cancer incidence in persons with Fanconi anemia. *Blood* 2003;101:822–6.
38. German, J. Bloom's syndrome. XX. The first 100 cancers. *Cancer Genet Cytogenet* 1997;93:100–6.
39. German, J.Constitutional hyperrecombinability and its consequences. *Genetics* 2004;168:1–8.
40. Futreal, P.A., Coin, L., Marshall, M., Down, T., Hubbard, T., Wooster, R., Rahman, N, and Stratton, M.R. A census of human cancer genes. Review. *Nat Rev Cancer* 2004;4:177–83.
41. Kinzler, K.W., and Vogelstein B. Cancer-susceptibility genes. Gatekeepers and caretakers. *Nature* 1997;386:761–63.
42. Kinzler, KW, and Vogelstein, B. Lessons from hereditary colorectal cancer. Review. *Cell* 1996;87:159–70.
43. Loeb, L.A., Loeb, K.R., and Anderson, J.P. Multiple mutations and cancer. Review. *Proc Natl Acad Sci U S A* 2003;100:776–81.
44. Loeb, L.A. Cancer cells exhibit a mutator phenotype. Review. *Adv Cancer Res* 1998;72:25–56.

45. Lehmann, A.R. The role of SMC proteins in the responses to DNA damage. *DNA Repair (Amst)* 2005;4:309–14.

46. Gollin, S.M. Mechanisms leading to chromosomal instability. *Semin Cancer Biol* 2005;15:33–42.

47. Gagos, S., and Irminger-Finger, I.Chromosome instability in neoplasia: chaotic roots to continuous growth. *Int J Biochem Cell Biol* 2005;37:1014–33.

48. McClintock B. The fusion of broken ends of sister half-chromatids following breakage at meiotic anaphases. *Mo Agric Exp Station Reseach Bull* 1938;2901–48.

49. Ioannides, C., and Lewis, D.F. Cytochromes P450 in the bioactivation of chemicals. Review. *Curr Top Med Chem* 2004;4:1767–88.

50. Wogan, G.N., Hecht, S.S., Felton, J.S., Conney, A.H., and Loeb, L.A. Environmental and chemical carcinogenesis. Review. *Semin Cancer Biol.* 2004;14:473–86.

51. Agundez, J.A. Cytochrome P450 gene polymorphism and cancer. Review. *Curr Drug Metab* 2004;5:211–24.

52. Ingelman-Sundberg, M. Polymorphism of cytochrome P450 and xenobiotic toxicity. Review. *Toxicology* 2002;181–182:447–52.

53. Patterson, L.H., and Murray, G.I. Tumour cytochrome P450 and drug activation. Review. *Curr Pharm Des* 2002;8:1335–47.

54. Poirier, M.C. Chemical-induced DNA damage and human cancer risk. Review. *Nat Rev Cancer* 2004;4:630–637. Erratum 2004;4:747.

55. Veglia, F., Matullo, G., and Vineis, P. Bulky DNA adducts and risk of cancer: a meta-analysis. Review. *Cancer Epidemiol Biomarkers Prev.* 2003;12:157–60.

56. De Bont, R., and van Larebeke, N. Endogenous DNA damage in humans: a review of quantitative data. Review. *Mutagenesis* 2004;19:169–85.

57. Cooke, M.S., Evans, M.D., Dizdaroglu, M., and Lunec, J. Oxidative DNA damage: mechanisms, mutation, and disease. Review. *FASEB J* 2003;17:1195–214.

58. Au, W.W., Oberheitmann, B., and Harms, C. Assessing DNA damage and health risk using biomarkers. Review. *Mutat Res* 2002;509:153–63.

59. Liu, L., Parekh-Olmedo, H., and Kmiec, E.B. The development and regulation of gene repair. Review. *Nat Rev Genet* 2003;4:679–89.

60. Thomas, C.E., Ehrhardt, A., and Kay, M.A. Progress and problems with the use of viral vectors for gene therapy. Review. *Nat Rev Genet* 2003;4:346–58.

61. Schrödinger, E. *What is life? The physical aspect of the living cell.* London: Folio Society, 2000.

# INDEX

Note: page numbers followed by *f* indicate a figure; *n*, footnote; and *t*, table